D0148434

Introductory Combinatorics

Fourth Edition

Richard A. Brualdi

University of Wisconsin – Madison

Upper Saddle River, New Jersey 07458

Library of Congress Cataloging–in–Publication Data

Brualdi, Richard A.
    Introductory combinatorics / Richard A. Brualdi. – 4th ed.
       p. cm.
    Includes bibliographical references and index.
    ISBN 0 - 13 - 100119 - 1
      1. Combinatorial analysis. 2. Computer science–Mathematics. I. Title.
QA164.B63 2004
511′.6–dc22   2004044455

Editor-in-Chief: *Sally Yagan*
Acquisition Editor: *George Lobell*
Production Editor: *Lynn Savino Wendel*
Vice President/Director of Production and Manufacturing: *David W. Riccardi*
Senior Managing Editor: *Linda Mihatov Behrens*
Assistant Managing Editor: *Bayani Mendoza DeLeon*
Executive Managing Editor: *Kathleen Schiaparelli*
Assistant Manufacturing Manager/Buyer: *Michael Bell*
Manufacturing Manager: *Trudy Pisciotti*
Marketing Manager: *Halee Dinsey*
Marketing Assistant: *Rachel Beckman*
Creative Director: *Jayne Conte*
Editorial Assistant: *Jennifer Brady*
Cover Designer: Kiwi Design
Cover Photo Credits: Kasimir Severinowich Malevich, (1878–1935), "Suprematism",
20th, 1917. Oil on canvas. Museum of Fine Arts, Krassnodar, Russia. Copyright Erich
Lessing/Art Resource, NY.

©2004, 1999, 1992, 1977 by Pearson Education, Inc.
Pearson Prentice Hall
Pearson Education, Inc.
Upper Saddle River, New Jersey, 07458

Pearson Prentice Hall® is a trademark of Pearson Education, Inc.

Printed in the United States of America
10 9 8 7 6 5 4

ISBN 0-13-100119-1

Pearson Education Ltd., London
Pearson Education Australia Pty. Limited, Sydney
Pearson Education Singapore Pte. Ltd.
Pearson Education North Asia, Ltd., Hong Kong
Pearson Education Canada, Ltd., Toronto
Pearson Educación de Mexico, S.A., de C.V.
Pearson Education, Japan, Tokyo
Pearson Education Malaysia, Pte. Ltd.

Contents

Chapter 5.   The Binomial Coefficients   **124**

Chapter 6.   The Inclusion–Exclusion Principle
and Applications   **160**

Chapter 7.   Recurrence Relations and
Generating Functions   **206**

Chapter 8.   Special Counting Sequences   **267**

Preface

In the Preface for the third edition of *Introductory Combinatorics* I related how I extensively rewrote some sections and included some new material and exercises. Some of the major changes from the second to third editions were given as:

> An introductory section on partial orders and equivalence relations has been added to Chapter 4.

> Chapter 5 contains a new section on partially ordered sets, where Dilworth's theorem and its dual are proved.

> A section on partitions of a positive integer has been added to Chapter 8.

> In Chapter 11, the first chapter on graph theory, a tree is now defined as a connected graph that becomes disconnected upon the removal of any edge, and the section on digraphs has been removed.

> Chapter 12, which is new, discusses digraphs and networks. This chapter includes a proof of the max-flow min-cut theorem of Ford and Fulkerson, from which Menger's theorem and König's theorem of Chapter 9 are deduced as corollaries.

> Fundamental numbers of graph theory, discussed in Chapter 12 in the second edition, are now Chapter 13.

> Pólya counting, formerly Chapter 13, is now Chapter 14.

For this fourth edition, I corrected all of the typos I knew of, made some minor adjustments in language (including, in the chapters on graph theory, replacing "chain" with "path"), inserted some small additions, and added more than 60 new, challenging exercises. I hesitated to change the book too much or add too many new topics. I don't like books that have "too many words" (and this preface will not have too many words), and I don't want to fall into that trap. Nevertheless, I added two new sections. In Chapter 6, I have added a new last section on Möbius inversion as a generalization of the inclusion-exclusion principle. In Chapter 8, I have added a new section on lattice paths and the small and large Schröder numbers.

As with earlier editions, one can use this book for either a one or two semester undergraduate course. A first semester could have an emphasis on counting and a second semester an emphasis on graph theory and designs. One could also put together a one semester course which does some counting and graph theory, or some counting and design theory. A brief commentary on each of the chapters and their interrelation follows:

Chapter 1 is an introductory chapter; I usually select one or two topics from it and spend at most two classes on this chapter. Chapter 2, on the pigeonhole principle, should be discussed at least in abbreviated form. But note that no use is made later of some of the harder applications of the pigeonhole principle and of the section on Ramsey's theorem. Chapters 3 to 8 are primarily concerned with counting techniques and properties of some of the resulting counting sequences. They should be covered in sequence. Chapter 4 is about schemes for generating permutations and combinations and, as mentioned above, includes an introduction to partial orders and equivalence relations. However, except for the section on partially ordered sets in Chapter 5, chapters beyond Chapter 4 are essentially independent of Chapter 4, and so this chapter can either be omitted or abbreviated. And one can decide not to cover partially ordered sets at all. Chapter 5 is on properties of the binomial coefficients, and Chapter 6 covers the inclusion-exclusion principle. The new section on Möbius inversion is not used in later sections. Chapter 7 is a long chapter on solving recurrence relations and the use of generating functions in counting. Chapter 8 is concerned mainly with the Catalan numbers, the Stirling numbers of the first and second kind, partition numbers and, now in this latest edition, the large and small Schröder numbers. The chapters that follow are independent of it.

Chapter 9 concerns matchings in bipartite graphs. I introduce bipartite graphs before graphs, but there is no essential dependence of this chapter on the later chapters on graph theory. Except for the application of matching theory to Latin squares, Chapter 10 on designs is independent of the rest of the book. Toward the end of section 10.4, I make use of the matching theory developed in Chapter 9. Chapters 11 and 13 contain an extensive discussion of graphs, with some emphasis on graph algorithms. Chapter 12 is concerned with digraphs and network flows. Chapter 14 deals with counting in the presence of the action of a permutation group and does make use of many of the earlier counting ideas. Except for the last example, it is independent of the chapters on graph theory and designs.

When I teach a one semester course out of this book, I like to conclude with Chaper 14 on Pólya counting as this theory enables one to solve many counting problems that can't be touched with the techniques of earlier chapters. Following Chapter 14, I give solutions and hints for some of the approximately 650 exercises in the book. A few of the exercises have a * symbol beside them, indicating that they are more challenging. The end of a proof and the end of an example are indicated by writing the symbol □.

It is difficult to assess the prerequisites for this book. As with all books intended as textbooks, having highly motivated and interested students helps. Perhaps the prerequisites can be best described as the mathematical maturity achieved by the successful completion of the calculus sequence and an elementary course on linear algebra. Use of calculus is minimal, and the references to linear algebra are few and should not cause any problem to those not familiar with it.

It is very gratifying for me that after more than 25 years since the first edition of *Introductory Combinatorics* was published, it continues to be well-received by the professional mathematical community.

I am very grateful to many individuals who encouraged me to do a fourth edition and who provided me with useful comments: Russ Rowlett (UNC, Chapel Hill), James Sellers (Penn State University), and Michael Buchner (Univ. of New Mexico). As in the third edition I want to especially acknowledge Leroy F. Meyers (Ohio State Univ.) and Tom Zaslavsky (SUNY at Binghamton), each of whom provided me with extensive and detailed comments (for the third edition). For the fourth edition, I received many useful comments from Nils Andersen (Univ. of Copenhagen), James Propp (Univ. of Wisconsin), and Louis Deaett (Univ. of Wisconsin), who read and commented on the new section on lattice paths. The book, I hope, continues to reflect my love of the subject of combinatorics, my enthusiasm for teaching it, and the way I teach it.

Finally, I want to again thank my dear wife, Mona, who continues to bring such happiness, spirit, and adventure into my life.

Richard A. Brualdi
brualdi@math.wisc.edu

Chapter 1

What Is Combinatorics?

It would be surprising indeed if a reader of this book had never solved a combinatorial problem. Have you ever counted the number of games n teams would play if each team played every other team exactly once? Have you ever constructed magic squares? Have you ever attempted to trace through a network without removing your pencil from the paper and without tracing any part of the network more than once? Have you ever counted the number of poker hands that are full houses in order to determine what the odds against a full house are? These are all combinatorial problems. As they might suggest, combinatorics has its historical roots in mathematical recreations and games. Many problems that were studied in the past, either for amusement or for their aesthetic appeal, are today of great importance in pure and applied science. Today, combinatorics is an important branch of mathematics, and its influence continues to expand. Part of the reason for the tremendous growth of combinatorics has been the major impact that computers have had and continue to have in our society. Because of their increasing speed, computers have been able to solve large-scale problems that previously would not have been possible. But computers do not function independently. They need to be programmed to perform. The bases for these programs often are combinatorial algorithms for the solutions of problems. Analysis of these algorithms for efficiency with regard to running time and storage requirements requires more combinatorial thinking.

Another reason for the continued growth of combinatorics is its applicability to disciplines that previously had little serious contact with mathematics. Thus, we find that the ideas and techniques of combinatorics are being used not only in the traditional area of mathematical application, namely the physical sciences, but also in the social sciences, the biological sciences, information theory, and so on. In addition, combinatorics and combinatorial thinking have become more and more important in many mathematical disciplines.

Combinatorics is concerned with arrangements of the objects of a set into patterns satisfying specified rules. Two general types of problems occur repeatedly:

- *Existence of the arrangement.* If one wants to arrange the objects of a set so that certain conditions are fulfilled, it may not be at all obvious whether such an arrangement is possible. This is the most basic of questions. If the arrangement is not always possible, it is then appropriate to ask under what conditions, both necessary and sufficient, the desired arrangement can be achieved.

- *Enumeration or classification of the arrangements.* If a specified arrangement is possible, there may be several ways of achieving it. If so, one may want to count their number or to classify them into types.

Although both existence and enumeration can be considered for any combinatorial problem, it often happens in practice that, if the existence question requires extensive study, the enumeration problem is very difficult. However, if the existence of a specified arrangement can be settled with reasonable ease, it may be possible to count the number of ways of achieving the arrangement. In exceptional cases (when their number is small), the arrangements can be listed. It is important to understand the distinction between listing all the arrangements and determining their number. Once the arrangements are listed, they can be counted by setting up a one-to-one correspondence between them and the set of integers $\{1, 2, 3, \ldots, n\}$ for some n. This is the way we count: one, two, three, ... However, we shall be concerned primarily with techniques for determining the number of arrangements of a particular type without first listing them. Of course the number of arrangements may be so large as to preclude listing them all. In sum, many combinatorial problems are of the following forms:

"Is it possible to arrange . . . ?"

"Does there exist a . . . ?"

"In how many ways can . . . ?"

"Count the number of . . . ."

Two other combinatorial problems that occur in conjunction with these forms are the following:

- *Study of a known arrangement.* After one has done the (possibly difficult) work of constructing an arrangement satisfying certain specified conditions, its properties and structure can then be investigated. Such structure may have implications for the classification problem and also for potential applications. It may also have implications for the next problem.

- *Construction of an optimal arrangement.* If more than one arrangement is possible, one may want to determine an arrangement that satisfies some optimality criterion—that is, to find a "best" or "optimal" arrangement in some prescribed sense.

Thus, a general description of combinatorics might be that *combinatorics is concerned with the existence, enumeration, analysis, and optimization of discrete structures.* In this book, discrete generally means finite, although some discrete structures are infinite.

One of the principal tools of combinatorics for verifying discoveries is *mathematical induction.* Induction is a powerful procedure, and it is especially so in combinatorics. It is often easier to prove a stronger result than a weaker result with mathematical induction. Although it is necessary to verify more in the inductive step, the inductive hypothesis is stronger. Part of the art of mathematical induction is to find the right *balance* of hypotheses to carry out the induction. We assume that the reader is familiar with induction; he or she will become more so as a result of working through this book.

The solutions of combinatorial problems often require *ad hoc* arguments sometimes coupled with use of general theory. One cannot always fall back onto application of formulas or known results. One must set up a mathematical model, study the model, do some computation for small cases, develop some insight, and use one's own ingenuity for the solution of the problem. I do not mean to imply that there are no general principles or methods that can be applied. For counting problems, the inclusion–exclusion principle, the so-called pigeonhole principle, the methods of recurrence relations and generating functions, Burnside's theorem, and Pólya counting are all examples of general principles and methods that we will consider in later chapters. But, often, to see that they can be applied and how to apply them requires cleverness. Thus, experience in solving combinatorial problems is very important. *The implication is that with combinatorics, as with mathematics in general, the more problems one solves, the more likely one is able to solve the next problem.*

In order to make the preceding discussion more meaningful, let us now turn to a few examples of combinatorial problems. They vary from relatively simple problems (but requiring ingenuity for solution) to problems whose solutions were a major achievement in combinatorics. Some of these problems will be considered in more detail in subsequent chapters.

1.1   Example: Perfect Covers of Chessboards

Consider an ordinary chessboard which is divided into 64 squares in 8 rows and 8 columns. Suppose there is available a supply of identically shaped dominoes, pieces which cover exactly two adjacent squares of the chessboard. Is it possible to arrange 32 dominoes on the chessboard so that no 2 dominoes overlap, every domino covers 2 squares, and all the squares of the chessboard are covered? We call such an arrangement a *perfect cover* of the chessboard by dominoes. This is an easy arrangement problem, and one quickly can construct many different perfect covers. It is difficult but nonetheless possible to count the number of different perfect covers. This number was found by Fischer[1] in 1961 to be $12,988,816 = 2^4 \times (901)^2$. The ordinary chessboard can be replaced by a more general chessboard divided into mn squares lying in m rows and n columns. A perfect cover need not exist now. Indeed, there is no perfect cover for the 3-by-3 board. For which values of m and n does the m-by-n chessboard have a perfect cover? It is not difficult to see that an m-by-n chessboard will have a perfect cover if and only if at least one of m and n is even or, equivalently, if and only if the number of squares of the chessboard is even. Fischer has derived general formulae involving trigonometric functions for the number of different perfect covers for the m-by-n chessboard. This problem is equivalent to a famous problem in molecular physics known as the *dimer problem*. It originated in the investigation of the absorption of diatomic atoms (dimers) on surfaces. The squares of the chessboard correspond to molecules, while the dominoes correspond to the dimers.

Consider once again the 8-by-8 chessboard and, with a pair of scissors, cut out two diagonally opposite corner squares, leaving a total of 62 squares. Is it possible to arrange 31 dominoes to obtain a perfect cover of this "pruned" board? Although the pruned board is very close to being the 8-by-8 chessboard, which has over twelve million perfect covers, it has no perfect cover. The proof of this is an example of simple but clever combinatorial reasoning. In an ordinary 8-by-8 chessboard the squares are alternately colored black and white, with

---

[1]M.E. Fischer: Statistical Mechanics of Dimers on a Plane Lattice, *Physical Review*, 124 (1961), 1664-1672.

32 of the squares white and 32 of the squares black. If we cut out two diagonally opposite corner squares, we have removed two squares of the same color, say white. This leaves 32 black and 30 white squares. But each domino covers one black and one white square, so that 31 nonoverlapping dominoes on the board cover 31 black and 31 white squares. Therefore the pruned board has no perfect cover, and the reasoning above can be summarized by

$$31 \boxed{\text{B} \mid \text{W}} \neq 32 \boxed{\text{B}} + 30 \boxed{\text{W}} .$$

More generally, one can take an m-by-n chessboard whose squares are alternately colored black and white and arbitrarily cut out some squares, leaving a pruned board. When does a pruned board have a perfect cover? For a perfect cover to exist the pruned board must have an equal number of black and white squares. But this is not sufficient, as the example in Figure 1.1 indicates.

W	×	W	B	W
×	W	B	×	B
W	B	×	B	W
B	W	B	W	B

Figure 1.1

Thus, we ask: What are necessary and sufficient conditions for a pruned board to have a perfect cover? We will return to this problem in Chapter 9 and will obtain a complete solution by applying the theory of matchings in bipartite graphs. There, a practical formulation of this problem is given in terms of assigning applicants to jobs for which they qualify.

There is another way to generalize the problem of a perfect cover of an m-by-n board by dominoes. Let b be a positive integer. In place of dominoes we consider 1-by-b pieces that consist of b 1-by-1 squares joined side by side consecutively. We call these pieces b-*ominoes*. Thus, a b-omino can cover b consecutive squares in a row or b consecutive squares in a column. In Figure 1.2, a 5-omino is illustrated. A 2-omino is simply a domino. A 1-omino is called a *monomino*.

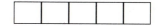

Figure 1.2. A 5-omino

A *perfect cover* of an m-by-n board by b-ominoes is an arrangement of b-ominoes on the board so that (i) no two b-ominoes overlap, (ii) every b-omino covers b squares of the board, and (iii) all the squares of the board are covered. *When does an m-by-n board have a perfect cover by b-ominoes?* Since each square of the board is covered by exactly one b-omino, in order for there to be a perfect cover b must be a factor of mn. Surely, a sufficient condition for the existence of a perfect cover is that b be a factor of m or b be a factor of n. For if b is a factor of m, we may perfectly cover the m-by-n board by arranging m/b b-ominoes in each of the n columns, while if b is a factor of n we may perfectly cover the board by arranging n/b b-ominoes in each of the m rows. Is this sufficient condition also necessary for there to be a perfect cover? Suppose for the moment that b is a prime number and that there is a perfect cover of the m-by-n board by b-ominoes. Then b is a factor of mn and, by a fundamental property of prime numbers, b is a factor of m or b is a factor of n. We conclude that, at least for the case of a prime number b, an m-by-n board can be perfectly covered by b-ominoes if and only if b is a factor of m or b is a factor of n.

In case b is not a prime number, we have to argue differently. So suppose we have the m-by-n board perfectly covered with b-ominoes. We want to show that either m or n has a remainder of 0 when divided by b. We divide m and n by b obtaining quotients p and q and remainders r and s, respectively:

$$
\begin{aligned}
m &= pb + r, \quad \text{where} \quad 0 \le r \le b - 1, \\
n &= qb + s, \quad \text{where} \quad 0 \le s \le b - 1.
\end{aligned}
$$

If $r = 0$, then b is a factor of m. If $s = 0$, then b is a factor of n. By interchanging the two dimensions of the board, if necessary, we may assume that $r \le s$. We then want to show that $r = 0$.

1	2	3	$\cdots$	$b-1$	b
b	1	2	$\cdots$	$b-2$	$b-1$
$b-1$	b	1	$\cdots$	$b-3$	$b-2$
$\vdots$	$\vdots$	$\vdots$		$\vdots$	$\vdots$
2	3	4	$\cdots$	b	1

Figure 1.3. Coloring of a b-by-b board with b colors

We now generalize the alternate black-white coloring used in the case of dominoes ($b = 2$) to b colors. We choose b colors which we label as 1, 2, ... , b. We color a b-by-b board in the manner indicated in Figure 1.3, and we extend this coloring to an m-by-n board in the manner illustrated in Figure 1.4 for the case $m = 10$, $n = 11$, and $b = 4$.

Each b-omino of the perfect covering covers one square of each of the b colors. It follows that there must be the same number of squares of each color on the board. We consider the board to be divided into three parts: the upper pb-by-n part, the lower left r-by-qb part, and the lower right r-by-s part. (For the 10-by-11 board in Figure 1.4, we would have the upper 8-by-11 part, the 2-by-8 part in the lower left, and the 2-by-3 part in the lower right.) In the upper part each color occurs p times in each column and hence pn times altogether. In the lower left part each color occurs q times in each row and hence rq times altogether. Since each color occurs the same number of times on the whole board, it now follows that each color occurs the same number of times in the lower right r-by-s part.

1	2	3	4	1	2	3	4	1	2	3
4	1	2	3	4	1	2	3	4	1	2
3	4	1	2	3	4	1	2	3	4	1
2	3	4	1	2	3	4	1	2	3	4
1	2	3	4	1	2	3	4	1	2	3
4	1	2	3	4	1	2	3	4	1	2
3	4	1	2	3	4	1	2	3	4	1
2	3	4	1	2	3	4	1	2	3	4
1	2	3	4	1	2	3	4	1	2	3
4	1	2	3	4	1	2	3	4	1	2

Figure 1.4. Coloring of a 10-by-11 board with four colors

How many times does color 1 (and, hence, each color) occur in the r-by-s part? Since $r \leq s$, the nature of the coloring is such that color 1 occurs once in each row of the r-by-s part and hence r times in the r-by-s part. Let us now count the number of squares in the r-by-s part. On the one hand there are rs squares; on the other hand, there are r squares of each of the b colors and so rb squares altogether. Equating we get $rs = rb$. If $r \neq 0$, we cancel to get $s = b$, contradicting $s \leq b-1$. So $r = 0$, as desired. We summarize as follows:

*An m-by-n board has a perfect cover by b-ominoes if and only if b
is a factor of m or b is a factor of n.*

A striking reformulation of the preceding statement is the following:
Call a perfect cover *trivial* if all the *b*-ominoes are horizontal or all the
b-ominoes are vertical. Then *an m-by-n board has a perfect cover by
b-ominoes if and only if it has a trivial perfect cover.* Note that this
does not mean that the only perfect covers are the trivial ones. It does
mean that if a perfect cover is possible, then a trivial perfect cover is
also possible.

1.2   Example: Cutting a Cube

Consider a block of wood in the shape of a cube, 3 feet on an edge. It is
desired to cut the cube into 27 smaller cubes, 1 foot on an edge. What
is the smallest number of cuts in which this can be accomplished? One
way of cutting the cube is to make a series of 6 cuts, 2 in each direction,
while keeping the cube in one block as shown in Figure 1.5. But is it
possible to use fewer cuts if the pieces can be rearranged between cuts?
An example is also given in Figure 1.5 where the second cut now cuts
through more wood than it would have if we had not rearranged the
pieces after the first cut. Since the number of pieces, and thus the
number of rearrangements, increases with each cut, this might appear
to be a difficult problem to analyze.

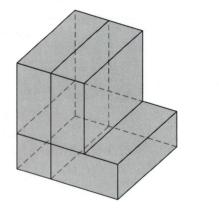 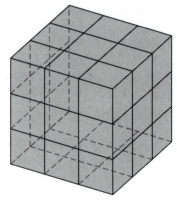

Figure 1.5

But let us look at it another way. Every one of the 27 small cubes
except the one in the middle has at least one face that was originally
part of one of the faces of the large cube. The cube in the middle has
every one of its faces formed by cuts. Since it has 6 faces, 6 cuts are
necessary to form it. Thus, at least 6 cuts are always necessary, and

rearranging between cuts does not help. An energetic student might wish to investigate the number of different ways in which the cube can be cut into 27 smaller cubes, using only 6 cuts.

Another example, which combines features of Example 1.1 and the cube-cutting example, is the following: Consider a 4-by-4 chessboard that is perfectly covered with 8 dominoes. *Show that it is always possible to cut the board into two nonempty horizontal pieces or two nonempty vertical pieces without cutting through one of the 8 dominoes.* The horizontal or vertical line of such a cut is called a *fault-line* of the perfect cover. Suppose there is a perfect cover of a 4-by-4 board such that none of the three horizontal lines and three vertical lines that cut the board into two nonempty pieces is a fault-line. Let x_1, x_2, and x_3 be, respectively, the number of dominoes that are cut by the horizontal lines (see Fig. 1.6).

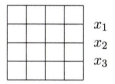

Figure 1.6

Because there is no fault-line, each of x_1, x_2, and x_3 is positive. A horizontal domino covers two squares in a row while a vertical domino covers one square in each of two rows. From these facts we conclude successively that x_1 is even, x_2 is even, and x_3 is even. Hence,

$$x_1 + x_2 + x_3 \geq 2 + 2 + 2 = 6,$$

and there are at least 6 vertical dominoes in the perfect cover. In a similar way, one concludes that there are at least 6 horizontal dominoes. Since $12 > 8$, we have a contradiction. Thus, it is impossible to perfectly cover a 4-by-4 board with dominoes without creating a fault-line.

1.3   Example: Magic Squares

Among the oldest and most popular forms of mathematical recreations are *magic squares*. A magic square of order n is an n-by-n array constructed out of the integers $1, 2, 3, \ldots, n^2$ in such a way that the sum of the integers in each row, in each column, and in each of the two diagonals is the same number s. The number s is called the *magic sum* of the magic square. Examples of magic squares of orders 3 and 4 are

$$
\begin{bmatrix} 8 & 1 & 6 \\ 3 & 5 & 7 \\ 4 & 9 & 2 \end{bmatrix}
\text{ and }
\begin{bmatrix} 16 & 3 & 2 & 13 \\ 5 & 10 & 11 & 8 \\ 9 & 6 & 7 & 12 \\ 4 & 15 & 14 & 1 \end{bmatrix},
\tag{1.1}
$$

with magic sums 15 and 34, respectively. In medieval times there was a certain mysticism associated with magic squares; they were worn for protection against evils. Benjamin Franklin was a magic square fan, and his papers contain many interesting examples.

The sum of all the integers in a magic square of order n is

$$
1 + 2 + 3 + \cdots + n^2 = \frac{n^2(n^2 + 1)}{2},
$$

using the formula for the sum of numbers in an arithmetic progression (see Section 7.1). Since a magic square of order n has n rows each with magic sum s, we obtain the relation $ns = n^2(n^2 + 1)/2$. Thus, any two magic squares of order n have the same magic sum, namely,

$$
s = \frac{n(n^2 + 1)}{2}.
$$

The combinatorial problem is to determine for which values of n there is a magic square of order n and to find general methods of construction. It is not difficult to verify that there can be no magic square of order 2 (the magic sum would have to be 5). But, for all other values of n, a magic square of order n can be constructed. There are many special methods of construction. We describe here a method found by de la Loubère in the seventeenth century for constructing magic squares of order n when n is odd. First a 1 is placed in the middle square of the top row. The successive integers are then placed in their natural order along a diagonal line that slopes upwards and to the right, with the following modifications:

(i) When the top row is reached, the next integer is put in the bottom row as if it came immediately above the top row.

(ii) When the right-hand column is reached, the next integer is put in the left-hand column as if it immediately succeeded the right-hand column.

(iii) When a square that has already been filled is reached or when the top right-hand square is reached, the next integer is placed in the square immediately below the last square that was filled.

The magic square of order 3 in (1.1), as well as the magic square

$$\begin{bmatrix} 17 & 24 & 1 & 8 & 15 \\ 23 & 5 & 7 & 14 & 16 \\ 4 & 6 & 13 & 20 & 22 \\ 10 & 12 & 19 & 21 & 3 \\ 11 & 18 & 25 & 2 & 9 \end{bmatrix}. \tag{1.2}$$

of order 5, was constructed by using de la Loubère's method.

Methods for constructing magic squares of even orders different from 2 and other methods for constructing magic squares of odd order can be found in a book by Rouse Ball.[2]

Three-dimensional analogs of magic squares have been considered. A *magic cube* of order n is an n-by-n-by-n cubical array constructed out of the integers $1, 2, \ldots, n^3$ in such a way that the sum s of the integers in the n cells of each of the following straight lines is the same:

(i) lines parallel to an edge of the cube;

(ii) the two diagonals of each plane cross section;

(iii) the four space diagonals.

The number s is called the *magic sum* of the magic cube and has the value $(n^4 + n)/2$. We leave it as an easy exercise to show that there is no magic cube of order 2, and we verify that there is no magic cube of order 3.

Suppose that there is a magic cube of order 3. Its magic sum would then be 42. Consider any 3-by-3 plane cross section

$$\begin{bmatrix} a & b & c \\ x & y & z \\ d & e & f \end{bmatrix},$$

with numbers as shown. Since the cube is magic,

[2]W.W. Rouse Ball: *Mathematical Recreations and Essays*; revised by H.S.M. Coxeter. Macmillan, New York (1962) 193-221.

$$a + y + f \;=\; 42$$
$$b + y + e \;=\; 42$$
$$c + y + d \;=\; 42$$
$$a + b + c \;=\; 42$$
$$d + e + f \;=\; 42.$$

Subtracting the last two equations from the first three, we get $3y = 42$ and, hence, $y = 14$. But this means that 14 has to be the center of each plane cross section of the magic cube and, thus, would have to occupy seven different places. But it can occupy only one place, and we conclude that there is no magic cube of order 3. It is more difficult to show that there is no magic cube of order 4. A magic cube of order 8 is given in an article by Gardner.[3]

Magic squares will not be studied furthered in this book.

1.4   Example: The 4-Color Problem

Consider a map on a plane or on the surface of a sphere where the countries are connected regions.[4] In order to be able to differentiate countries quickly, it is required to color them so that two countries which have a common boundary receive different colors (a corner does not count as a common boundary). What is the smallest number of colors necessary to guarantee that every map can be so colored? Until fairly recently, this was one of the famous unsolved problems in mathematics. Its appeal to the layperson is due to the fact that it can be simply stated and understood. Except for the well-known angle-trisection problem, it has probably intrigued more amateur mathematicians than any other problem. First posed by Francis Guthrie about 1850 when he was a graduate student, it has also stimulated a large body of mathematical research. Some maps require four colors. An example is the map in Figure 1.7. Since each pair of the four countries of this map have a common boundary, it is clear that four colors are necessary to

---

[3]M. Gardner: Mathematical Games, *Scientific American*, January (1976), 118-123.

[4]Thus, the state of Michigan would not be allowed as a country for such a map, unless one takes into account that the upper and lower peninsulas of Michigan are connected by the Straits of Mackinac Bridge. Kentucky would also not be allowed, since its westernmost tip of Fulton County is completely surrounded by Missouri and Tennessee.

color the map. It was proven by Heawood[5] in 1890 that five colors are always enough to color any map. It is not too difficult to show that it is impossible to have a map in the plane which has five countries, every pair of which have a boundary in common. Such a map, if it had existed, would have required five colors. But not having five countries every two of which have a common boundary does not mean that four colors suffice. It might be that some map in the plane requires five colors for other more subtle reasons.

In 1976 Appel and Haken[6] announced that they had proven that any map in the plane could be colored with four colors. Their proof required about 1200 hours of computer calculations, nearly 10 billion separate, logical decisions! A complete description of their proof appears in their book.[7] Recently, the Appel–Haken proof was simplified by N. Robertson, D.P. Sanders, P.D. Seymour, and R. Thomas,[8] although the proof still requires very substantial computer verification.

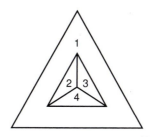

Figure 1.7

1.5 Example: The Problem of the 36 Officers

Given 36 officers of 6 ranks and from 6 regiments, can they be arranged in a 6-by-6 formation so that in each row and column there is one officer of each rank and one officer from each regiment? This problem, which was posed in the eighteenth century by the Swiss mathematician L. Euler as a problem in recreational mathematics, has important repercussions in statistics, especially in the design of experiments (see

---

[5]P.J. Heawood: Map-colour theorems, *Quarterly J. Mathematics*, Oxford ser., 24 (1890), 332-338.

[6]K. Appel and W. Haken: Every planar map is four colorable, *Bulletin of the American Mathematical Society*, 82 (1976), 711-712.

[7]K. Appel and W. Haken: *Every planar map is four colorable*, American Math. Society, Providence (1989).

[8]N. Robertson, D.P. Sanders, P.D. Seymour, and R. Thomas: The four-colour theorem, *J. Combin. Theory Ser. B*, 70 (1997), 2-44.

Chapter 10). An officer can be designated by an ordered pair (i, j), where i denotes his rank $(i = 1, 2, \ldots, 6)$ and j denotes his regiment $(j = 1, 2, \ldots, 6)$. Thus, the problem asks the following question:

Can the 36 ordered pairs (i, j) $(i = 1, 2, \ldots, 6; j = 1, 2, \ldots, 6)$ be arranged in a 6-by-6 array so that in each row and each column the integers $1, 2, \ldots, 6$ occur in some order in the first positions and in some order in the second positions of the ordered pairs?

Such an array can be split into two 6-by-6 arrays, one corresponding to the first positions of the ordered pairs (the *rank array*) and the other to the second positions (the *regiment array*). Thus, the problem can be stated as follows:

Do there exist two 6-by-6 arrays whose entries are taken from the integers $1, 2, \ldots, 6$ such that

(i) in each row and in each column of these arrays the integers $1, 2, \ldots, 6$ occur in some order, and

(ii) when the two arrays are juxtaposed, all of the 36 ordered pairs (i, j) $(i = 1, 2, \ldots, 6; j = 1, 2, \ldots, 6)$ occur?

To make this concrete, suppose instead that there are 9 officers of 3 ranks and from 3 different regiments. Then a solution for the problem in this case is

$$\begin{bmatrix} 1 & 2 & 3 \\ 3 & 1 & 2 \\ 2 & 3 & 1 \end{bmatrix}, \quad \begin{bmatrix} 1 & 2 & 3 \\ 2 & 3 & 1 \\ 3 & 1 & 2 \end{bmatrix} \quad \longrightarrow \quad \begin{bmatrix} (1,1) & (2,2) & (3,3) \\ (3,2) & (1,3) & (2,1) \\ (2,3) & (3,1) & (1,2) \end{bmatrix}.$$

$$\text{rank array} \qquad \text{regiment array} \qquad\qquad \text{juxtaposed array}$$

$$(1.3)$$

The preceding rank and regiment arrays are examples of what are called *Latin squares* of order 3; each of the integers 1, 2, and 3 occurs once in each row and once in each column. The following are Latin squares of orders 2 and 4:

$$\begin{bmatrix} 1 & 2 \\ 2 & 1 \end{bmatrix} \quad \text{and} \quad \begin{bmatrix} 1 & 2 & 3 & 4 \\ 4 & 1 & 2 & 3 \\ 3 & 4 & 1 & 2 \\ 2 & 3 & 4 & 1 \end{bmatrix}. \qquad (1.4)$$

The two Latin squares of order 3 in (1.3) are called *orthogonal* because when they are juxtaposed, all of the 9 possible ordered pairs (i, j), with $i = 1, 2, 3$ and $j = 1, 2, 3$, result. We can thus rephrase Euler's question:

Do there exist two orthogonal Latin squares of order 6?

Euler investigated the more general problem of orthogonal Latin squares of order n. It is easy to see that there is no pair of orthogonal Latin squares of order 2, since, besides the Latin square of order 2 given in (1.4), the only other one is

$$\begin{bmatrix} 2 & 1 \\ 1 & 2 \end{bmatrix},$$

and these are not orthogonal. Euler showed how to construct a pair of orthogonal Latin squares of order n whenever n is odd or has 4 as a factor. Notice that this does not include $n = 6$. On the basis of many trials he concluded, but did not prove, that there is no pair of orthogonal Latin squares of order 6, and he conjectured that no such pair existed for any of integers $6, 10, 14, 18, \ldots, 4k + 2, \ldots$. By exhaustive enumeration, Tarry[9] in 1901 proved that Euler's conjecture was true for $n = 6$. Around 1960, three mathematician–statisticians, R.C Bose, E.T. Parker, and S.S. Shrikhande,[10] succeeded in proving that Euler's conjecture was false for all $n > 6$. That is, they showed how to construct a pair of orthogonal Latin squares of order n for every n of the form $4k+2$, $k = 2, 3, 4, \ldots$. This was a major achievement and put Euler's conjecture to rest. Later we shall explore how to construct orthogonal Latin squares using finite number systems called finite fields and how they can be applied in *experimental design*.

1.6   Example: Shortest-Route Problem

Consider a system of streets and intersections. A person wishes to travel from one intersection A to another intersection B. In general, there are many available routes from A to B. The problem is to determine a route for which the distance traveled is as small as possible—a *shortest route*. This is an example of a combinatorial *optimization* problem. One possible way of solving this problem is to list in a systematic way all possible routes from A to B. It is not necessary to

---

[9]G. Tarry: Le problème de 36 officeurs, *Compte Rendu de l'Association Francaise pour l'Advancement de Science Naturel*, 1 (1900), 122-123; 2 (1901), 170-203.

[10]R.C. Bose, E.T. Parker and S.S. Shrikhande: Further results on the construction of mutually orthogonal Latin squares and the falsity of the Euler's conjecture, *Canadian Journal of Mathematics*, 12 (1960), 189-203.

travel over any street more than once; thus, there are only a finite number of such routes. Then, compute the distance traveled for each and select a shortest route. But this is not a very efficient procedure and, when the system is large, the amount of work may be too great to permit a solution in a reasonable amount of time. What is needed is an algorithm for determining a shortest route in which the work involved in carrying out the algorithm does not increase too rapidly as the system increases in size. The more precise way of saying this is that the amount of work should be bounded by a polynomial function (as opposed to, say, an exponential function) of the size of the problem. We shall later describe such an algorithm. This algorithm will actually find a shortest route from A to every other intersection in the system.

The problem of finding a shortest route between two intersections can be viewed abstractly. Let X be a finite set of objects called *vertices* (which correspond to the intersections and the ends of dead-end streets), and let E be a set of unordered pairs of vertices called *edges* (which correspond to the streets). Thus, some pairs of vertices are joined by edges, while others are not. The pair (X, E) is called a *graph*. A *walk* in the graph joining vertices x and y is a sequence of vertices such that the first vertex is x and the last vertex is y, and any two consecutive vertices are joined by an edge. Now associate with each edge a nonnegative real number—the *length* of the edge. The *length of a walk* is the sum of the lengths of the edges that join consecutive vertices of the walk. Given two vertices x and y, the shortest-route problem is to find a walk from x to y that has the smallest length. In the graph depicted in Figure 1.8, there are 6 vertices and 10 edges. The numbers on the edges denote their lengths. One walk joining x and y is x, a, b, d, y, and it has length 4. Another is x, b, d, y, and it has length 3. It is not difficult to see that the latter walk gives a shortest route joining x and y.

A graph is an example of a discrete structure, which has been and continues to be extensively studied in combinatorics. The generality of the notion allows for its wide applicability in such diverse fields as psychology, sociology, chemistry, genetics, and communications science. Thus, the vertices of a graph might correspond to people, with two vertices joined by an edge if the corresponding people distrust each other; or the vertices might represent atoms, and the edges represent the bonds between atoms. You can probably imagine other ways in which graphs can be used to model phenomena. Some important concepts and properties of graphs are studied in Chapters 9, 11, and 12.

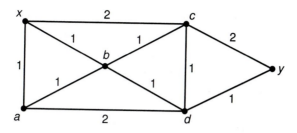

Figure 1.8

1.7   Example: The Game of Nim

We close this introductory chapter by returning to the roots of combinatorics in recreational mathematics and investigate the ancient game of Nim.[11] Its solution depends on *parity*, an important problem-solving concept in combinatorics. We used a simple parity argument in investigating perfect covers of chessboards when we showed that a board had to have an even number of squares in order that it have a perfect cover with dominoes.

Nim is a game played by two players with heaps of coins (or stones or beans or . . .). Suppose that there are $k \geq 1$ heaps of coins that contain, respectively, $n_1, n_2, \ldots, n_k$ coins. The *object* of the game is to select the last coin. The *rules* of the game are the following:

(i)   The players alternate turns (let us call the player who makes the first move I and then call the other player II).

(ii)   Each player, when it is his or her turn, selects one of the heaps and removes at least one of the coins from the selected heap. (The player may take all of the coins from the selected heap, thereby leaving an empty heap, which is now "out of play.")

The game ends when all the heaps are empty. The last player to make a move, that is, the player who takes the last coin(s), is the *winner*.

The variables in this game are the number k of heaps and the numbers $n_1, n_2, \ldots, n_k$ of coins in the heaps. The combinatorial problem is to determine whether the first or second player wins[12] and how that player should move in order to guarantee a win—a *winning strategy*.

---

[11]Nim derives from the German *Nimm!*, meaning *Take!*.
[12]With intelligent play.

In order to develop some understanding of Nim, we consider some special cases.[13] If there is initially only one heap, then player I wins by removing all its coins. Now suppose that there are $k = 2$ heaps, with n_1 and n_2 coins, respectively. Whether or not player I can win depends not on the actual values of n_1 and n_2 but on whether or not they are equal. Suppose that $n_1 \neq n_2$. Player I can remove enough coins from the larger heap in order to leave two heaps of equal size for player II. Now player I, when it is her turn, can mimic player II's moves. Thus if player II takes c coins from one of the heaps, then player I takes the same number c of coins from the other heap. Such a strategy guarantees a win for player I. If $n_1 = n_2$, then player II can win by mimicking player I's moves. Thus, we have completely solved 2-heap Nim. An example of play in the 2-heap game of Nim with heaps of sizes 8 and 5, respectively, is

$$8, 5 \xrightarrow{\text{I}} 5, 5 \xrightarrow{\text{II}} 5, 2 \xrightarrow{\text{I}} 2, 2 \xrightarrow{\text{II}} 0, 2 \xrightarrow{\text{I}} 0, 0.$$

The preceding idea in solving 2-heap Nim, namely, moving in such a way as to leave two equal heaps, can be generalized to any number k of heaps. The insight one needs is provided by the concept of the base 2 numeral of an integer. Recall that each positive integer n can be expressed as a base 2 numeral by repeatedly removing the largest power of 2 which does not exceed the number. For instance, to express the decimal number 57 in base 2, we observe that

$$\begin{aligned}
2^5 &\leq 57 < 2^6, & 57 - 2^5 &= 25 \\
2^4 &\leq 25 < 2^5, & 25 - 2^4 &= 9 \\
2^3 &\leq 9 < 2^4, & 9 - 2^3 &= 1 \\
2^0 &\leq 1 < 2^1, & 1 - 2^0 &= 0.
\end{aligned}$$

Thus,

$$57 = 2^5 + 2^4 + 2^3 + 2^0,$$

and the base 2 numeral for 57 is

$$111001.$$

Each digit in a base 2 numeral is either 0 or 1. The digit in the ith position, the one corresponding to 2^i, is called the ith bit[14] ($i \geq 0$). We can think of each heap of coins as consisting of *subheaps* of powers

---

[13]This is an important principle to follow in general: Consider small or special cases in order to develop understanding and intuition. Then try to extend your ideas in order to solve the problem in general.

[14]The word *bit* is short for *binary digit*.

of 2, according to its base numeral. Thus a heap of size 53 consists of subheaps of sizes $2^5, 2^4, 2^2$, and 2^0. In the case of 2-heap Nim, the total number of subheaps of each size is either 0, 1, or 2. There is exactly one subheap of a particular size if and only if the two heaps have different sizes. Put another way, the total number of subheaps of each size is even if and only if the two heaps have the same size—that is, if and only if player II can win the Nim game.

Now consider a general Nim game with heaps of sizes $n_1, n_2, \ldots, n_k$. Express each of the numbers n_i as base 2 numerals:

$$
\begin{aligned}
n_1 &= a_s \cdots a_1 a_0 \\
n_2 &= b_s \cdots b_1 b_0 \\
&\cdots \\
n_k &= e_s \cdots e_1 e_0.
\end{aligned}
$$

(By including leading 0's we can assume that all of the heap sizes have base 2 numerals with the same number of digits.) We call a Nim game *balanced*, provided that the number of subheaps of each size is even. Thus, a Nim game is balanced if and only if

$$a_s + b_s + \cdots + e_s \text{ is even,}$$
$$\vdots$$
$$a_i + b_i + \cdots + e_i \text{ is even,}$$
$$\vdots$$
$$a_0 + b_0 + \cdots + e_0 \text{ is even.}$$

A Nim game that is not balanced is called *unbalanced*. We say that the ith bit is *balanced*, provided that the sum $a_i + b_i + \cdots + e_i$ is even, and is *unbalanced* otherwise. Thus, a balanced game is one in which all bits are balanced, while an unbalanced game is one in which there is at least one unbalanced bit.

We then have the following:

> Player I can win in unbalanced Nim games, and player II can win in balanced Nim games.

To see this, we generalize the strategies used in 2-pile Nim. Suppose the Nim game is unbalanced. Let the largest unbalanced bit be the jth bit. Then player I moves in such a way as to leave a balanced game for player II. She does this by selecting a heap whose jth bit is 1 and removing a number of coins from it so that the resulting game is balanced (see also Exercise 33). No matter what player II does, she leaves for player I an unbalanced game again, and player I once again balances it. Continuing like this ensures player I a win. If the game starts out balanced, then player I's first move unbalances it, and now player II adopts the strategy of balancing the game whenever it is her move.

For example, consider a 4-pile Nim game with heaps of sizes 7, 9, 12, and 15. The base 2 numerals for these heap sizes are, respectively, 0111, 1001, 1100, and 1111. In terms of subheaps of powers of 2 we have:

	$2^3 = 8$	$2^2 = 4$	$2^1 = 2$	$2^0 = 1$
Heap of size 7	0	1	1	1
Heap of size 9	1	0	0	1
Heap of size 12	1	1	0	0
Heap of size 15	1	1	1	1

This game is unbalanced with the 3rd, 2nd and 0th bits unbalanced. Player I can remove 11 coins from the pile of size 12, leaving 1 coin. Since the base 2 numeral of 1 is 0001, the game is now balanced. Alternatively, player I can remove 5 coins from the pile of size 9, leaving 4 coins, or player I can remove 13 coins from the pile of size 15, leaving 2 coins.

1.8   Exercises

1. Show that an m-by-n chessboard has a perfect cover by dominoes if and only if at least one of m and n is even.

2. Consider an m-by-n chessboard with m and n both odd. To fix the notation, suppose that the square in the upper left-hand corner is colored white. Show that if a white square is cut out anywhere on the board, the resulting pruned board has a perfect cover by dominoes.

3. Imagine a prison consisting of 64 cells arranged like the squares of an 8-by-8 chessboard. There are doors between all adjoining cells. A prisoner in one of the corner cells is told that he will be released, provided he can get into the diagonally opposite corner cell after passing through every other cell exactly once. Can the prisoner obtain his freedom?

4. (a) Let $f(n)$ count the number of different perfect covers of a 2-by-n chessboard by dominoes. Evaluate $f(1), f(2), f(3), f(4)$, and $f(5)$. Try to find (and verify) a simple relation that the counting function f satisfies. Use this relation to compute $f(12)$.

* (b) Let $g(n)$ be the number of different perfect covers of a 3-by-n chessboard by dominoes. Evaluate $g(1), g(2), \ldots, g(6)$.

5. Find the number of different perfect covers of a 3-by-4 chessboard by dominoes.

6. Show how to cut a cube, 3 feet on an edge, into 27 cubes, 1 foot on an edge, using exactly 6 cuts, but making a nontrivial rearrangement of the pieces between two of the cuts.

7. Consider the following three-dimensional version of the chessboard problem: A *three-dimensional domino* is defined to be the geometric figure that results when two cubes, 1 unit on an edge, are joined along a face. Show that it is possible to construct a cube n units on an edge from dominoes if and only if n is even. If n is odd, is it possible to construct a cube n units on an edge with a 1-by-1 hole in the middle? (Hint: Think of a cube n units on an edge as being composed of n^3 cubes 1 unit on an edge. Color the cubes alternately black and white.)

8. Let a and b be positive integers with a a factor of b. Show that an m-by-n board has a perfect cover by a-by-b pieces if and only if a is a factor of both m and n and b is a factor of either m or n. (Hint: Partition the a-by-b pieces into a 1-by-b pieces.)

9. Use Exercise 8 to conclude that when a is a factor of b, an m-by-n board has a perfect cover by a-by-b pieces if and only if it has a trivial perfect cover in which all the pieces are oriented the same way.

10. Show that the conclusions of Exercises 8 and 9 need not hold when a is not a factor of b.

11. Verify that there is no magic square of order 2.

12. Use de la Loubère's method to construct a magic square of order 7.

13. Use de la Loubère's method to construct a magic square of order 9.

14. Construct a magic square of order 6.

15. Show that a magic square of order 3 must have a 5 in the middle position. Deduce that there are exactly 8 magic squares of order 3.

16. Can the partial square below be completed to a magic square of order 4?

$$\begin{bmatrix} 2 & 3 & & \\ 4 & & & \\ & & & \\ & & & \end{bmatrix}$$

17. Show that the result of replacing every integer a in a magic square of order n with $n^2 + 1 - a$ is a magic square of order n.

18. Let n be a positive integer divisible by 4, say $n = 4m$. Consider the following construction of an n-by-n array:

    (i) Proceeding from left to right and from first row to nth row, fill in the places of the array with the integers $1, 2, \ldots, n^2$ in order.

    (ii) Partition the resulting square array into m^2 4-by-4 smaller arrays. Replace each number a on the two diagonals of each of the 4-by-4 arrays with its "complement" $n^2 + 1 - a$.

    Verify that this construction produces a magic square of order n when $n = 4$ and $n = 8$. (Actually it produces a magic square for each n divisible by 4.)

19. Show that there is no magic cube of order 2.

20. * Show there is no magic cube of order 4.

21. Show that the following map of 10 countries $\{1, 2, \ldots, 10\}$ can be colored with three but no fewer colors. If the colors used are red, white, and blue, determine the number of different colorings.

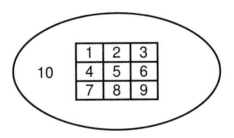

22. (a) Does there exist a *magic hexagon* of order 2? That is, is it possible to arrange the numbers $1, 2, \ldots, 7$ in the hexagonal array below so that all of the nine "line" sums (the sum of the numbers in the hexagonal boxes penetrated by a line through midpoints of opposite sides) are the same?

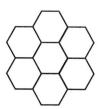

* (b) Construct a magic hexagon of order 3; that is, arrange the integers $1, 2, \ldots, 19$ in a hexagonal array (three integers on a side) in such a way that all of the fifteen "line" sums are the same (namely, 38).

23. Construct a pair of orthogonal Latin squares of order 4.

24. Construct Latin squares of orders 5 and 6.

25. Find a general method for constructing a Latin square of order n.

26. A 6-by-6 chessboard is perfectly covered with 18 dominoes. Prove that it is possible to cut it either horizontally or vertically into two nonempty pieces without cutting through a domino; that is, prove that there must be a fault-line.

27. Construct a perfect cover of an 8-by-8 chessboard with dominoes having no fault-line.

28. Determine all shortest routes from A to B in the system of intersections and streets (graph) in the following diagram. The numbers on the streets represent the lengths of the streets measured in terms of some unit.

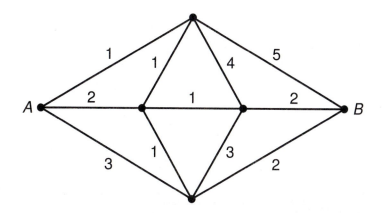

29. Consider 3-heap Nim with piles of sizes 1, 2, and 4. Show that this game is unbalanced and determine a first move for player I.

30. Is 4-pile Nim with heaps of sizes 22, 19, 14, and 11 balanced or unbalanced? Player I's first move is to remove 6 coins from the heap of size 19. What should player II's first move be?

31. Consider 5-pile Nim with heaps of sizes 10, 20, 30, 40, and 50. Is this game balanced? Determine a first move for player I.

32. Show that player I can always win a Nim game in which the number of heaps with an odd number of coins is odd.

33. Show that in an unbalanced game of Nim in which the largest unbalanced bit is the jth bit, player I can always balance the game by removing coins from any heap the base 2 numeral of whose number has a 1 in the jth bit.

34. Suppose we change the object of Nim so that the player who takes the last coin loses (the *misère* version). Show that the following is a winning strategy: Play as in ordinary Nim until all but exactly one heap contains a single coin. Then remove either all or all but one of the coins of the exceptional heap so as to leave an *odd* number of heaps of size 1.

35. A game is played between two players, alternating turns as follows: The game starts with an empty pile. When it is his turn a player may add either 1, 2, 3, or 4 coins to the pile. The person who adds the 100th coin to the pile is the winner. Determine whether it is the first or second player who can guarantee a win in this game. What is the winning strategy to follow?

36. Suppose that in Exercise 35, the player who adds the 100th coin loses. Now who wins and how?

37. Eight people are at a party and pair off to form four teams of two. In how many ways can this be done? (This is sort of an "unstructured" domino covering problem.)

38. A Latin square of order n is *idempotent* provided the integers $1, 2, \ldots, n$ occur, in this order, in the diagonal positions $(1, 1), (2, 2)$, $\ldots, (n, n)$, and is *symmetric* provided the integer in position (i, j) equals the integer in position (j, i) whenever $i \neq j$. There is no symmetric, idempotent Latin square of order 2. Construct a symmetric, idempotent Latin square of order 3. Show that there is

no symmetric, idempotent Latin square of order 4. What about order n in general, where n is even?

39. Take any set of $2n$ points in the plane with no three collinear, and then arbitrarily color each point red or blue. Prove that it is always possible to pair up the red points with the blue points by drawing line segments connecting them so that no two of the line segments intersect.

40. Consider an n-by-n board and L-tetrominoes (4 squares joined in the shape of an L). Show that if there is a perfect cover of the n-by-n board with L-tetrominoes, then n is divisible by 4. What about m-by-n-boards?

Chapter 2

The Pigeonhole Principle

We consider in this chapter an important, but elementary, combinatorial principle that can be used to solve a variety of interesting problems, often with surprising conclusions. This principle is known under a variety of names, the most common of which are the *pigeonhole principle*, the *Dirichlet drawer principle*, and the *shoebox principle*.[1] Formulated as a principle about pigeonholes, it says roughly that if a lot of pigeons fly into not too many pigeonholes, then at least one pigeonhole will be occupied by two or more pigeons. A more precise statement is given below.

2.1 Pigeonhole Principle: Simple Form

The simplest form of the pigeonhole principle is the following:

Theorem 2.1.1 *If $n+1$ objects are put into n boxes, then at least one box contains two or more of the objects.*

Proof. If each of the n boxes contains at most one of the objects, then the total number of objects is at most n. Since we start with $n + 1$ objects, some box contains at least two of the objects. $\square$

Notice that neither the pigeonhole principle nor its proof gives any help in finding a box that contains two or more of the objects. They simply assert that if one examines each of the boxes, one will come upon a box that contains more than one object. The pigeonhole principle merely guarantees the existence of such a box. Thus, whenever the

---

[1] The word *shoebox* is a mistranslation and folk etymology for the German *Schubfach*, which means *pigeonhole* (in a desk).

pigeonhole principle is applied to prove the existence of an arrangement or some phenomenon, it will give no indication of how to construct the arrangement or find an instance of the phenomenon other than to examine all possibilities.

Notice also that the conclusion of the pigeonhole principle cannot be *guaranteed* if there are only n (or fewer) objects. This is because we may put a different object in each of the n boxes. Of course, it is possible to distribute as few as two objects among the boxes in such a way that a box contains two objects, but there is no guarantee. The pigeonhole principle asserts that, no matter how one distributes $n + 1$ objects among n boxes, one cannot avoid putting two objects in the same box.

Instead of putting objects into boxes one may think of coloring each object with one of n colors. The pigeonhole principle then asserts that if $n + 1$ objects are colored with n colors, then two objects have the same color.

We begin with two simple applications:

Application 1. Among 13 people there are two who have their birthdays in the same month. □

Application 2. There are n married couples. How many of the $2n$ people must be selected in order to guarantee that one has selected a married couple?

To apply the pigeonhole principle in this case, think of n boxes, one corresponding to each of the n couples. If we select $n + 1$ people and put each of them in the box corresponding to the couple to which they belong, then some box contains two people; that is, we have selected a married couple. Two of the ways to select n people without getting a married couple are to select all the husbands or all the wives. Therefore $n + 1$ is the smallest number that will guarantee a married couple has been selected. □

There are other principles related to the pigeonhole principle that are worth stating formally:

- *If n objects are put into n boxes and no box is empty, then each box contains exactly one object.*

- *If n objects are put into n boxes and no box gets more than one object, then each box has an object in it.*

Referring to Application 2, if we select n people in such a way that we have selected at least one person from each married couple, then we have selected exactly one person from each couple. Also, if we select n people without selecting more than one person from each married couple, then we have selected at least one (and, hence, exactly one) person from each couple.

More abstract formulations of the three principles enunciated thus far are the following:

Let X and Y be finite sets and let $f : X \to Y$ be a function from X to Y.

- *If X has more elements than Y, then f is not one to one.*

- *If X and Y have the same number of elements and f is onto, then f is one to one.*

- *If X and Y have the same number of elements and f is one to one, then f is onto.*

$\square$

Application 3. Given m integers $a_1, a_2, \ldots, a_m$, there exist integers k and l with $0 \le k < l \le m$ such that $a_{k+1} + a_{k+2} + \cdots + a_l$ is divisible by m. Less formally, there exist consecutive a's in the sequence $a_1, a_2, \ldots, a_m$ whose sum is divisible by m.

To see this, consider the m sums

$$a_1, a_1 + a_2, a_1 + a_2 + a_3, \ldots, a_1 + a_2 + a_3 + \cdots + a_m.$$

If any of these sums is divisible by m, then the conclusion holds. Thus, we may suppose that each of these sums has a nonzero remainder when divided by m, and so a remainder equal to one of $1, 2, \ldots, m-1$. Since there are m sums and only $m-1$ remainders, two of the sums have the same remainder when divided by m. Therefore, there are integers k and l with $k < l$ such that $a_1 + a_2 + \cdots + a_k$ and $a_1 + a_2 + \cdots + a_l$ have the same remainder r when divided by m:

$$a_1 + a_2 + \cdots + a_k = bm + r, \qquad a_1 + a_2 + \cdots + a_l = cm + r.$$

Subtracting, we find that $a_{k+1} + \cdots + a_l = (c-b)m$; thus, $a_{k+1} + \cdots + a_l$ is divisible by m.

To illustrate this argument,[2] let $m = 7$ and let our integers be $2, 4, 6, 3, 5, 5$, and 6. Computing the sums as above, we get $2, 6, 12, 15, 20,$ 25, and 31 whose remainders when divided by 7 are, respectively, $2, 6, 5, 1,$ $6, 4,$ and 3. We have two remainders equal to 6, and this implies the conclusion that $6 + 3 + 5 = 14$ is divisible by 7. □

Application 4. A chess master who has 11 weeks to prepare for a tournament decides to play at least one game every day but, in order not to tire himself, he decides not to play more than 12 games during any calendar week. Show that there exists a succession of (consecutive) days during which the chess master will have played *exactly* 21 games.

Let a_1 be the number of games played on the first day, a_2 the total number of games played on the first and second days, a_3 the total number of games played on the first, second, and third days, and so on. The sequence of numbers $a_1, a_2, \ldots, a_{77}$ is a strictly increasing sequence[3] since at least one game is played each day. Moreover, $a_1 \geq 1$, and since at most 12 games are played during any one week, $a_{77} \leq 12 \times 11 = 132$.[4] Hence, we have

$$1 \leq a_1 < a_2 < \cdots < a_{77} \leq 132.$$

The sequence $a_1 + 21, a_2 + 21, \ldots, a_{77} + 21$ is also a strictly increasing sequence:

$$22 \leq a_1 + 21 < a_2 + 21 < \cdots < a_{77} + 21 \leq 132 + 21 = 153.$$

Thus each of the 154 numbers

$$a_1, a_2, \ldots, a_{77}, a_1 + 21, a_2 + 21, \ldots, a_{77} + 21$$

is an integer between 1 and 153. It follows that two of them are equal. Since no two of the numbers $a_1, a_2, \ldots, a_{77}$ are equal and no two of the numbers $a_1 + 21, a_2 + 21, \ldots, a_{77} + 21$ are equal, there must be an i and a j such that $a_i = a_j + 21$. Therefore on days $j + 1, j + 2, \ldots, i$ the chess master played a total of 21 games. □

Application 5. From the integers $1, 2, \ldots, 200$, we choose 101 integers. Show that, among the integers chosen, there are two such that one of them is divisible by the other.

---

[2]The argument actually contains a nice algorithm, whose validity relies on the pigeonhole principle, for finding the consecutive a's, which is more efficient than examining all sums of consecutive a's.

[3]Each term of the sequence is larger than the one that precedes it.

[4]This is the only place where the assumption that at most 12 games are played during any of the 11 calendar weeks is used. Thus, this assumption could be replaced by the assumption that at most 132 games are played in 77 days.

By factoring out as many 2's as possible, we see that any integer can be written in the form $2^k \times a$, where $k \geq 0$ and a is odd. For an integer between 1 and 200, a is one of the 100 numbers $1, 3, 5, \ldots, 199$. Thus among the 101 integers chosen, there are two having a's of equal value when written in this form. Let these two numbers be $2^r \times a$ and $2^s \times a$. If $r < s$, then the second number is divisible by the first. If $r > s$, then the first is divisible by the second.   □

Let us note that the result of Application 5 is the best possible in the sense that one may select 100 integers from $1, 2, \ldots, 200$ in such a way that no one of the selected integers is divisible by any other, for instance, the 100 integers $101, 102, \ldots, 199, 200$.

We conclude this section with another application from number theory. First, we recall that two positive integers m and n are said to be *relatively prime* if their greatest common divisor[5] is 1. Thus 12 and 35 are relatively prime, but 12 and 15 are not since 3 is a common divisor of 12 and 15.

Application 6. (*Chinese remainder theorem*) Let m and n be relatively prime positive integers, and let a and b be integers where $0 \leq a \leq m - 1$ and $0 \leq b \leq n - 1$. Then there is a positive integer x such that the remainder when x is divided by m is a, and the remainder when x is divided by n is b; that is, x can be written in the form $x = pm + a$ and also in the form $x = qn + b$ for some integers p and q.

To show this, we consider the n integers

$$a, m + a, 2m + a, \ldots, (n - 1)m + a.$$

Each of these integers has remainder a when divided by m. Suppose that two of them had the same remainder r when divided by n. Let the two numbers be $im + a$ and $jm + a$ where $0 \leq i < j \leq n - 1$. Then there are integers q_i and q_j such that

$$im + a = q_i n + r$$

and

$$jm + a = q_j n + r.$$

Subtracting the first equation from the second, we get

$$(j - i)m = (q_j - q_i)n.$$

---

[5] Also called *greatest common factor* or *highest common factor*.

The preceding equation tells us that n is a factor of the number $(j-i)m$. Since n has no common factor other than 1 with m, it follows that n is a factor of $j-i$. However, $0 \le i < j \le n-1$ implies that $0 < j-i \le n-1$, and hence n cannot be a factor of $j - i$. This contradiction arises from our supposition that two of the numbers

$$a, m + a, 2m + a, \ldots, (n - 1)m + a$$

had the same remainder when divided by n. We conclude that each of these n numbers has a different remainder when divided by n. By the pigeonhole principle each of the n numbers $0, 1, \ldots, n - 1$ occurs as a remainder; in particular, the number b does. Let p be the integer with $0 \le p \le n-1$ such that the number $x = pm + a$ has remainder b when divided by n. Then, for some integer q,

$$x = qn + b.$$

So $x = pm + a$ and $x = qn + b$, and x has the required properties. □

The fact that a rational number a/b has a decimal expansion that eventually repeats is a consequence of the pigeonhole principle, and we leave a proof of this fact for the exercises.

For further applications we will need a stronger form of the pigeonhole principle.

2.2   Pigeonhole Principle: Strong Form

The following theorem contains Theorem 2.1.1 as a special case:

Theorem 2.2.1 *Let $q_1, q_2, \ldots, q_n$ be positive integers. If*

$$q_1 + q_2 + \cdots + q_n - n + 1$$

objects are put into n boxes, then either the first box contains at least q_1 objects, or the second box contains at least q_2 objects, ..., or the nth box contains at least q_n objects.

Proof. Suppose that we distribute $q_1 + q_2 + \cdots + q_n - n + 1$ objects among n boxes. If for each $i = 1, 2, \ldots, n$ the ith box contains fewer than q_i objects, then the total number of objects in all boxes does not exceed

$$(q_1 - 1) + (q_2 - 1) + \cdots + (q_n - 1) = q_1 + q_2 + \cdots + q_n - n.$$

Since this number is one less than the number of objects distributed, we conclude that for some $i = 1, 2, \ldots, n$ the ith box contains at least q_i objects.   □

Notice that it is possible to distribute $q_1 + q_2 + \cdots + q_n - n$ objects among n boxes in such a way that for no $i = 1, 2, \ldots, n$ is it true that the ith box contains q_i or more objects. We do this by putting $q_1 - 1$ objects into the first box, $q_2 - 1$ objects into the second box, and so on.

The simple form of the pigeonhole principle is obtained from the strong form by taking $q_1 = q_2 = \cdots = q_n = 2$. Then

$$q_1 + q_2 + \cdots + q_n - n + 1 = 2n - n + 1 = n + 1.$$

In terms of coloring, the strong form of the pigeonhole principle asserts that if each of $q_1 + q_2 + \cdots + q_n - n + 1$ objects is assigned one of n colors, then there is an i such that there are (at least) q_i objects of the ith color.

In elementary mathematics the strong form of the pigeonhole principle is most often applied in the special case when $q_1, q_2, \ldots, q_n$ are all equal to some integer r. In this case the principle reads as follows:

- If $n(r - 1) + 1$ objects are put into n boxes, then at least one of the boxes contains r or more of the objects. Equivalently,

- If the average of n nonnegative integers $m_1, m_2, \ldots, m_n$ is greater than $r - 1$, that is,

$$\frac{m_1 + m_2 + \cdots + m_n}{n} > r - 1,$$

  then at least one of the integers is greater than or equal to r.

The connection between these two formulations is obtained by taking $n(r - 1) + 1$ objects and putting them into n boxes. For $i = 1, 2, \ldots, n$, let m_i be the number of objects in the ith box. Then the average of the numbers $m_1, m_2, \ldots, m_n$ is

$$\frac{m_1 + m_2 + \cdots + m_n}{n} = \frac{n(r - 1) + 1}{n} = (r - 1) + \frac{1}{n}.$$

Since this average is greater than $r - 1$, one of the integers m_i is at least r. In other words, one of the boxes contains at least r objects.

Another averaging principle is the following:

- If the average of n nonnegative integers $m_1, m_2, \ldots, m_n$ is less than $r + 1$, that is,

$$\frac{m_1 + m_2 + \cdots + m_n}{n} < r + 1,$$

then at least one of the integers is less than $r + 1$.

Application 7. A basket of fruit is being arranged out of apples, bananas, and oranges. What is the smallest number of pieces of fruit that should be put in the basket in order to guarantee that either there are at least 8 apples or at least 6 bananas or at least 9 oranges?

By the strong form of the pigeonhole principle, $8 + 6 + 9 - 3 + 1 = 21$ pieces of fruit, no matter how selected, will guarantee a basket of fruit with the desired properties. But 7 apples, 5 bananas, and 8 oranges, a total of 20 pieces of fruit, will not. □

The following is yet another averaging principle:

- If the average of n nonnegative integers $m_1, m_2, \ldots, m_n$ is at least equal to r, then at least one of the integers $m_1, m_2, \ldots, m_n$ satisfies $m_i \geq r$.

Application 8. Two disks, one smaller than the other, are each divided into 200 congruent sectors.[6] In the larger disk, 100 of the sectors are chosen arbitrarily and painted red; the other 100 sectors are painted blue. In the smaller disk, each sector is painted either red or blue with no stipulation on the number of red and blue sectors. The small disk is then placed on the larger disk so that their centers coincide. Show that it is possible to align the two disks so that the number of sectors of the small disk whose color matches the corresponding sector of the large disk is at least 100.

To see this, we observe that if the large disk is fixed in place, there are 200 possible positions for the small disk such that each sector of the small disk is contained in a sector of the large disk. We first count the total number of color matches over all of the 200 possible positions of the disks. Since the large disk has 100 sectors of each of the two colors, each sector of the small disk will match in color the corresponding sector of the large disk in exactly 100 of the 200 possible positions. Thus, the total number of color matches over all the positions equals the number of sectors of the small disk multiplied by 100, and this

---

[6] 200 equal slices of a pie.

equals 20,000. Therefore, the average number of color matches per position is $20,000/200 = 100$. So there must be some position with at least 100 color matches. □

We next present an ~~application that was first discovered by~~ Erdös and Szekeres.[7]

Application 9. Show that every sequence $a_1, a_2, \ldots, a_{n^2+1}$ of $n^2 + 1$ real numbers contains either an increasing subsequence of length $n + 1$ or a decreasing subsequence of length $n + 1$.

We first clarify the notion of a subsequence. If $b_1, b_2, \ldots, b_m$ is a sequence, then $b_{i_1}, b_{i_2}, \ldots, b_{i_k}$ is a *subsequence*, provided that $1 \le i_1 < i_2 < \cdots < i_k \le m$. Thus b_2, b_4, b_5, b_6 is a subsequence of $b_1, b_2, \ldots, b_8$, but b_2, b_6, b_5 is not. The subsequence $b_{i_1}, b_{i_2}, \ldots, b_{i_k}$ is *increasing* (more properly *not decreasing*) if $b_{i_1} \le b_{i_2} \le \cdots \le b_{i_k}$ and *decreasing* if $b_{i_1} \ge b_{i_2} \ge \cdots \ge b_{i_k}$.

We now prove the assertion. We suppose that there is no increasing subsequence of length $n + 1$ and show that there must be a decreasing subsequence of length $n + 1$. For each $k = 1, 2, \ldots, n^2 + 1$, let m_k be the length of the longest increasing subsequence that begins with a_k. Suppose $m_k \le n$ for each $k = 1, 2, \ldots, n^2 + 1$, so that there is no increasing subsequence of length $n + 1$. Since $m_k \ge 1$ for each $k = 1, 2, \ldots, n^2 + 1$, the numbers $m_1, m_2, \ldots, m_{n^2+1}$ are $n^2 + 1$ integers each between 1 and n. By the strong form of the pigeonhole principle, $n + 1$ of the numbers $m_1, m_2, \ldots, m_{n^2+1}$ are equal. Let

$$m_{k_1} = m_{k_2} = \cdots = m_{k_{n+1}},$$

where $1 \le k_1 < k_2 < \cdots < k_{n+1} \le n^2 + 1$. Suppose that for some $i = 1, 2, \ldots, n$, $a_{k_i} < a_{k_{i+1}}$. Then, since $k_i < k_{i+1}$ we could take a longest increasing subsequence beginning with $a_{k_{i+1}}$ and put a_{k_i} in front to obtain an increasing subsequence beginning with a_{k_i}. Since this implies that $m_{k_i} > m_{k_{i+1}}$, we conclude that $a_{k_i} \ge a_{k_{i+1}}$. Since this is true for each $i = 1, 2, \ldots, n$, we have

$$a_{k_1} \ge a_{k_2} \ge \cdots \ge a_{k_{n+1}},$$

and we conclude that $a_{k_1}, a_{k_2}, \ldots, a_{k_{n+1}}$ is a decreasing subsequence of length $n + 1$. □

An amusing formulation of Application 9 is the following: Suppose that $n^2 + 1$ people are lined up shoulder to shoulder in a straight line.

---

[7]P. Erdös and A. Szekeres: A combinatorial problem in geometry, *Compositio Mathematica*, 2 (1935), 463-470.

Then it is always possible to choose $n+1$ of the people to take one step forward so that going from left to right their heights are increasing (or decreasing). It is instructive to read through the proof of Application 9 in these terms.

2.3 A Theorem of Ramsey

We now discuss without proof a profound and important generalization of the pigeonhole principle which is called Ramsey's theorem,[8] after the English logician Frank Ramsey.[9]

The following is the most popular and easily understood instance of Ramsey's theorem:

> Of six (or more) people, either there are three, each pair of whom are acquainted, or there are three, each pair of whom are unacquainted.

One way to prove this result is to examine all the different ways in which 6 people can be acquainted and unacquainted. This is a tedious task, but nonetheless one that can be accomplished with a little fortitude. There is, however, a simple and elegant proof that avoids consideration of cases. Before giving this proof, we formulate the result more abstractly as

$$K_6 \to K_3, K_3 \qquad (\text{read } K_6 \text{ arrows } K_3, K_3). \qquad (2.1)$$

What does this mean? First, by K_6 we mean a set of 6 objects (e.g., people) and all of the 15 (unordered) pairs of these objects. We can picture K_6 by choosing 6 points in the plane, no 3 of which are collinear, and then drawing the edge or line segment connecting each pair of points (the edges now represent the pairs). In general, we mean by K_n a set of n objects and all of the pairs of these objects.[10] Illustrations for K_n ($n = 1, 2, 3, 4, 5$) are given in Figure 2.1. Notice that the picture of K_3 is that of a triangle, and we often refer to K_3 as a *triangle*.

---

[8]For a proof, see H.J. Ryser: *Combinatorial Mathematics*, Mathematical Association of America, Providence (1963) or R.L. Graham, B.L. Rothschild and J.H. Spencer: *Ramsey Theory*, second edition, Wiley, New York (1990).

[9]Frank Ramsey was born in 1903 and died in 1930 when he was not quite 27 years of age. In spite of his premature death, he laid the foundation for what is now called *Ramsey theory*.

[10]In later chapters K_n is called the *complete graph* of order n.

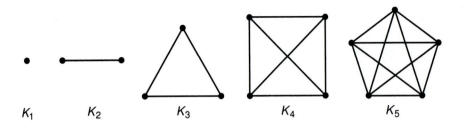

K_1 K_2 K_3 K_4 K_5

Figure 2.1

We distinguish between acquainted pairs and unacquainted pairs by coloring edges red for acquainted and blue for unacquainted. "Three mutually acquainted people" now means "a K_3 each of whose edges is colored red: a *red K_3*." Similarly, three mutually unacquainted people form a *blue K_3*. We can now explain the expression (2.1):

$K_6 \rightarrow K_3, K_3$ is the assertion that *no matter how the edges of K_6 are colored with the colors red and blue, there is always a red K_3 (3 of the original 6 points with the 3 line segments between them all colored red) or a blue K_3 (3 of the original 6 points with the 3 line segments between them all colored blue), in short a monochromatic triangle.*

To prove that $K_6 \rightarrow K_3, K_3$, we argue as follows: Suppose the edges of K_6 have been colored red or blue in any way. Consider one of the points p of K_6. It meets 5 edges. Since each of these 5 edges is colored red or blue, it follows (from the strong form of the pigeonhole principle) that either at least 3 of them are colored red or at least 3 of them are colored blue. We suppose that 3 of the 5 edges meeting the point p are red. (If 3 are blue a similar argument works.) Let the 3 red edges meeting p join p to points a, b, and c, respectively. Consider the edges which join a, b, c in pairs. If all of these are blue, then a, b, c determine a blue K_3. If one of them, say the one joining a and b is red, then p, a, b determine a red K_3. Thus, we are guaranteed either a red K_3 or a blue K_3.

We observe that the assertion $K_5 \rightarrow K_3, K_3$ is false. This is because there is *some* way to color the edges of K_5 without creating a red K_3 or a blue K_3. This is shown in Figure 2.2, where the edges of the pentagon (the solid edges) are the red edges and the edges of the inscribed pentagram (the dashed edges) are the blue edges.

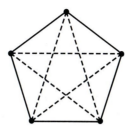

Figure 2.2

More generally, Ramsey's theorem, still not in its full generality, asserts

if $m \geq 2$ and $n \geq 2$ are integers, then there is a positive integer p such that

$$K_p \to K_m, K_n.$$

In words, given m and n there is a positive integer p such that, if the edges of K_p are colored red or blue, then either there is a red K_m or there is a blue K_n. The existence of either a red K_m or a blue K_n is guaranteed, no matter how the edges of K_p are colored.

If $K_p \to K_m, K_n$, then $K_q \to K_m, K_n$ for any integer $q \geq p$. The *Ramsey number* $r(m,n)$ is the smallest integer p such that $K_p \to K_m, K_n$. Ramsey's theorem asserts the existence of the number $r(m,n)$. Earlier we proved that

$$r(3,3) = 6.$$

The Ramsey numbers $r(2,n)$ and $r(m,2)$ can be determined. We show that $r(2,n) = n$ by the following argument:

$(r(2,n) \leq n)$: If we color the edges of K_n either red or blue, then either some edge is colored red (and so we have a red K_2) or all edges are blue (and so we have a blue K_n).

$(r(2,n) > n-1)$: If we color all the edges of K_{n-1} blue, then we have neither a red K_2 nor a blue K_n.

In a similar way, one shows that $r(m,2) = m$. These are the *trivial Ramsey numbers*. In general, by interchanging the colors red and blue, we see that

$$r(m,n) = r(n,m).$$

With this observation in mind, the following table[11] contains known facts about nontrivial Ramsey numbers $r(m,n)$:

---

[11]The paper "Small Ramsey Numbers" by S.P. Radziszowski, *Electronic Journal of Combinatorics*, Dynamic Survey #1 contains this and other information.

$$r(3,3) = 6,$$
$$r(3,4) = r(4,3) = 9,$$
$$r(3,5) = r(5,3) = 14,$$
$$r(3,6) = r(6,3) = 18,$$
$$r(3,7) = r(7,3) = 23,$$
$$r(3,8) = r(8,3) = 28,$$
$$r(3,9) = r(9,3) = 36,$$
$$40 \leq r(3,10) = r(10,3) \leq 43,$$
$$r(4,4) = 18,$$
$$r(4,5) = r(5,4) = 25,$$
$$43 \leq r(5,5) \leq 49.$$

Notice that the fact that $r(3,10)$ lies between 40 and 43 implies that

$$K_{43} \to K_3, K_{10}$$

and

$$K_{39} \not\to K_3, K_{10}.$$

Thus, there is no way to color the edges of K_{43} without creating either a red K_3 or a blue K_{10}; there is a way to color the edges of K_{39} without creating either a red K_3 or a blue K_{10}, but neither of these conclusions is known to be true for K_{40}, K_{41}, and K_{42}. The assertion $43 \leq r(5,5) \leq 55$ implies that $K_{55} \to K_5, K_5$ and that there is a way to color the edges of K_{42} without creating a monochromatic K_5.

Ramsey's theorem generalizes to any number of colors. Thus, if n_1, n_2, and n_3 are integers greater than or equal to 2, then there exists an integer p such that

$$K_p \to K_{n_1}, K_{n_2}, K_{n_3}.$$

In words, if each of the edges of K_p is colored red, blue, or green, then either there is a red K_{n_1} or a blue K_{n_2} or a green K_{n_3}. The smallest integer p for which this assertion holds is the Ramsey number $r(n_1, n_2, n_3)$. The only nontrivial Ramsey number of this type that is known is

$$r(3,3,3) = 17.$$

The Ramsey numbers $r(n_1, n_2, \ldots, n_k)$ are defined in a similar way, and Ramsey's theorem in its full generality for pairs asserts that these numbers exist; that is, there is an integer p such that

$$K_p \to K_{n_1}, K_{n_2}, \ldots, K_{n_k}.$$

There is a more general form of Ramsey's theorem in which pairs (subsets of two elements) are replaced by subsets of t elements for some fixed integer $t \geq 1$. Let

$$K_n^t$$

denote the collection of all subsets of t elements of a set of n elements. Generalizing our preceding notation, we obtain the general form of Ramsey's theorem: Given integers $t \geq 2$ and integers $q_1, q_2, \ldots, q_k \geq t$, there exists an integer p such that

$$K_p^t \to K_{q_1}^t, K_{q_2}^t, \ldots, K_{q_k}^t.$$

In words, there exists an integer p such that if each of the t-element subsets of a p-element set is assigned one of k colors $c_1, c_2, \ldots, c_k$, then either there are q_1 elements, all of whose t element subsets are assigned the color c_1, or there are q_2 elements, all of whose t-element subsets are assigned the color c_2, ..., or there are q_k elements, all of whose t-element subsets are assigned the color c_k. The smallest such integer p is the *Ramsey number*

$$r_t(q_1, q_2, \ldots, q_k).$$

Suppose $t = 1$. Then $r_1(q_1, q_2, \ldots, q_k)$ is the smallest number p such that, if the elements of a set of p elements are colored with one of the colors $c_1, c_2, \ldots, c_k$, then either there are q_1 elements of color c_1, or q_2 elements of color c_2, or . . . , or q_k elements of color c_k. Thus, by the strong form of the pigeonhole principle,

$$r_1(q_1, q_2, \ldots, q_k) = q_1 + q_2 + \cdots q_k - k + 1.$$

This demonstrates that Ramsey's theorem is a generalization of the strong form of the pigeonhole principle.

The determination of the general Ramsey numbers $r_t(q_1, q_2, \ldots, q_k)$ is a difficult problem. Very little is known about their exact values. It is not difficult to see that

$$r_t(t, q_2, \ldots, q_k) = r_t(q_2, \ldots, q_k),$$

and that the order in which $q_1, q_2, \ldots, q_k$ are listed does not affect the value of the Ramsey number.

2.4 Exercises

1. Concerning Application 4, show that there is a succession of days during which the chess master will have played exactly k games,

for each $k = 1, 2, \ldots, 21$. (The case $k = 21$ is the case treated in Application 4.) Is it possible to conclude that there is a succession of days during which the chess master will have played exactly 22 games?

2. * Concerning Application 5, show that if 100 integers are chosen from $1, 2, \ldots, 200$, and one of the integers chosen is less than 16, then there are two chosen numbers such that one of them is divisible by the other.

3. Generalize Application 5 by choosing (how many?) integers from the set
$$\{1, 2, \ldots, 2n\}.$$

4. Show that if $n+1$ integers are chosen from the set $\{1, 2, \ldots, 2n\}$, then there are always two which differ by 1.

5. Show that if $n+1$ integers are chosen from the set $\{1, 2, \ldots, 3n\}$, then there are always two which differ by at most 2.

6. Generalize Exercises 4 and 5.

7. * Show that for any given 52 integers there exist two of them whose sum, or else whose difference, is divisible by 100.

8. Use the pigeonhole principle to prove that the decimal expansion of a rational number m/n eventually is repeating. For example,
$$34,478/99,900 = .34512512512512512 \cdots.$$

9. In a room there are 10 people, none of whom are older than 60 (ages are given in whole numbers only) but each of whom is at least 1 year old. Prove that one can always find two groups of people (with no common person) the sum of whose ages is the same. Can 10 be replaced by a smaller number?

10. A child watches TV at least one hour each day for 7 weeks but never more than 11 hours in any one week. Prove that there is some period of consecutive days in which the child watches exactly 20 hours of TV. (It is assumed that the child watches TV for a whole number of hours each day.)

11. A student has 37 days to prepare for an examination. From past experience she knows that she will require no more than 60 hours of study. She also wishes to study at least 1 hour per day.

Show that no matter how she schedules her study time (a whole number of hours per day, however), there is a succession of days during which she will have studied exactly 13 hours.

12. Show by example that the conclusion of the Chinese remainder theorem (Application 6) need not hold when m and n are not relatively prime.

13. ∗ Let S be a set of 6 points in the plane, with no 3 of the points collinear. Color either red or blue each of the 15 line segments determined by the points of S. Show that there are at least two triangles determined by points of S which are either red triangles or blue triangles. (Both may be red, or both may be blue, or one may be red and the other blue.)

14. A bag contains 100 apples, 100 bananas, 100 oranges, and 100 pears. If I pick one piece of fruit out of the bag every minute, how long will it be before I am assured of having picked at least a dozen pieces of fruit of the same kind?

15. Prove that, for any $n+1$ integers $a_1, a_2, \ldots, a_{n+1}$, there exist two of the integers a_i and a_j with $i \neq j$ such that $a_i - a_j$ is divisible by n.

16. Prove that in a group of $n > 1$ people there are two who have the same number of acquaintances in the group. (It is assumed that no one is acquainted with him or herself.)

17. There are 100 people at a party. Each person has an even number (possibly zero) of acquaintances. Prove that there are three people at the party with the same number of acquaintances.

18. Prove that of any five points chosen within a square of side length 2, there are two whose distance apart is at most $\sqrt{2}$.

19. (a) Prove that of any five points chosen within an equilateral triangle of side length 1, there are two whose distance apart is at most $\frac{1}{2}$.

    (b) Prove that of any ten points chosen within an equilateral triangle of side length 1, there are two whose distance apart is at most $\frac{1}{3}$.

    (c) Determine an integer m_n such that if m_n points are chosen within an equilateral triangle of side length 1, there are two whose distance apart is at most $1/n$.

20. Prove that $r(3, 3, 3) \leq 17$.

21. * Prove that $r(3, 3, 3) \geq 17$ by exhibiting a coloring, with colors red, blue, and green, of the line segments joining 16 points with the property that there do not exist 3 points such that the 3 line segments joining them are all colored the same.

22. Prove that

$$r(\underbrace{3, 3, \ldots, 3}_{k+1}) \leq (k+1)(r(\underbrace{3, 3, \ldots, 3}_{k}) - 1) + 2.$$

Use this result to obtain an upper bound for

$$r(\underbrace{3, 3, \ldots, 3}_{n}).$$

23. The line segments joining 10 points are arbitrarily colored red or blue. Prove that there must exist 3 points such that the 3 line segments joining them are all red, or 4 points such that the 6 line segments joining them are all blue (that is, $r(3, 4) \leq 10$).

24. Let q_3 and t be positive integers with $q_3 \geq t$. Determine the Ramsey number $r_t(t, t, q_3)$.

25. Let $q_1, q_2, \ldots, q_k, t$ be positive integers, where $q_1 \geq t$, $q_2 \geq t, \ldots,$ $q_k \geq t$. Let m be the largest of $q_1, q_2, \ldots, q_k$. Show that

$$r_t(m, m, \ldots, m) \geq r_t(q_1, q_2, \ldots, q_k).$$

Conclude that, to prove Ramsey's theorem, it is enough to prove it in the case that $q_1 = q_2 = \cdots = q_k$.

26. Suppose that the mn people of a marching band are standing in a rectangular formation of m rows and n columns in such a way that in each row each person is taller than the one to her or his left. Suppose that the leader rearranges the people in each column in increasing order of height from front to back. Show that the rows are still arranged in increasing order of height from left to right.

27. A collection of subsets of $\{1, 2, \ldots, n\}$ has the property that each pair of subsets has at least one element in common. Prove that there are at most 2^{n-1} subsets in the collection.

28. At a dance-hop there are 100 men and 20 women. For each i from $1, 2, \ldots, 100$, the ith man selects a group of a_i women as potential dance partners (his dance list), but in such a way that given any group of 20 men, it is always possible to pair the 20 men up with the 20 women with each man paired up with a woman on his dance list. What is the smallest sum $a_1 + a_2 + \cdots + a_{100}$ that will *guarantee* this?

Chapter 3

Permutations and
Combinations

Most readers of this book will have had some experience with simple counting problems, so that the concepts "permutations" and "combinations" are probably familiar. But the experienced counter knows that even rather simple-looking problems can pose difficulties in their solutions. While it is generally true that in order to learn mathematics one must *do* mathematics, it is especially so here—the serious student should attempt to solve a large number of problems.

In this chapter, we explore four general principles and some of the counting formulas that they imply. Each of these principles gives a "complementary" principle which we also discuss.

3.1   Four Basic Counting Principles

The first principle[1] is very basic. It is one formulation of the principle that the whole is equal to the sum of its parts.

A *partition* of a set S is a collection $S_1, S_2, \ldots, S_m$ of subsets of S such that each element of S is in exactly one of those subsets:

$$S = S_1 \cup S_2 \cup \cdots \cup S_m,$$

$$S_i \cap S_j = \emptyset, \quad (i \neq j).$$

---

[1]According to the *The Random House College Dictionary, Revised Edition*, 1975, a *principle* is 1. an accepted or professed rule of action or conduct. 2. a basic law, axiom, or doctrine. Our principles are basic *laws* of mathematics, and important *rules of action* for solving counting problems.

The subsets $S_1, S_2, \ldots, S_m$ are called the *parts* of the partition. We note that by this definition a part of a partition may be empty, but usually there is no advantage in considering partitions with one or more empty parts. The *number of objects of a set* S is denoted by $|S|$ and is sometimes called the *size* of S.

Addition Principle. *Suppose that a set S is partitioned into parts $S_1, S_2, \ldots, S_m$. The number of objects in S can be determined by finding the number of objects in each of the parts, and adding the numbers so obtained:*

$$|S| = |S_1| + |S_2| + \cdots + |S_m|.$$

If the sets $S_1, S_2, \ldots, S_m$ are allowed to overlap, then a more profound principle, the inclusion–exclusion principle of Chapter 6, can be used to count the number of objects in S.

In applying the addition principle, we usually define the parts descriptively. In other words, we break the problem up into mutually exclusive cases that exhaust all possibilities. The art of applying the addition principle is to partition the set S to be counted into "manageable parts"—that is, parts which one can readily count. But this statement needs to be qualified. If we partition S into too many parts, then we may have defeated ourselves. For instance, if we partition S into parts each containing only one element, then applying the addition principle is the same as counting the number of parts, and this is basically the same as listing all the objects of S. Thus, a more appropriate description is that *the art of applying the addition principle is to partition the set S into "not too many manageable parts."*

Example. Suppose we wish to find the number of different courses offered by the University of Wisconsin–Madison. We partition the courses according to the department in which they are listed. *Provided there is no cross-listing* (cross-listing occurs when the same course is listed by more than one department), the number of courses offered by the University equals the sum of the number of courses offered by each department. □

Another formulation of the addition principle in terms of choices is the following: *If an object can be selected from one pile in p ways and also an object can be selected from a separate pile in q ways, then the selection of one object chosen from either of the two piles can be made in $p + q$ ways.* This formulation has an obvious generalization to more than two piles.

Example. A student wishes to take either a mathematics course or a biology course, but not both. If there are 4 mathematics courses and 3 biology courses for which the student has the necessary prerequisites, then the student can choose a course to take in $4 + 3 = 7$ ways. □

The second principle is a little more complicated. We state it for two sets, but it can also be generalized to any finite number of sets.

Multiplication Principle. *Let S be a set of ordered pairs (a, b) of objects, where the first object a comes from a set of size p, and for each choice of object a there are q choices for object b. Then the size of S is $p \times q$:*

$$|S| = p \times q.$$

The multiplication principle is actually a consequence of the addition principle. Let $a_1, a_2, \ldots, a_p$ be the p different choices for the object a. We partition S into parts $S_1, S_2, \ldots, S_p$ where S_i is the set of ordered pairs in S with first object a_i, $(i = 1, 2, \ldots, p)$. The size of each S_i is q; hence, by the addition principle,

$$\begin{aligned} |S| &= |S_1| + |S_2| + \cdots + |S_p| \\ &= q + q + \cdots + q \quad (p \ q\text{'s}) \\ &= p \times q. \end{aligned}$$

Note how the basic fact – multiplication of whole numbers is just repeated addition – enters into the above derivation.

A second useful formulation of the multiplication principle is as follows: *If a first task has p outcomes and, no matter what the outcome of the first task, a second task has q outcomes, then the two tasks performed consecutively have $p \times q$ outcomes.*

Example. A student is to take two courses. The first meets at any one of 3 hours in the morning, and the second at any one of 4 hours in the afternoon. The number of schedules that are possible for the student is $3 \times 4 = 12$. □

As already remarked, the multiplication principle can be generalized to $3, 4$, or any finite number of sets. Rather than formulate it in terms of n sets, we give examples for $n = 3$ and $n = 4$.

Example. Chalk comes in 3 different lengths, 8 different colors, and 4 different diameters. How many different kinds of chalk are there?

To determine a piece of chalk we carry out 3 different tasks (it matters not in which order we take these tasks): choose a length, choose a color, choose a diameter. By the multiplication principle there are $3 \times 8 \times 4 = 96$ different kinds of chalk. □

Example. The number of ways a man, woman, boy, and girl can be selected from 5 men, 6 women, 2 boys, and 4 girls is $5 \times 6 \times 2 \times 4 = 240$.

The reason is that we have 4 different tasks to carry out: select a man (5 ways), select a woman (6 ways), select a boy (2 ways), select a girl (4 ways). If, in addition, we ask for the number of ways one person can be selected, the answer is $5 + 6 + 2 + 4 = 17$. This follows from the addition principle for 4 piles. □

Example. Determine the number of positive integers that are factors of the number

$$3^4 \times 5^2 \times 11^7 \times 13^8.$$

The numbers $3, 5, 11$, and 13 are prime numbers. By the *fundamental theorem of arithmetic*, each factor is of the form

$$3^i \times 5^j \times 11^k \times 13^l,$$

where $0 \leq i \leq 4$, $0 \leq j \leq 2$, $0 \leq k \leq 7$, and $0 \leq l \leq 8$. There are 5 choices for i, 3 for j, 8 for k, and 9 for l. By the multiplication principle, the number of factors is

$$5 \times 3 \times 8 \times 9 = 1080.$$

□

In the multiplication principle the q choices for object b may vary with the choice of a. The only requirement is that there be the *same number q* of choices, not necessarily the same choices.

Example. How many two-digit numbers have distinct and nonzero digits?

A two-digit number ab can be regarded as an ordered pair (a, b) where a is the tens digit and b is the units digit. Neither of these digits is allowed to be 0 in the problem, and the two digits are to be different. There are 9 choices for a, namely $1, 2, \ldots, 9$. Once a is chosen, there are 8 choices for b. If $a = 1$, these 8 choices are $2, 3, \ldots, 9$, if $a = 2$, the 8 choices are $1, 3, \ldots, 9$, and so on. What is important for application of the multiplication principle is that the number of choices is always 8. The answer to the questions is, by the multiplication principle, $9 \times 8 = 72$.

We can arrive at the answer 72 in another way. There are 90 two-digit numbers, $10, 11, 12, \ldots, 99$. Of these numbers, 9 have a 0, (namely, $10, 20, \ldots, 90$) and 9 have identical digits (namely, $11, 22, \ldots, 99$). Thus the number of two-digit numbers with distinct and nonzero digits equals $90 - 9 - 9 = 72$. □

The preceding example illustrates two ideas. One is that there may be more than one way to arrive at the answer to a counting question. The other idea is that to find the number of objects in a set A (in this case the set of two-digit numbers with distinct and nonzero digits) it may be easier to find the number of objects in a larger set U containing S (the set of all two-digit numbers in the example above) and then subtract the number of objects of U that do not belong to A (the two-digit numbers containing a 0 or identical digits). We formulate this idea as our third principle.

Subtraction Principle. Let A be a set and let U be a larger set containing A. Let

$$\overline{A} = \{x \in U : x \notin A\}$$

be the *complement of A in U*. Then the number $|A|$ of objects in A is given by the rule

$$|A| = |U| - |\overline{A}|.$$

In applying the subtraction principle, the set U is usually some natural set consisting of all the objects under discussion (the so-called *universal set*). Using the subtraction principle makes sense only if it is easier to count the number of objects in U and in $|\overline{A}|$ than to count the number of objects in A.

Example. Computer passwords are to consist of a string of 6 symbols taken from the digits $0, 1, 2, \ldots, 9$ and the lowercase letters $a, b, c, \ldots, z$. How many computer passwords have a repeated symbol?

We want to count the number of objects in the set A of computer passwords with a repeated symbol. Let U be the set of all computer passwords. Taking the complement of A in U we get the set $\overline{A}$ of computer passwords with no repeated symbol. By two applications of the multiplication principle, we get

$$|U| = 36^6 = 2,176,782,336$$

and

$$|\overline{A}| = 36 \cdot 35 \cdot 34 \cdot 33 \cdot 32 \cdot 31 = 1,402,410,240.$$

Therefore,

$$|A| = |U| - |\overline{A}| = 2,176,782,336 - 1,402,410,240 = 774,372,096.$$

$\square$

We now formulate the final principle of this section.

Division Principle. Let S be a finite set that is partitioned into k parts in such a way that each part contains the same number of objects. Then the number of parts in the partition is given by the rule

$$k = \frac{|S|}{\text{number of objects in a part}}.$$

Thus, we can determine the number of parts if we know the number of objects in S and the common value of the number of objects in the parts.

Example. There are 740 pigeons in a collection of pigeonholes. If each pigeonhole contains 5 pigeons, the number of pigeonholes equals

$$\frac{740}{5} = 148.$$

□

More profound applications of the division principle will occur later in this book. Now consider the next example.

Example. You wish to give your Aunt Mollie a basket of fruit. In your refrigerator you have 6 oranges and 9 apples. The only requirement is that there must be at least one piece of fruit in the basket (that is, an empty basket of fruit is not allowed). How many different baskets of fruit are possible?

One way to count the number of baskets is the following: First, ignore the requirement that the basket cannot be empty. We can compensate for that later. What distinguishes one basket of fruit from another is the number of oranges and number of apples in the basket. There are 7 choices for the number of oranges $(0, 1, \ldots, 6)$ and 10 choices for the number of apples $(0, 1, \ldots, 9)$. By the multiplication principle, the number of different baskets is $7 \times 10 = 70$. Subtracting the empty basket, the answer is 69. Notice that if we had not (temporarily) ignored the requirement that the basket be nonempty, then there would have been 9 or 10 choices for the number of apples depending on whether or not the number of oranges was 0, and we could not have applied the multiplication principle directly. But an alternative solution is the following. Partition the nonempty baskets into two parts, S_1 and S_2, where S_1 consists of those baskets with no oranges and S_2 consists of those baskets with at least one orange. The size of S_1 is 9 $(1, 2, \ldots, 9$ apples) and the size of S_2 by the reasoning above is $6 \times 10 = 60$. The number of possible baskets of fruit is, by the addition principle, $9 + 60 = 69$. □

We made an implicit assumption in the preceding example, which we should now bring into the open. It was assumed in the solution that the oranges were indistinguishable from one another (an orange is an orange is an orange is ...), and that the apples were indistinguishable from one another. Thus, what mattered in making up a basket of fruit was *not* which apples and which oranges went into it but only the *number* of each type of fruit. If we distinguished between the various oranges and the various apples (one orange is perfectly round, another is bruised, a third very juicy, and so on), then the number of baskets would be larger. We will return to this example in Section 3.5.

Before continuing with more examples, we discuss some general ideas.

A great many counting problems are of the following types:

(i) Count the number of *ordered* arrangements or *ordered* selections of objects

    (a) without repeating any object,

    (b) with repetition of objects permitted (but perhaps limited).

(ii) Count the number of *unordered* arrangements or *unordered* selections of objects

    (a) without repeating any object,

    (b) with repetition of objects permitted (but perhaps limited).

Instead of distinguishing between nonrepetition and repetition of objects, it is sometimes more convenient to distinguish between selections from a set and a multiset. A *multiset* is like a set except that its members need not be distinct.[2] For example, the multiset

$$M = \{a, a, a, b, c, c, d, d, d, d\}$$

has 10 elements of 4 different types: 3 of type a, 1 of type b, 2 of type c, and 4 of type d. We shall also indicate a multiset by specifying the number of times different types of elements occur. Thus, M shall also be denoted by $\{3 \cdot a, 1 \cdot b, 2 \cdot c, 4 \cdot d\}$. The numbers $3, 1, 2$, and 4 are the *repetition numbers* of the multiset M. A set is a multiset that has all repetition numbers equal to 1. In order to include the

---

[2] Thus a multiset breaks one of the cardinal rules of sets, namely, elements are not repeated in sets; they are either in the set or not in the set. The set $\{a, a, b\}$ is the same as the set $\{a, b\}$ but not so for multisets.

listed case (b) when there is no limit on the number of times an object of each type can occur (except for that imposed by the size of the arrangement), we allow infinite repetition numbers.[3] Thus, a multiset in which a and c each have an infinite repetition number and b and d have repetition numbers 2 and 4, respectively, is denoted by $\{\infty \cdot a, 2 \cdot b, \infty \cdot c, 4 \cdot d\}$. Arrangements or selections in (i) in which order is taken into consideration are called *permutations*, whereas arrangements or selections in (ii) in which order is irrelevant are called *combinations*. In the next two sections we will develop some general formulas for the number of permutations and combinations of sets and multisets. But not all permutation and combination problems can be solved by using these formulas. It is often necessary to return to the basic addition and multiplication principles.

Example. How many odd numbers between 1000 and 9999 have distinct digits?

A number between 1000 and 9999 is an *ordered* arrangement of 4 digits. Thus we are asked to count a certain collection of permutations. We have 4 choices to make: a units, a tens, a hundreds, and a thousands digit. Since the numbers we want to count are odd, the units digit can be any one of $1, 3, 5, 7, 9$. The tens and the hundreds digit can be any one of $0, 1, \ldots, 9$, while the thousands digit can be any one of $1, 2, \ldots, 9$. Thus, there are 5 choices for the units digit. Since the digits are to be distinct, we have 8 choices for the thousands digit, whatever the choice of the units digit. Then, there are 8 choices for the hundreds digit, whatever the first 2 choices were, and 7 choices for the tens digit, whatever the first 3 choices were. Thus, by the multiplication principle, the answer to the question is $5 \times 8 \times 8 \times 7 = 2240$. □

Suppose in the previous example we made the choices in a different order: first choose the thousands digit, then the hundreds, tens, and units. There are 9 choices for the thousands digit, then 9 choices for the hundreds digit (since we are allowed to use 0), 8 choices for the tens digit, but now the number of choices for the units digit (which has to be odd) *depends* on the previous choices. If we had chosen no odd digits, the number of choices for the units digit would be 5; if we had chosen one odd digit, the number of choices for the units digit would be 4; and so on. Thus, we cannot invoke the multiplication principle, if we carry out our choices in the reverse order. There are two lessons to learn from this example. One is that as soon as your answer for the

---

[3]There are no circumstances in which we will have to worry about different sizes of infinity.

number of choices of one of the tasks is "it depends" (or some such words), the multiplication principle cannot be applied. The second is that there may not be a fixed order in which the tasks have to be taken, and by changing the order a problem may be more readily solved by the multiplication principle. A *rule of thumb* to keep in mind is to "make the most restrictive choice" first.

Example. How many integers strictly between 0 and 10,000 have exactly one digit equal to 5?

Let S be the set of integers between 0 and 10,000 with exactly one digit equal to 5.

First solution: One digit equal to 5. It is natural to partition S into the set S_1 of one-digit numbers in S, the set S_2 of two-digit numbers in S, the set S_3 of three-digit numbers in S, and the set S_4 of four-digit numbers in S. There are no five-digit numbers in S. We clearly have

$$|S_1| = 1.$$

The numbers in S_2 naturally fall into two types: (i) the units digit is 5, and (ii) the tens digit is 5. The number of the first type is 8 (the tens digit cannot be 0 nor can it be 5). The number of the second type is 9 (the units digit cannot be 5). Hence,

$$|S_2| = 8 + 9 = 17.$$

Reasoning in a similar way, we obtain

$$|S_3| = 8 \times 9 + 8 \times 9 + 9 \times 9 = 225, \text{ and}$$

$$|S_4| = 8 \times 9 \times 9 + 8 \times 9 \times 9 + 8 \times 9 \times 9 + 9 \times 9 \times 9 = 2,673.$$

Thus,

$$|S| = 1 + 17 + 225 + 2,673 = 2,916.$$

Second solution: By including leading zeros (e.g., think of 6 as 0006, 25 as 0025, 352 as 0352) we can regard each number in S as a four-digit number. Now we partition S into the sets S_1', S_2', S_3', S_4' according to whether the 5 is in the first, second, third, or fourth position. Each of the four sets in the partition contains $9 \times 9 \times 9 = 729$ integers, and so the number of integers in S equals

$$4 \times 729 = 2,916.$$

$\square$

Example. How many different five-digit numbers can be constructed out of the digits 1, 1, 1, 3, 8?

Here we are asked to count permutations of a multiset with 3 objects of one type, 1 of another, and 1 of a third. We really have only two choices to make: which position is to be occupied by the 3 (5 choices) and then which position is to be occupied by the 8 (4 choices). The remaining three places are occupied by 1's. By the multiplication principle the answer is $5 \times 4 = 20$.

If the five digits are 1, 1, 1, 3, 3, the answer is 10, half as many. $\square$

These examples clearly demonstrate that mastery of the addition and multiplication principles is essential for becoming an expert counter.

3.2 Permutations of Sets

Let r be a positive integer. By an *r-permutation* of a set S of n elements we understand an ordered arrangement of r of the n elements. If $S = \{a, b, c\}$, then the three 1-permutations of S are

$$a \quad b \quad c,$$

the six 2-permutations of S are

$$ab \quad ac \quad ba \quad bc \quad ca \quad cb,$$

and the six 3-permutations of S are

$$abc \quad acb \quad bac \quad bca \quad cab \quad cba.$$

There are no 4-permutations of S since S has fewer than 4 elements.

We denote by $P(n, r)$ the number of r-permutations of an n-element set. If $r > n$, then $P(n, r) = 0$. Clearly $P(n, 1) = n$ for each positive integer n. An n-permutation of an n-element set S will be more simply called a *permutation of S* or a *permutation of n elements*. Thus, a *permutation of a set S can be thought of as a listing of the elements of S in some order.* Previously we saw that $P(3, 1) = 3, P(3, 2) = 6$, and $P(3, 3) = 6$.

Theorem 3.2.1 *For n and r positive integers with $r \leq n$,*

$$P(n, r) = n \times (n - 1) \times \cdots \times (n - r + 1).$$

Proof. In constructing an r-permutation of an n-element set, we can choose the first item in n ways, the second item in $n-1$ ways, whatever the choice of the first item, . . . , and the rth item in $n - (r - 1)$ ways, whatever the choice of the first $r-1$ items. By the multiplication principle the r items can be chosen in $n \times (n - 1) \times \cdots \times (n - r + 1)$ ways. □

For a nonnegative integer n, we define $n!$ (read n *factorial*) by

$$n! = n \times (n - 1) \times \cdots \times 2 \times 1,$$

with the convention that $0! = 1$. We may then write

$$P(n, r) = \frac{n!}{(n - r)!}.$$

For $n \geq 0$, we define $P(n, 0)$ to be 1, and this agrees with the formula when $r = 0$. The number of permutations of n elements is

$$P(n, n) = \frac{n!}{0!} = n!.$$

□

Example. The number of 4-letter "words" that can be formed by using each of the letters a, b, c, d, e at most once is $P(5, 4)$, and this equals $5!/(5 - 4)! = 120$. The number of 5-letter words equals $P(5, 5)$, which is also 120. □

Example. The so-called "15 puzzle" consists of 15 sliding unit squares labeled with the numbers 1 through 15 and mounted in a 4-by-4 square frame as shown in Figure 3.1. The challenge of the puzzle is to move from the initial position shown to any specified position. By a position, we mean an arrangement of the 15 numbered squares in the frame with one empty unit square. What is the number of positions in the puzzle (ignoring whether it is possible to move to the position from the initial one)?

1	2	3	4
5	6	7	8
9	10	11	12
13	14	15	

Figure 3.1

The problem is equivalent to determining the number of ways to assign the numbers $1, 2, \ldots, 15$ to the 16 squares of a 4-by-4 grid, leaving one square empty. Since we can assign the number 16 to the empty square, the problem is also equivalent to determining the number of assignments of the numbers $1, 2, \ldots, 16$ to the 16 squares, and this is $P(16, 16) = 16!$

What is the number of ways to assign the numbers $1, 2, \ldots, 15$ to the squares of a 6-by-6 grid, leaving 21 squares empty? These assignments correspond to the 15-permutations of the 36 squares as follows: To an assignment of $1, 2, \ldots, 15$ to 15 of the squares, we associate the 15-permutation of the 36 squares obtained by putting the square labeled 1 first, the square labeled 2 second, and so on. Hence the total number of assignments is $P(36, 15) = 36!/21!$. $\square$

Example. What is the number of ways to order the 26 letters of the alphabet so that no two of the vowels $a, e, i, o,$ and u occur consecutively?

The solution to this problem (like so many counting problems) is straightforward once one sees how to do it. We think of two main tasks to be accomplished. The first task is to decide how to order the consonants among themselves. There are 21 consonants, and so 21! permutations of the consonants. Since we cannot have two consecutive vowels in our final arrangement, the vowels must be in 5 of the 22 spaces before, between, and after the consonants. Our second task is to put the vowels in these places. There are 22 places for the a, then 21 for the e, 20 for the i, 19 for the o, and 18 for the u. That is, the second task can be accomplished in

$$P(22, 5) = \frac{22!}{17!}$$

ways. By the multiplication principle, we determine that the number of ordered arrangements of the letters of the alphabet with no two vowels consecutive is

$$21! \times \frac{22!}{17!}.$$

$\square$

Example. How many seven-digit numbers are there such that the digits are distinct integers taken from $\{1, 2, \ldots, 9\}$ and such that the digits 5 and 6 do not appear consecutively in either order?

We want to count certain 7-permutations of the set $\{1, 2, \ldots, 9\}$, and we partition these 7-permutations into 4 types: (i) neither 5 nor

6 appears as a digit; (ii) 5, but not 6, appears as a digit; (iii) 6, but not 5, appears as a digit; (iv) both 5 and 6 appears as digits. The permutations of type (i) are the 7-permutations of $\{1, 2, 3, 4, 7, 8, 9\}$, and hence their number is $P(7,7) = 7! = 5040$. The permutations of type (ii) can be counted as follows: The digit equal to 5 can be any one of the 7 digits. The remaining 6 digits are a 6-permutation of $\{1, 2, 3, 4, 7, 8, 9\}$. Hence there are $7P(7,6) = 7(7!) = 35,280$ numbers of type (ii). In a similar way we see that there are 35,280 numbers of type (iii). To count the number of permutations of type (iv), we partition the permutations of types (iv) into three parts:

First digit equal to 5, and so second digit not equal to 6:

$$5 \quad \neq 6$$

$$- \quad - \quad - \quad - \quad - \quad - \quad \text{--}$$

There are 5 places for the 6. The other 5 digits constitute a 5-permutation of the 7 digits $\{1, 2, 3, 4, 7, 8, 9\}$. Hence, there are

$$5 \times P(7,5) = \frac{5 \times 7!}{2!} = 12,600$$

numbers in this part.

Last digit equal to 5, and so next to last digit not equal to 6:

$$\neq 6 \quad 5$$

$$- \quad - \quad - \quad - \quad - \quad - \quad \text{--}$$

By an argument similar to the preceding, we conclude there are also 12,600 numbers in this part.

A digit other than the first or last is equal to 5:

$$\neq 6 \quad 5 \quad \neq 6$$

$$- \quad - \quad - \quad - \quad - \quad - \quad \text{--}$$

The place occupied by 5 is any one of 5 places. The place for the 6 can then be chosen in 4 ways. The remaining 5 digits constitute a 5-permutation of the 7 digits $\{1, 2, 3, 4, 7, 8, 9\}$. Hence, there are $5 \times 4 \times P(7,5) = 50,400$ numbers in this category. Thus, there are

$$2(12{,}600) + 50{,}400 = 75{,}600$$

numbers of types (iv). By the addition principle, the answer to the problem posed is

$$5{,}040 + 2(35{,}280) + 75{,}600 = 151{,}200.$$

The solution just given was arrived at by partitioning the set of objects we wanted to count into manageable parts, parts the number of whose objects we could calculate, and then using the addition principle. But the addition principle can be used in a more subtle way which leads to an alternative, and computationally easier, solution to the problem. Let us consider the entire collection T of seven-digit numbers that can be formed by using distinct integers from $\{1, 2, \ldots, 9\}$. The set T then contains

$$P(9, 7) = \tfrac{9!}{2!} = 181,440$$

numbers. We partition T into two subsets: S, which consists of those numbers in T in which 5 and 6 do not occur consecutively, and its complement[4] $\overline{S}$ which consists of those numbers in T in which 5 and 6 do occur consecutively. We wish to determine the size of S. By the addition principle the size of T equals the size of S plus the size of $\overline{S}$. If we can find the size of $\overline{S}$, then our problem is solved. How many numbers are there in $\overline{S}$? In $\overline{S}$, the digits 5 and 6 occur consecutively. There are 6 ways to position a 5 followed by a 6, and 6 ways to position a 6 followed by a 5. The remaining digits constitute a 5-permutation of $\{1, 2, 3, 4, 7, 8, 9\}$. So the number of numbers in $\overline{S}$ is

$$2 \times 6 \times P(7, 5) = 30,240.$$

But then S contains

$$181,440 - 30,240 = 151,200$$

numbers.   □

The permutations that we have just considered are more properly called *linear permutations*. We think of the objects as being arranged in a line. If instead of arranging objects in a line, we arrange them in a circle, the number of permutations is smaller. Think of it this way: Suppose 6 children are marching in a circle. In how many different ways can they form their circle? Since the children are moving, what matters are their positions relative to each other and not to their environment. Thus, it is natural to regard two circular permutations as being the same provided one can be brought to the other by a rotation, that is, by a circular shift. There are 6 linear permutations for each circular permutation. For example, the circular permutation

1

---

[4]The *complement* of S (relative to T) consists of all elements of T that do not belong to S.

$$2 \qquad\qquad\qquad 6$$

$$3 \qquad\qquad\qquad 5$$

$$4$$

arises from each of the linear permutations

$$123456 \qquad 234561 \qquad 345612$$

$$456123 \qquad 561234 \qquad 612345$$

by regarding the last digit as coming before the first digit. Thus, there is a 6-to-1 correspondence between the linear permutations of 6 children and the circular permutations of the 6 children. Therefore, to find the number of circular permutations we divide the number of linear permutations by 6. Hence, the number of circular permutations of the 6 children equals $6!/6 = 5!$.

Theorem 3.2.2 *The number of circular r-permutations of a set of n elements is given by*

$$\frac{P(n,r)}{r} = \frac{n!}{r \cdot (n-r)!}.$$

In particular, the number of circular permutations of n elements is $(n-1)!$.

Proof. A proof is essentially contained in the preceding paragraph and uses the division principle. The set of linear r-permutations can be partitioned into parts in such a way that two linear r-permutations correspond to the same circular r-permutation if and only if they are in the same part. Thus, the number of circular r-permutations equals the number of parts. Since each part contains r linear r-permutations, the number of parts is given by

$$\frac{P(n,r)}{r} = \frac{n!}{r \cdot (n-r)!}.$$

$\square$

For emphasis, we remark that the preceding argument worked because each part contained the same number of r-permutations so that we could apply the division principle. If, for example, we partition a set of 10 objects into 3 parts of sizes $2, 4$, and 4, respectively, the number of parts is *not* $10/3$.

Another way to view the counting of circular permutations is the following: Suppose we wish to count the number of circular permutations of A, B, C, D, E, and F (the number of ways to seat A, B, C, D, E, and F around a table). Since we are free to rotate the people, any circular permutation can be rotated so that A is in a fixed position; think of it as the "head" of the table:

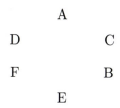

Now that A is fixed, the circular permutations of A, B, C, D, E, and F can be identified with the linear permutations of B, C, D, E, and F. (The preceding circular permutation is identified with the linear permutation $DFEBC$.) There are 5! linear permutations of B, C, D, E, and F, and hence 5! circular permutations of A, B, C, D, E, and F.

This way of looking at circular permutations is also useful when the formula for circular permutations cannot be applied directly.

Example. Ten people, including two who do not wish to sit next to one another, are to be seated at a round table. How many circular seating arrangements are there?

To solve this problem, let the 10 people be $P_1, P_2, P_3, \ldots, P_{10}$, where P_1 and P_2 are the two who do not wish to sit together. Consider seating arrangements for 9 people $X, P_3, \ldots, P_{10}$ at a round table. There are 8! such arrangements. If we replace X by either P_1, P_2 or by P_2, P_1 in each of these arrangements, we obtain a seating arrangement for the 10 people in which P_1 and P_2 are next to one another. Hence the number of such arrangements in which P_1 and P_2 are not together is $9! - 2 \times 8! = 7 \times 8!$.

Another way to analyze this problem is the following: First seat P_1 at the "head" of the table. Then P_2 cannot be on either side of P_1. There are 8 choices for the person on P_1's left, 7 choices for the person on P_1's right, and the remaining seats can be filled in 7! ways. Thus, the number of seating arrangements in which P_1 and P_2 are not together is

$$8 \times 7 \times 7! = 7 \times 8!.$$

$\square$

As we did before our discussion of circular permutations, we will continue to use "permutation" to mean "linear permutation."

Example. The number of ways 12 different markings can be placed on a rotating drum is $P(12, 12)/12 = 11!$. □

Example. What is the number of necklaces that can be made from 20 beads, each of a different color?

There are 20! permutations of the 20 beads. Since each necklace can be rotated without changing the arrangement of the beads, the number of necklaces is at most $20!/20 = 19!$. Since a necklace can also be turned over without changing the arrangement of the beads, the total number of necklaces is $19!/2$. □

Circular permutations and necklaces are counted again in Chapter 14, in a more general context.

3.3   Combinations of Sets

Let r be a nonnegative integer. By an r-*combination* of a set S of n elements, we understand an unordered selection of r of the n objects of S. In other words, an r-combination of S is a subset of S consisting of r of the n objects of S—that is, an r-element subset of S. If $S = \{a, b, c, d\}$, then

$$\{a, b, c\}, \{a, b, d\}, \{a, c, d\}, \{b, c, d\}$$

are the four 3-combinations of S. We denote by $\binom{n}{r}$ the number of r-combinations of an n-element set.[5] Obviously,

$$\binom{n}{r} = 0 \qquad \text{if } r > n.$$

Also,

$$\binom{0}{r} = 0 \qquad \text{if } r > 0.$$

The following additional facts are readily seen to be true for each non-negative integer n:

$$\binom{n}{0} = 1, \quad \binom{n}{1} = n, \quad \binom{n}{n} = 1.$$

In particular, $\binom{0}{0} = 1$. The basic formula for combinations is given in the next theorem.

---

[5] Other common notations for these numbers are $C(n, r)$ and $_nC_r$.

Theorem 3.3.1 *For $0 \le r \le n$,*

$$P(n,r) = r! \binom{n}{r}.$$

Hence,

$$\binom{n}{r} = \frac{n!}{r!(n-r)!}.$$

Proof. Let S be an n-element set. Each r-permutation of S arises in exactly one way as a result of carrying out the following two tasks:

(i) Choose r elements from S.

(ii) Arrange the chosen r elements in some order.

The number of ways to carry out the first task is, by definition, the combination number $\binom{n}{r}$. The number of ways to carry out the second task is $P(r,r) = r!$. By the multiplication principle, we have $P(n,r) = r! \binom{n}{r}$. We now use our formula $P(n,r) = \frac{n!}{(n-r)!}$ and obtain

$$\binom{n}{r} = \frac{P(n,r)}{r!} = \frac{n!}{r!(n-r)!}. \qquad \Box$$

Example. Twenty-five points are chosen in the plane so that no three of them are collinear. How many straight lines do they determine? How many triangles do they determine?

Since no three of the points lie on a line, every pair of points determines a unique straight line. Thus, the number of straight lines determined equals the number of 2-combinations of a 25-element set, and this is given by

$$\binom{25}{2} = \frac{25!}{2!23!} = 300.$$

Similarly, every three points determine a unique triangle, so that the number of triangles determined is given by

$$\binom{25}{3} = \frac{25!}{3!22!}. \qquad \Box$$

Example. There are 15 people enrolled in a mathematics course, but exactly 12 attend on any given day. The number of different ways that 12 students can be chosen is

$$\binom{15}{12} = \frac{15!}{12!3!}.$$

If there are 25 seats in the classroom, the 12 students could seat themselves in $P(25, 12) = 25!/13!$ ways. Thus, there are

$$\binom{15}{12} P(25, 12) = \frac{15!25!}{12!3!13!}$$

ways in which an instructor might see the 12 students in the classroom.
□

Example. How many 8-letter words can be constructed by using the 26 letters of the alphabet if each word contains $3, 4$, or 5 vowels? It is understood that there is no restriction on the number of times a letter can be used in a word.

We count the number of words according to the number of vowels they contain and then use the addition principle.

First, consider words with 3 vowels. The 3 positions occupied by the vowels can be chosen in $\binom{8}{3}$ ways; the other 5 positions are occupied by consonants. The vowel positions can then be completed in 5^3 ways and the consonant positions in 21^5 ways. Thus, the number of words with 3 vowels is

$$\binom{8}{3} 5^3 21^5 = \frac{8!}{3!5!} 5^3 21^5.$$

In a similar way, we see that the number of words with 4 vowels is

$$\binom{8}{4} 5^4 21^4 = \frac{8!}{4!4!} 5^4 21^4,$$

and the number of words with 5 vowels is

$$\binom{8}{5} 5^5 21^3 = \frac{8!}{5!3!} 5^5 21^3.$$

Hence, the total number of words is

$$\frac{8!}{3!5!} 5^3 21^5 + \frac{8!}{4!4!} 5^4 21^4 + \frac{8!}{5!3!} 5^5 21^3.$$

□

The following important property is immediate from Theorem 3.3.1:

Corollary 3.3.2 *For $0 \le r \le n$,*

$$\binom{n}{r} = \binom{n}{n-r}.$$

□

The numbers $\binom{n}{r}$ have many important and fascinating properties, and Chapter 5 is devoted to some of these. For the moment, we mention only one more property.

Theorem 3.3.3 *We have*

$$\binom{n}{0} + \binom{n}{1} + \binom{n}{2} + \cdots + \binom{n}{n} = 2^n,$$

and the common value equals the number of combinations of an n-element set.

Proof. We prove this theorem by showing that both sides of the preceding equation count the number of combinations of an n-element set S, but in different ways. First we observe that every combination of S is an r-combination of S for some $r = 0, 1, 2, \ldots, n$. Since $\binom{n}{r}$ equals the number of r-combinations of S, it follows from the addition principle that

$$\binom{n}{0} + \binom{n}{1} + \binom{n}{2} + \cdots + \binom{n}{n}$$

equals the number of combinations of S.

We can also count the number of combinations of S as follows, by breaking down the choice of a combination into n tasks: Let the elements of S be $x_1, x_2, \ldots, x_n$. In choosing a combination of S we have two choices to make for each of the n elements: x_1 either goes into the combination or it doesn't, x_2 either goes into the combination or it doesn't, $\ldots$, x_n either goes into the combination or it doesn't. Thus, by the multiplication principle, there are 2^n ways we can form a combination of S. We now equate the two counts and complete the proof. □

The proof of Theorem 3.3.3 is an instance of obtaining an identity by counting the objects of a set (in this case the combinations of a set of n elements) in two different ways and setting the results equal to one another. This technique of "double counting" is a powerful one in combinatorics, and we will see several other applications of it.

Example. The number of 2-combinations of the set $\{1, 2, \ldots, n\}$ of the first n positive integers is $\binom{n}{2}$. Partition the 2-combinations according to the largest integer they contain. For each $i = 1, 2, \ldots, n$, the number of 2-combinations in which i is the largest integer is $i - 1$ (the other integer can be any of $1, 2, \ldots, i - 1$). Equating the two counts, we obtain the identity

$$0 + 1 + 2 + \cdots + (n - 1) = \binom{n}{2} = \frac{n(n-1)}{2}.$$

$\square$

3.4   Permutations of Multisets

If S is a multiset, an *r-permutation* of S is an ordered arrangement of r of the objects of S. If the total number of objects of S is n (counting repetitions), then an n-permutation of S will also be called a *permutation of S*. For example, if $S = \{2 \cdot a, 1 \cdot b, 3 \cdot c\}$, then

$$acbc \qquad cbcc$$

are 4-permuations of S, while

$$abccca$$

is a permutation of S. The multiset S has no 7-permutations since $7 > 2 + 1 + 3 = 6$, the number of objects of S. We first count the number of r-permutations of a multiset S, each of whose repetition number is infinite.

Theorem 3.4.1 *Let S be a multiset with objects of k different types, where each has an infinite repetition number. Then the number of r-permutations of S is k^r.*

Proof. In constructing an r-permutation of S, we can choose the first item to be an object of any one of the k types. Similarly, the second item can be an object of any one of the k types, and so on. Since all repetition numbers of S are infinite, the number of different choices for any item is always k and does not depend on the choices of any previous items. By the multiplication principle, the r items can be chosen in k^r ways. $\square$

An alternative phrasing of the theorem is the following: The number of r-permutations of k distinct objects, each available in unlimited supply, equals k^r. We also note that the conclusion of the theorem remains true if the repetition numbers of the k different types of objects of S are all at least r. The assumption that the repetition numbers are infinite is a simple way of ensuring that we never run out of objects of any type.

Example. What is the number of ternary numerals[6] with at most 4 digits.

The answer to this question is the number of 4-permutations of the multiset $\{\infty \cdot 0, \infty \cdot 1, \infty \cdot 2\}$ or of the multiset $\{4 \cdot 0, 4 \cdot 1, 4 \cdot 2\}$. By Theorem 3.4.1, this number equals $3^4 = 81$. □

We now count permutations of a multiset with objects of k different types, each with a finite repetition number.

Theorem 3.4.2 *Let S be a multiset with objects of k different types with finite repetition numbers $n_1, n_2, \ldots, n_k$, respectively. Let the size of S be $n = n_1 + n_2 + \cdots + n_k$. Then the number of permutations of S equals*

$$\frac{n!}{n_1! n_2! \cdots n_k!}.$$

Proof. We are given a multiset S having objects of k types, say $a_1, a_2, \ldots, a_k$, with repetition numbers $n_1, n_2, \ldots, n_k$, respectively, for a total of $n = n_1 + n_2 + \cdots + n_k$ objects. We want to determine the number of permutations of these n objects. We can think of it this way. There are n places, and we want to put exactly one of the objects of S in each of the places. We first decide which places are to be occupied by the a_1's. Since there are n_1 a_1's in S, we must choose a subset of n_1 places from the set of n places. We can do this in $\binom{n}{n_1}$ ways. We next decide which places are to be occupied by the a_2's. There are $n - n_1$ places left, and we must choose n_2 of them. This can be done in $\binom{n-n_1}{n_2}$ ways. We next find that there are $\binom{n-n_1-n_2}{n_3}$ ways to choose the places for the a_3's. We continue like this, and invoke the multiplication principle and find that the number of permutations of S equals

$$\binom{n}{n_1} \binom{n-n_1}{n_2} \binom{n-n_1-n_2}{n_3} \cdots \binom{n-n_1-n_2-\cdots-n_{k-1}}{n_k}.$$

Using Theorem 3.3.1, we see that this number equals

---

[6]A *ternary numeral*, or base 3 numeral, is one arrived at by representing a number in terms of powers of 3. For instance, $46 = 1 \times 3^3 + 2 \times 3^2 + 0 \times 3^1 + 1 \times 3^0$, and so its ternary numeral is 1201.

$$\frac{n!}{n_1!(n-n_1)!}\frac{(n-n_1)!}{n_2!(n-n_1-n_2)!}\frac{(n-n_1-n_2)!}{n_3!(n-n_1-n_2-n_3)!}\cdots$$
$$\cdots\frac{(n-n_1-n_2-\cdots-n_{k-1})!}{n_k!(n-n_1-n_2-\cdots-n_k)!},$$

which, after cancellation, reduces to

$$\frac{n!}{n_1!n_2!n_3!\cdots n_k!0!}=\frac{n!}{n_1!n_2!n_3!\cdots n_k!}.$$

□

Example. The number of permutations of the letters in the word MISSISSIPPI is

$$\frac{11!}{1!4!4!2!},$$

since this number equals the number of permutations of the multiset $\{1\cdot M, 4\cdot I, 4\cdot S, 2\cdot P\}$.   □

If the multiset S has only two types, a_1 and a_2, of objects with repetition numbers n_1 and n_2, respectively, where $n = n_1 + n_2$, then according to Theorem 3.4.2, the number of permutations of S is

$$\frac{n!}{n_1!n_2!}=\frac{n!}{n_1!(n-n_1)!}=\binom{n}{n_1}.$$

Thus we may regard $\binom{n}{n_1}$ as the number of n_1-combinations of a set of n objects or as the number of permutations of an multiset with two types of objects with repetition numbers n_1 and $n-n_1$, respectively.

There is another interpretation of the numbers $\frac{n!}{n_1!n_2!\cdots n_k!}$ that occur in Theorem 3.4.2. This concerns the problem of partitioning a set of objects into parts of prescribed sizes *where the parts now have labels assigned to them*. To understand the implications of the last phrase, we offer the next example.

Example. Consider a set of the 4 objects $\{a, b, c, d\}$ that is to be partitioned into two sets, each of size 2. If the parts are not labeled, then there are 3 different partitions:

$$\{a, b\}, \{c, d\}; \{a, c\}, \{b, d\}; \{a, d\}, \{b, c\}.$$

However, if the parts are labeled, let us label them with colors red and blue. Then the number of partitions is greater; indeed, there are 6, since we can assign the labels red and blue to each part of in 2 ways. For instance, for the particular partition $\{a,b\},\{c,d\}$ we have

$$\text{red box}\{a,b\}, \text{blue box}\{c,d\}$$

and

$$\text{blue box}\{a,b\}, \text{red box}\{c,d\}.$$

□

In the general case, we label the parts $B_1, B_2, \ldots, B_k$ (think of color 1, color 2, ..., color k), and we also think of the parts as boxes.

Theorem 3.4.3 *Let n be a positive integer and let $n_1, n_2, \ldots, n_k$ be positive integers with $n = n_1 + n_2 + \cdots + n_k$. The number of ways to partition a set of n objects into k labeled boxes $B_1, B_2, \ldots, B_k$ in which Box 1 contains n_1 objects, Box 2 contains n_2 objects, ..., Box k contains n_k objects equals*

$$\frac{n!}{n_1! n_2! \cdots n_k!}.$$

If the boxes are not labeled, and $n_1 = n_2 = \cdots = n_k$, then the number of partitions equals

$$\frac{n!}{k! n_1! n_2! \cdots n_k!}.$$

Proof. The proof is a direct application of the multiplication principle. We have to choose which objects go into which boxes, subject to the size restrictions. We first choose n_1 objects for the first box, then n_2 of the remaining $n - n_1$ objects for the second box, then n_3 of the remaining $n - n_1 - n_2$ objects for the third box, ..., finally $n - n_1 - \cdots - n_{k-1} = n_k$ objects for the kth box. By the multiplication principle, the number of ways to make these choices is

$$\binom{n}{n_1}\binom{n-n_1}{n_2}\binom{n-n_1-n_2}{n_3}\cdots\binom{n-n_1-n_2-\cdots-n_{k-1}}{n_k}.$$

As in the proof of Theorem 3.4.2, this gives

$$\frac{n!}{n_1! n_2! \cdots n_k!}.$$

If boxes are not labeled and $n_1 = n_2 = \cdots = n_k$, then the result has to be divided by $k!$. This is so because, for each way of distributing the objects into the k unlabeled boxes there are $k!$ ways in which we can now attach the labels $1, 2, \ldots, k$. Hence, using the division principle, we find that the number of partitions with unlabeled boxes is

$$\frac{n!}{k!n_1!n_2!\cdots n_k!}.$$

$\square$

The more difficult problem of counting partitions in which the sizes of the parts are not prescribed is studied in Section 8.2.

We conclude this section with an example of a kind that we shall refer to many times in the remainder of the text.[7] The example concerns nonattacking rooks on a chessboard. Lest the reader be concerned that knowledge of chess is a prerequisite for the rest of the book, let us say at the outset that the only fact one needs to know about the game of chess is that *two rooks can attack one another if and only if they lie in the same row or the same column of the chessboard.* No other knowledge of chess is necessary (nor does it help!). Thus, a set of nonattacking rooks on a chessboard simply means a collection of "pieces" called rooks that occupy certain squares of the board, and no two of the rooks lie in the same row or in the same column.

Example. How many possibilities are there for 8 nonattacking rooks on an 8-by-8 chessboard?

An example of 8 nonattacking rooks on an 8-by-8 board is the following:

We give each square on the board a pair (i, j) of coordinates. The integer i designates the row number of the square, and the integer j designates the column number of the square. Thus, i and j are integers between 1 and 8. Since the board is 8-by-8 and there are to be 8 rooks on the board that cannot attack one another, there must be exactly one rook in each row. Thus, the rooks must occupy 8 squares with coordinates

$$(1, j_1), (2, j_2), \ldots, (8, j_8).$$

---

[7]It is the author's favorite kind of example to illustrate many ideas.

But there must also be exactly one rook in each column so that no two of the numbers $j_1, j_2, \ldots, j_8$ can be equal. More precisely,

$$j_1, j_2, \ldots, j_8$$

must be a permutation of $\{1, 2, \ldots, 8\}$. Conversely, if $j_1, j_2, \ldots, j_8$ is a permutation of $\{1, 2, \ldots, 8\}$, then putting rooks in the squares with co-ordinates $(1, j_1), (2, j_2), \ldots, (8, j_8)$, we arrive at 8 nonattacking rooks on the board. Thus, we have a one-to-one correspondence between sets of 8 nonattacking rooks on the 8-by-8 board and permutations of $\{1, 2, \ldots, 8\}$. Since there are 8! permutations of $\{1, 2, \ldots, 8\}$, there are 8! ways to place 8 rooks on an 8-by-8 board so that they are nonattacking.

We implicitly assumed in the preceding argument that the rooks were *indistinguishable* from one another. Therefore, the only thing that mattered was which squares were occupied by rooks. If we have 8 distinct rooks, say 8 rooks each colored with one of 8 different colors, then we have also to take into account which rook is in each of the 8 occupied squares. Let us thus suppose that we have 8 rooks of 8 different colors. Having decided which 8 squares are to be occupied by the rooks (8! possibilities), we now have also to decide what the color is of the rook in each of the occupied squares. As we look at the rooks from row 1 to row 8, we see a permutation of the 8 colors. Hence, having decided which 8 squares are to be occupied (8! possibilities), we then have to decide which permutation of the 8 colors (8! permutations) we shall assign. Thus, the number of ways to have 8 nonattacking rooks of 8 different colors on an 8-by-8 board equals

$$8! 8! = (8!)^2.$$

Now suppose that, instead of rooks of 8 different colors, we have 1 red (R) rook, 3 blue (B) rooks, and 4 (Y) yellow rooks. It is assumed that rooks of the same color are indistinguishable from one another.[8] Now, as we look at the rooks from row 1 to row 8, we see a permutation of the colors of the multiset

$$\{1 \cdot R, 3 \cdot B, 4 \cdot Y\}.$$

The number of permutations of this multiset equals, by Theorem 3.4.2,

$$\frac{8!}{1! 3! 4!}.$$

---

[8]Put another way, the only way we can tell one rook from another is by color.

Thus, the number of ways to place 1 red, 3 blue, and 4 yellow rooks on an 8-by-8 board so that no rook can attack another equals

$$8! \frac{8!}{1!3!4!} = \frac{(8!)^2}{1!3!4!}.$$

□

The reasoning in the preceding example is quite general and leads to the next theorem.

Theorem 3.4.4 *There are n rooks of k colors with n_1 rooks of the first color, n_2 rooks of the second color, . . . , and n_k rooks of the kth color. The number of ways to arrange these rooks on an n-by-n board so that no rook can attack another equals*

$$n! \frac{n!}{n_1! n_2! \cdots n_k!} = \frac{(n!)^2}{n_1! n_2! \cdots n_k!}.$$

Note that if the rooks all have different colors ($k = n$ and all $n_i = 1$), the formula gives $(n!)^2$ as an answer. If the rooks are all colored the same ($k = 1$ and $n_1 = n$), the formula gives $n!$ as an answer.

Let S be an n-element multiset with repetition numbers equal to $n_1, n_2, \ldots, n_k$, so that $n = n_1 + n_2 + \cdots + n_k$. Theorem 3.4.2 furnishes a simple formula for the number of n-permutations of S. If $r < n$, there is, in general, no simple formula for the number of r-permutations of S. Nonetheless a solution can be obtained by the technique of generating functions, and we discuss this in Chapter 7. In certain cases, we can argue as in the next example.

Example. Consider the multiset $S = \{3 \cdot a, 2 \cdot b, 4 \cdot c\}$ of 9 objects of 3 types. Find the number of 8-permutations of S.

The 8-permutations of S can be partitioned into three parts:

(i) 8-permutations of $\{2 \cdot a, 2 \cdot b, 4 \cdot c\}$, of which there are

$$\frac{8!}{2!2!4!} = 420;$$

(ii) 8-permutations of $\{3 \cdot a, 1 \cdot b, 4 \cdot c\}$, of which there are

$$\frac{8!}{3!1!4!} = 280;$$

(iii) 8-permutations of $\{3 \cdot a, 2 \cdot b, 3 \cdot c\}$, of which there are

$$\frac{8!}{3!2!3!} = 560.$$

Thus, the number of 8-permutations of S is

$$420 + 280 + 560 = 1260.$$

$\square$

3.5 Combinations of Multisets

If S is a multiset, then an *r-combination* of S is an unordered selection of r of the objects of S. Thus, an r-combination of S is itself a multiset, a *submultiset* of S. If S has n objects, then there is only one n-combination of S, namely, S itself. If S contains objects of k different types, then there are k 1-combinations of S.

Example. If $S = \{2 \cdot a, 1 \cdot b, 3 \cdot c\}$, then the 3-combinations of S are

$$\{2 \cdot a, 1 \cdot b\}, \quad \{2 \cdot a, 1 \cdot c\}, \quad \{1 \cdot a, 1 \cdot b, 1 \cdot c\},$$

$$\{1 \cdot a, 2 \cdot c\}, \quad \{1 \cdot b, 2 \cdot c\}, \quad \{3 \cdot c\}.$$

$\square$

We first count the number of r-combinations of a multiset all of whose repetition numbers are infinite.

Theorem 3.5.1 *Let S be a multiset with objects of k types, each with an infinite repetition number. Then the number of r-combinations of S equals*

$$\binom{r+k-1}{r} = \binom{r+k-1}{k-1}.$$

Proof. Let the k types of objects of S be $a_1, a_2, \ldots, a_k$ so that

$$S = \{\infty \cdot a_1, \infty \cdot a_2, \ldots, \infty \cdot a_k\}.$$

Any r-combination of S is of the form $\{x_1 \cdot a_1, x_2 \cdot a_2, \ldots, x_k \cdot a_k\}$ where $x_1, x_2, \ldots, x_k$ are nonnegative integers with $x_1 + x_2 + \cdots + x_k = r$. Conversely, every sequence $x_1, x_2, \ldots, x_k$ of nonnegative integers with $x_1 + x_2 + \cdots + x_k = r$ corresponds to an r-combination of S. Thus, the number of r-combinations of S equals the number of solutions of the equation

$$x_1 + x_2 + \cdots + x_k = r,$$

where $x_1, x_2, \ldots, x_k$ are nonnegative integers. We show that the number of these solutions equals the number of permutations of the multiset

$$T = \{r \cdot 1, (k-1) \cdot *\}$$

of objects of two different types.[9] Given a permutation of T, the $k-1$ *'s divide the r 1's into k groups. Let there be x_1 1's to the left of the first $*$, x_2 1's between the first and the second $*$, $\ldots$, and x_k 1's to the right of the last $*$. Then $x_1, x_2, \ldots, x_k$ are nonnegative integers with $x_1 + x_2 + \cdots + x_k = r$. Conversely, given nonnegative integers $x_1, x_2, \ldots, x_k$ with $x_1 + x_2 + \cdots + x_k = r$, we can reverse the preceding steps and construct a permutation of T.[10] Thus, the number of r-combinations of the multiset S is equal to the number of permutations of the multiset T, which by Theorem 3.4.2 equals

$$\frac{(r+k-1)!}{r!(k-1)!} = \binom{r+k-1}{r}.$$

$\square$

Another way of phrasing Theorem 3.5.1 is as follows:

> *The number of r-combinations of k distinct objects, each available in unlimited supply, equals*
>
> $$\binom{r+k-1}{r}.$$

We note that Theorem 3.5.1 remains true if the repetition numbers of the k distinct objects of S are all at least r.

Example. A bakery boasts 8 varieties of doughnuts. If a box of doughnuts contains 1 dozen, how many different options are there for a box of doughnuts?

It is assumed that the bakery has on hand a large number (at least 12) of each variety. This is a combination problem, since we assume

---

[9] Equivalently, the number of sequences of 0's and 1's of length $r+k-1$ in which there are r 1's and $k-1$ 0's.

[10] For example, if $k = 4$ and $r = 5$, then the permutation of $T = \{5 \cdot 1, 3 \cdot *\}$ given by $*111 * *11$ corresponds to the solution of $x_1 + x_2 + x_3 + x_4 = 5$ given by $x_1 = 0, x_2 = 3, x_3 = 0, x_4 = 2$.

the order of the doughnuts in a box is irrelevant for the purchaser's purpose. The number of different options for boxes equals the number of 12-combinations of a multiset with objects of 8 types, each having an infinite repetition number. By Theorem 3.5.1, this number equals

$$\binom{12 + 8 - 1}{12} = \binom{19}{12}.$$

□

Example. What is the number of nondecreasing sequences of length r whose terms are taken from $1, 2, \ldots, k$?

The nondecreasing sequences to be counted can be obtained by first choosing an r-combination of the multiset

$$S = \{\infty \cdot 1, \infty \cdot 2, \ldots, \infty \cdot k\}$$

and then arranging the elements in increasing order. Thus, the number of such sequences equals the number of r-combinations of S, and hence, by Theorem 3.5.1, equals

$$\binom{r + k - 1}{r}.$$

□

In the proof of Theorem 3.5.1, we have seen that there is a one-to-one correspondence between r-combinations of a multiset S with objects of k different types and the nonnegative integral solutions of the equation

$$x_1 + x_2 + \cdots + x_k = r.$$

In this correspondence, x_i represents the number of objects of the ith type that are used in the r-combination. Putting restrictions on the number of times each type of object is to occur in the r-combination is equivalent to putting restrictions on the x_i.

Example. Let S be the multiset $\{10 \cdot a, 10 \cdot b, 10 \cdot c, 10 \cdot d\}$ with objects of four types, a, b, c and d. What is the number of 10-combinations of S which have the property that each of the four types of objects occurs at least once?

The answer is the number of *positive* integral solutions of

$$x_1 + x_2 + x_3 + x_4 = 10,$$

where x_1 represents the number of a's in a 10-combination, x_2 the number of b's, x_3 the number of c's, and x_4 the number of d's. Since the repetition numbers all equal 10, and 10 is the size of the combinations being counted, we can ignore the repetition numbers. We then perform the changes of variable:

$$y_1 = x_1 - 1, \ y_2 = x_2 - 1, \ y_3 = x_3 - 1, \ y_4 = x_4 - 1$$

to get

$$y_1 + y_2 + y_3 + y_4 = 6,$$

where the y_i's are to be nonnegative. The number of nonnegative integral solutions of the new equation is, by Theorem 3.5.1,

$$\binom{6 + 4 - 1}{6} = \binom{9}{6} = 84. \qquad \square$$

Example. Continuing with the doughnut example following Theorem 3.5.1, we see that the number of different options for boxes of doughnuts containing at least one doughnut of each of the 8 varieties equals

$$\binom{4 + 8 - 1}{4} = \binom{11}{4} = 330.$$

$$\square$$

General lower bounds on the number of times each type of object occurs in the combination also can be handled by a change of variable.

Example. What is the number of integral solutions of the equation

$$x_1 + x_2 + x_3 + x_4 = 20,$$

in which

$$x_1 \geq 3, \ x_2 \geq 1, \ x_3 \geq 0 \ \text{ and } \ x_4 \geq 5?$$

We introduce the new variables

$$y_1 = x_1 - 3, \ y_2 = x_2 - 1, \ y_3 = x_3, \ y_4 = x_4 - 5,$$

and our equation becomes

$$y_1 + y_2 + y_3 + y_4 = 11.$$

The lower bounds on the x_i's are satisfied if and only if the y_i's are nonnegative. The number of nonnegative integral solutions of the new equation is

$$\binom{11 + 4 - 1}{11} = \binom{14}{11} = 364.$$

$\Box$

It is more difficult to count the number of r-combinations of a multiset

$$S = \{n_1 \cdot a_1, n_2 \cdot a_2, \ldots, n_k \cdot a_k\}$$

with k types of objects and general repetition numbers $n_1, n_2, \ldots, n_k$. The number of r-combinations of S is the same as the number of integral solutions of

$$x_1 + x_2 + \cdots + x_k = r$$

where

$$0 \le x_1 \le n_1, \quad 0 \le x_2 \le n_2, \quad \ldots, \quad 0 \le x_k \le n_k.$$

We now have upper bounds on the x_i's, and these cannot be handled in the same way as lower bounds. In Chapter 6 we show how the inclusion–exclusion principle provides a satisfactory method for this case.

3.6 Exercises

1. For each of the four combinations of the two properties (a) and (b), count the number of four-digit numbers whose digits are either $1, 2, 3, 4,$ or 5:

   (a) The digits are distinct.
   (b) The number is even.

   Note that there are four problems here: $\emptyset$ (no further restriction), $\{a\}$ (property (a) holds), $\{b\}$ (property (b) holds), $\{a, b\}$ (both properties (a) and (b) hold).

2. How many orderings are there for a deck of 52 cards if all the cards of the same suit are together?

3. In how many ways can a poker hand (5 cards) be dealt? How many different poker hands are there?

4. How many distinct positive divisors do each of the following numbers have?

   (a) $3^4 \times 5^2 \times 7^6 \times 11$

   (b) 620

   (c) 10^{10}

5. Determine the largest power of 10 that is a factor of the following numbers (equivalently, the number of terminal 0's, using ordinary base 10 representation):

   (a) 50!.

   (b) 1000!.

6. How many integers greater than 5400 have both of the following properties?

   (a) The digits are distinct.

   (b) The digits 2 and 7 do not occur.

7. Determine the number of poker hands of the following types:

   (a) full houses (3 cards of one rank and 2 of a different rank).

   (b) straights (5 consecutive ranks).

   (c) flushes (5 cards of the same suit).

   (d) straight flushes (5 consecutive cards of the same suit).

   (e) exactly two pairs (2 cards of one rank, 2 cards of another rank, and 1 card of a third rank).

   (f) exactly one pair (2 cards of one rank, and 3 cards of three other and different ranks).

8. In how many ways can six men and six women be seated at a round table if the men and women are to sit in alternate seats?

9. In how many ways can 15 people be seated at a round table if B refuses to sit next to A? What if B only refuses to sit on A's right?

10. A committee of 5 is to be chosen from a club that boasts a membership of 10 men and 12 women. How many ways can the committee be formed if it is to contain at least 2 women? How many ways if, in addition, one particular man and one particular woman who are members of the club refuse to serve together on the committee?

11. How many sets of 3 numbers each can be formed from the numbers $\{1, 2, 3, \ldots, 20\}$ if no 2 consecutive numbers are to be in a set?

12. A football team of 11 players is to be selected from a set of 15 players, 5 of whom can play only in the backfield, 8 of whom can play only on the line, and 2 of whom can play either in the backfield or on the line. Assuming a football team has 7 men on the line and 4 in the backfield, determine the number of football teams possible.

13. There are 100 students at a school and three dormitories, A, B, and C, with capacities 25, 35 and 40 respectively.

    (a) How many ways are there to fill the dormitories?

    (b) Suppose that, of the 100 students, 50 are men and 50 are women and that A is an all-men's dorm, B is an all-women's dorm, and C is co-ed. How many ways are there to fill the dormitories?

14. A classroom has 2 rows of 8 seats each. There are 14 students, 5 of whom always sit in the front row and 4 of whom always sit in the back row. In how many ways can the students be seated?

15. At a party there are 15 men and 20 women.

    (a) How many ways are there to form 15 couples consisting of one man and one woman?

    (b) How many ways are there to form 10 couples consisting of one man and one woman?

16. Prove that
$$\binom{n}{r} = \binom{n}{n-r}$$
by using a combinatorial argument and not the values of these numbers as given in Theorem 3.3.1.

17. In how many ways can 6 indistinguishable rooks be placed on a 6-by-6 board so that no two rooks can attack one another? In how many ways if there are 2 red and 4 blue rooks?

18. In how many ways can 2 red and 4 blue rooks be placed on an 8-by-8 board so that no two rooks can attack one another?

19. We are given 8 rooks, 5 of which are red and 3 of which are blue.

   (a) In how many ways can the 8 rooks be placed on an 8-by-8 chessboard so that no two rooks can attack one another?

   (b) In how many ways can the 8 rooks be placed on a 12-by-12 chessboard so that no two rooks can attack one another?

20. Determine the number of circular permutations of $\{0, 1, 2, \ldots, 9\}$ in which 0 and 9 are not opposite. (Hint: Count those in which 0 and 9 are opposite.)

21. How many permutations are there of the letters of the word AD-DRESSES? How many 8-permutations are there of these 9 letters?

22. A footrace takes place between 4 runners. If ties are allowed (even all 4 runners finishing at the same time), how many ways are there for the race to finish?

23. Bridge is played with 4 players and an ordinary deck of 52 cards. Each player begins with a hand of 13 cards. In how many ways can a bridge game start?

24. A roller coaster has 5 cars, each containing 4 seats, two in front and two in back. There are 20 people ready for a ride. In how many ways can the ride begin? Same question but now a certain two people want to sit in different cars.

25. A ferris wheel has 5 cars, each containing 4 seats in a row. There are 20 people ready for a ride. In how many ways can the ride begin? Same question but now a certain two people want to sit in different cars.

26. A group of mn people are to be arranged into m teams each with n players.

   (a) Determine the number of ways if each team has a different name.

   (b) Determine the number of ways if the teams don't have names.

27. In how many ways can 5 indistinguishable rooks be placed on an 8-by-8 chessboard so that no rook can attack another and neither the first row nor the first column is empty?

28. A secretary works in a building located 9 blocks east and 8 blocks north of his home. Every day he walks 17 blocks to work. (See the map that follows.)

(a) How many different routes are possible for him?

(b) How many different routes if the street in the easterly direction, which begins 4 blocks east and 3 blocks north of his home, is under water (and he can't swim)? (Hint: Count the routes that use the block under water.)

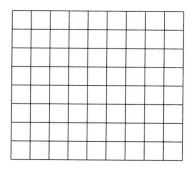

29. Let S be a multiset with repetition numbers $n_1, n_2, \ldots, n_k$ where $n_1 = 1$. Let $n = n_2 + \cdots + n_k$. Prove that the number of circular permutations of S equals

$$\frac{n!}{n_2! \cdots n_k!}.$$

30. We are to seat 5 men, 5 women, and 1 dog in a circular arrangement around a table. In how many ways can this be done if no man is to sit next to a man and no woman is to sit next to a woman?

31. In a soccer tournament of 15 teams, the top 3 teams are awarded gold, silver, and bronze cups, and the last 3 teams are dropped to a lower league. We regard two outcomes of the tournament as the same if the teams which receive the gold, silver, and bronze cups, respectively, are identical and the teams which drop to a lower league are also identical. How many different possible outcomes are there for the tournament?

32. Determine the number of 11-permutations of the multiset

$$S = \{3 \cdot a, 4 \cdot b, 5 \cdot c\}.$$

33. Determine the number of 10-permutations of the multiset

$$S = \{3 \cdot a, 4 \cdot b, 5 \cdot c\}.$$

34. Determine the number of 11-permutations of the multiset
$$\{3 \cdot a, 3 \cdot b, 3 \cdot c, 3 \cdot d\}.$$

35. List all 3-combinations and 4-combinations of the multiset
$$\{2 \cdot a, 1 \cdot b, 3 \cdot c\}.$$

36. Determine the total number of combinations (of any size) of a multiset of objects of k different types with finite repetition numbers $n_1, n_2, \ldots, n_k$, respectively.

37. A bakery sells 6 different kinds of pastry. If the bakery has at least a dozen of each kind, how many different options for a dozen of pastry are there? What if a box is to contain at least one of each kind of pastry?

38. How many integral solutions of
$$x_1 + x_2 + x_3 + x_4 = 30$$
satisfy $x_1 \geq 2$, $x_2 \geq 0$, $x_3 \geq -5$, and $x_4 \geq 8$?

39. There are 20 identical sticks lined up in a row occupying 20 distinct places as follows:

Six of them are to be chosen.

   (a) How many choices are there?
   (b) How many choices are there if no two of the chosen sticks can be consecutive?
   (c) How many choices are there if there must be at least two sticks between each pair of chosen sticks?

40. There are n sticks lined up in a row and k of them are to be chosen.

   (a) How many choices are there?
   (b) How many choices are there if no two of the chosen sticks can be consecutive?
   (c) How many choices are there if there must be at least l sticks between each pair of chosen sticks?

41. In how many ways can 12 indistinguishable apples and 1 orange be distributed among three children in such a way that each child gets at least one piece of fruit?

42. Determine the number of ways to distribute 10 orange drinks, 1 lemon drink, and 1 lime drink to 4 thirsty students so that each student gets at least 1 drink, and the lemon and lime drinks go to different students.

43. Determine the number of r-combinations of the multiset

$$\{1 \cdot a_1, \infty \cdot a_2, \ldots, \infty \cdot a_k\}.$$

44. Prove that the number of ways to distribute n different objects among k children equals k^n.

45. Twenty different books are to be put on five book shelves, each of which holds at least twenty books.

    (a) How many different arrangements are there if you only care about the number of books on the shelves (and not which book is where)?

    (b) How many different arrangements are there if you care about which books are where but the order of the books on the shelves doesn't matter?

    (c) How many different arrangements are there if the order on the shelves does matter?

46. (a) There is an even number $2n$ of people at a party, and they talk together in pairs with everyone talking with someone (so n pairs). In how many different ways can the $2n$ people be talking like this?

    (b) Now suppose that there is an odd number $2n + 1$ of people at the party with everyone but one person talking with someone. How many different pairings are there?

47. There are $2n + 1$ identical books to be put in a bookcase with three shelves. In how many ways can this be done if each pair of shelves together contain more books than the other shelf?

48. Prove that the number of permutations of m A's and at most n B's equals

$$\binom{m+n+1}{m+1}.$$

49. Prove that the number of permutations of at most m A's and at most n B's equals
$$\binom{m+n+2}{m+1} - 2.$$

50. In how many ways can five identical rooks be placed on the squares of an 8-by-8 board so that four of them form the corners of a rectangle?

51. Consider the multiset $\{n \cdot a, 1, 2, 3, \ldots, n\}$ of size $2n$. Determine the number of its n-combinations.

52. Consider the multiset $\{n \cdot a, n \cdot b, 1, 2, 3, \ldots, n+1\}$ of size $3n+1$. Determine the number of its n-combinations.

53. Establish a one-to-one correspondence between the permutations of the set $\{1, 2, \ldots, n\}$ and the towers $A_0 \subset A_1 \subset A_2 \subset \cdots \subset A_n$ where $|A_k| = k$ for $k = 0, 1, 2, \ldots, n$.

54. Determine the number of towers of the form $\emptyset \subseteq A \subseteq B \subseteq \{1, 2, \ldots, n\}$.

55. How many permutations are there of the letters in the word

PNEUMONOULTRAMICROSCOPICSILICOVOLCANOCONIOSIS?

This word is, by some accounts, the longest word in the English language.

Chapter 4

Generating Permutations and Combinations

In this chapter we explore some features of permutations and combinations that are not directly related to counting. We discuss some ordering schemes for permutations and combinations, and algorithms for carrying them out. We also introduce the idea of a relation on a set and discuss two important instances, those of partial order and equivalence relation.

4.1   Generating Permutations

The set $\{1, 2, \ldots, n\}$ consisting of the first n positive integers has $n!$ permutations which, even if n is only moderately large, is quite enormous. For instance, 15! is more than $1,000,000,000,000$. A useful and readily computable approximation to $n!$ is given by *Stirling's formula*,

$$n! \sim \sqrt{2\pi n}\left(\frac{n}{e}\right)^n,$$

where $\pi = 3.141\ldots$, and $e = 2.718\ldots$ is the base of the natural logarithm. As n grows without bound, the ratio of $n!$ to $\sqrt{2\pi n}\left(\frac{n}{e}\right)^n$ approaches 1. A proof of this can be found in many texts on advanced calculus and in an article by Feller.[1]

Permutations are of importance in many different circumstances, both theoretical and applied. For sorting techniques in computer science they correspond to the unsorted input data. We consider in this

---

[1]W. Feller: A direct proof of Stirling's formula, *American Mathematical Monthly*, 74 (1967), 1223-1225.

section a simple but elegant algorithm for generating all the permutations of $\{1, 2, \ldots, n\}$.

Because of the large number of permutations of a set of n elements, in order for such an algorithm to be effective on a computer the individual steps must be simple to perform. The result of the algorithm should be a list containing each of the permutations of $\{1, 2, \ldots, n\}$ exactly once. The algorithm to be described momentarily has these features. It was independently discovered by Johnson[2] and Trotter[3], and was described by Gardner in a popular article.[4] The algorithm is based on the following observation:

> If the integer n is deleted from a permutation of $\{1, 2, \ldots, n\}$,
> the result is a permutation of $\{1, 2, \ldots, n - 1\}$.

The same permutation of $\{1, 2, \ldots, n - 1\}$ can result from different permutations of $\{1, 2, \ldots, n\}$. For instance, if $n = 5$ and we delete 5 from the permutation $3, 4, 1, 5, 2$, the result is $3, 4, 1, 2$. However $3, 4, 1, 2$ also results when 5 is deleted from $3, 5, 4, 1, 2$. Indeed there are exactly 5 permutations of $\{1, 2, 3, 4, 5\}$ which yield $3, 4, 1, 2$ upon the deletion of 5, namely,

$$5, 3, 4, 1, 2$$
$$3, 5, 4, 1, 2$$
$$3, 4, 5, 1, 2$$
$$3, 4, 1, 5, 2$$
$$3, 4, 1, 2, 5.$$

More generally, each permutation of $\{1, 2, \ldots, n - 1\}$ results from exactly n permutations of $\{1, 2, \ldots, n\}$ upon the deletion of n. Looked at from the opposite viewpoint, given a permutation of $\{1, 2, \ldots, n - 1\}$ there are exactly n ways to insert n into this permutation to obtain a permutation of $\{1, 2, \ldots, n\}$. Thus, given a list of the $(n - 1)!$ permutations of $\{1, 2, \ldots, n - 1\}$ we can obtain a list of the $n!$ permutations of $\{1, 2, \ldots, n\}$ by systematically inserting n into each permutation of $\{1, 2, \ldots n - 1\}$ in all possible ways. We now give an inductive description of such an algorithm. It generates the permutations of $\{1, 2, \ldots, n\}$ from the permutations of $\{1, 2, \ldots, n - 1\}$. Thus, starting with the unique permutation of $\{1\}$, we build up the permutations of

[2]S.M. Johnson: Generation of permutations by adjacent transpositions, *Mathematics of Computation*, 17 (1963), 282-285.

[3]H.F. Trotter: Algorithm 115, *Communications of the Association for Computing Machinery*, 5 (1962), 434-435.

[4]M. Gardner: Mathematical Games, *Scientific American*, November (1974), 122-125.

$\{1,2\}$, then the permutations of $\{1,2,3\}$, and so on until finally we obtain the permutations of $\{1,2,\ldots,n\}$.

$n = 2$: To generate the permutations of $\{1,2\}$, write the unique permutation of $\{1\}$ twice and "interlace" the 2:

$$
\begin{array}{cc}
1 & \mathbf{2} \\
\mathbf{2} & 1
\end{array}
$$

The second permutation is obtained from the first by switching the two numbers.

$n = 3$: To generate the permutations of $\{1,2,3\}$, write down each of the permutations of $\{1,2\}$ three times in the order generated above, and interlace the 3 with them as shown:

$$
\begin{array}{ccc}
1 & 2 & \mathbf{3} \\
1 & \mathbf{3} & 2 \\
\mathbf{3} & 1 & 2 \\
\mathbf{3} & 2 & 1 \\
2 & \mathbf{3} & 1 \\
2 & 1 & \mathbf{3}
\end{array}
$$

It is seen that each permutation other than the first is obtained from the preceding one by switching two adjacent numbers. When the 3 is fixed, as it is from the third to the fourth permutation in the sequence of generation, the switch comes from a corresponding switch for $n = 2$. We note that by switching 1 and 2 in the last permutation generated, we obtain the first one, namely, 123.

$n = 4$: To generate the permutations of $\{1,2,3,4\}$, write down each of the permutations of $1,2,3$ four times in the order generated above, and interlace the 4 with them as shown:

```
    1     2      3 4
    1     2  4 3
    1  4 2       3
  4 1     2      3
  4 1     3      2
    1  4 3       2
    1      3 4 2
    1      3      2 4
    3      1      2 4
    3      1  4 2
    3  4 1        2
  4 3     1      2
  4 3     2      1
    3  4 2        1
    3      2 4 1
    3      2      1 4
    2      3      1 4
    2      3 4 1
    2  4 3        1
  4 2     3      1
  4 2     1      3
    2  4 1        3
    2      1  4 3
    2      1      3 4
```

One again observes that each permutation is obtained from the preceding one by switching two adjacent numbers. When the 4 is fixed, as it is between the 4th and 5th, the 8th and 9th, the 12th and 13th, the 16th and 17th, and the 20th and 21st permutations in the sequence of generation, the switch comes from a corresponding switch for $n = 3$. Also, by switching 1 and 2 in the last permutation generated we obtain the first permutation 1234.

It should now be clear how to proceed for any n. It follows by induction on n, using our earlier remarks, that the algorithm generates all permutations of $\{1, 2, \ldots, n\}$ exactly once. Moreover, each permutation other than the first is obtained from the preceding one by switching two adjacent numbers. The first permutation generated is $12 \cdots n$. This is so for $n = 1$ and follows by induction, since, in the algorithm, n is first put on the extreme right. Provided that $n \geq 2$, the last permutation generated is always $213 \cdots n$. This observation can be verified by induction on n as follows: If $n = 2$, the last permutation generated is 21. Now suppose that $n \geq 3$ and that $213 \cdots (n-1)$ is the last permutation generated for $\{1, 2, \ldots, n-1\}$. There are $(n-1)!$, an even number, of permutations of $\{1, 2, \ldots n - 1\}$, and it follows that, in applying the algorithm, the integer n ends on the extreme right. Hence, $213 \cdots n$ is the last permutation generated. Since the last permutation is $213 \cdots n$, by switching 1 and 2 in the last permutation the first permutation results. Thus the algorithm is cyclical in nature.

To generate the permutations of $\{1, 2, \ldots, n\}$ in the manner just described, we must first generate the permutations of $\{1, 2, \ldots, n-1\}$. To generate the permutations of $\{1, 2, \ldots, n-1\}$, we must first generate the permutations of $\{1, 2, \ldots, n - 2\}$, and so on. What we would like be able to do is to generate the permutations one at a time, using only the current permutation in order to generate the next one. We next show how it is possible to generate in this way the permutations of $\{1, 2, \ldots, n\}$ in the same order as above. Thus, rather than having to retain a list of all the permutations, we can simply overwrite the current permutation with the one that follows it. To do this, one needs to determine which two adjacent integers are to be switched as the permutations appear on the list. The particular description we give is taken from Even.[5]

Given an integer k, we assign a *direction* to it by writing an arrow above it pointing to the left or to the right: $\overleftarrow{k}$ or $\overrightarrow{k}$. Consider a permutation of $\{1, 2, \ldots, n\}$ in which each of the integers is given a direction. The integer k is called *mobile* if its arrow points to a smaller integer adjacent to it. For example, for

$$\overrightarrow{2}\ \overrightarrow{6}\ \overrightarrow{3}\ \overleftarrow{1}\ \overrightarrow{5}\ \overrightarrow{4}$$

only $3, 5$, and 6 are mobile. It follows that the integer 1 can never be mobile since there is no integer in $\{1, 2, \ldots, n\}$ smaller than 1. The integer n is mobile, except in two cases:

(i) n is the first integer and its arrow points to the left: $\overleftarrow{n} \cdots$,

(ii) n is the last integer and its arrow points to the right: $\cdots \overrightarrow{n}$.

This is because n, being the largest integer in the set $\{1, 2, \ldots, n\}$, is mobile provided its arrow points to an integer. We can now describe the algorithm for generating the permutations of $\{1, 2, \ldots, n\}$ directly.

---

[5] S. Even: *Algorithmic Combinatorics*, Macmillan, New York, 1973.

Algorithm for generating the permutations of $\{1, 2, \ldots, n\}$

Begin with $\overleftarrow{1}\ \overleftarrow{2}\ \cdots\ \overleftarrow{n}$.

While there exists a mobile integer, do the following:

(1) Find the largest mobile integer m.

(2) Switch m and the adjacent integer its arrow points to.

(3) Switch the direction of all the arrows above integers p with $p > m$.

We illustrate the algorithm for $n = 4$. The results are displayed in two columns, with the first column giving the first 12 permutations:

$\overleftarrow{1}$	$\overleftarrow{2}$	$\overleftarrow{3}$	$\overleftarrow{4}$	$\overrightarrow{4}$	$\overrightarrow{3}$	$\overleftarrow{2}$	$\overleftarrow{1}$
$\overleftarrow{1}$	$\overleftarrow{2}$	$\overleftarrow{4}$	$\overleftarrow{3}$	$\overrightarrow{3}$	$\overrightarrow{4}$	$\overleftarrow{2}$	$\overleftarrow{1}$
$\overleftarrow{1}$	$\overleftarrow{4}$	$\overleftarrow{2}$	$\overleftarrow{3}$	$\overrightarrow{3}$	$\overleftarrow{2}$	$\overrightarrow{4}$	$\overleftarrow{1}$
$\overleftarrow{4}$	$\overleftarrow{1}$	$\overleftarrow{2}$	$\overleftarrow{3}$	$\overrightarrow{3}$	$\overleftarrow{2}$	$\overleftarrow{1}$	$\overrightarrow{4}$
$\overrightarrow{4}$	$\overleftarrow{1}$	$\overleftarrow{3}$	$\overleftarrow{2}$	$\overleftarrow{2}$	$\overrightarrow{3}$	$\overleftarrow{1}$	$\overrightarrow{4}$
$\overleftarrow{1}$	$\overrightarrow{4}$	$\overleftarrow{3}$	$\overleftarrow{2}$	$\overleftarrow{2}$	$\overrightarrow{3}$	$\overleftarrow{4}$	$\overleftarrow{1}$
$\overleftarrow{1}$	$\overleftarrow{3}$	$\overrightarrow{4}$	$\overleftarrow{2}$	$\overleftarrow{2}$	$\overrightarrow{4}$	$\overrightarrow{3}$	$\overleftarrow{1}$
$\overleftarrow{1}$	$\overleftarrow{3}$	$\overleftarrow{2}$	$\overrightarrow{4}$	$\overleftarrow{4}$	$\overleftarrow{2}$	$\overrightarrow{3}$	$\overleftarrow{1}$
$\overleftarrow{3}$	$\overleftarrow{1}$	$\overleftarrow{2}$	$\overrightarrow{4}$	$\overrightarrow{4}$	$\overleftarrow{2}$	$\overleftarrow{1}$	$\overrightarrow{3}$
$\overleftarrow{3}$	$\overleftarrow{1}$	$\overleftarrow{4}$	$\overleftarrow{2}$	$\overleftarrow{2}$	$\overrightarrow{4}$	$\overleftarrow{1}$	$\overrightarrow{3}$
$\overleftarrow{3}$	$\overleftarrow{4}$	$\overleftarrow{1}$	$\overleftarrow{2}$	$\overleftarrow{2}$	$\overleftarrow{1}$	$\overrightarrow{4}$	$\overrightarrow{3}$
$\overleftarrow{4}$	$\overleftarrow{3}$	$\overleftarrow{1}$	$\overleftarrow{2}$	$\overleftarrow{2}$	$\overleftarrow{1}$	$\overrightarrow{3}$	$\overrightarrow{4}$

Since no integer is mobile in $\overleftarrow{2}\ \overleftarrow{1}\ \overrightarrow{3}\ \overrightarrow{4}$, the algorithm stops.

That this algorithm generates the permutations of $\{1, 2, \ldots, n\}$, and in the same order as our previous method, follows by induction on n. We illustrate the inductive step from $n = 3$ to $n = 4$. We begin with $\overleftarrow{1}\ \overleftarrow{2}\ \overleftarrow{3}\ \overleftarrow{4}$, with 4 the largest mobile integer. The integer 4 remains mobile until it reaches the extreme left. At that point 4 has been inserted in all possible ways in the permutation 123 of $\{1, 2, 3\}$. Now 4 is no longer mobile. The largest mobile integer is 3, which is the same as the largest mobile integer in $\overleftarrow{1}\ \overleftarrow{2}\ \overleftarrow{3}$. Then 3 and 2 switch places and 4 changes direction. The switch is the same switch that would have occurred in $\overleftarrow{1}\ \overleftarrow{2}\ \overleftarrow{3}$. The result is now $\overrightarrow{4}\ \overleftarrow{1}\ \overleftarrow{3}\ \overleftarrow{2}$;

now 4 is mobile again and remains mobile until it reaches the extreme right. Again a switch takes place, which is the same switch that would have occurred in $\overset{\leftarrow}{1}\,\overset{\leftarrow}{3}\,\overset{\leftarrow}{2}$. The algorithm continues like this, and 4 is interlaced in all possible ways with each permutation of $\{1, 2, 3\}$.

It is possible to determine, for a given permutation of $\{1, 2, \ldots, n\}$, at which step it occurs in the preceding algorithm. Conversely, it is possible to determine which permutation occurs at a given step. For a clear analysis of this, we refer to the book by Even.[6]

4.2 Inversions in Permutations

In this section we discuss a method of describing a permutation by means of its inversions discovered by Hall.[7] The notion of an inversion is an old one, and it plays an important role in the theory of determinants of matrices.

Let $i_1 i_2 \ldots i_n$ be a permutation of the set $\{1, 2, \ldots, n\}$. The pair (i_k, i_l) is called an *inversion* if $k < l$ and $i_k > i_l$. Thus, an inversion in a permutation corresponds to a pair of numbers that are out of their natural order. For example, the permutation 31524 has four inversions, namely $(3, 1), (3, 2), (5, 2), (5, 4)$. The only permutation of $\{1, 2, \ldots, n\}$ with no inversions is $12 \ldots n$. For a permutation $i_1 i_2 \ldots i_n$ we let a_j denote the number of inversions whose second component is j. In other words,

> a_j equals the number of integers that precede j in the permutation, but are greater than j; it measures how much j is out of order.

The sequence of numbers

$$a_1, a_2, \ldots, a_n$$

is called the *inversion sequence* of the permutation $i_1 i_2 \ldots i_n$. The number $a_1 + a_2 + \cdots + a_n$ measures the *disorder* of a permutation.

Example. The inversion sequence of the permutation 31524 is

$$1, 2, 0, 1, 0.$$

□

---

[6]Op. cit.

[7]M. Hall, Jr.: *Proceedings Symposium in Pure Mathematics*, American Mathematical Society, Providence, 6 (1963), 203.

The inversion sequence $a_1, a_2, \ldots, a_n$ of the permutation $i_1 i_2 \ldots i_n$ satisfies the conditions

$$0 \leq a_1 \leq n - 1, \ 0 \leq a_2 \leq n - 2, \ \ldots, \ 0 \leq a_{n-1} \leq 1, \ a_n = 0.$$

This is so because for each $k = 1, 2, \ldots, n$ there are $n - k$ integers in the set $\{1, 2, \ldots, n\}$ which are greater than k. Using the multiplication principle, we see that the number of sequences of integers $b_1, b_2, \ldots, b_n$, with

$$0 \leq b_1 \leq n - 1, \ 0 \leq b_2 \leq n - 2, \ \ldots, \ 0 \leq b_{n-1} \leq 1, \ b_n = 0, \quad (4.1)$$

equals $n \times (n - 1) \times \cdots \times 2 \times 1 = n!$.

Thus, there are as many permutations of $\{1, 2, \ldots, n\}$ as there are possible inversion sequences. This suggests (but does not yet prove!) that different permutations of $\{1, 2, \ldots, n\}$ have different inversion sequences. If we can show that each sequence of integers $b_1, b_2, \ldots, b_n$ satisfying (4.1) is the inversion sequence of a permutation of $\{1, 2, \ldots, n\}$, then it follows (from the pigeonhole principle) that different permutations have different inversion sequences.

Theorem 4.2.1 *Let $b_1, b_2, \ldots, b_n$ be a sequence of integers satisfying*

$$0 \leq b_1 \leq n - 1, \ 0 \leq b_2 \leq n - 2, \ \ldots \ 0 \leq b_{n-1} \leq 1, \ b_n = 0.$$

Then there exists a unique permutation of $\{1, 2, \ldots, n\}$ whose inversion sequence is $b_1, b_2, \ldots, b_n$.

Proof. We describe two methods for uniquely constructing a permutation whose inversion sequence is $b_1, b_2, \ldots, b_n$.

Algorithm I

Construction of a permutation from its inversion sequence

n: Write down n.

n − 1: Consider b_{n-1}. We are given that $0 \leq b_{n-1} \leq 1$. If $b_{n-1} = 0$, then $n - 1$ *must* be placed before n. If $b_{n-1} = 1$, then $n - 1$ *must* be placed after n.

n − 2: Consider b_{n-2}. We are given that $0 \leq b_{n-2} \leq 2$. If $b_{n-2} = 0$, then $n-2$ *must* be placed before the two numbers from step $n-1$. If $b_{n-2} = 1$, then $n - 2$ *must* be placed between the two numbers from step $n - 1$. If $b_{n-2} = 2$, then $n - 2$ *must* be placed after the two numbers from step $n - 1$.

$\vdots$

n − k: (*general step*) Consider b_{n-k}. We are given that $0 \leq b_{n-k} \leq k$. In steps n through $n-k+1$, the k numbers $n, n-1, \ldots, n-k+1$ have already been placed in the required order. If $b_{n-k} = 0$, then $n - k$ *must* be placed before all the numbers from step $n-k+1$. If $b_{n-k} = 1$, then $n - k$ *must* be placed between the first two numbers. . . . If $b_{n-k} = k$, then $n - k$ *must* be placed after all the numbers.

⋮

1: We *must* place 1 after the b_1st number in the sequence constructed in step $n - 1$.

Steps $n, n-1, n-2, \ldots, 1$, when carried out, determine the unique permutation of $\{1, 2, \ldots, n\}$ whose inversion sequence is $b_1, b_2, \ldots, b_n$. The disadvantage of this algorithm is that the location of each integer in the permutation is not known until the very end; only the relative positions of the integers remain fixed throughout the algorithm.

In the second algorithm,[8] it is the positions of the integers $1, 2, \ldots, n$ in the permutation that are determined.

Algorithm II

Construction of a permutation from its inversion sequence

We begin with n empty locations, which we label $1, 2, \ldots, n$ from left to right.

1: Since there are to be b_1 integers that precede 1 in the permutation, we must put 1 in location number $b_1 + 1$.

2: Since there are to be b_2 integers that precede 2 and are larger than 2 in the permutation, and since these integers have not yet been inserted, we must leave exactly b_2 empty locations for them. Thus, counting from the left, we put 2 in the $(b_2 + 1)$st empty location.

⋮

k: (*general step*) Since there are to be b_k integers that precede k in the permutation, and since these integers have not yet been inserted, we must leave exactly b_k empty locations for them. We observe that the number of empty locations at the beginning of

---

[8]This algorithm was brought to my attention by J. Csima.

this step is $n - (k - 1) = n - k + 1$. Counting from the left, we put k in the $(b_k + 1)$st empty location. Since $b_k \leq n - k$, we have $b_k + 1 \leq n - k + 1$ and such an empty location can be determined.

$\vdots$

n: We put n in the one remaining empty location.

Carrying out the steps $1, 2, \ldots, n$ in the order described, we obtain the unique permutation of $\{1, 2, \ldots, n\}$ whose inversion sequence is $b_1, b_2, \ldots, b_n$. $\qquad\square$

Example. Determine the permutation of $\{1, 2, 3, 4, 5, 6, 7, 8\}$ whose inversion sequence is $5, 3, 4, 0, 2, 1, 1, 0$.

The steps in the two algorithms in the proof of Theorem 4.2.1, when carried out for the given inversion sequence, yield the following results:

Algorithm I

$$
\begin{array}{rl}
8: & 8 \\
7: & 87 \\
6: & 867 \\
5: & 8657 \\
4: & 48657 \\
3: & 486537 \\
2: & 4862537 \\
1: & 48625137
\end{array}
$$

Thus, the permutation is 48625137.

Algorithm II

	(1)	(2)	(3)	(4)	(5)	(6)	(7)	(8)
1:						1		
2:				2		1		
3:				2		1	3	
4:	4			2		1	3	
5:	4			2	5	1	3	
6:	4		6	2	5	1	3	
7:	4		6	2	5	1	3	7
8:	4	8	6	2	5	1	3	7

Again, the permutation is 48625137. $\qquad\square$

It follows from Theorem 4.2.1 that the correspondence which associates the inversion sequence to each permutation is a one-to-one correspondence between the permutations of $\{1, 2, \ldots, n\}$ and the sequences of integers $b_1, b_2, \ldots, b_n$ satisfying

$$0 \le b_1 \le n - 1, \ 0 \le b_2 \le n - 2, \ \ldots \ , 0 \le b_{n-1} \le 1, b_n = 0.$$

Thus, a permutation is uniquely specified by specifying its inversion sequence. Think of it as a "code" for the permutation. In the proof of Theorem 4.2.1, we have given two methods to "break this code."

There is a subtle distinction worth making between a permutation and its inversion sequence. In choosing a permutation of $\{1, 2, \ldots, n\}$ we have to make n choices, one for each term of the permutation. We choose the first term, in any one of n ways, then the second term, in any one of $n - 1$ ways, but notice that while the *number* of choices for the second term is $n - 1$, independent of the choice of the first term, the *choices* for the second term are *not* independent of the first term (one cannot choose whatever has already been chosen for the first term). A similar situation occurs for the choice of the kth term. One has $n - (k - 1)$ choices for the kth term, but the actual choices depend on what has already been chosen for the first $k - 1$ terms.

The preceding description can be contrasted with choosing an inversion sequence $b_1, b_2, \ldots, b_n$ for a permutation of $\{1, 2, \ldots, n\}$. For b_1, we can choose any of the n integers $0, 1, \ldots, n - 1$. For b_2, we can choose any of the $n - 1$ integers $0, 1, \ldots, n - 2$, and *it does not matter what our choice for b_1 is*. In general, for b_k, we can choose any of the $n - (k - 1)$ integers $0, 1, \ldots, n - k$, and *it does not matter what our choices for $b_1, b_2, \ldots, b_{k-1}$ are*. Thus, the inversion sequence replaces dependent choices by independent choices.

It is customary to call a permutation $i_1 i_2 \ldots i_n$ of $\{1, 2, \ldots, n\}$ *even* or *odd* according as its number of inversions is even or odd. The *sign* of the permutation is then defined to be $+1$ or -1 according to whether it is even or odd. The sign of a permutation is important in the theory of determinants of matrices, where the determinant of an $n \times n$ matrix

$$A = [a_{ij}] \qquad (i, j = 1, 2, \ldots, n)$$

is defined to be

$$\det(A) = \sum \epsilon(i_1 i_2 \ldots i_n) a_{1 i_1} a_{2 i_2} \cdots a_{n i_n},$$

where the summation extends over all permutations $i_1 i_2 \ldots i_n$ of the set $\{1, 2, \ldots, n\}$, and $\epsilon(i_1 i_2 \ldots i_n)$ is equal to the sign of $i_1 i_2 \ldots i_n$.[9]

---

[9]Thinking of an $n \times n$ matrix as an n-by-n chessboard in which the squares

If the permutation $i_1 i_2 \ldots i_n$ has inversion sequence $b_1, b_2, \ldots, b_n$ and $k = b_1 + b_2 + \cdots + b_n$ is the number of inversions, then $i_1 i_2 \ldots i_n$ can be brought to $12 \ldots n$ by k successive switches of adjacent numbers. We first switch 1 successively with the b_1 numbers to its left. We then switch 2 successively with the b_2 numbers to its left which are greater than 2, and so on. In this way, we arrive at $12 \ldots n$ after $b_1 + b_2 + \cdots + b_n$ switches.

Example. Bring the permutation 361245 to 123456 by successive switches of adjacent numbers.

The inversion sequence is 220110. The results of successive switches are as follows:

$$
\begin{array}{cccccc}
3 & 6 & 1 & 2 & 4 & 5 \\
3 & 1 & 6 & 2 & 4 & 5 \\
1 & 3 & 6 & 2 & 4 & 5 \\
1 & 3 & 2 & 6 & 4 & 5 \\
1 & 2 & 3 & 6 & 4 & 5 \\
1 & 2 & 3 & 4 & 6 & 5 \\
1 & 2 & 3 & 4 & 5 & 6 \\
\end{array}
$$

$\square$

This procedure is one instance of a sorting procedure common in computer science. The elements of a permutation $i_1 i_2 \ldots i_n$ correspond to the unsorted data. For more efficient sorting techniques and their analysis, the reader may consult Knuth.[10]

4.3 Generating Combinations

Let S be a set of n elements. For reasons that will be clear shortly, we take S to be the set

$$ S = \{x_{n-1}, \ldots, x_1, x_0\}. $$

We now seek an algorithm that generates all of the 2^n combinations (subsets) of S. What this means is that we want a systematic procedure that lists the combinations of S. The resulting list should contain all the combinations of S (and only combinations of S) with no duplications. Thus according to Theorem 3.3.3 there will be 2^n combinations on the list.

---

are occupied by numbers, the terms in the summation for the formula for the determinant correspond to the $n!$ ways to place n nonattacking rooks on the board!

[10]D.E. Knuth: *Sorting and Searching.* Volume 3 of *The Art of Computer Programming,* Addison-Wesley, Reading, MA (1973).

Given a combination A of S, then each element x either belongs or does not belong to A. If we use 1 to denote that an element belongs and 0 to denote that an element does not belong, then we can identify the 2^n combinations of S with the 2^n n-tuples

$$(a_{n-1}, \ldots, a_1, a_0) = a_{n-1} \cdots a_1 a_0$$

of 0's and 1's.[11] We let the ith term a_i of the n-tuple correspond to the element x_i for each $i = 0, 1, \ldots, n-1$. For example, when $n = 3$, the $2^3 = 8$ combinations and their corresponding 3-tuples are given as follows:

	a_2	a_1	a_0
$\emptyset$	0	0	0
$\{x_0\}$	0	0	1
$\{x_1\}$	0	1	0
$\{x_1, x_0\}$	0	1	1
$\{x_2\}$	1	0	0
$\{x_2, x_0\}$	1	0	1
$\{x_2, x_1\}$	1	1	0
$\{x_2, x_1, x_0\}$	1	1	1

Example. Let $S = \{x_6, x_5, x_4, x_3, x_2, x_1, x_0\}$. The 7-tuple corresponding to the combination $\{x_5, x_4, x_2, x_0\}$ is 0110101. The combination corresponding to the 7-tuple 1010001 is $\{x_6, x_4, x_0\}$.   □

Because of this identification of combinations of a set of n elements with n-tuples of 0's and 1's, in order to generate the combinations of a set of n elements, it suffices to describe a systematic procedure for writing in a list the 2^n n-tuples of 0's and 1's. Now, each such n-tuple can be regarded as a base 2 numeral.[12] For example, 10011 is the binary numeral for the integer 19, since

$$19 = 1 \times 2^4 + 0 \times 2^3 + 0 \times 2^2 + 1 \times 2^1 + 1 \times 2^0.$$

In general, given an integer m from 0 up to $2^n - 1$, it can be expressed in the form

$$m = a_{n-1} \times 2^{n-1} + a_{n-2} \times 2^{n-2} + \cdots + a_1 \times 2^1 + a_0 \times 2^0,$$

where each a_i is 0 or 1. Its binary numeral is

$$a_{n-1}a_{n-2} \cdots a_1 a_0.$$

---

[11]In the language of Section 3.3, we are identifying the combinations with the n-permutations of the multiset $\{n \cdot 0, n \cdot 1\}$.

[12]See also Section 1.7.

Conversely, since

$$2^{n-1} + 2^{n-2} + \cdots + 2^1 + 2^0 = 2^n - 1,$$

every expression of the preceding form has value equal to an integer between 0 and $2^n - 1$. The n-tuples of 0's and 1's are thus in one-to-one correspondence with the integers $0, 1, \ldots, 2^n - 1$. Note that, in writing the binary numeral for an integer between 0 and $2^n - 1$, our convention is to use exactly n digits and thus to include, if necessary, some initial 0's that are not normally included.

Example. Let $n = 7$. The number 29 is between 0 and $2^7 - 1 = 127$ and can be expressed as

$$29 = 0 \times 2^6 + 0 \times 2^5 + 1 \times 2^4 + 1 \times 2^3 + 1 \times 2^2 + 0 \times 2^1 + 1 \times 2^0.$$

Thus, 29 has a binary numeral of 7 digits given by 0011101 and corresponds to the combination $\{x_4, x_3, x_2, x_0\}$ of the set

$$S = \{x_6, x_5, x_4, x_3, x_2, x_1, x_0\}.$$

$\square$

How do we generate the 2^n combinations of $S = \{x_{n-1}, \ldots, x_1, x_0\}$? Equivalently, how do we generate the 2^n n-tuples of 0's and 1's? The answer is now simple. We write down the numbers from 0 to $2^n - 1$ in increasing order by size, *but in binary form, adding 1 each time, using base 2 arithmetic!* This is how the 3-tuples of 0's and 1's were generated earlier.

Example. Generate the 4-tuples of 0's and 1's.

Number	Binary Numeral
0	0 0 0 0
1	0 0 0 1
2	0 0 1 0
3	0 0 1 1
4	0 1 0 0
5	0 1 0 1
6	0 1 1 0
7	0 1 1 1
8	1 0 0 0

$$
\begin{array}{rl}
9 & 1\,0\,0\,1 \\
10 & 1\,0\,1\,0 \\
11 & 1\,0\,1\,1 \\
12 & 1\,1\,0\,0 \\
13 & 1\,1\,0\,1 \\
14 & 1\,1\,1\,0 \\
15 & 1\,1\,1\,1
\end{array}
$$

□

Example. Using the scheme just described, what is the combination of $\{x_6, x_5, x_4, x_3, x_2, x_1, x_0\}$ that immediately follows the combination $\{x_6, x_4, x_2, x_1, x_0\}$?

The combination $\{x_6, x_4, x_2, x_1, x_0\}$ corresponds to the binary numeral 1010111. Using base 2 arithmetic, we see that the next combination corresponds to

$$
\begin{array}{r}
1\ \ 0\ \ 1\ \ 0\ \ 1\ \ 1\ \ 1 \\
+ \hspace{4.5em} 1 \\
\hline
1\ \ 0\ \ 1\ \ 1\ \ 0\ \ 0\ \ 0
\end{array}
$$

and thus is the combination $\{x_6, x_4, x_3\}$. Since

$$
1 \times 2^6 + 0 \times 2^5 + 1 \times 2^4 + 0 \times 2^3 + 1 \times 2^2 + 1 \times 2^1 + 1 \times 2^0 = 87,
$$

the combination $\{x_6, x_4, x_2, x_1, x_0\}$ is the 87th on the list. The combination that is 88th on the list is $\{x_6, x_4, x_3\}$. Note that the places on the list are numbered beginning with 0. The combination occupying the 0th place is always the empty set. When we say, for instance, the 5th combination on the list, we mean the combination on the list corresponding to the number 5, and not the combination corresponding to the number 4. Five combinations precede the 5th combination on the list. If this is not yet clear, the next example should clarify our convention. □

Example. Which combination of $S = \{x_6, x_5, x_4, x_3, x_2, x_1, x_0\}$ is 108th on the list?

We first find the base 2 numeral for 108:

$$
108 = 1 \times 2^6 + 1 \times 2^5 + 0 \times 2^4 + 1 \times 2^3 + 1 \times 2^2 + 0 \times 2^1 + 0 \times 2^0.
$$

Hence, the base 2 numeral for 108 is

1101100.

Thus, the combination is $\{x_6, x_5, x_3, x_2\}$. Which combination immediately precedes this one? We simply subtract in base 2:

$$
\begin{array}{cccccccc}
 & 1 & 1 & 0 & 1 & 1 & 0 & 0 \\
- & & & & & & & 1 \\
\hline
 & 1 & 1 & 0 & 1 & 0 & 1 & 1
\end{array}
$$

This corresponds to the combination $\{x_6, x_5, x_3, x_1, x_0\}$. □

We now describe in compact form our algorithm for generating the combinations of a set of n elements. The description is in terms of n-tuples of 0's and 1's. The *rule of succession* given in the algorithm is a consequence of addition using base 2 arithmetic.

Base 2 Algorithm for Generating the Combinations of $\{x_{n-1}, \ldots, x_1, x_0\}$

Begin with $a_{n-1} \cdots a_1 a_0 = 0 \cdots 00$.

While $a_{n-1} \cdots a_1 a_0 \neq 1 \cdots 11$, do the following:

(1) Find the smallest integer j (between $n-1$ and 0) such that $a_j = 0$.

(2) Replace a_j by 1 and each of $a_{j-1}, \ldots, a_0$ (which by our choice of j all equal 1) by 0.

The algorithm comes to an end when $a_{n-1} \cdots a_1 a_0 = 1 \cdots 11$, which is the last binary n-tuple on the resulting list.

The ordering of the n-tuples of 0's and 1's produced by the base 2 generation scheme is called the *lexicographic ordering of n-tuples*. In this ordering, an n-tuple $a_{n-1} \cdots a_1 a_0$ occurs earlier on the list than another n-tuple $b_{n-1} \cdots b_1 b_0$, provided that, starting at the left, the first position in which they disagree, say position j, we have $a_j = 0$ and $b_j = 1$. (This is equivalent to saying that the number whose base 2 numeral is given by $a_{n-1} \cdots a_1 a_0$ is smaller than the number whose base 2 numeral is given by $b_{n-1} \cdots b_1 b_0$.) Thinking of the n-tuples as "words" of length n in an alphabet of two "letters," 0 and 1, in which 0 is the first letter of the alphabet and 1 is the second letter, the lexicographic ordering is the order in which these words would occur in a dictionary.

Viewing the n-tuples as combinations of the set $\{x_{n-1}, \ldots, x_1, x_0\}$, for each j with $n - 1 > j$ all the combinations of $\{x_j, \ldots, x_1, x_0\}$ precede those combinations which contain at least one of the elements $x_{n-1}, \ldots, x_{j+1}$. For this reason, the lexicographic ordering on n-tuples of 0's and 1's, when viewed as an ordering of the combinations of $\{x_{n-1}, \ldots, x_1, x_0\}$, is sometimes called the *squashed ordering of combinations*. In the squashed ordering we list all the combinations of the current elements before introducing a new element. The squashed ordering of the combinations of $\{x_3 = 4, x_2 = 3, x_1 = 2, x_0 = 1\}$ is as follows and corresponds to our earlier (lexicographic) listing of the binary 4-tuples:

$\emptyset$	4
1	1, 4
2	2, 4
1, 2	1, 2, 4
3	3, 4
1, 3	1, 3, 4
2, 3	2, 3, 4
1, 2, 3	1, 2, 3, 4

Notice how, in this ordering, all the combinations that do not contain 4 come before those that do. Of the combinations that do not contain 4, those that do not contain 3 come before those that do. Of the combinations that contain neither 4 nor 3, those that do not contain 2 come before those that do.

The immediate successor of a combination in the squashed ordering of combinations (equivalently, the immediate successor of an n-tuple in the lexicographic ordering of n-tuples) may differ greatly from the combination itself. The combination $A = \{x_6, x_4, x_3\}$ (equivalently, the 7-tuple 1011000) which follows the combination $B = \{x_6, x_4, x_2, x_1, x_0\}$ (equivalently, the 7-tuple 1010111) differs from B in four instances, since A contains x_3 (and B doesn't) while B contains x_2, x_1 and x_0 (and A doesn't). This suggests consideration of the following question: Is it possible to generate the combinations of a set of n elements in a different order so that the immediate successor of a combination differs from it as little as possible? *As little as possible* means here that the immediate successor of a combination is obtained by either including a new element or deleting an old element, but not both; in short, one in or one out. Such a generation scheme can be important for many reasons, not the least of which is that there would be a smaller chance of error in generating all the combinations.

Example. Let $S = \{x_{n-1}, \ldots, x_1, x_0\}$, and consider the following lists of the combinations of S and the corresponding n-tuples for $n = 1, 2, 3$.

$n = 1$	
$\emptyset$	0
$\{x_0\}$	1

$n = 2$	
$\emptyset$	0 0
$\{x_0\}$	0 1
$\{x_1, x_0\}$	1 1
$\{x_1\}$	1 0

$n = 3$	
$\emptyset$	0 0 0
$\{x_0\}$	0 0 1
$\{x_1, x_0\}$	0 1 1
$\{x_1\}$	0 1 0
$\{x_2, x_1\}$	1 1 0
$\{x_2, x_1, x_0\}$	1 1 1
$\{x_2, x_0\}$	1 0 1
$\{x_2\}$	1 0 0

In each list, the transition from one combination to the next is obtained by inserting a new element or removing an element already present, but not both. In terms of n-tuples of 0's and 1's, we change a 0 to a 1 or a 1 to a 0, but not both.

□

We now make a further identification, this time a geometric one. We regard an n-tuple of 0's and 1's as the coordinates of a point in n-dimensional space. Thus, for $n = 1$, the identification is with points on a line; for $n = 2$, it is with points in 2-space or a plane; for $n = 3$, it is with points in three-dimensional space.

Figure 4.1

Example. Let $n = 1$. The 1-tuples of 0's and 1's correspond to the endpoints or corners of a unit line segment as shown above in Figure 4.1.

□

Example. Let $n = 2$. The 2-tuples of 0's and 1's correspond to the corners of a unit square as shown below in Figure 4.2.

□

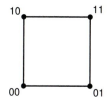

Figure 4.2

Example. Let $n = 3$. The 3-tuples of 0's and 1's correspond to the corners of a unit cube as shown in Figure 4.3.   □

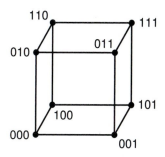

Figure 4.3

Notice that in all three examples there is an edge between two corners precisely when their coordinates differ in only one place. This is precisely the feature we are looking for in generating the n-tuples of 0's and 1's!

We can generalize to any n. The *unit n-cube* (a 1-cube is a line segment, a 2-cube is a square, a 3-cube is an ordinary cube) has 2^n corners whose coordinates are the 2^n n-tuples of 0's and 1's. There is an edge of the n-cube joining two corners precisely when the coordinates of the corners differ in only one place. An algorithm for generating the n-tuples of 0's and 1's which has the property that the successor of an n-tuple differs from it in only one place corresponds to a walk along the edges of an n-cube that visits every corner exactly once. Any such walk (or the resulting list of n-tuples) is called a *Gray code of order* n.[13] If it is possible to traverse over one more edge to get from the terminal corner to the initial corner of the walk, then the Gray code is called *cyclic*. The lists for $n = 1, 2$, and 3 in the examples are cyclic Gray codes. They have an additional property that makes them quite special and we now investigate it.

---

[13]In 1878, the French engineer Émile Baudot demonstrated the use of a Gray code in a telegraph. It was the Bell Labs researcher Frank Gray who first patented these codes in 1953.

Figure 4.4

Let us begin with the unit 1-cube and the Gray code, which starts at 0 and ends at 1 as shown in Figure 4.4. We build a unit 2-cube by taking two copies of the 1-cube and joining corresponding corners. We attach a 0 to the coordinates of one copy and a 1 to the coordinates of the other: We obtain a cyclic Gray code for the 2-cube by first following the Gray code on one copy of the 1-cube, crossing over to the other copy, and then following the Gray code for the 1-cube in reverse direction, as shown on the left in Figure 4.5.

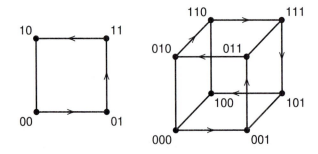

Figure 4.5

We build a unit 3-cube in a similar way from the unit 2-cube. We take two copies of the 2-cube and join corresponding corners. We attach a 0 to the coordinates of one copy and a 1 to the coordinates of the other. We obtain a cyclic Gray code for the 3-cube by first following the Gray code on one copy of the 2-cube, crossing over to the other copy, and then following the Gray code for the 2-cube in the reverse direction as shown on the right in Figure 4.5.

We may continue in this manner to construct inductively a Gray code of order n for any integer $n \geq 1$. The Gray code constructed in this way is called the *reflected Gray code*. The n-cube is only a visual device and needn't be introduced in order to obtain the reflected Gray code of order n. Of course, for $n > 3$ we can only get a visual picture of the corners and edges of the n-cube. The reflected Gray code for

$n = 4$ is as follows:

$$
\begin{array}{cccc}
0 & 0 & 0 & 0 \\
0 & 0 & 0 & 1 \\
0 & 0 & 1 & 1 \\
0 & 0 & 1 & 0 \\
0 & 1 & 1 & 0 \\
0 & 1 & 1 & 1 \\
0 & 1 & 0 & 1 \\
0 & 1 & 0 & 0 \\
1 & 1 & 0 & 0 \\
1 & 1 & 0 & 1 \\
1 & 1 & 1 & 1 \\
1 & 1 & 1 & 0 \\
1 & 0 & 1 & 0 \\
1 & 0 & 1 & 1 \\
1 & 0 & 0 & 1 \\
1 & 0 & 0 & 0.
\end{array}
$$

The general inductive definition of the reflected Gray code of order n is the following:

(1) The reflected Gray code of order 1 is $\begin{smallmatrix} 0 \\ 1 \end{smallmatrix}$.

(2) Suppose $n > 1$ and the reflected Gray code of order $n - 1$ has been constructed. To construct the reflected Gray code of order n, we first list the $(n-1)$-tuples of 0's and 1's in the order given by the reflected Gray code of order $n - 1$, and attach a 0 at the beginning of each $(n-1)$-tuple. We then list the $(n-1)$-tuples in the order which is the reverse of that given by the reflected Gray code of order $n - 1$, and attach a 1 at the beginning.

It follows from this inductive definition that the reflected Gray code of order n begins with the n-tuple $00\cdots0$ and ends with the n-tuple $10\cdots0$. It is therefore cyclic, since $00\cdots0$ and $10\cdots0$ differ in only one place.

Since the reflected Gray codes have been defined inductively, to construct the reflected Gray code of order n, we first construct the reflected Gray code of order $n - 1$. So, for instance, to construct the reflected Gray code of order 6, we first construct the reflected Gray code of order 5. To do that we construct the reflected Gray code of order 4, and so on. Therefore, to construct the reflected Gray code of order 6, using the inductive definition we must construct sequentially the reflected Gray codes of orders 1, 2, 3, 4, and 5. We now describe an algorithm that enables us to construct the reflected Gray code of order n directly. In order to do this we need a *rule of succession*, which tells us which place to change (from a 0 to a 1 or a 1 to a 0) in going from one n-tuple to the next in the reflected Gray code. This rule of succession is provided in the next algorithm.

If $a_{n-1}a_{n-2} \cdots a_0$ is an n-tuple of 0's and 1's, then

$$\sigma(a_{n-1}a_{n-2} \cdots a_0) = a_{n-1} + a_{n-2} + \cdots + a_0$$

is the number of its 1's (and thus equals the size of the combination to which it corresponds).

Algorithm for generating the n-tuples of 0's and 1's in the reflected Gray code order

Begin with the n-tuple $a_{n-1}a_{n-2} \cdots a_0 = 00 \cdots 0$.

While the n-tuple $a_{n-1}a_{n-2} \cdots a_0 \neq 10 \cdots 0$, do the following:

(1) Compute $\sigma(a_{n-1}a_{n-2} \cdots a_0) = a_{n-1} + a_{n-2} + \cdots + a_0$.

(2) If $\sigma(a_{n-1}a_{n-2} \cdots a_0)$ is even, change a_0 (from 0 to 1 or 1 to 0).

(3) Else, determine j such that $a_j = 1$ and $a_i = 0$ for all i with $j > i$, and then change a_{j+1} (from 0 to 1 or 1 to 0).

We note that if, in step (3), we have $a_{n-1}a_{n-2} \cdots a_0 \neq 10 \cdots 0$, then $j \leq n-2$, so that $j+1 \leq n-1$ and a_{j+1} is defined. We also note that in step (3) we may have $j = 0$–that is, $a_0 = 1$; in this case there is no i with $i < j$, and we change a_1 as instructed in step (3).

The reader may wish to check that this algorithm does give the Gray code of order 4 as already presented.

Theorem 4.3.1 *The algorithm for generating the n-tuples of 0's and 1's previously described produces the reflected Gray code of order n for each positive integer n.*

Proof. We prove the theorem by induction on n. It is clear that the algorithm applied to $n = 1$ produces the reflected Gray code of order 1. Let $n > 1$ and assume that the algorithm applied to $n-1$ produces the reflected Gray of order $n-1$. The first 2^{n-1} n-tuples of the reflected Gray code of order n consist of the $(n-1)$-tuples of the reflected Gray code of order $n-1$ with a 0 attached at the beginning of each $(n-1)$-tuple. Since the $(n-1)$-tuple $10\cdots0$ occurs last in the reflected Gray code of order $n-1$, it follows that the rule of succession applied to the first $(2^{n-1}-1)$ n-tuples of the reflected Gray code of order n has the same effect as applying the rule of succession to all but the last $(n-1)$-tuple of the reflected Gray code of order $n-1$, and then attaching a 0. Hence it is a consequence of the inductive hypothesis that the rule of succession produces the first half of the reflected Gray code of order n. The 2^{n-1}st n-tuple of the reflected Gray code of order n is $010\cdots0$. Since $\sigma(010\cdots0)=1$, an odd number, the rule of succession applied to $010\cdots0$ gives $110\cdots0$, which is the $(2^{n-1}+1)$st n-tuple of the reflected Gray code of order n.

Consider now two consecutive n-tuples in the second half of the reflected Gray code of order n:

$$1 \ a_{n-2}\cdots a_0$$
$$1 \ b_{n-2}\cdots b_0$$

Then $a_{n-2}\cdots a_0$ immediately follows $b_{n-2}\cdots b_0$ in the reflected Gray code of order $n-1$:

$$b_{n-2}\cdots b_0$$
$$a_{n-2}\cdots a_0$$

Now $\sigma(a_{n-2}\cdots a_0)$ and $\sigma(b_{n-2}\cdots b_0)$ are of opposite parity. One is even and the other is odd. Also, $\sigma(1a_{n-2}\cdots a_0)$ and $\sigma(a_{n-2}\cdots a_0)$ are of opposite parity and so are $\sigma(1b_{n-2}\cdots b_0)$ and $\sigma(b_{n-2}\cdots b_0)$. Suppose that $\sigma(b_{n-2}\cdots b_0)$ is even. Then, $\sigma(a_{n-2}\cdots a_0)$ is odd and $\sigma(1a_{n-2}\cdots a_0)$ is even. By the induction assumption, $a_{n-2}\cdots a_0$ is obtained from $b_{n-2}\cdots b_0$ by changing b_0. The rule of succession applied to $1a_{n-2}\cdots a_0$ instructs us to change a_0, and this gives $1b_{n-2}\cdots b_0$ as desired. Now suppose that $\sigma(b_{n-2}\cdots b_0)$ is odd. Then $\sigma(a_{n-2}\cdots a_0)$ is even and $\sigma(1a_{n-2}\cdots a_0)$ is odd. The rule of succession applied to $1a_{n-2}\cdots a_0$ has the opposite effect from the rule of succession applied to $b_{n-2}\cdots b_0$. Hence, it also follows by the induction assumption that the rule of succession applied to $1a_{n-2}\cdots a_0$ gives $1b_{n-2}\cdots b_0$, as desired. Therefore, the theorem holds by induction. $\square$

Example. Determine the 8-tuples that are successors of 10100110, 00011111, and 01010100 in the reflected Gray code of order 8.

Since $\sigma(10100110)=4$, an even number, 10100111 follows 10100110. Since $\sigma(00011111)=5$, an odd number, then in step (3) of the algorithm $j=1$ so that 00011101 follows 00011111. Since $\sigma(01010100)=3$, 01011100 follows 01010100. $\square$

We have described two linear orderings of the 2^n binary n-tuples: the lexicographic order obtained, starting with $00\cdots0$, by using base 2 arithmetic, and the reflected Gray code order which also starts with $00\cdots0$. The lexicographic order corresponds to the integers from 0 to $2^n - 1$ in base 2, and we can think of the reflected Gray code order as listing the binary n-tuples in a specified order from 0 to $2^n - 1$. Let $a_{n-1}\cdots a_1 a_0$ be a binary n-tuple. We can say explicitly in what place this binary n-tuple occurs on the list in Gray code order. For $i = 0, 1, \ldots, n - 1$, let

$$b_i = \begin{cases} 0 & \text{if } a_{n-1} + \cdots + a_i \text{ is even, and} \\ 1 & \text{if } a_{n-1} + \cdots + a_i \text{ is odd.} \end{cases}$$

Then $a_{n-1}\cdots a_1 a_0$ is in the same place on the Gray code order list as $b_{n-1}\cdots b_1 b_0$ is on the lexicographic order list. Put another way, $a_{n-1}\cdots a_1 a_0$ is in place

$$k = b_{n-1} \times 2^{n-1} + \cdots + b_1 \times 2 + b_0 \times 2^0$$

on the Gray code order list. We leave this verification as an exercise.

4.4   Generating r-Combinations

In Section 4.3, we have described two orderings for the combinations of a set of n elements and corresponding algorithms based on a rule of succession for generating the combinations. We now consider only the combinations of a fixed size r and seek a method to generate these combinations. One way to do this is to generate *all* combinations and then go through the list and select those that contain exactly r elements. But this is obviously a very inefficient approach.

Example. In Section 4.3, we listed all the 4-combinations of $\{1, 2, 3, 4\}$ in the squashed ordering. Selecting the 2-combinations from among them, we get the squashed ordering of the 2-combinations of $\{1, 2, 3, 4\}$:

$$1, 2$$
$$1, 3$$
$$2, 3$$
$$1, 4$$
$$2, 4$$
$$3, 4$$

$\square$

In this section, we develop an algorithm for a lexicographic ordering of the r-combinations of a set of n elements, where r is a fixed integer with $1 \le r \le n$. We now take our set to be the set

$$S = \{1, 2, \ldots, n\}$$

consisting of the first n positive integers. This gives us a natural order,

$$1 < 2 < \cdots < n,$$

on the elements of S. Let A and B be two r-combinations of the set $\{1, 2, \ldots, n\}$. Then we say that A *precedes* B *in the lexicographic order* provided that the smallest integer which is in their union $A \cup B$, but not in their intersection $A \cap B$ (that is, in one but not both of the sets), is in A.

Example. Let 5-combinations A and B of $\{1, 2, 3, 4, 5, 6, 7, 8\}$ be given by

$$A = \{2, 3, 4, 7, 8\}, \qquad B = \{2, 3, 5, 6, 7\}.$$

The smallest element that is in one, but not both, of the sets is 4 (which belongs to A). Hence A precedes B in the lexicographic order. □

How is this a lexicographic order in the sense used in the preceding section and in the sense used in a dictionary? We think of the elements of S as the letters of an alphabet, where 1 is the first letter of the alphabet, 2 is the second letter, and so on. We want to think of the r-combinations as "words" of length r over the alphabet S and then impose a dictionary-type order on the words. But the letters in a word form an ordered sequence (e.g., *part* is not the same word as *trap*) and for combinations, as we have learned, order doesn't matter. Well, since order doesn't matter in a combination, let us agree that, whenever we write a combination of $\{1, 2, \ldots, n\}$, we write the integers in it from smallest to largest. Thus, we agree that an r-combination of $S = \{1, 2, \ldots, n\}$ is to be written in the form

$$a_1, a_2, \ldots, a_r, \text{ where } 1 \le a_1 < a_2 < \cdots < a_r \le n.$$

Let us also agree, for convenience, to write this r-combination as

$$a_1 a_2 \cdots a_r$$

without commas, that is, as a word of length r. We now have established a convention for writing combinations that allows us to regard a combination as a word. But note that not all words are allowed. The only words that will be in our dictionary are those that have r letters from our alphabet $1, 2, \ldots, n$ *and* for which the letters are in strictly increasing order (in particular, there are no repeated letters in our words).

Example. We return to our previous example and now, with our established conventions, write $A = 23478$ and $B = 23567$. We see that A and B agree in their first two letters and disagree in their third letter. Since $4 < 5$ (4 comes earlier in our alphabet than 5), A precedes B in the lexicographic order. □

Example. We consider the lexicographic order of the 5-combinations of $\{1, 2, 3, 4, 5, 6, 7, 8, 9\}$. The first 5-combination is clearly 12345; the last 5-combination is 56789. What 5-combination immediately follows 12389 (in our dictionary)? Among the 5-combinations that begin with 123, 12389 is the last. Among the 5-combinations that begin with 12 and don't have a 3 in the third position, 12456 is the first. Thus, 12456 immediately follows 12389. □

We generalize this example and determine, for all but the last word in our dictionary, the word that immediately follows it.

Theorem 4.4.1 *Let $a_1 a_2 \cdots a_r$ be an r-combination of $\{1, 2, \ldots, n\}$. The first r-combination in the lexicographic ordering is $12 \cdots r$. The last r-combination in the lexicographic ordering is $(n - r + 1)(n - r + 2) \cdots n$. Assume that $a_1 a_2 \cdots a_r \neq (n - r + 1)(n - r + 2) \cdots n$. Let k be the largest integer such that $a_k < n$ and $a_k + 1$ is different from each of $a_1, a_2, \ldots, a_r$. Then the r-combination that is the immediate successor of $a_1 a_2 \cdots a_r$ in the lexicographic ordering is*

$$a_1 \cdots a_{k-1}(a_k + 1)(a_k + 2) \cdots (a_k + r - k + 1).$$

Proof. It follows from the definition of the lexicographic order that $12 \cdots r$ is the first and $(n-r+1)(n-r+2) \cdots n$ is the last r-combination in the lexicographic ordering. Now let $a_1 a_2 \cdots a_r$ be any r-combination other than the last, and determine k as indicated in the theorem. Then

$$a_1 a_2 \cdots a_r = a_1 \cdots a_{k-1} a_k (n - r + k + 1)(n - r + k + 2) \cdots (n)$$

where

$$a_k + 1 < n - r + k + 1.$$

Thus $a_1 a_2 \cdots a_r$ is the last r-combination that begins with $a_1 \cdots a_{k-1} a_k$. The r-combination

$$a_1 \cdots a_{k-1}(a_k + 1)(a_k + 2) \cdots (a_k + r - k + 1)$$

is the first r-combination that begins $a_1 \cdots a_{k-1} a_k + 1$ and hence is the immediate successor of $a_1 a_2 \cdots a_r$. $\square$

From Theorem 4.4.1, we conclude that the next algorithm generates the r-combinations of $\{1, 2, \ldots, n\}$ in lexicographic order.

Algorithm for generating the r-combinations of $\{\{1, 2, \ldots, n\}$
in lexicographic order

Begin with the r-combination $a_1 a_2 \cdots a_r = 12 \cdots r$.
While $a_1 a_2 \cdots a_r \neq (n - r + 1)(n - r + 2) \cdots n$, do the following:

(1) Determine the largest integer k such that $a_k + 1 \leq n$ and $a_k + 1$ is not one of $a_1, a_2, \ldots, a_r$.

(2) Replace $a_1 a_2 \cdots a_r$ with the r-combination

$$a_1 \cdots a_{k-1}(a_k + 1)(a_k + 2) \cdots (a_k + r - k + 1).$$

Example. We apply the algorithm to generate the 4-combinations of $S = \{1, 2, 3, 4, 5, 6\}$ and obtain the following:

1234	1256	2345
1235	1345	2346
1236	1346	2356
1245	1356	2456
1246	1456	3456.

$\square$

Combining the algorithm for generating permutations of a set with that for generating r-combinations of an n-element set, we obtain an algorithm for generating r-permutations of an n-element set.

Example. Generate the 3-permutations of $\{1, 2, 3, 4\}$. We first generate the 3-combinations in lexicographic order: 123, 124, 134, 234. For each 3-combination, we then generate all of its permutations:

123	124	134	234
132	142	143	243
312	412	413	423
321	421	431	432
231	241	341	342
312	214	314	324.

$\square$

We conclude by determining the position of each r-combination in the lexicographic order of the r-combinations of $\{1, 2, \ldots, n\}$.

Theorem 4.4.2 *The r-combination $a_1 a_2 \cdots a_r$ of $\{1, 2, \ldots, n\}$ occurs in place number*

$$\binom{n}{r} - \binom{n-a_1}{r} - \binom{n-a_2}{r-1} - \cdots - \binom{n-a_{r-1}}{2} - \binom{n-a_r}{1}$$

in the lexicographic order of the r-combinations of $\{1, 2, \ldots, n\}$.

Proof. We first count the number of r-combinations that come *after* $a_1 a_2 \cdots a_r$:

(1) There are $\binom{n-a_1}{r}$ r-combinations whose first element is greater than a_1 that come after $a_1 a_2 \cdots a_r$.

(2) There are $\binom{n-a_2}{r-1}$ r-combinations whose first element is a_1 but whose second element is greater than a_2 that come after $a_1 a_2 \cdots a_r$.

$\vdots$

$(r-1)$ There are $\binom{n-a_{r-1}}{2}$ r-combinations that begin $a_1 \cdots a_{r-2}$ but whose $(r-1)$st element is greater than a_{r-1} that come after $a_1 a_2 \cdots a_r$.

(r) There are $\binom{n-a_r}{1}$ r-combinations that begin $a_1 \cdots a_{r-1}$ but whose rth element is greater than a_r that come after $a_1 a_2 \cdots a_r$.

Subtracting the number of r-combinations that come after $a_1 a_2 \cdots a_r$ from the total number $\binom{n}{r}$ of r-combinations, we find that the place of $a_1 a_2 \cdots a_r$ is as given in the theorem. $\qquad \square$

Example. In which place is the combination 1258 among the 4-combinations of $\{1, 2, 3, 4, 5, 6, 7, 8\}$ in lexicographic order? We apply Theorem 4.4.2 and find that 1258 is in place

$$\binom{8}{4} - \binom{7}{4} - \binom{6}{3} - \binom{3}{2} - \binom{0}{1} = 12.$$

4.5 Partial Orders and Equivalence Relations

In this chapter we have defined various "natural" orders on the sets of permutations, combinations, and r-combinations of a finite set, namely, the orders determined by the generating schemes. These orders are "total orders" in the sense that there is a first object, a second object, a third object, ... , a last object. There is a more general notion of order, called "partial order," which is extremely important and useful in mathematics. Perhaps the two partial orders which are not total orders that are most familiar are those defined by containment of one set in another and divisibility of one integer by another. These are *partial* orders in the sense that, given any two sets, neither may be a subset of the other, and given any two integers, neither may be divisible by the other.

In order to give a precise definition of a partial order, it is important to know what is meant in mathematics by a *relation*. Let X be a set. A relation R on X is a "property" that may or may not hold between any two given elements of X. More formally, a relation on X is a subset R of the set $X \times X$ of ordered pairs of elements of X. We write $a\,R\,b$, provided that the ordered pair (a, b) belongs to R; we also write $a\,\not\!R\,b$ whenever (a, b) is not in R.

Example. Let $X = \{1, 2, 3, 4, 5, 6\}$. Write $a \mid b$ to mean that a is a divisor of b (equivalently, b is divisible by a). This defines a partial order on X and we have, for example, $2 \mid 6$ and $3 \nmid 5$.

Now consider the collection $\mathcal{P}(X)$ of all subsets (i.e., combinations) of X. For A and B in $\mathcal{P}(X)$, we write as usual $A \subseteq B$, read A is *contained in* B, provided that every element of A is also an element of B. This defines a relation on $\mathcal{P}(X)$ and we have, for example, $\{1\} \subseteq \{1, 3\}$ and $\{1, 2\} \not\subseteq \{2, 3\}$.   □

The following are special properties that a relation R on a set X may have:

1. R is *reflexive*, provided that $x\,R\,x$ for all x in X.

2. R is *irreflexive*, provided that $x\,\not\!R\,x$ for all x in X.

3. R is *symmetric*, provided that, for all x and y in X, whenever we have $x\,R\,y$ we also have $y\,R\,x$.

4. R is *antisymmetric*, provided that, for all x and y in X with $x \neq y$, whenever we have $x\,R\,y$, we also have $y\,\not\!R\,x$. Equivalently, for all x and y in X, $x\,R\,y$ and $y\,R\,x$ together imply that $x = y$.

5. R is *transitive*, provided that, for all x, y, z in X, whenever we have $x\,R\,y$ and $y\,R\,z$, we also have $x\,R\,z$.

Example. The relations of subset, $\subseteq$, and divisibility, $\mid$, as used in the previous example are reflexive and transitive. The relation of subset is also antisymmetric, as is that of divisibility provided we consider only positive integers.

The relation of *proper subset*, $\subset$, defined by $A \subset B$, provided that every element of A is also an element of B and $A \neq B$, is irreflexive, antisymmetric, and transitive. The relation of *less than or equal*, $\leq$, on a set of numbers, is reflexive, antisymmetric, and transitive, while the relation of *less than*, $<$, is irreflexive, antisymmetric, and transitive. □

A *partial order* on a set X is a reflexive, antisymmetric, and transitive relation R. A *strict partial order* on a set X is an irreflexive, antisymmetric, and transitive relation. Thus, $\subseteq$, $\leq$, and $\mid$ are partial orders, while $\subset$ and $<$ are strict partial orders.[14] If a relation R is a partial order, we generally use the usual inequality symbol "$\leq$" instead of R;[15] the relation $<$ defined by $a < b$ if and only if $a \leq b$ and $a \neq b$ is then a strict partial order. (Conversely, starting from a strict partial order $<$ on X, the relation $\leq$ defined by $a \leq b$ if and only if $a < b$ or $a = b$ is a partial order.)

A set X on which a partial order $\leq$ is defined is sometimes referred to as a *partially ordered set* (or sometimes simply as a *poset*) and denoted by $(X, \leq)$.

If R is a relation on a set X, then for x and y in X, x and y are *comparable*, provided that either $x\,R\,y$ or $y\,R\,x$; x and y are *incomparable* otherwise.[16] A partial order R on a set X is a *total order*, provided that every pair of elements of X is comparable. The standard relation $\leq$ on a set of numbers is a total order.[17]

If X is a finite set and we list the elements of X in some linear order $a_1, a_2, \ldots, a_n$ (a permutation of X), then by defining $a_i \leq a_j$ provided that $i \leq j$ (that is, provided that a_i comes before a_j in the permutation), it can be checked that we obtain a total order on X. We now show that every total order on X arises in this way.

Theorem 4.5.1 *Let X be a finite set with n elements. Then there is a one-to-one correspondence between the total orders on X and the permutations of X. In particular, the number of different total orders on X is $n!$.*

Proof. We show by induction on n that each total order $\leq$ on X corresponds to a permutation $a_1, a_2, \ldots, a_n$ of X with $a_1 < a_2 < \cdots <$

---

[14]The relation *is divisible by but does not equal* is also a strict partial order.

[15]It is important, then, to be aware that $a \leq b$ does not mean that a and b are numbers with a no bigger than b. The symbol "$\leq$" now becomes an abstract symbol for a partial order.

[16]Think of the phrase "x and y are incomparable" as an abstract version of the common phrase "one cannot compare apples and oranges" and so apples and oranges are incomparable.

[17]This is one reason why one should be careful to distinguish between the abstract symbol "$\leq$" for a partial order and the standard relation "$\leq$" on numbers; the latter is a total order where any two numbers a and b are comparable (either $a \leq b$ or $b \leq a$), but this property does not hold for a general partial order.

a_n. If $n = 1$, this is trivial. Let $n > 1$. We first show that there is a *minimal element* of X; that is, an element a_1 such that $b \le a_1$ implies that $b = a_1$ (equivalently, there is no element x with $x < a_1$). Let a be any element of X. If a is not a minimal element, then there is an element b such that $b < a$. If b is not a minimal element, there is an element c such that $c < b$ so that $c < b < a$. Continuing like this and using the fact that X is a finite set, eventually we locate a minimal element a_1. Suppose there is an element $x \ne a_1$ of X such that $a_1 \not< x$. Since we have a total order, we must have $x < a_1$, contradicting the minimality of a_1. Hence, $a_1 < x$ for all x in X different from a_1. Applying induction to the set of $n - 1$ elements of X different from a_1, we conclude that these elements can be ordered $a_2, a_3, \ldots, a_n$ with $a_2 < a_3 < \cdots < a_n$. Hence, $a_1, a_2, a_3, \ldots, a_n$ is a permutation of the elements of X with $a_1 < a_2 < a_3 < \cdots < a_n$. $\qquad\square$

As a result of Theorem 4.5.1, a finite totally ordered set is often denoted as $a_1 < a_2 < \cdots < a_n$, or simply as a permutation $a_1, a_2, \ldots, a_n$.

A partially ordered set can be represented geometrically. In order to illustrate this we need to define the cover relation of a partially ordered set $(X, \le)$. Let a and b be in X. Then a is *covered by* b (also expressed as b *covers* a), denoted $a <_c b$, provided that $a < b$ and no element c can be squeezed between a and b; that is, there is no element c such that both $a < c$ and $c < b$ hold. If X is a finite set, then, by transitivity, the partial order $\le$ is uniquely determined by its cover relation. Thus, the cover relation is an efficient way to describe a partial order. It follows from Theorem 4.5.1 that, if $(X, \le)$ is a totally ordered set, then the elements of X can be listed as $x_1, x_2, \ldots, x_n$ such that $x_1 <_c x_2 <_c \cdots <_c x_n$. It is for this reason that a totally ordered set is also called a *linearly ordered set*.

A *diagram* of a finite partially ordered set $(X, \le)$ is obtained by taking a point in the plane for each element of X, being careful to put the point for x below the point for y if $x <_c y$, and connecting x and y by a line segment if and only if x is covered by y. (We put x below y in order to know that x is covered by y rather than the other way around.)

Example. A totally ordered set of 5 elements is represented by the diagram, shown in Figure 4.6, of 5 vertical points, with 4 vertical line segments connecting the points. $\qquad\square$

Figure 4.6

Example. The partially ordered set of subsets of the set $\{1,2,3\}$ is represented by the diagram, shown in Figure 4.7, of a cube "sitting" on one of its corners.   □

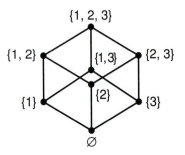

Figure 4.7

Example. The set of the first eight positive integers, partially ordered by "is a divisor of," is represented by the diagram in Figure 4.8.   □

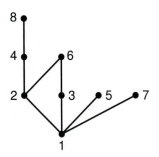

Figure 4.8

Let $\leq_1$ and $\leq_2$ be two partial orders on the same set X. Then the partially ordered set $(X, \leq_2)$ is an *extension* of the partially ordered set $(X, \leq_1)$, provided that whenever $a \leq_1 b$ holds, $a \leq_2 b$ also holds. In particular, an extension of a partially ordered set has more comparable pairs. We show that every finite partially ordered set $(X, \leq)$ has a *linear extension*—that is, an extension which is a linearly ordered set. This means that it is possible to list the elements of X in a linear order

$x_1, x_2, \ldots, x_n$ so that x_i is listed before x_j whenever $x_i < x_j$; that is, if $x_i < x_j$, then $i < j$ (here $i < j$ means that i is a smaller integer than j).

Theorem 4.5.2 *Let* $(X, \leq)$ *be a finite partially ordered set. Then there is a linear extension of* $(X, \leq)$.

Proof. There is a very simple algorithm for listing the elements of X in a linear order $x_1, x_2, \ldots, x_n$ to obtain a linear extension of $(X, \leq)$:

Algorithm for a linear extension of a partially ordered set

(1) Choose a minimal element x_1 of X (with respect to the partial order $\leq$).

(2) Delete x_1 from X and choose a minimal element x_2 from among the remaining $n - 1$ elements.

(3) Delete, in addition, x_2 from X, and choose a minimal element x_3 from among the remaining $n - 2$ elements.

(4) Delete, in addition, x_3 from X, and choose a minimal element x_4 from among the remaining $n - 3$ elements.

$\vdots$

(n) Delete x_{n-1} from X, leaving exactly one element x_n.

We show that $x_1, x_2, \ldots, x_n$ is a linear extension of $(X, \leq)$ by arguing by contradiction. Suppose there are x_i and x_j such that $x_i < x_j$ but $j < i$. Then, in step (j) in the preceding algorithm, when we chose x_j, x_i was among the remaining elements, and since $x_i < x_j$, x_j was not a minimal element as required by the algorithm. Thus, $x_1, x_2, \ldots, x_n$ is a linear extension of $(X, \leq)$. $\square$

Example. Let $X = \{1, 2, \ldots, n\}$ be the set consisting of the first n positive integers, and consider the partially ordered set $(X, |)$, where, as before, $|$ means "is a divisor of." Since, if $i \mid j$, then i is smaller than j, it follows that $1, 2, \ldots, n$ is a linear extension of $(X, \leq)$. $\square$

Example. Let X be a set of n elements, and consider the partially ordered set $(\mathcal{P}(X), \subseteq)$ of all subsets of X partially ordered by containment. Since, if $A \subseteq B$ implies that $|A| \leq |B|$, it follows that, if we start with the empty set, list all the one-element subsets in some order, then the two-element subsets in some order, then the three-element subsets in some order, and so on, we obtain a linear extension of $(\mathcal{P}(X), \subseteq)$. For instance, if $n = 3$ and $X = \{1, 2, 3\}$, then

$$\emptyset, \{1\}, \{2\}, \{3\}, \{1,2\}, \{1,3\}, \{2,3\}, \{1,2,3\}$$

is a linear extension of $(\mathcal{P}(X), \subseteq)$.   □

We now define another special class of relations. Let X be a set. A relation R on X is an *equivalence relation*, provided it is reflexive, symmetric, and transitive. (Thus, an equivalence relation differs from a partial order only in that an equivalence relation is symmetric and a partial order is antisymmetric.) A relation that is an equivalence relation is usually denoted by $\sim$. If $a \sim b$, then we say that a is *equivalent* to b. Just as a partial order can be considered as a generalization of the usual order $\leq$ of numbers, an equivalence relation can be considered as a generalization of equality $=$ of numbers. We now show that equivalence relations on X naturally correspond to partitions of X into nonempty sets.

Let $\sim$ be an equivalence relation on X. For each a in X, the *equivalence class* of a is the set

$$[a] = \{x : x \sim a\}$$

of all elements equivalent to a. Since $a \sim a$, the equivalence class of a contains a and thus is nonempty.

Theorem 4.5.3 *Let $\sim$ be an equivalence relation on a set X. Then the distinct equivalence classes partition X into nonempty parts. Conversely, given any partition of X into nonempty parts, there is an equivalence relation on X whose equivalence classes are the parts of the partition.*

Proof. First let $\sim$ be an equivalence relation on X. We need to show that the different equivalence classes are pairwise disjoint and that their union is X. Each equivalence class is nonempty, and each element of X is contained in an equivalence class (the equivalence class of a contains a). It remains only to show that the distinct equivalence classes are pairwise disjoint, or equivalently, that, if two equivalence classes have a nonempty intersection, then they are identical sets. Suppose $[a] \cap [b] \neq \emptyset$, and let c be an element common to both $[a]$ and $[b]$. Then $c \sim a$ (and so $a \sim c$) and $c \sim b$ (and so $b \sim c$). Let x be contained in $[a]$. Then $x \sim a$. Since $a \sim c$ and $c \sim b$, transitivity implies that $x \sim b$ and hence x is contained in $[b]$. We conclude that $[a] \subseteq [b]$. In a similar way we conclude that $[b] \subseteq [a]$ and hence that $[a] = [b]$.

Conversely, let $A_1, A_2, \ldots, A_s$ be a partition of X into nonempty sets. For x and y in X, define $x \sim y$ if and only if x and y are in the same part of the partition. Then it is straightforward to check that $\sim$ is an equivalence relation on X whose distinct equivalence classes are $A_1, A_2, \ldots, A_s$. See Exercise 44. $\qquad\square$

4.6   Exercises

1. Which permutation of $\{1, 2, 3, 4, 5\}$ follows 31524 in using the algorithm described in Section 4.1? Which permutation comes before 31524?

2. Determine the mobile integers in

$$\overset{\rightarrow}{4}\ \overset{\leftarrow}{8}\ \overset{\rightarrow}{3}\ \overset{\leftarrow}{1}\ \overset{\rightarrow}{6}\ \overset{\leftarrow}{7}\ \overset{\leftarrow}{2}\ \overset{\rightarrow}{5}\ .$$

3. Use the algorithm of Section 4.1 to generate the permutations $\{1, 2, 3, 4, 5\}$, starting with $\overset{\leftarrow}{1}\ \overset{\leftarrow}{2}\ \overset{\leftarrow}{3}\ \overset{\leftarrow}{4}\ \overset{\leftarrow}{5}$.

4. Prove, that in the algorithm of Section 4.1, which generates directly the permutations of $\{1, 2, \ldots, n\}$, the directions of 1 and 2 never change.

5. Let $i_1 i_2 \cdots i_n$ be a permutation of $\{1, 2, \ldots, n\}$ with inversion sequence $b_1, b_2, \ldots, b_n$, and let $k = b_1 + b_2 + \cdots + b_n$. Show by induction that one cannot bring $i_1 i_2 \cdots i_n$ to $12 \cdots n$ by fewer than k successive switches of adjacent numbers.

6. Determine the inversion sequences of the following permutations of $\{1, 2, \ldots, 8\}$:

   (a) 35168274

   (b) 83476215

7. Construct the permutations of $\{1, 2, \ldots, 8\}$ whose inversion sequences are

   (a) $2, 5, 5, 0, 2, 1, 1, 0$

   (b) $6, 6, 1, 4, 2, 1, 0, 0$

8. How many permutations of $\{1, 2, 3, 4, 5, 6\}$ have

   (a) exactly 15 inversions?

   (b) exactly 14 inversions?

(c) exactly 13 inversions?

9. Show that the largest number of inversions of a permutation of $\{1, 2, \ldots, n\}$ equals $n(n-1)/2$. Determine the unique permutation with $n(n-1)/2$ inversions. Also determine all those permutations with one fewer inversion.

10. Bring the permutations 256143 and 436251 to 123456 by successive switches of adjacent numbers.

11. Let $S = \{x_7, x_6, \ldots, x_1, x_0\}$. Determine the 8-tuples of 0's and 1's corresponding to the following combinations of S:

   (a) $\{x_5, x_4, x_3\}$

   (b) $\{x_7, x_5, x_3, x_1\}$

   (c) $\{x_6\}$

12. Let $S = \{x_7, x_6, \ldots, x_1, x_0\}$. Determine the combinations of S corresponding to the following 8-tuples:

   (a) 00011011

   (b) 01010101

   (c) 00001111

13. Generate the 5-tuples of 0's and 1's by using the base 2 arithmetic generating scheme and identify them with combinations of the set $\{x_4, x_3, x_2, x_1, x_0\}$.

14. Repeat Exercise 13 for the 6-tuples of 0's and 1's.

15. For each of the following combinations of $\{x_7, x_6, \ldots, x_1, x_0\}$, determine the combination that immediately follows it by using the base 2 arithmetic generating scheme:

   (a) $\{x_4, x_1, x_0\}$

   (b) $\{x_7, x_5, x_3\}$

   (c) $\{x_7, x_5, x_4, x_3, x_2, x_1, x_0\}$

   (d) $\{x_0\}$

16. For each of the combinations (a), (b), (c), and (d) in the preceding exercise, determine the combination that immediately *precedes* it in the base 2 arithmetic generating scheme.

17. Which combination of $\{x_7, x_6, \ldots, x_1, x_0\}$ is 150th on the list of combinations of S when the base 2 arithmetic generating scheme is used? 200th? 250th? [As in Section 4.3, the places on the list are numbered beginning with 0.]

18. Build (the corners and edges of) the 4-cube, and indicate the reflected Gray code on it.

19. Give an example of a noncyclic Gray code of order 3.

20. Give an example of a cyclic Gray code of order 3 that is not the reflected Gray code.

21. Construct the reflected Gray code of order 5 by

    (a) using the inductive definition, and

    (b) using the Gray code algorithm.

22. Determine the reflected Gray code of order 6.

23. Determine the immediate successors of the following 9-tuples in the reflected Gray code of order 9:

    (a) 010100110

    (b) 110001100

    (c) 111111111

24. Determine the predecessors of each of the 9-tuples of the previous Exercise 23 in the reflected Gray code of order 9.

25. * The reflected Gray code of order n is properly called the reflected *binary* Gray code, since it is a listing of the n-tuples of 0's and 1's. It can be generalized to any base system, in particular the ternary and decimal system. Thus, the reflected decimal Gray code of order n is a listing of all the decimal numbers of n digits such that consecutive numbers in the list differ in only one place and the absolute value of the difference is 1. Determine the reflected decimal Gray codes of orders 1 and 2. (Note we have not said precisely what a reflected decimal Gray code is. Part of the problem is to discover what it is.) Also, determine the reflected ternary Gray codes of orders 1, 2, and 3.

26. Generate the 2-combinations of $\{1, 2, 3, 4, 5\}$ in lexicographic order by using the algorithm described in Section 4.4.

27. Generate the 3-combinations of $\{1, 2, 3, 4, 5, 6\}$ in lexicographic order by using the algorithm described in Section 4.4.

28. Determine the 6-combination of $\{1, 2, \ldots, 10\}$ that immediately follows $2, 3, 4, 6, 9, 10$ in the lexicographic order. Determine the 6-combination that immediately precedes $2, 3, 4, 6, 9, 10$?

29. Determine the 7-combination of $\{1, 2, \ldots, 15\}$ that immediately follows $1, 2, 4, 6, 8, 14, 15$ in the lexicographic order. Determine the 7-combination that immediately precedes $1, 2, 4, 6, 8, 14, 15$.

30. Generate the inversion sequences of the permutations of $\{1, 2, 3\}$ in the lexicographic order, and write down the corresponding permutations. Repeat for the inversion sequences of permutations of $\{1, 2, 3, 4\}$.

31. Generate the 3-permutations of $\{1, 2, 3, 4, 5\}$.

32. Generate the 4-permutations of $\{1, 2, 3, 4, 5, 6\}$.

33. In which position does the combination 2489 occur in the lexicographic order of the 4-combinations of $\{1, 2, 3, 4, 5, 6, 7, 8, 9\}$?

34. Consider the r-combinations of $\{1, 2, \ldots, n\}$ in lexicographic order.

    (a) What are the first $(n - r + 1)$ r-combinations?
    (b) What are the last $(r + 1)$ r-combinations?

35. The *complement* of an r-combination A of $\{1, 2, \ldots, n\}$ is the $(n - r)$-combination $\overline{A}$ of $\{1, 2, \ldots, n\}$, consisting of all those elements that do not belong to A. Let $M = \binom{n}{r}$, the number of r-combinations, and the number of $(n - r)$-combinations of $\{1, 2, \ldots, n\}$. Prove that, if

$$A_1, A_2, A_3, \ldots, A_M$$

are the r-combinations in lexicographic order, then

$$\overline{A_M}, \ldots, \overline{A_3}, \overline{A_2}, \overline{A_1}$$

are the $(n - r)$-combinations in lexicographic order.

36. Let X be a set of n elements. How many different relations on X are there? How many of these are reflexive? Symmetric? Antisymmetric? Reflexive and symmetric? Reflexive and antisymmetric?

37. Let R' and R'' be two partial orders on a set X. Define a new relation R on X by $x\,R\,y$ if and only if both $x\,R'\,y$ and $x\,R''\,y$ hold. Prove that R is also a partial order on X. (R is called the *intersection* of R' and R''.)

38. Let $(X_1, \leq_1)$ and $(X_2, \leq_2)$ be partially ordered sets. Define a relation T on the set

$$X_1 \times X_2 = \{(x_1, x_2) : x_1 \text{ in } X_1, x_2 \text{ in } X_2\}$$

by

$$(x_1, x_2)\, T\, (x_1', x_2') \text{ if and only if } x_1 \leq_1 x_1' \text{ and } x_2 \leq_2 x_2'.$$

Prove that $(X_1 \times X_2, T)$ is a partially ordered set. $(X_1 \times X_2, T)$ is called the *direct product* of $(X_{,1} \leq_1)$ and $(X_2, \leq_2)$ and is also denoted by $(X_1, \leq_1) \times (X_2, \leq_2)$. More generally, prove that the direct product $(X_1, \leq_1) \times (X_2, \leq_2) \times \cdots \times (X_m, \leq_m)$ of partially ordered sets is also a partially ordered set.

39. Let $(J, \leq)$ be the partially ordered set with $J = \{0, 1\}$ and with $0 < 1$. By identifying the combinations of a set X of n elements with the n-tuples of 0's and 1's, prove that the partially ordered set $(X, \subseteq)$ can be identified with the n-fold direct product $(J, \leq) \times (J, \leq) \times \cdots \times (J, \leq)$ (n factors).

40. Generalize Exercise 39 to the multiset of all combinations of the multiset $X = \{n_1 \cdot a_1, n_2 \cdot a_2, \ldots, n_m \cdot a_m\}$. (Part of this exercise is to determine the "natural" partial order on these multisets.)

41. Prove that a partial order on a finite set is uniquely determined by its cover relation.

42. Describe the cover relation for the partial order $\subseteq$ on the collection $\mathcal{P}(X)$ of all subsets of a set X.

43. Let $X = \{a, b, c, d, e, f\}$ and let the relation R on X be defined by $a\,R\,b$, $b\,R\,c$, $c\,R\,d$, $a\,R\,e$, $e\,R\,f$, $f\,R\,d$. Verify that R is the cover relation of a partially ordered set, and determine all the linear extensions of this partial order.

44. Let $A_1, A_2, \ldots, A_s$ be a partition of a set X. Define a relation R on X by $x\,R\,y$ if and only if x and y belong to the same part of the partition. Prove that R is an equivalence relation.

45. Define a relation R on the set Z of all integers by $a \, R \, b$ if and only if $a = \pm b$. Is R an equivalence relation on Z?

46. Let m be a positive integer and define a relation R on the set X of all nonnegative integers by: $a \, R \, b$ if and only if a and b have the same remainder when divided by m. Prove that R is an equivalence relation on X. How many different equivalence classes does this equivalence relation have?

47. Let Π_n denote the set of all partitions of the set $\{1, 2, \ldots, n\}$. Given two partitions π and σ in Π_n, define $\pi \leq \sigma$, provided each part of π is contained in a part of σ. Thus, the partition π can be obtained by partitioning the parts of σ. This relation is usually expressed by saying that π is a *refinement* of σ.

    (a) Prove that this relation is a partial order on Π_n.

    (b) By Theorem 4.5.3, we know that there is a one-to-one correspondence between Π_n and the set Λ_n of all equivalence relations on $\{1, 2, \ldots, n\}$. What is the partial order on Λ_n that corresponds to this partial order on Π_n?

    (c) Construct the diagram of $(\Pi_n, \leq)$ for $n = 1, 2, 3,$ and 4.

48. Consider the partial order $\leq$ on the set X of positive integers given by "is a divisor of." Let a and b be two integers. Let c be the largest integer such that $c \leq a$ and $c \leq b$, and let d be the smallest integer such that $a \leq d$ and $b \leq d$. What are c and d?

49. Prove that the intersection $R \cap S$ of two equivalence relations R and S on a set X is also an equivalence relation on X. Is the union of two equivalence relations on X always an equivalence relation?

50. Consider the partially ordered set $(X, \subseteq)$ of subsets of the set $X = \{a, b, c\}$ of 3 elements. How many linear extensions are there?

51. Let n be a positive integer, and let X_n be the set of $n!$ permutations of $\{1, 2, \ldots, n\}$. Let π and σ be two permutations in X_n, and define $\pi \leq \sigma$ provided the set of inversions of π is a subset of the set of inversions of σ. Verify that this defines a partial order on X_n, called the *inversion poset*. Describe the cover relation for this partial order and then draw the diagram for the inversion poset $(H_4, \leq)$.

52. Verify that a binary n-tuple $a_{n-1} \cdots a_1 a_0$ is in place k in the Gray code order list where k is determined as follows: For $i = 0, 1, \ldots, n-1$, let

$$b_i = \begin{cases} 0 & \text{if } a_{n-1} + \cdots + a_i \text{ is even, and} \\ 1 & \text{if } a_{n-1} + \cdots + a_i \text{ is odd.} \end{cases}$$

Then
$$k = b_{n-1} \times 2^{n-1} + \cdots + b_1 \times 2 + b_0 \times 2^0.$$

Thus, $a_{n-1} \cdots a_1 a_0$ is in the same place in the Gray code order list of binary n-tuples as $b_{n-1} \cdots b_1 b_0$ is in the lexicographic order list of binary n-tuples.

53. Referring to Exercise 52, show that $a_{n-1} \cdots a_1 a_0$ can be recovered from $b_{n-1} \cdots b_1 b_0$ by $a_{n-1} = b_{n-1}$, and for $i = 0, 1, \ldots, n-1$,

$$a_i = \begin{cases} 0 & \text{if } b_i + b_{i+1} \text{ is even, and} \\ 1 & \text{if } b_i + b_{i+1} \text{ is odd.} \end{cases}$$

54. Let $(X, \leq)$ be a finite partially ordered set. By Theorem 4.5.2 we know that $(X, \leq)$ has a linear extension. Let a and b be incomparable elements of X. Modify the proof of Theorem 4.5.2 to obtain a linear extension of $(X, \leq)$ such that $a < b$. Hint: First find a partial order $\leq'$ on X such that whenever $x \leq y$ then $x \leq' y$ and, in addition, $a \leq' b$.

55. Use Exercise 54 to prove that a finite partially ordered set is the intersection of all its linear extensions (see Exercise 37.)

56. The *dimension* of a finite partially ordered set $(X, \leq)$ is the smallest number of its linear extensions whose intersection is $(X, \leq)$. By Exercise 55, every partially ordered set has a dimension. Those that have dimension 1 are the linear orders. Let n be a positive integer and let $i_1, i_2, \ldots, i_n$ be a permutation σ of $\{1, 2, \ldots, n\}$ that is different from $1, 2, \ldots, n$. Let $X = \{(1, i_1), (2, i_2), \ldots, (n, i_n)\}$. Now define a relation R on X by $(k, i_k) \, R \, (l, i_l)$ if and only if $k \leq l$ (ordinary integer inequality) and $i_k \leq i_l$ (again ordinary inequality); that is, (i_k, i_l) is not an inversion of σ. Thus, for instance, if $n = 3$ and $\sigma = 2, 3, 1$, then $X = \{(1, 2), (2, 3), (3, 1)\}$, and $(1, 2) \, R \, (2, 3)$, but $(1, 2) \, \cancel{R} \, (3, 1)$. Prove that R is a partial order on X and that the dimension of the partially ordered set (X, R) is 2, provided that $i_1, i_2, \ldots, i_n$ is not the identity permutation $1, 2, \ldots, n$.

Chapter 5

The Binomial Coefficients

The numbers $\binom{n}{k}$ count the number of k-combinations of a set of n elements. They have many fascinating properties and satisfy a number of interesting identities. Because of their appearance in the binomial theorem (see Section 5.2), they are called the *binomial coefficients*. In formulas arising in the analysis of algorithms in theoretical computer science, the binomial coefficients occur over and over again, so a facility for manipulating with them is important. In this chapter, we discuss some of their elementary properties and identities. We prove a useful theorem of Sperner, and then continue our study of partially ordered sets and prove an important theorem of Dilworth.

5.1 Pascal's Formula

The binomial coefficients $\binom{n}{k}$ have been defined in Section 3.3 for all nonnegative integers k and n. Recall that $\binom{n}{k} = 0$ if $k > n$ and that $\binom{n}{0} = 1$ for all n. If n is positive and $1 \leq k \leq n$, then

$$\binom{n}{k} = \frac{n!}{k!(n-k)!} = \frac{n(n-1)\cdots(n-k+1)}{k(k-1)\cdots 1}. \qquad (5.1)$$

In Section 3.3, we noted that

$$\binom{n}{k} = \binom{n}{n-k}.$$

This relation is valid for all integers k and n with $0 \leq k \leq n$.

Theorem 5.1.1 (Pascal's formula) *For all integers* n *and* k *with* $1 \leq k \leq n - 1$,

$$\binom{n}{k} = \binom{n-1}{k} + \binom{n-1}{k-1}.$$

Proof. One way to prove this identity is to substitute the values of the binomial coefficients and then check that both sides are equal. We leave this straightforward verification to the reader.

A *combinatorial proof* can be obtained as follows: Let S be a set of n elements. We distinguish one of the elements of S and denote it by x. We then partition the set X of k-combinations of S into two parts, A and B. In A we put all those k-combinations which do not contain x. In B we put all the k-combinations which do contain x. The size of X is $|X| = \binom{n}{k}$; hence, by the addition principle,

$$\binom{n}{k} = |A| + |B|.$$

The k-combinations in A are exactly the k-combinations of the set $S - \{x\}$ of $n - 1$ elements; thus, the size of A is

$$|A| = \binom{n-1}{k}.$$

A k-combination in B is obtained by adjoining the element x to a $(k-1)$-combination of $S - \{x\}$. Hence, the size of B satisfies

$$|B| = \binom{n-1}{k-1}.$$

Combining these facts, we obtain

$$\binom{n}{k} = \binom{n-1}{k} + \binom{n-1}{k-1}.$$

$\square$

To illustrate the proof, let $n = 5$, $k = 3$, and $S = \{x, a, b, c, d\}$. Then the 3-combinations of S in A are

$$\{a, b, c\}, \{a, b, d\}, \{a, c, d\}, \{b, c, d\}.$$

These are the 3-combinations of the set $\{a, b, c, d\}$. The 3-combinations of S in B are

$$\{x, a, b\}, \{x, a, c\}, \{x, a, d\}, \{x, b, c\}, \{x, b, d\}, \{x, c, d\}.$$

Upon deletion of the element x in these 3-combinations, we obtain

$$\{a,b\},\{a,c\},\{a,d\},\{b,c\},\{b,d\},\{c,d\},$$

the 2-combinations of $\{a,b,c,d\}$. Thus,

$$\binom{5}{3} = 10 = 4 + 6 = \binom{4}{3} + \binom{4}{2}.$$

By using the relation

$$\binom{n}{k} = \binom{n-1}{k} + \binom{n-1}{k-1}$$

and the initial information

$$\binom{n}{0} = 1 \text{ and } \binom{n}{n} = 1, \qquad (n \geq 0),$$

the binomial coefficients can be calculated without recourse to the formula (5.1). When the binomial coefficients are calculated in this way, the results are often displayed in an array known as *Pascal's triangle*. This array, which appeared in Blaise Pascal's *Traité du triangle arithmétique* in 1653, is illustrated in Figure 5.1.

$n\backslash k$	0	1	2	3	4	5	6	7	8	$\cdots$
0	1									
1	1	1								
2	1	2	1							
3	1	3	3	1						
4	1	4	6	4	1					
5	1	5	10	10	5	1				
6	1	6	15	20	15	6	1			
7	1	7	21	35	35	21	7	1		
8	1	8	28	56	70	56	28	8	1	
$\vdots$	$\vdots$	$\vdots$	$\vdots$	$\vdots$	$\vdots$	$\vdots$	$\vdots$	$\vdots$	$\vdots$	$\ddots$

Figure 5.1: Pascal's triangle

Each entry in the triangle, other than those equal to 1 occurring on the left side and hypotenuse, is obtained by adding together two entries in the row above: the one directly above and the one immediately to the left. This is in accordance with Pascal's formula as given in Theorem 5.1.1. For instance, in row $n = 8$, we have

$$\binom{8}{3} = 56 = 35 + 21 = \binom{7}{3} + \binom{7}{2}.$$

Many of the relations involving binomial coefficients can be discovered by careful examination of Pascal's triangle. The symmetry relation

$$\binom{n}{k} = \binom{n}{n-k}$$

is readily noticed in the triangle. The identity

$$\binom{n}{0} + \binom{n}{1} + \cdots + \binom{n}{n} = 2^n$$

of Theorem 3.3.2 is discovered by adding the numbers in a row of Pascal's triangle. The numbers $\binom{n}{1} = n$ in column $k = 1$ are the counting numbers. The numbers $\binom{n}{2} = n(n-1)/2$ in column $k = 2$ are the so-called *triangular numbers*, which equal the number of dots in the triangular arrays of dots illustrated in Figure 5.2.

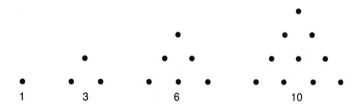

Figure 5.2

The numbers $\binom{n}{3} = n(n-1)(n-2)/3!$ in column $k = 3$ are the so-called *tetrahedral numbers*, and they equal the number of dots in tetrahedral arrays of dots (think of stacked cannon balls). The reader is encouraged now to examine Pascal's triangle for other relations involving binomial coefficients.

Another interpretation can be given to the entries of Pascal's triangle. Let n be a nonnegative integer and let k be an integer with $0 \le k \le n$. Define

$$p(n, k)$$

as the number of paths from the top left corner (the entry $\binom{0}{0} = 1$) to the entry $\binom{n}{k}$ where in each path we move from one entry to the entry in the next row immediately below it or immediately to its right. The two types of moves allowed in going from one entry to the next on the path are illustrated in Figure 5.3.

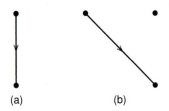

(a)   (b)

Figure 5.3

We define $p(0,0)$ to be 1, and, for each nonnegative integer n, we have

$$p(n,0) = 1 \qquad \text{(we must move straight down to reach } \binom{n}{0} \text{)}$$

and

$$p(n,n) = 1 \qquad \text{(we must move diagonally to reach } \binom{n}{n} \text{).}$$

We note that each path from $\binom{0}{0}$ to $\binom{n}{k}$ is either

(i) a path from $\binom{0}{0}$ to $\binom{n-1}{k}$ followed by one vertical move (a) or

(ii) a path from $\binom{0}{0}$ to $\binom{n-1}{k-1}$ followed by one diagonal move (b).

Thus, by the addition principle, we have

$$p(n,k) = p(n-1,k) + p(n-1,k-1),$$

a Pascal-type relation for the numbers $p(n,k)$. The numbers $p(n,k)$ are computed in exactly the same way as the binomial coefficients $\binom{n}{k}$, starting with the same initial values. Hence, for all integers n and k with $0 \le k \le n$,

$$p(n,k) = \binom{n}{k}.$$

Consequently, the value of an entry $\binom{n}{k}$ of Pascal's triangle represents the number of paths from the top left corner to that entry, using only moves of types (a) and (b). Therefore, we have another combinatorial interpretation of the numbers $\binom{n}{k}$.

5.2   The Binomial Theorem

The binomial coefficients receive their name from their appearance in the binomial theorem. The first few cases of this theorem should be familiar algebraic identities.

Theorem 5.2.1 *Let n be a positive integer. Then, for all x and y,*

$$(x + y)^n = x^n + \binom{n}{1} x^{n-1} y + \binom{n}{2} x^{n-2} y^2 + \cdots$$

$$+ \binom{n}{n-1} x^1 y^{n-1} + y^n.$$

In summation notation,

$$(x + y)^n = \sum_{k=0}^{n} \binom{n}{k} x^{n-k} y^k.$$

First proof. Write $(x + y)^n$ as the product

$$(x + y)(x + y) \cdots (x + y)$$

of n factors each equal to $x + y$. We completely expand this product, using the distributive law, and group together like terms. Since, for each factor $(x + y)$, we can choose either x or y in multiplying out $(x + y)^n$, there are 2^n terms that result and each can be arranged in the form $x^{n-k} y^k$ for some $k = 0, 1, \ldots, n$. We obtain the term $x^{n-k} y^k$ by choosing y in k of the n factors and x (by default) in the remaining $n - k$ factors. Thus, the number of times the term $x^{n-k} y^k$ occurs in the expanded product equals the number $\binom{n}{k}$ of k-combinations of the set of n factors. Therefore,

$$(x + y)^n = \sum_{k=0}^{n} \binom{n}{k} x^{n-k} y^k.$$

Second proof. The proof is by induction on n. It's more cumbersome and helps one appreciate the combinatorial viewpoint given in the first proof. If $n = 1$, the formula becomes

$$(x + y)^1 = \sum_{k=0}^{1} \binom{1}{k} x^{1-k} y^k = \binom{1}{0} x^1 y^0 + \binom{1}{1} x^0 y^1 = x + y,$$

and this is clearly true. We now assume that the formula is true for a positive integer n and prove that it is true when n is replaced by $n+1$. We write

$$(x + y)^{n+1} = (x + y)(x + y)^n,$$

which, by the induction assumption, becomes

$$(x + y)^{n+1} = (x + y)\left(\sum_{k=0}^{n} \binom{n}{k} x^{n-k} y^k\right)$$

$$= x\left(\sum_{k=0}^{n} \binom{n}{k} x^{n-k} y^k\right) + y\left(\sum_{k=0}^{n} \binom{n}{k} x^{n-k} y^k\right)$$

$$= \sum_{k=0}^{n} \binom{n}{k} x^{n+1-k} y^k + \sum_{k=0}^{n} \binom{n}{k} x^{n-k} y^{k+1}$$

$$= \binom{n}{0} x^{n+1} + \sum_{k=1}^{n} \binom{n}{k} x^{n+1-k} y^k$$

$$+ \sum_{k=0}^{n-1} \binom{n}{k} x^{n-k} y^{k+1} + \binom{n}{n} y^{n+1}.$$

Replacing k by $k-1$ in the last summation, we obtain

$$\sum_{k=1}^{n} \binom{n}{k-1} x^{n+1-k} y^k.$$

Hence,

$$(x + y)^{n+1} = x^{n+1} + \sum_{k=1}^{n} \left[\binom{n}{k} + \binom{n}{k-1}\right] x^{n+1-k} y^k + y^{n+1},$$

which, using Pascal's identity, becomes

$$(x + y)^{n+1} = x^{n+1} + \sum_{k=1}^{n} \binom{n+1}{k} x^{n+1-k} y^k + y^{n+1}.$$

Since $\binom{n+1}{0} = \binom{n+1}{n+1} = 1$, we may rewrite this last equation and obtain

$$(x + y)^{n+1} = \sum_{k=0}^{n+1} \binom{n+1}{k} x^{n+1-k} y^k.$$

This is the binomial theorem with n replaced by $n+1$, and the theorem holds by induction. $\qquad\square$

The binomial theorem can be written in several other equivalent forms:

$$(x+y)^n = \sum_{k=0}^{n} \binom{n}{n-k} x^{n-k} y^k,$$

$$(x+y)^n = \sum_{k=0}^{n} \binom{n}{n-k} x^k y^{n-k},$$

$$(x+y)^n = \sum_{k=0}^{n} \binom{n}{k} x^k y^{n-k}.$$

The first of these follows from Theorem 5.2.1 and the fact that

$$\binom{n}{k} = \binom{n}{n-k}, \quad (k = 0, 1, \ldots, n).$$

The other two follow by interchanging x with y.

The case $y = 1$ occurs sufficiently often to record it now as a special case.

Theorem 5.2.2 *Let n be a positive integer. Then, for all x,*

$$(1+x)^n = \sum_{k=0}^{n} \binom{n}{k} x^k = \sum_{k=0}^{n} \binom{n}{n-k} x^k.$$

The special cases $n = 2, 3, 4$ of the binomial theorem are

$$
\begin{aligned}
(x+y)^2 &= x^2 + 2xy + y^2, \\
(x+y)^3 &= x^3 + 3x^2y + 3xy^2 + y^3, \text{ and} \\
(x+y)^4 &= x^4 + 4x^3y + 6x^2y^2 + 4xy^3 + y^4.
\end{aligned}
$$

We note that the coefficients that occur in these expansions are the numbers in the row of Pascal's triangle. From Theorem 5.2.1 and the construction of Pascal's triangle, this is always the case.

5.3 Identities

We now consider some additional identities satisfied by the binomial coefficients. The identity

$$k\binom{n}{k} = n\binom{n-1}{k-1}, \quad (n \text{ and } k \text{ positive integers}) \qquad (5.2)$$

follows immediately from the fact that $\binom{n}{k} = 0$ if $k > n$ and

$$\binom{n}{k} = \frac{n(n-1)\cdots(n-k+1)}{k(k-1)\cdots 1} \qquad \text{for} \quad 1 \le k \le n.$$

The identity

$$\binom{n}{0} + \binom{n}{1} + \binom{n}{2} + \cdots + \binom{n}{n} = 2^n, \quad (n \ge 0) \qquad (5.3)$$

has already been proved as Theorem 3.3.2, but it also follows from the binomial theorem by setting $x = y = 1$. If we set $x = 1$, $y = -1$ in the binomial theorem, then we obtain

$$\binom{n}{0} - \binom{n}{1} + \binom{n}{2} - \cdots + (-1)^n \binom{n}{n} = 0, \quad (n \ge 1). \qquad (5.4)$$

We can also write this as

$$\binom{n}{0} + \binom{n}{2} + \cdots = \binom{n}{1} + \binom{n}{3} + \cdots, \quad (n \ge 1). \qquad (5.5)$$

This identity can be interpreted as follows: If S is a set of n elements, then the number of combinations of S with an even number of elements equals the number of combinations of S with an odd number of elements. Indeed, both have the value 2^{n-1}; that is,

$$\binom{n}{0} + \binom{n}{2} + \cdots = 2^{n-1}, \text{ and} \qquad (5.6)$$

$$\binom{n}{1} + \binom{n}{3} + \cdots = 2^{n-1}. \qquad (5.7)$$

We can verify these identities by combinatorial reasoning as follows: Let $S = \{x_1, x_2, \ldots, x_n\}$ be a set of n elements. We can think of combinations of S as resulting from the following decision process:

(1) we consider x_1 and decide either to put it in or leave it out (2 choices);

(2) we consider x_2 and decide either to put it in or leave it out (2 choices);

$\vdots$

(n) we consider x_n and decide either to put it in or leave it out (2 choices).

We have n decisions to make each with two choices. Thus, there are 2^n combinations as we know by (5.3).

Now, suppose we want to choose a combination with an even number of elements. Then we have two choices for each of $x_1, \ldots, x_{n-1}$. But when we get to x_n, we have only one choice. For, if we have chosen an even number of the elements $x_1, x_2, \ldots, x_{n-1}$, we must leave x_n out; if we have chosen an odd number of the elements $x_1, x_2, \ldots, x_{n-1}$, we must put x_n in. Hence, the number of combinations of S with an even number of elements equals 2^{n-1}. Since the left side of (5.6) also counts the number of combinations of S with an even number of elements, (5.6) holds. In a similar way one verifies (5.7). (However, now that we know that both (5.3) and (5.6) hold, so does (5.7).)

Using identities (5.2) and (5.3), we can derive the following identity:

$$1\binom{n}{1} + 2\binom{n}{2} + \cdots + n\binom{n}{n} = n2^{n-1}, (n \geq 1). \tag{5.8}$$

To see this, we first note that it follows from (5.2) that (5.8) is equivalent to

$$n\binom{n-1}{0} + n\binom{n-1}{1} + \cdots + n\binom{n-1}{n-1} = n2^{n-1}, (n \geq 1). \tag{5.9}$$

But now, by (5.3), with n replaced by $n-1$,

$$n\binom{n-1}{0} + n\binom{n-1}{1} + \cdots + n\binom{n-1}{n-1}$$

$$= n\left(\binom{n-1}{0} + \binom{n-1}{1} + \cdots + \binom{n-1}{n-1}\right)$$

$$= n2^{n-1}.$$

Thus, (5.9) and hence (5.8) hold. Another way to verify (5.8) is the following: By the binomial theorem,

$$(1+x)^n = \binom{n}{0} + \binom{n}{1}x + \binom{n}{2}x^2 + \binom{n}{3}x^3 + \cdots + \binom{n}{n}x^n.$$

If we differentiate both sides with respect to x, we obtain

$$n(1+x)^{n-1} = \binom{n}{1} + 2\binom{n}{2}x + 3\binom{n}{3}x^2 + \cdots + n\binom{n}{n}x^{n-1}.$$

Substituting $x = 1$, we get (5.8).

A number of interesting identities can be derived by successive differentiation and multiplication of the binomial expansion. For brevity we use the summation notation now. We begin with

$$(1+x)^n = \sum_{k=0}^{n} \binom{n}{k}x^k.$$

Differentiating both sides with respect to x, we get

$$n(1+x)^{n-1} = \sum_{k=1}^{n} k\binom{n}{k}x^{k-1}.$$

If we now multiply both sides by x, we get

$$nx(1+x)^{n-1} = \sum_{k=1}^{n} k\binom{n}{k}x^k.$$

Differentiating both sides with respect to x again, we now get

$$n\left[(1+x)^{n-1} + (n-1)x(1+x)^{n-2}\right] = \sum_{k=1}^{n} k^2\binom{n}{k}x^{k-1}.$$

Substituting $x = 1$, we obtain

$$n\left[2^{n-1} + (n-1)2^{n-2}\right] = \sum_{k=1}^{n} k^2\binom{n}{k};$$

hence,

$$n(n+1)2^{n-2} = \sum_{k=1}^{n} k^2\binom{n}{k}, \quad (n \geq 1). \qquad (5.10)$$

By successively multiplying by x and differentiating with respect to x, we can obtain an identity for

$$\sum_{k=1}^{n} k^p \binom{n}{k}$$

for any positive integer p.

An identity for the sum of the squares of the numbers in a row of Pascal's triangle is

$$\sum_{k=0}^{n} \binom{n}{k}^2 = \binom{2n}{n}, \quad (n \geq 0). \tag{5.11}$$

Identity (5.11) can be verified by combinatorial reasoning. Let S be a set with $2n$ elements. The right side of (5.11) counts the number of n-combinations of S. We partition S into two subsets, A and B, of n elements each. We use this partition of S to partition the n-combinations of S. Each n-combination of S contains a number k of elements of A, and the remaining $n - k$ elements come from B. Here, k may be any integer between 0 and n. We partition the n-combinations of S into $n + 1$ parts,

$$C_0, C_1, C_2, \ldots, C_n,$$

where C_k consists of those n-combinations which contain k elements from A and $n - k$ elements from B. By the addition principle,

$$\binom{2n}{n} = |C_0| + |C_1| + |C_2| + \cdots + |C_n|. \tag{5.12}$$

An n-combination in C_k is obtained by choosing k elements from A (there are $\binom{n}{k}$ choices) and then $(n - k)$ elements from B (there are $\binom{n}{n-k}$ choices). Hence, by the multiplication principle,

$$|C_k| = \binom{n}{k}\binom{n}{n-k} = \binom{n}{k}^2, \quad (k = 0, 1, \ldots, n).$$

Substituting this into (5.12), we obtain

$$\binom{2n}{n} = \binom{n}{0}^2 + \binom{n}{1}^2 + \binom{n}{2}^2 + \cdots + \binom{n}{n}^2,$$

and this proves (5.11). (A generalization of this identity, called the *Vandermonde convolution*, is given in Exercise 25.)

We now extend the domain of definition of the numbers $\binom{n}{k}$ to allow n to be any real number and k to be any integer (positive, negative, or zero).

Let r be a real number and let k be an integer. We then define the binomial coefficient $\binom{r}{k}$ by

$$\binom{r}{k} = \begin{cases} \dfrac{r(r-1)\cdots(r-k+1)}{k!} & \text{if } k \geq 1 \\ 1 & \text{if } k = 0 \\ 0 & \text{if } k \leq -1. \end{cases}$$

For instance,

$$\binom{5/2}{4} = \frac{(5/2)(3/2)(1/2)(-1/2)}{4!} = \frac{-5}{128},$$

$$\binom{-8}{2} = \frac{(-8)(-9)}{2} = 36,$$

$$\binom{3.2}{0} = 1, \text{ and}$$

$$\binom{3}{-2} = 0.$$

Pascal's formula and formula (5.2), namely,

$$\binom{r}{k} = \binom{r-1}{k} + \binom{r-1}{k-1} \text{ and } k\binom{r}{k} = r\binom{r-1}{k-1},$$

are now valid for all r and k. Each of these formulas can be verified by direct substitution. By iteration of Pascal's formula, we can obtain two summation formulas for the binomial coefficients.

Consider Pascal's formula,

$$\binom{r}{k} = \binom{r-1}{k} + \binom{r-1}{k-1},$$

with k equal to a positive integer. We can apply Pascal's formula to either of the binomial coefficients on the right and obtain an expression for $\binom{r}{k}$ as a sum of three binomial coefficients. Suppose we repeatedly apply Pascal's formula to the second binomial coefficient that appears in it (the one with the smaller lower argument). We then obtain

$$\binom{r}{k} = \binom{r-1}{k} + \binom{r-1}{k-1}$$

$$\binom{r}{k} = \binom{r-1}{k} + \binom{r-2}{k-1} + \binom{r-2}{k-2}$$

$$\binom{r}{k} = \binom{r-1}{k} + \binom{r-2}{k-1} + \binom{r-3}{k-2} + \binom{r-3}{k-3}$$

$$\vdots$$

$$\binom{r}{k} = \binom{r-1}{k} + \binom{r-2}{k-1} + \binom{r-3}{k-2} + \cdots + \binom{r-k}{1} + \binom{r-k-1}{0} + \binom{r-k-1}{-1}.$$

The last term $\binom{r-k-1}{-1}$ has value 0 and can be deleted. If we replace r by $r+k+1$ in the preceding summation and transpose terms, we obtain

$$\binom{r}{0} + \binom{r+1}{1} + \cdots + \binom{r+k}{k} = \binom{r+k+1}{k}. \qquad (5.13)$$

Identity (5.13) is valid for all real numbers r and all integers k. Notice that in (5.13) the upper argument starts with r, the lower argument starts with 0, and these arguments are successively increased by 1.

We now apply Pascal's formula repeatedly to the first binomial coefficient that appears in it. Let k be a positive integer. We then get

$$\binom{r}{k} = \binom{r-1}{k} + \binom{r-1}{k-1}$$

$$\binom{r}{k} = \binom{r-2}{k} + \binom{r-2}{k-1} + \binom{r-1}{k-1}$$

$$\binom{r}{k} = \binom{r-3}{k} + \binom{r-3}{k-1} + \binom{r-2}{k-1} + \binom{r-1}{k-1}$$

$$\vdots$$

$$\binom{r}{k} = \binom{r-t}{k} + \binom{r-t}{k-1} + \binom{r-t+1}{k-1} + \cdots + \binom{r-2}{k-1} + \binom{r-1}{k-1}.$$

Here t denotes an integer equal to the number of applications of Pascal's formula. Let us now assume that $r = n$ is a positive integer. Then, after $t = n$ applications of Pascal's formula, we arrive at a binomial coefficient whose upper argument is 0. Since $\binom{0}{k} = 0$, we get

$$\binom{n}{k} = \binom{0}{k-1} + \binom{1}{k-1} + \cdots + \binom{n-2}{k-1} + \binom{n-1}{k-1}.$$

We now replace k by $k+1$ and n by $n+1$ and transpose terms to obtain

$$\binom{0}{k} + \binom{1}{k} + \cdots + \binom{n-1}{k} + \binom{n}{k} = \binom{n+1}{k+1}. \qquad (5.14)$$

The identity (5.14) is valid for all nonnegative integers k and n. It is important to understand that this identity is just an iterated form of Pascal's formula.

If we take $k = 1$ in (5.14), we obtain

$$1 + 2 + \cdots + (n-1) + n = \frac{(n+1)n}{2},$$

the formula for the sum of the first n positive integers.

The identities (5.13) and (5.14) can be proved formally by mathematical induction and Pascal's formula. These are left as exercises. Some other identities for the binomial coefficients are given in the exercises.

5.4   Unimodality of Binomial Coefficients

If one examines the binomial coefficients in a row of Pascal's triangle, one notices that the numbers increase for a while and then decrease. A sequence of numbers with this property is called *unimodal*. Thus, the sequence $s_0, s_1, s_2, \ldots, s_n$ is unimodal, provided that there is an integer t with $0 \le t \le n$, such that

$$s_0 \overset{\cdot}{\le} s_1 \le \cdots \le s_t, \qquad s_t \ge s_{t+1} \ge \cdots \ge s_n.$$

The number s_t is the largest number in the sequence. The integer t is not necessarily unique because the largest number may occur in the sequence more than once. For instance, if $s_0 = 1$, $s_1 = 3$, $s_2 = 3$, and $s_3 = 2$, then

$$s_0 \le s_1 \le s_2, \qquad s_2 \ge s_3, \qquad (t = 2)$$

but also

$$s_0 \leq s_1, \qquad s_1 \geq s_2 \geq s_3 \qquad (t = 1).$$

Theorem 5.4.1 *Let n be a positive integer. The sequence of binomial coefficients*

$$\binom{n}{0}, \binom{n}{1}, \binom{n}{2}, \ldots, \binom{n}{n}$$

is a unimodal sequence. More precisely, if n is even,

$$\binom{n}{0} < \binom{n}{1} < \cdots < \binom{n}{n/2},$$

$$\binom{n}{n/2} > \cdots > \binom{n}{n-1} > \binom{n}{n},$$

and if n is odd,

$$\binom{n}{0} < \binom{n}{1} < \cdots < \binom{n}{(n-1)/2} = \binom{n}{(n+1)/2},$$

$$\binom{n}{(n+1)/2} > \cdots > \binom{n}{n-1} > \binom{n}{n}.$$

Proof. We consider the quotient of successive binomial coefficients in the sequence. Let k be an integer with $1 \leq k \leq n$. Then

$$\frac{\binom{n}{k}}{\binom{n}{k-1}} = \frac{\frac{n!}{k!(n-k)!}}{\frac{n!}{(k-1)!(n-k+1)!}} = \frac{n-k+1}{k}.$$

Hence,

$$\binom{n}{k-1} < \binom{n}{k}, \quad \binom{n}{k-1} = \binom{n}{k} \quad \text{or} \quad \binom{n}{k-1} > \binom{n}{k},$$

according as

$$k < n - k + 1, \qquad k = n - k + 1 \quad \text{or} \quad k > n - k + 1.$$

Now, $k < n - k + 1$ if and only if $k < (n+1)/2$. If n is even, then $k < (n+1)/2$ is equivalent to $k \leq n/2$. If n is odd, then $k < (n+1)/2$ is equivalent to $k \leq (n-1)/2$. Hence, the binomial coefficients increase as indicated in the statement of the theorem. We now observe that

$k = n - k + 1$ if and only if $2k = n + 1$. If n is even, $2k \neq n + 1$ for any k. If n is odd, then $2k = n + 1$, for $k = (n + 1)/2$. Thus, for n even, no two consecutive binomial coefficients in the sequence are equal. For n odd, the only two consecutive binomial coefficients of equal value are

$$\binom{n}{(n-1)/2} \quad \text{and} \quad \binom{n}{(n+1)/2}.$$

That the binomial coefficients decrease as indicated in the statement of the theorem follows in a similar way. □

For any real number x, let $\lfloor x \rfloor$ denote the greatest integer that is less than or equal to x. The integer $\lfloor x \rfloor$ is called the *floor* of x. Similarly, the *ceiling* of x is the smallest integer $\lceil x \rceil$ that is greater than or equal to x. For instance,

$$\lfloor 2\tfrac{1}{2} \rfloor = 2, \quad \lfloor 3 \rfloor = 3, \quad \lfloor -1\tfrac{1}{2} \rfloor = -2$$

and

$$\lceil 2\tfrac{1}{2} \rceil = 3, \quad \lceil 3 \rceil = 3, \quad \lceil -1\tfrac{1}{2} \rceil = -1.$$

We also have

$$\left\lfloor \frac{n}{2} \right\rfloor = \left\lceil \frac{n}{2} \right\rceil = \frac{n}{2}, \text{ if } n \text{ is even,}$$

and

$$\left\lfloor \frac{n}{2} \right\rfloor = \frac{n-1}{2} \text{ and } \left\lceil \frac{n}{2} \right\rceil = \frac{n+1}{2}, \text{ if } n \text{ is odd.}$$

Corollary 5.4.2 *For n a positive integer, the largest of the binomial coefficients*

$$\binom{n}{0}, \binom{n}{1}, \binom{n}{2}, \dots, \binom{n}{n}$$

is

$$\binom{n}{\lfloor n/2 \rfloor} = \binom{n}{\lceil n/2 \rceil}.$$

Proof. The corollary follows from Theorem 5.4.1 and the preceding observations about the floor and ceiling functions. □

To conclude this section we discuss a generalization of Theorem 5.4.1 called Sperner's Theorem.[1] Let S be a set of n elements. A

---

[1]E. Sperner: Ein Satz über Untermengen einer endlichen Menger [A theorem about subsets of finite sets], *Math. Zeitschrift*, 27 (1928), 544-548.

clutter (called an *antichain* in the more general context of a partially ordered set) of S is a collection $\mathcal{C}$ of combinations of S with the property that no combination in $\mathcal{C}$ is contained in another. For example, if $S = \{a, b, c, d\}$, then

$$\mathcal{C} = \{\{a, b\}, \{b, c, d\}, \{a, d\}, \{a, c\}\}$$

is a clutter. One way to obtain a clutter on a set S is to choose an integer $k \leq n$ and then take $\mathcal{C}_k$ to be the collection of all k-combinations of S. Since each combination in $\mathcal{C}_k$ has k elements, no combination in $\mathcal{C}_k$ can contain another; hence, $\mathcal{C}_k$ is a clutter. It follows from the previous corollary, that a clutter constructed in this way contains at most

$$\binom{n}{\lfloor \frac{n}{2} \rfloor}$$

sets. For example, if $n = 4$ and $S = \{a, b, c, d\}$, the 2-combinations of S give the clutter

$$\mathcal{C}_2 = \{\{a, b\}, \{a, c\}, \{a, d\}, \{b, c\}, \{b, d\}, \{c, d\}\}$$

of size 6. Can we do better by choosing combinations of more than one size? The negative answer to this question is the conclusion of Sperner's Theorem.

Theorem 5.4.3 *Let S be a set of n elements. Then a clutter on S contains at most $\binom{n}{\lfloor \frac{n}{2} \rfloor}$ sets.*

Proof. The proof is not hard, but it will seem a little long since we introduce and illustrate some new concepts in it. We take the set S to be the set $\{1, 2, \ldots, n\}$ of the first n positive integers and prove the theorem by induction on n. Actually, we shall prove the following stronger result by induction on n:

> The collection of all 2^n combinations of S can be partitioned into $\binom{n}{\lfloor \frac{n}{2} \rfloor}$ parts such that *any* clutter contains at most one combination from each part.

How can we be sure that a clutter contains at most one combination from each part? That's easy! We insist that our parts have the property that for any two combinations in a part, one is contained in the other. A collection of combinations with this property is called a *chain*. Here are partitions into chains for $n = 1, 2, 3$:

$n = 1$:

$$\emptyset \subset \{1\};$$

$n = 2$:

$$\emptyset \subset \{1\} \subset \{1,2\}$$

$$\{2\};$$

$n = 3$:

$$\emptyset \subset \{1\} \subset \{1,2\} \subset \{1,2,3\}$$

$$\{2\} \subset \{2,3\}$$

$$\{3\} \subset \{1,3\}.$$

For instance, for $n = 3$, a clutter contains at most 3 combinations, since it can contain at most 1 combination from each of the 3 chains shown. We can obtain a chain partition for the combinations of $\{1,2,3,4\}$ from that shown for $\{1,2,3\}$ as follows: We take each chain with more than one combination in it (for $n = 3$ all chains shown have this property) and make two chains for $n = 4$: one obtained by attaching a new combination at the end obtained by appending 4 to the last combination of the chain, and the other obtained by appending 4 to all but the last combination of the chain. Thus, the chain

$$\emptyset \subset \{1\} \subset \{1,2\} \subset \{1,2,3\}$$

becomes

$$\emptyset \subset \{1\} \subset \{1,2\} \subset \{1,2,3\} \subset \{1,2,3,4\} \text{ and}$$

$$\{4\} \subset \{1,4\} \subset \{1,2,4\};$$

the chain

$$\{2\} \subset \{2,3\}$$

becomes

$$\{2\} \subset \{2,3\} \subset \{2,3,4\} \text{ and}$$

$$\{2,4\};$$

and the chain

$$\{3\} \subset \{1,3\}$$

becomes

$$\{3\} \subset \{1,3\} \subset \{1,3,4\} \text{ and}$$

$$\{3,4\}.$$

Consequently, we have a chain partition of $6 = \binom{4}{2}$ chains of the combinations of $\{1,2,3,4\}$. The chains in this partition for $n = 4$ have two special properties. Each combination in a chain has one more element than the combination that precedes it (when there is a preceding combination). The size of the first combination in a chain plus the size of the last combination in the chain is $n = 4$. Similar properties hold for the chain partitions given for $n = 1, 2$ and 3. A chain partition of the combinations of $\{1,2,\ldots,n\}$ is a *symmetric chain partition*, provided that:

(i) each combination in a chain has one more element than the combination that precedes it in the chain; and

(ii) the size of the first combination in a chain plus the size of the last combination in the chain equals n.

Each chain in a symmetric chain partition contains exactly one $\lfloor n/2 \rfloor$-combination (and exactly one $\lceil n/2 \rceil$-combination); hence, the number of chains in a symmetric chain partition is

$$\binom{n}{\lfloor \frac{n}{2} \rfloor} = \binom{n}{\lceil \frac{n}{2} \rceil}.$$

A symmetric chain decomposition for $\{1,2,\ldots,n\}$ can be obtained inductively from a symmetric chain decomposition of $\{1,2,\ldots,n-1\}$, previously illustrated for $n = 3$. We take each chain

$$A_1 \subset A_2 \subset \cdots \subset A_k, \text{ where } |A_1| + |A_k| = n - 1$$

in a symmetric chain partition for $\{1,2,\ldots,n-1\}$ and, depending on whether $k = 1$ or > 1, obtain one or two chains for $\{1,2,\ldots,n\}$:

$$A_1 \subset A_2 \subset \cdots \subset A_k \subset A_k \cup \{n\}, \text{ where } |A_1| + |A_k \cup \{n\}| = n,$$

and

$$A_1 \cup \{n\} \subset \cdots \subset A_{k-1} \cup \{n\} \text{ where } |A_1 \cup \{n\}| + |A_{k-1} \cup \{n\}| = n.$$

(If $k = 1$, the second chain does not occur.) Every combination of $\{1,2,\ldots,n\}$ occurs in exactly one of the chains constructed in this way; hence, the resulting collection of chains forms a symmetric chain partition for $\{1, 2, \ldots, n\}$.

The number of chains in a symmetric chain partition of $\{1, 2, \ldots, n\}$ is

$$\binom{n}{\lfloor n/2 \rfloor}.$$

Thus, the number of combinations in a clutter of $\{1, 2, \ldots, n\}$ is at most equal to

$$\binom{n}{\lfloor n/2 \rfloor}.$$

$\square$

If n is even, it can be shown that the only clutter of size $\binom{n}{\lfloor \frac{n}{2} \rfloor}$ is the clutter of all $\frac{n}{2}$-combinations of S. If n is odd, the only clutters of this size are the clutter of all $\frac{n-1}{2}$-combinations of S and the clutter of all $\frac{n+1}{2}$-combinations of S. See Exercise 29.

5.5   The Multinomial Theorem

The binomial theorem gives a formula for $(x + y)^n$ for each positive integer n. It can be generalized to give a formula for $(x + y + z)^n$ or more generally for the nth power of the sum of t real numbers: $(x_1 + x_2 + \cdots + x_t)^n$. In the general formula, the role of the binomial coefficients is taken over by numbers called the *multinomial coefficients*, which are defined by

$$\binom{n}{n_1 \ n_2 \ \cdots \ n_t} = \frac{n!}{n_1! n_2! \cdots n_t!}. \tag{5.15}$$

Here, $n_1, n_2, \ldots, n_t$ are nonnegative integers with

$$n_1 + n_2 + \cdots + n_t = n.$$

Recall from Section 3.4 that (5.15) represents the number of permutations of a multiset of objects of t different types with repetition numbers $n_1, n_2, \ldots, n_t$, respectively. The binomial coefficient $\binom{n}{k}$, for nonnegative n and k and having the value

$$\frac{n!}{k!(n-k)!}, \quad (k = 0, 1, \ldots, n)$$

in this notation becomes

$$\left(\begin{array}{cc} & n \\ k & n-k \end{array}\right)$$

and represents the number of permutations of a multiset of objects of two types with repetition numbers k and $n-k$, respectively.

In the same notation, Pascal's formula for the binomial coefficients with n and k positive is

$$\left(\begin{array}{cc} & n \\ k & n-k \end{array}\right) = \left(\begin{array}{cc} & n-1 \\ k & n-k-1 \end{array}\right) + \left(\begin{array}{cc} & n-1 \\ k-1 & n-k \end{array}\right).$$

Pascal's formula for the multinomial coefficients is

$$\left(\begin{array}{cccc} & & n & \\ n_1 & n_2 & \cdots & n_t \end{array}\right) = \left(\begin{array}{cccc} & & n-1 & \\ n_1-1 & n_2 & \cdots & n_t \end{array}\right)$$

$$+ \left(\begin{array}{cccc} & & n-1 & \\ n_1 & n_2-1 & \cdots & n_t \end{array}\right) + \cdots + \left(\begin{array}{cccc} & & n-1 & \\ n_1 & n_2 & \cdots & n_t-1 \end{array}\right). \qquad (5.16)$$

Formula (5.16) can be verified by direct substitution, using the value of the multinomial coefficients in (5.15). For instance, let $t = 3$ and let n_1, n_2, and n_3 be positive integers with $n_1 + n_2 + n_3 = n$. Then

$$\left(\begin{array}{ccc} & n-1 & \\ n_1-1 & n_2 & n_3 \end{array}\right) + \left(\begin{array}{ccc} & n-1 & \\ n_1 & n_2-1 & n_3 \end{array}\right) + \left(\begin{array}{ccc} & n-1 & \\ n_1 & n_2 & n_3-1 \end{array}\right)$$

$$= \frac{(n-1)!}{(n_1-1)!n_2!n_3!} + \frac{(n-1)!}{n_1!(n_2-1)!n_3!} + \frac{(n-1)!}{n_1!n_2!(n_3-1)!}$$

$$= \frac{n_1 \times (n-1)!}{n_1!n_2!n_2!} + \frac{n_2 \times (n-1)!}{n_1!n_2!n_3!} + \frac{n_3 \times (n-1)!}{n_1!n_2!n_3!}$$

$$= (n_1 + n_2 + n_3) \times \frac{(n-1)!}{n_1!n_2!n_3!} = n \times \frac{(n-1)!}{n_1!n_2!n_3!}$$

$$= \frac{n!}{n_1!n_2!n_3!} = \left(\begin{array}{ccc} & n & \\ n_1 & n_2 & n_3 \end{array}\right).$$

In the exercises, a hint is given for a combinatorial verification of (5.16).

Before stating the general theorem, we first consider a special case. Let x_1, x_2, x_3 be real numbers. If we completely multiply out

$$(x_1 + x_2 + x_3)^3$$

and collect like terms (the reader is urged to do so), we obtain the sum

$$x_1^3 + x_2^3 + x_3^3 + 3x_1^2 x_2 + 3x_1 x_2^2 + 3x_1^2 x_3 + 3x_1 x_3^2 + 3x_2^2 x_3 + 3x_2 x_3^2 + 6x_1 x_2 x_3.$$

The terms that appear in the preceding sum are all the terms of the form $x_1^{n_1} x_2^{n_2} x_3^{n_3}$, where n_1, n_2, n_3 are nonnegative integers with $n_1 + n_2 + n_3 = 3$. The coefficient of $x_1^{n_1} x_2^{n_2} x_3^{n_3}$ in this expression equals

$$\binom{3}{n_1\ n_2\ n_3} = \frac{3!}{n_1! n_2! n_3!}.$$

More generally, we have the following *multinomial theorem*:

Theorem 5.5.1 *Let n be a positive integer. For all $x_1, x_2, \ldots, x_t$,*

$$(x_1 + x_2 + \cdots + x_t)^n = \sum \binom{n}{n_1\ n_2\ \cdots\ n_t} x_1^{n_1} x_2^{n_2} \cdots x_t^{n_t}$$

where the summation extends over all nonnegative integral solutions $n_1, n_2, \ldots, n_t$ of $n_1 + n_2 + \cdots + n_t = n$.

Proof. We generalize the first proof of the binomial theorem. We write $(x_1 + x_2 + \cdots + x_t)^n$ as a product of n factors, each equal to $(x_1 + x_2 + \cdots + x_t)$. We completely expand this product, using the distributive law and collect like terms. For each of the n factors we choose one of the t numbers $x_1, x_2, \ldots, x_t$ and form their product. There are t^n terms that result in this way, and each can be arranged in the form $x_1^{n_1} x_2^{n_2} \cdots x_t^{n_t}$, where $n_1, n_2, \ldots, n_t$ are nonnegative integers summing to n. We obtain the term $x_1^{n_1} x_2^{n_2} \cdots x_t^{n_t}$ by choosing x_1 in n_1 of the n factors, x_2 in n_2 of the remaining $n - n_1$ factors, $\ldots$, x_t in n_t of the remaining $n - n_1 - \cdots - n_{t-1}$ factors. Thus, by the multiplication principle, the number of times the term $x_1^{n_1} x_2^{n_2} \cdots x_t^{n_t}$ occurs is given by

$$\binom{n}{n_1}\binom{n - n_1}{n_2} \cdots \binom{n - n_1 - \cdots - n_{t-1}}{n_t}.$$

We have already seen in Section 3.4 that this number equals the multi-nomial coefficient

$$\frac{n!}{n_1! n_2! \cdots n_t!},$$

and this proves the theorem. □

Example. When $(x_1 + x_2 + x_3 + x_4 + x_5)^7$ is expanded, the coefficient of $x_1^2 x_3 x_4^3 x_5$ equals

$$\binom{7}{2\ 0\ 1\ 3\ 1} = \frac{7!}{2!0!1!3!1!} = 420.$$

□

Example. When $(2x_1 - 3x_2 + 5x_3)^6$ is expanded, the coefficient of $x_1^3 x_2 x_3^2$ equals

$$\binom{6}{3\ 1\ 2} 2^3(-3)(5)^2 = -36,000.$$

□

The number of different terms that occur in the multinomial expansion of $(x_1 + x_2 + \cdots + x_t)^n$ equals the number of nonnegative integral solutions of

$$n_1 + n_2 + \cdots + n_t = n.$$

It follows from Section 3.5 that the number of these solutions equals

$$\binom{n + t - 1}{n}.$$

For instance, $(x_1 + x_2 + x_3 + x_4)^6$ contains

$$\binom{6 + 4 - 1}{6} = \binom{9}{6} = 84$$

different terms if multiplied out completely.

5.6 Newton's Binomial Theorem

In 1676, Newton generalized the binomial theorem given in Section 5.2 to obtain an expansion for $(x + y)^\alpha$, where α is any real number. For general exponents, however, the expansion becomes an infinite series and questions of convergence need to be considered. We shall be satisfied with stating the theorem and considering some special cases. A proof of the theorem can be found in most advanced calculus texts.

Theorem 5.6.1 *Let α be a real number. Then, for all x and y with* $0 \leq |x| < |y|$,

$$(x + y)^\alpha = \sum_{k=0}^{\infty} \binom{\alpha}{k} x^k y^{\alpha-k},$$

where

$$\binom{\alpha}{k} = \frac{\alpha(\alpha - 1) \cdots (\alpha - k + 1)}{k!}.$$

If α is a positive integer n, then for $k > n$, $\binom{n}{k} = 0$, and the preceding expansion becomes

$$(x + y)^n = \sum_{k=0}^{n} \binom{n}{k} x^k y^{n-k}.$$

This agrees with the binomial theorem of Section 5.2.

If we set $z = x/y$, then $(x+y)^\alpha = y^\alpha (z+1)^\alpha$. Thus, Theorem 5.6.1 can be stated in the equivalent form: For any z with $|z| < 1$,

$$(1 + z)^\alpha = \sum_{k=0}^{\infty} \binom{\alpha}{k} z^k.$$

Suppose that n is a positive integer and we choose α to be the negative integer $-n$. Then

$$\binom{\alpha}{k} = \binom{-n}{k} = \frac{-n(-n-1)\cdots(-n-k+1)}{k!}$$

$$= (-1)^k \frac{n(n+1)\cdots(n+k-1)}{k!}$$

$$= (-1)^k \binom{n+k-1}{k}.$$

Thus, for $|z| < 1$,

$$(1 + z)^{-n} = \frac{1}{(1+z)^n} = \sum_{k=0}^{\infty} (-1)^k \binom{n+k-1}{k} z^k.$$

Replacing z by $-z$, we obtain

$$(1 - z)^{-n} = \frac{1}{(1-z)^n} = \sum_{k=0}^{\infty} \binom{n+k-1}{k} z^k. \tag{5.17}$$

If $n = 1$, then $\binom{n+k-1}{k} = \binom{k}{k} = 1$, and we obtain

$$\frac{1}{1+z} = \sum_{k=0}^{\infty} (-1)^k z^k \qquad (|z| < 1)$$

and

$$\frac{1}{1-z} = \sum_{k=0}^{\infty} z^k \qquad (|z| < 1). \qquad (5.18)$$

The binomial coefficient $\binom{n+k-1}{k}$ that occurs in the expansion (5.17) is of a type that has occurred before in counting problems, and this suggests a possible combinatorial derivation of (5.17). We start with the infinite geometric series (5.18). Then

$$\frac{1}{(1-z)^n} = (1 + z + z^2 + \cdots) \cdots (1 + z + z^2 + \cdots) \quad (n \text{ factors}). \quad (5.19)$$

We obtain a term z^k in this product by choosing z^{k_1} from the first factor, z^{k_2} from the second factor, $\ldots$, z^{k_n} from the nth factor, where $k_1, k_2, \ldots, k_n$ are nonnegative integers summing to k:

$$z^{k_1} z^{k_2} \cdots z^{k_n} = z^{k_1 + k_2 + \cdots + k_n} = z^k.$$

Thus, the number of different ways to get z^k, that is, the coefficient of z^k in (5.19), equals the number of nonnegative integral solutions of

$$k_1 + k_2 + \cdots + k_n = k,$$

and we know this to be

$$\binom{n+k-1}{k}.$$

The binomial theorem can be used to obtain square roots to any desired accuracy. If we take $\alpha = \frac{1}{2}$, then

$$\binom{\alpha}{0} = 1,$$

while, for $k > 0$,

$$\begin{aligned}
\binom{\alpha}{k} = \binom{1/2}{k} &= \frac{\frac{1}{2}(\frac{1}{2} - 1) \cdots (\frac{1}{2} - k + 1)}{k!} \\
&= \frac{(-1)^{k-1}}{2^k} \frac{1 \times 2 \times 3 \times 4 \times \cdots \times (2k-3) \times (2k-2)}{2 \times 4 \times \cdots \times (2k-2) \times (k!)} \\
&= \frac{(-1)^{k-1}}{k \times 2^{2k-1}} \frac{(2k-2)!}{(k-1)!^2} \\
&= \frac{(-1)^{k-1}}{k \times 2^{2k-1}} \binom{2k-2}{k-1}.
\end{aligned}$$

Thus, for $|z| < 1$,

$$\sqrt{1+z} \;=\; (1+z)^{1/2} = 1 + \sum_{k=1}^{\infty} \frac{(-1)^{k-1}}{k \times 2^{2k-1}} \binom{2k-2}{k-1} z^k$$

$$= \; 1 + \frac{1}{2}z - \frac{1}{2 \times 2^3}\binom{2}{1}z^2 + \frac{1}{3 \times 2^5}\binom{4}{2}z^3 - \cdots.$$

For example,

$$\sqrt{20} \;=\; \sqrt{16+4} = 4\sqrt{1+0.25}$$

$$= \; 4\left(1 + \frac{1}{2}(0.25) - \frac{1}{8}(0.25)^2 + \frac{1}{16}(0.25)^3 - \cdots\right)$$

$$= \; 4.472\ldots.$$

In Chapter 7 we shall apply the general binomial theorem in the solution of certain recurrence relations by generating functions.

5.7   More on Partially Ordered Sets

In Section 5.4, we discussed the notions of antichain (or clutter) and chain in the partially ordered set $\mathcal{P}(X)$ of all subsets of a set X. In the current section we extend these notions to partially ordered sets in general, and prove some basic theorems.

Let $(X, \leq)$ be a finite partially ordered set. An *antichain* is a subset A of X no pair of whose elements are comparable. In contrast, a *chain* is a subset C of X each pair of whose elements is comparable. Thus, a chain C is a totally ordered subset of X, and hence, by Theorem 4.5.2, the elements of a chain can be linearly ordered: $x_1 < x_2 < \cdots < x_t$. We usually present a chain by writing it in a linear order in this way. It follows immediately from definitions that a subset of a chain is also a chain, and that a subset of an antichain is also an antichain. The important connection between antichains and chains is that

$$|A \cap C| \leq 1 \text{ if } A \text{ is an antichain and } C \text{ is a chain.}$$

Example. Let $X = \{1, 2, \ldots, 10\}$, and consider the partially ordered set $(X, |)$ whose partial order $|$ is "is divisible by." Then $\{4, 6, 7, 9, 10\}$ is an antichain of size 5, while $1 \mid 2 \mid 4 \mid 8$ is a chain of size 4. There are no antichains of size 6 and no chains of size 5. $\square$

Let $(X, \leq)$ be a finite partially ordered set. We consider partitions of X into chains and also into antichains. Surely, if there is a chain

C of size r, then since no two elements of C can belong to the same antichain, X cannot be partitioned into fewer than r antichains. Similarly, if there is an antichain A of size s, then since no two elements of A can belong to the same chain, X cannot be partitioned into fewer that s chains. Our primary goal in this section is to prove two theorems that makes more precise this connection between antichains and chains. In spite of the "duality" between chains and antichains,[2] the proof of one of these theorems is quite short and simple while that of the other is less so.

Recall that a *minimal element* of a partially ordered set is an element a such that no element x different from a satisfies $x \leq a$. A *maximal element* is an element b such that no element y different from b satisfies $b \leq y$. The set of all minimal elements of a partially ordered set forms an antichain as does the set of all maximal elements.

Theorem 5.7.1 *Let $(X, \leq)$ be a finite partially ordered set, and let r be the largest size of a chain. Then X can be partitioned into r but no fewer antichains.*

Proof. As already noted, X cannot be partitioned into fewer than r antichains. Thus, it suffices to show that X can be partitioned into r antichains. Let $X_1 = X$ and let A_1 be the set of minimal elements of X. Delete the elements of A_1 from X_1 to get X_2, and let A_2 be the set of minimal elements of X_2. Note that for each element a_2 of A_2, there is an element a_1 of A_1 such that $a_1 < a_2$. Delete the elements of A_2 from X_2 to get X_3, and let A_3 be the set of minimal elements of X_3. Continue like this until the first integer p such that $X_p \neq \emptyset$ and $X_{p+1} = \emptyset$. Then $A_1, A_2, \ldots, A_p$ is a partition of X into antichains. Moreover, there is a chain

$$a_1 < a_2 < \cdots < a_p$$

where a_1 is in A_1, a_2 is in A_2, $\ldots$, a_p is in A_p. Since r is the largest size of a chain, $r \geq p$. Since X is partitioned into p antichains, $r \leq p$. Hence $r = p$ and the theorem is proved. □

The "dual" theorem is generally known as *Dilworth's Theorem*.

Theorem 5.7.2 *Let $(X, \leq)$ be a finite partially ordered set, and let m be the largest size of an antichain. Then X can be partitioned into m but no fewer chains.*

---

[2] In a chain every pair of elements is comparable; in an antichain every pair of elements is incomparable.

Proof.[3] As already noted above, X cannot be partitioned into fewer than m chains. Thus it suffices to show that X can be partitioned into m chains. We prove this by induction on the number n of elements in X. If $n = 1$, then the conclusion holds trivially. Assume that $n > 1$.

We consider two cases:

Case 1. There is an antichain A of size m that is neither the set of all maximal elements nor the set of all minimal elements of X.

In this case, let

$$A^+ = \{x : x \text{ in } X \text{ with } a \le x \text{ for some } a \text{ in } A\}$$

$$A^- = \{x : x \text{ in } X \text{ with } x \le a \text{ for some } a \text{ in } A\}.$$

Thus, A^+ consists of all elements "above" A, and A^- consists of all elements "below" A. The following properties hold:

1. $A^+ \ne X$ (and thus $|A^+| < |X|$), since there is a minimal element not in A;

2. $A^- \ne X$ (and thus $|A^-| < |X|$), since there is a maximal element not in A;

3. $A^+ \cap A^- = A$, since, if there were an element x in $A^+ \cap A^-$ not in A, then we would have $a_1 < x < a_2$ for some elements a_1 and a_2 in A, contradicting the assumption that A is an antichain;

4. $A^+ \cup A^- = X$, since, if there were an element x not in $A^+ \cup A^-$, $A \cup \{x\}$ would be an antichain of larger size than A.

We apply the induction assumption to the smaller partially ordered sets A^+ and A^- and conclude that A^+ can be partitioned into m chains $E_1, E_2, \ldots, E_m$, and A^- can be partitioned into m chains $F_1, F_2, \ldots, F_m$. The elements of A are the maximal elements of A^- and so the last elements on the chains $F_1, F_2, \ldots, F_m$; the elements of A are also the minimal elements of A^+ and so the first elements on the chains $E_1, E_2, \ldots, E_m$. We "glue" the chains together in pairs to form m chains that partition X.

Case 2. There are at most two antichains of size m, one or both of the set of all maximal elements and the set of all minimal elements. Let x be a minimal element and y a maximal element with $x \le y$ (x may equal y). Then the largest size of an antichain of $X - \{x, y\}$ is $m - 1$.

---

[3]This particularly simple proof is taken from M.A. Perles: A proof of Dilworth's decomposition theorem for partially ordered sets, *Israel J. Math.*, 1 (1963), 105-7.

By the induction hypothesis, $X - \{x, y\}$ can be partitioned into $m - 1$ chains. These chains, together with the chain $x \le y$, give a partition of X into m chains.   □

5.8   Exercises

1. Prove Pascal's formula by substituting the values of the binomial coefficients as given in equation (5.1).

2. Fill in the rows of Pascal's triangle corresponding to $n = 9$ and 10.

3. Consider the sum of the binomial coefficients along the diagonals of Pascal's triangle running upward from the left. The first few are: $1, 1, 1 + 1 = 2, 1 + 2 = 3, 1 + 3 + 1 = 5, 1 + 4 + 3 = 8$. Compute several more of these diagonal sums, and determine how these sums are related. (Compare them with the values of the counting function f in Exercise 4 of Chapter 1.)

4. Expand $(x + y)^5$ and $(x + y)^6$, using the binomial theorem.

5. Expand $(2x - y)^7$, using the binomial theorem.

6. What is the coefficient of $x^5 y^{13}$ in the expansion of $(3x - 2y)^{18}$? What is the coefficient of $x^8 y^9$? (There is not a misprint in this last question!)

7. Use the binomial theorem to prove that

$$3^n = \sum_{k=0}^{n} \binom{n}{k} 2^k.$$

Generalize to find the sum

$$\sum_{k=0}^{n} \binom{n}{k} r^k$$

for any real number r.

8. Use the binomial theorem to prove that

$$2^n = \sum_{k=0}^{n} (-1)^k \binom{n}{k} 3^{n-k}.$$

9. Evaluate the sum

$$\sum_{k=0}^{n} (-1)^k \binom{n}{k} 10^k.$$

10. Use *combinatorial* reasoning to prove the identity (5.2). (Hint: Think of choosing a team with one person designated as captain.)

11. Use *combinatorial* reasoning to prove the identity (in the form given)

$$\binom{n}{k} - \binom{n-3}{k} = \binom{n-1}{k-1} + \binom{n-2}{k-1} + \binom{n-3}{k-1}.$$

(Hint: Let S be a set with three distinguished elements $a, b,$ and c and count certain k-combinations of S.)

12. Let n be a positive integer. Prove that

$$\sum_{k=0}^{n} (-1)^k \binom{n}{k}^2 = \begin{cases} 0 & \text{if } n \text{ is odd} \\ (-1)^m \binom{2m}{m} & \text{if } n = 2m. \end{cases}$$

(Hint: For $n = 2m$, consider the coefficient of x^n in $(1 - x^2)^n = (1 + x)^n (1 - x)^n$.)

13. Find one binomial coefficient equal to the following expression

$$\binom{n}{k} + 3\binom{n}{k-1} + 3\binom{n}{k-2} + \binom{n}{k-3}.$$

14. Prove that

$$\binom{r}{k} = \frac{r}{r-k}\binom{r-1}{k}$$

for r a real number and k an integer with $r \neq k$.

15. Prove, that for every integer $n > 1$,

$$\binom{n}{1} - 2\binom{n}{2} + 3\binom{n}{3} + \cdots + (-1)^{n-1} n \binom{n}{n} = 0.$$

16. By integrating the binomial expansion, prove that, for a positive integer n,

$$1 + \frac{1}{2}\binom{n}{1} + \frac{1}{3}\binom{n}{2} + \cdots + \frac{1}{n+1}\binom{n}{n} = \frac{2^{n+1} - 1}{n+1}.$$

17. Prove the identity in the previous exercise by using (5.2) and (5.3).

18. Evaluate the sum

$$1 - \frac{1}{2}\binom{n}{1} + \frac{1}{3}\binom{n}{2} - \frac{1}{4}\binom{n}{3} + \cdots + (-1)^n \frac{1}{n+1}\binom{n}{n}.$$

19. Sum the series $1^2 + 2^2 + 3^2 + \cdots + n^2$ by observing that

$$m^2 = 2\binom{m}{2} + \binom{m}{1}$$

and using the identity (5.14).

20. Find integers a, b, and c such that

$$m^3 = a\binom{m}{3} + b\binom{m}{2} + c\binom{m}{1}$$

for all m. Then sum the series $1^3 + 2^3 + 3^3 + \cdots + n^3$.

21. Prove that, for all real numbers r and all integers k,

$$\binom{-r}{k} = (-1)^k \binom{r+k-1}{k}.$$

22. Prove that, for all real numbers r and all integers k and m,

$$\binom{r}{m}\binom{m}{k} = \binom{r}{k}\binom{r-k}{m-k}.$$

23. Every day a student walks from her home to school, which is located 10 blocks east and 14 blocks north from home. She always takes a shortest walk of 24 blocks.

    (a) How many different walks are possible?

    (b) Suppose that 4 blocks east and 5 blocks north of her home lives her best friend, whom she meets each day on her way to school. Now how many different walks are possible?

    (c) Suppose, in addition, that 3 blocks east and 6 blocks north of her friend's house there is a park where the two girls stop each day to rest and play. Now how many different walks are there?

(d) Stopping at a park to rest and play, the two students often get to school late. To avoid the temptation of the park, our two students decide never to pass the intersection where the park is. Now how many different walks are there?

24. Consider a three-dimensional grid whose dimensions are 10 by 15 by 20. You are at the front lower left corner of the grid and wish to get to the back upper right corner 45 "blocks" away. How many different routes are there in which you walk exactly 45 blocks?

25. Use a combinatorial argument to prove the *Vandermonde convolution* for the binomial coefficients: For all positive integers m_1, m_2, and n,

$$\sum_{k=0}^{n} \binom{m_1}{k}\binom{m_2}{n-k} = \binom{m_1+m_2}{n}.$$

Deduce the identity (5.11) as a special case.

26. Let n be a positive integer. Verify by substitution that

$$\binom{2n}{n+1} + \binom{2n}{n} = \frac{1}{2}\binom{2n+2}{n+1}.$$

Then give a combinatorial proof.

27. Let n and k be integers with $1 \le k \le n$. Prove that

$$\sum_{k=1}^{n} \binom{n}{k}\binom{n}{k-1} = \frac{1}{2}\binom{2n+1}{n+1} - \binom{2n}{n}.$$

28. Let n and k be positive integers. Give a combinatorial proof of the identity (5.10):

$$n(n+1)2^{n-2} = \sum_{k=1}^{n} k^2 \binom{n}{k}.$$

29. Let n and k be positive integers. Give a combinatorial proof that

$$\sum_{k=1}^{n} k\binom{n}{k}^2 = n\binom{2n-1}{n-1}.$$

30. Find and prove a formula for

$$\sum_{\substack{r,s,t \geq 0 \\ r+s+t=n}} \binom{m_1}{r}\binom{m_2}{s}\binom{m_3}{t},$$

where the summation extends over all nonnegative integers r, s and t with sum $r+s+t = n$.

31. Prove that the only clutter of $S = \{1,2,3,4\}$ of size 6 is the clutter of all 2-combinations of S.

32. Prove that there are only two clutters of $S = \{1,2,3,4,5\}$ of size 10 (10 is maximum by Sperner's Theorem), namely, the clutter of all 2-combinations of S and the clutter of all 3-combinations.

33. * Let S be a set of n elements. Prove that, if n is even, the only clutter of size $\binom{n}{\lfloor \frac{n}{2} \rfloor}$ is the clutter of all $\frac{n}{2}$-combinations; if n is odd, prove that the only clutters of this size are the clutter of all $\frac{n-1}{2}$-combinations and the clutter of all $\frac{n+1}{2}$-combinations.

34. Construct a partition of the combinations of $\{1,2,3,4,5\}$ into symmetric chains.

35. In a partition of the combinations of $\{1,2,\ldots,n\}$ into symmetric chains, how many chains have only one combination in them? two combinations? k combinations?

36. A talk show host has just bought 10 new jokes. Each night he tells some of the jokes. What is the largest number of nights on which you can tune in so that you never hear on one night at least all the jokes you heard on *one* of the other nights? (Thus, for instance, it is acceptable that you hear jokes 1, 2, and 3 on one night, jokes 3 and 4 on another, and jokes 1, 2, and 4 on a third. It is not acceptable that you hear jokes 1 and 2 on one night and joke 2 on another night.)

37. Prove the identity of Exercise 25, using the binomial theorem and the relation $(1+x)^{m_1}(1+x)^{m_2} = (1+x)^{m_1+m_2}$.

38. Use the multinomial theorem to show that, for positive integers n and t,

$$t^n = \sum \binom{n}{n_1\ n_2\cdots n_t},$$

where the summation extends over all nonnegative integral solutions $n_1, n_2, \ldots, n_t$ of $n_1 + n_2 + \cdots + n_t = n$.

39. Use the multinomial theorem to expand $(x_1 + x_2 + x_3)^4$.

40. Determine the coefficient of $x_1^3 x_2 x_3^4 x_5^2$ in the expansion of

$$(x_1 + x_2 + x_3 + x_4 + x_5)^{10}.$$

41. What is the coefficient of $x_1^3 x_2^3 x_3 x_4^2$ in the expansion of

$$(x_1 - x_2 + 2x_3 - 2x_4)^9?$$

42. Expand $(x_1 + x_2 + x_3)^n$ by observing that

$$(x_1 + x_2 + x_3)^n = ((x_1 + x_2) + x_3)^n$$

and then using the binomial theorem.

43. Prove the identity (5.16) by a combinatorial argument. (Hint: Consider the permutations of a multiset of objects of t different types with repetition numbers $n_1, n_2, \ldots, n_t$, respectively. Partition these permutations according to what type of object is in the first position.)

44. Prove by induction on n that, for n a positive integer,

$$\frac{1}{(1-z)^n} = \sum_{k=0}^{\infty} \binom{n+k-1}{k} z^k, \qquad |z| < 1.$$

Assume the validity of

$$\frac{1}{1-z} = \sum_{k=0}^{\infty} z^k, \qquad |z| < 1.$$

45. Use Newton's binomial theorem to approximate $\sqrt{30}$.

46. Use Newton's binomial theorem to approximate $10^{1/3}$.

47. Use Theorem 5.7.1 to show that, if m and n are positive integers, then a partially ordered set of $mn + 1$ elements has a chain of size $m + 1$ or an antichain of size $n + 1$.

48. Use the result of the previous exercise to show that a sequence of $mn + 1$ real numbers either contains an increasing subsequence of $m + 1$ numbers or a decreasing subsequence of $n + 1$ numbers (see Application 9 of Section 2.2).

49. Consider the partially ordered set $(X, |)$ on the set $X = \{1, 2, \ldots, 12\}$ of the first 12 positive integers, partially ordered by "is divisible by."

    (a) Determine a chain of largest size and a partition of X into the smallest number of antichains.

    (b) Determine an antichain of largest size and a partition of X into the smallest number of chains.

50. Let R and S be two partial orders on the same set X. Considering R and S as subsets of $X \times X$, we assume that $R \subseteq S$ but $R \neq S$. Show that there exists an ordered pair (p, q) where $(p, q) \in S$ and $(p, q) \notin R$ such that $R' = R \cup \{(p, q)\}$ is also a partial order on X. Show by example that not every such (p, q) has the property that R' is a partial order on X.

Chapter 6

The Inclusion–Exclusion Principle and Applications

In this chapter we derive and apply an important counting formula called the inclusion–exclusion principle. Recall that the addition principle gives a simple formula for counting the number of objects in a union of sets, *provided that the sets do not overlap* (i.e., provided that the sets determine a partition). The inclusion–exclusion principle gives a formula for the most general of circumstances in which the sets are free to overlap without restriction. The formula is necessarily more complicated but, as a result, is more widely applicable.

6.1   The Inclusion–Exclusion Principle

In Chapter 3 we have seen several examples in which it is easier to make an indirect count of the number of objects in a set rather than to count the objects directly. We now give two more examples.

Example. Count the permutations $i_1 i_2 \ldots i_n$ of $\{1, 2, \ldots, n\}$ in which 1 is not in the first position (that is, $i_1 \neq 1$).

We could make a direct count by observing that the permutations with 1 not in the first position can be divided into $n-1$ parts according to which of the $n-1$ integers k from $\{2, 3, \ldots, n\}$ is in the first position. A permutation with k in the first position consists of k followed by a permutation of the $(n-1)$-element set $\{1, \ldots, k-1, k+1, \ldots, n\}$. Hence, there are $(n-1)!$ permutations of $\{1, 2, \ldots, n\}$ with k in the first position. By the addition principle, there are $(n-1)!(n-1)$ permutations of $\{1, 2, \ldots, n\}$ with 1 not in the first position.

160

Alternatively, we could make an indirect count by observing that the number of permutations of $\{1, 2, \ldots, n\}$ with 1 in the first position is the same as the number $(n-1)!$ of permutations of $\{2, 3, \ldots, n\}$. Since the total number of permutations of $\{1, 2, \ldots, n\}$ is $n!$, the number of permutations of $\{1, 2, \ldots, n\}$ in which 1 is not in the first position is $n! - (n-1)! = (n-1)!(n-1)$. $\qquad\square$

Example. Count the number of integers between 1 and 600, inclusive, which are not divisible by 6.

We can do this indirectly as follows. The number of integers between 1 and 600 which are divisible by 6 is $600/6 = 100$ since every sixth integer is divisible by 6. Hence $600 - 100 = 500$ of the integers between 1 and 600 are not divisible by 6. $\qquad\square$

The rule used to obtain an indirect count in these examples is the following. If A is a subset of a set S, then the number of objects in A equals the number of objects in S minus the number not in A. Recall that

$$\overline{A} = S - A = \{x : x \text{ in } S \text{ but } x \text{ not in } A\}$$

is the *complement* of A in S—that is, the set consisting of those objects in S which are not in A. The rule can then be written as

$$|A| = |S| - |\overline{A}| \text{ or, equivalently, } |\overline{A}| = |S| - |A|.$$

This formula is the simplest instance of the inclusion–exclusion principle.

We shall formulate the inclusion–exclusion principle in a manner in which it is convenient to apply. As a first generalization of the preceding rule, let S be a finite set of objects, and let P_1 and P_2 be two "properties" that each object in S may or may not possess. We wish to count the number of objects in S that have neither property P_1 nor property P_2. We can do this by first including all objects of S in our count, then excluding all objects that have property P_1 and excluding all objects which have property P_2, and then, noting that we have excluded objects having both properties P_1 and P_2 twice, readmitting all such objects once. We can write this symbolically as follows: Let A_1 be the subset of objects of S that have property P_1, and let A_2 be the subset of objects of S that have property P_2. Then $\overline{A}_1$ consists of those objects of S not having property P_1, and $\overline{A}_2$ consists of those objects of S not having property P_2. The objects of the set $\overline{A}_1 \cap \overline{A}_2$ are those having neither property P_1 nor property P_2. We then have

$$|\overline{A}_1 \cap \overline{A}_2| = |S| - |A_1| - |A_2| + |A_1 \cap A_2|.$$

Since the left side of the preceding equation counts the number of objects of S that have neither property P_1 nor property P_2, we can establish the validity of this equation by showing that an object with neither of the two properties P_1 and P_2 makes a net contribution of 1 to the right side, and every other object makes a net contribution of 0. If x is an object with neither of the properties P_1 and P_2, it is counted among the objects of S, not counted among the objects of A_1 or of A_2, and not counted among the objects of $A_1 \cap A_2$. Hence, its net contribution to the right side of the equation is

$$1 - 0 - 0 + 0 = 1.$$

If x has only the property P_1, it contributes

$$1 - 1 - 0 + 0 = 0$$

to the right side, while if it has only the property P_2, it contributes

$$1 - 0 - 1 + 0 = 0$$

to the right side. Finally, if x has both properties P_1 and P_2, it contributes

$$1 - 1 - 1 + 1 = 0$$

to the right side of the equation. Thus, the right side of the equation also counts the number of objects of S with neither property P_1 nor property P_2.

More generally, let $P_1, P_2, \ldots, P_m$ be m properties referring to the objects in S, and let

$$A_i = \{x : x \text{ in } S \text{ and } x \text{ has property } P_i\}, \quad (i = 1, 2, \ldots, m)$$

be the subset of objects of S that have property P_i (and possibly other properties). Then $A_i \cap A_j$ is the subset of objects that have both properties P_i and P_j (and possibly others), $A_i \cap A_j \cap A_k$ is the subset of objects which have properties P_i, P_j, and P_k, and so on. The subset of objects having none of the properties is $\overline{A}_1 \cap \overline{A}_2 \cap \cdots \cap \overline{A}_m$. The inclusion–exclusion principle shows how to count the number of objects in this set by counting objects according to the properties they *do* have. Thus, in this sense, it "inverts" the counting process.

Theorem 6.1.1 *The number of objects of S that have none of the properties $P_1, P_2, \ldots, P_m$ is given by*

$$\begin{aligned}
|\overline{A}_1 \cap \overline{A}_2 \cap \cdots \cap \overline{A}_m| \;=\;& |S| - \Sigma|A_i| + \Sigma|A_i \cap A_j| - \Sigma|A_i \cap A_j \cap A_k| \\
& + \cdots + (-1)^m |A_1 \cap A_2 \cap \cdots \cap A_m| \quad (6.1)
\end{aligned}$$

where the first sum is over all 1-combinations $\{i\}$ of $\{1, 2, \ldots, m\}$, the second sum is over all 2-combinations $\{i, j\}$ of $\{1, 2, \ldots, m\}$, the third sum is over all 3-combinations $\{i, j, k\}$ of $\{1, 2, \ldots, m\}$, and so on.

If $m = 3$, (6.1) becomes

$$
\begin{aligned}
|\overline{A}_1 \cap \overline{A}_2 \cap \overline{A}_3| &= |S| - (|A_1| + |A_2| + |A_3|) + \\
&\quad (|A_1 \cap A_2| + |A_1 \cap A_3| + |A_2 \cap A_3|) \\
&\quad - |A_1 \cap A_2 \cap A_3|.
\end{aligned}
$$

Note that there are $1 + 3 + 3 + 1 = 8$ terms on the right side. If $m = 4$, then equation (6.1) becomes

$$
\begin{aligned}
|\overline{A}_1 \cap \overline{A}_2 \cap \overline{A}_3 \cap \overline{A}_4| &= |S| - (|A_1| + |A_2| + |A_3| + |A_4|) \\
&\quad + (|A_1 \cap A_2| + |A_1 \cap A_3| + |A_1 \cap A_4| \\
&\quad + |A_2 \cap A_3| + |A_2 \cap A_4| + |A_3 \cap A_4|) \\
&\quad - (|A_1 \cap A_2 \cap A_3| + |A_1 \cap A_2 \cap A_4| \\
&\quad + |A_1 \cap A_3 \cap A_4| + |A_2 \cap A_3 \cap A_4|) \\
&\quad + |A_1 \cap A_2 \cap A_3 \cap A_4|.
\end{aligned}
$$

In this case there are $1 + 4 + 6 + 4 + 1 = 16$ terms on the right side. In the general case, the number of terms on the right side of (6.1) is

$$
\binom{m}{0} + \binom{m}{1} + \binom{m}{2} + \binom{m}{3} + \cdots + \binom{m}{m} = 2^m.
$$

Proof of Theorem 6.1.1. The left side of equation (6.1) counts the number of objects of S with none of the properties. We can establish the validity of the equation by showing that an object with none of the properties $P_1, P_2, \ldots, P_m$ makes a net contribution of 1 to the right side, and an object with at least one of the properties makes a net contribution of 0. First, consider an object x with none of the properties. Its contribution to the right side of (6.1) is

$$
1 - 0 + 0 - 0 + \cdots + (-1)^m 0 = 1,
$$

since it is in S, but in none of the other sets. Now consider an object y with exactly $n \geq 1$ of the properties. The contribution of y to $|S|$ is $1 = \binom{n}{0}$. Its contribution to $\Sigma |A_i|$ is $n = \binom{n}{1}$ since it has exactly n of the properties and so is a member of exactly n of the sets $A_1, A_2, \ldots, A_m$. The contribution of y to $\Sigma |A_i \cap A_j|$ is $\binom{n}{2}$ since we may select a pair of the properties y has in $\binom{n}{2}$ ways, and so y is a member of exactly $\binom{n}{2}$ of the sets $A_i \cap A_j$. The contribution of y to $\Sigma|A_i \cap A_j \cap A_k|$ is $\binom{n}{3}$, and so on. Thus, the net contribution of y to the right side of (6.1) is

$$\binom{n}{0} - \binom{n}{1} + \binom{n}{2} - \binom{n}{3} + \cdots + (-1)^m \binom{n}{m},$$

which equals

$$\binom{n}{0} - \binom{n}{1} + \binom{n}{2} - \binom{n}{3} + \cdots + (-1)^n \binom{n}{n},$$

since $n \leq m$ and $\binom{n}{k} = 0$ if $k > n$. Since this last expression equals 0 according to the identity (5.4), the net contribution of y to the right side of (6.1) is 0 if y has at least one of the properties. $\qquad \square$

Theorem 6.1.1 implies a formula for the number of objects in the union of sets that are free to overlap.

Corollary 6.1.2 *The number of objects of S which have at least one of the properties $P_1, P_2, \ldots, P_m$ is given by*

$$|A_1 \cup A_2 \cup \cdots \cup A_m| = \Sigma|A_i| - \Sigma|A_i \cap A_j| + \Sigma|A_i \cap A_j \cap A_k| - \cdots$$
$$+ (-1)^{m+1}|A_1 \cap A_2 \cap \cdots \cap A_m|, \qquad (6.2)$$

where the summations are as specified in Theorem 6.1.1.

Proof. The set $A_1 \cup A_2 \cup \cdots \cup A_m$ consists of all those objects in S which possess at least one of the properties. Also,

$$|A_1 \cup A_2 \cup \cdots \cup A_m| = |S| - |\overline{A_1 \cup A_2 \cup \cdots \cup A_m}|.$$

Since, as is readily verified,

$$\overline{A_1 \cup A_2 \cup \cdots \cup A_m} = \overline{A}_1 \cap \overline{A}_2 \cap \cdots \cap \overline{A}_m,$$

we have

$$|A_1 \cup A_2 \cup \cdots \cup A_m| = |S| - |\overline{A}_1 \cap \overline{A}_2 \cap \cdots \cap \overline{A}_m|.$$

Combining this equation with equation (6.1), we obtain equation (6.2). $\qquad \square$

Example. Find the number of integers between 1 and 1000, inclusive, that are not divisible by 5, 6, and 8.

To solve this problem we introduce some notation. For a real number r, recall that $\lfloor r \rfloor$ stands for the largest integer that does not exceed r. Also, we shall abbreviate the least common multiple of two integers, a, b, or three integers, a, b, c, by $\operatorname{lcm}\{a, b\}$ and $\operatorname{lcm}\{a, b, c\}$, respectively. Let P_1 be the property that an integer is divisible by 5, P_2 the property that an integer is divisible by 6, and P_3 the property that an integer is divisible by 8. Let S be the set consisting of the first thousand positive integers. For $i = 1, 2, 3$, let A_i be the set consisting of those integers in S with property P_i. We wish to find the number of integers in $\overline{A}_1 \cap \overline{A}_2 \cap \overline{A}_3$.

We first see that

$$|A_1| = \lfloor \tfrac{1000}{5} \rfloor = 200,$$

$$|A_2| = \lfloor \tfrac{1000}{6} \rfloor = 166,$$

$$|A_3| = \lfloor \tfrac{1000}{8} \rfloor = 125.$$

Integers in the set $A_1 \cap A_2$ are divisible by both 5 and 6. But an integer is divisible by both 5 and 6 if and only if it is divisible by $\operatorname{lcm}\{5, 6\}$. Since $\operatorname{lcm}\{5, 6\} = 30$, $\operatorname{lcm}\{5, 8\} = 40$, and $\operatorname{lcm}\{6, 8\} = 24$, we see that

$$|A_1 \cap A_2| = \lfloor \tfrac{1000}{30} \rfloor = 33,$$

$$|A_1 \cap A_3| = \lfloor \tfrac{1000}{40} \rfloor = 25,$$

$$|A_2 \cap A_3| = \lfloor \tfrac{1000}{24} \rfloor = 41.$$

Because $\operatorname{lcm}\{5, 6, 8\} = 120$, we conclude that

$$|A_1 \cap A_2 \cap A_3| = \left\lfloor \frac{1000}{120} \right\rfloor = 8.$$

Thus, by the inclusion–exclusion principle, the number of integers between 1 and 1000 that are not divisible by $5, 6$, and 8 equals

$$
\begin{aligned}
|\overline{A}_1 \cap \overline{A}_2 \cap \overline{A}_3| &= 1000 - (200 + 166 + 125) + (33 + 25 + 41) - 8 \\
&= 600.
\end{aligned}
$$

$\square$

Example. How many permutations of the letters

$$M, A, T, H, I, S, F, U, N$$

are there such that none of the words MATH, IS, and FUN occur as consecutive letters? (Thus, for instance, the permutation MATH-ISFUN is not allowed nor are the permutations INUMATHSF and ISMATHFUN.)

We apply the inclusion–exclusion principle (6.1). First, we identify the set S as the set of all permutations of the 9 letters given. We then let P_1 be the property that a permutation in S contains the word MATH as consecutive letters, let P_2 be the property that a permutation contains the word IS, and let P_3 be the property that a permutation contains the word FUN. For $i = 1, 2, 3$, let A_i be the set of those permutations in S satisfying property P_i. We wish to find the number of permutations in $\overline{A_1} \cap \overline{A_2} \cap \overline{A_3}$.

We have $|S| = 9! = 362,880$. The permutations in A_1 can be thought of as permutations of the six symbols

$$MATH, I, S, F, U, N.$$

Hence,

$$|A_1| = 6! = 720.$$

Similarly, the permutations in A_2 are permutations of the eight symbols

$$M, A, T, H, IS, F, U, N,$$

so

$$|A_2| = 8! = 40,320,$$

and the permutations in A_3 are permutations of the seven symbols

$$M, A, T, H, I, S, FUN,$$

so

$$|A_3| = 7! = 5040.$$

The permutations in $A_1 \cap A_2$ are permutations of the five symbols

$$MATH, IS, F, U, N;$$

the permutations in $A_1 \cap A_3$ are permutations of the four symbols

$$MATH, I, S, FUN;$$

and the permutations in $A_2 \cap A_3$ are permutations of the six symbols

$$M, A, T, H, IS, FUN.$$

Hence, we have

$$|A_1 \cap A_2| = 5! = 120, |A_1 \cap A_3| = 4! = 24, \text{ and } |A_2 \cap A_3| = 6! = 720.$$

Finally, $A_1 \cap A_2 \cap A_3$ consists of the permutations of the three symbols $MATH, IS, FUN$; therefore,

$$|A_1 \cap A_2 \cap A_3| = 3! = 6.$$

Substituting into (6.1), we obtain

$$|\overline{A_1} \cap \overline{A_2} \cap \overline{A_3}| = 362,880 - 720 - 40,320 - 5040$$

$$+120 + 24 + 720 - 6 = 317,658.$$

□

In subsequent sections we consider applications of the inclusion–exclusion principle to some general problems. The following special case of the inclusion–exclusion principle will be useful:

Assume that the size of the set $A_{i_1} \cap A_{i_2} \cap \cdots \cap A_{i_k}$ that occurs in the inclusion–exclusion principle depends only on k and not on which k sets are used in the intersection. Thus, there are constants $\alpha_0, \alpha_1, \alpha_2, \ldots, \alpha_n$ such that

$$\alpha_0 = |S|$$
$$\alpha_1 = |A_1| = |A_2| = \cdots = |A_m|$$
$$\alpha_2 = |A_1 \cap A_2| = \cdots = |A_{m-1} \cap A_m|$$
$$\alpha_3 = |A_1 \cap A_2 \cap A_3| = \cdots = |A_{m-2} \cap A_{m-1} \cap A_m|$$
$$\vdots$$
$$\alpha_m = |A_1 \cap A_2 \cap \cdots \cap A_m|.$$

In this case, the inclusion–exclusion principle simplifies to

$$|\overline{A_1} \cap \overline{A_2} \cap \cdots \cap \overline{A_m}| = \alpha_0 - \binom{m}{1}\alpha_1 + \binom{m}{2}\alpha_2 - \binom{m}{3}\alpha_3 + \cdots +$$

$$(-1)^k \binom{m}{k}\alpha_k + \cdots + (-1)^m \alpha_m. \qquad (6.3)$$

This is because the kth summation that occurs in the inclusion–exclusion principle contains $\binom{m}{k}$ summands, each equal to α_k.

Example. How many integers between 0 and 99,999 (inclusive) have among their digits each of 2, 5, and 8?

Let S be the set of integers between 0 and 99,999. Each integer in S has 5 digits including possible leading 0's. (Thus we think of the integers in S as the 5-permutations of the multiset in which each digit $0, 1, 2, \ldots, 9$ has repetition number 5 or greater.) Let P_1 be the property that an integer does not contain the digit 2, let P_2 be the property that an integer does not contain the digit 5, and let P_3 be the property that an integer does not contain the digit 8. For $i = 1, 2, 3$, let A_i be the set consisting of those integers in S with property P_i. We wish to count the number of integers in $\overline{A_1} \cap \overline{A_2} \cap \overline{A_3}$.

Using the notation in the preceding example, we have

$$\alpha_0 = 10^5$$
$$\alpha_1 = 9^5$$
$$\alpha_2 = 8^5$$
$$\alpha_3 = 7^5.$$

For instance, the number of integers between 0 and 99,999 that do not contain the digit 2 and that do not contain the digit 5, the size of $|A_1 \cap A_2|$, equals the number of 5-permutations of the multiset

$$\{5 \cdot 0, 5 \cdot 1, 5 \cdot 3, 5 \cdot 4, 5 \cdot 6, 5 \cdot 7, 5 \cdot 8, 5 \cdot 9\},$$

and this equals 8^5. By (6.2), we obtain the answer

$$10^5 - 3 \times 9^5 + 3 \times 8^5 - 7^5.$$

$\square$

6.2   Combinations with Repetition

In Sections 3.3 and 3.5 we have shown that the number of r-combinations of a set of n distinct elements is

$$\binom{n}{r} = \frac{n!}{r!(n-r)!}$$

and that the number of r-combinations of a multiset with k distinct objects, each with an infinite repetition number, equals

$$\binom{r+k-1}{r}.$$

In this section we show how the latter formula, in conjunction with the inclusion–exclusion principle, gives a method for finding the number of

r-combinations of a multiset without any restrictions on its repetition numbers.

Suppose T is a multiset and an object x of T of a certain type has a repetition number that is greater than r. Then the number of r-combinations of T equals the number of r-combinations of the multiset obtained from T by replacing the repetition number of x by r. This is so because the number of times x can be used in an r-combination of T cannot exceed r. Therefore, any repetition number that is greater than r can be replaced by r. For example, the number of 8-combinations of the multiset $\{3 \cdot a, \infty \cdot b, 6 \cdot c, 10 \cdot d, \infty \cdot e\}$ is the same as the number of 8-combinations of the multiset $\{3 \cdot a, 8 \cdot b, 6 \cdot c, 8 \cdot d, 8 \cdot e\}$. We can summarize by saying that we have determined the number of r-combinations of a multiset $T = \{n_1 \cdot a_1, n_2 \cdot a_2, \ldots, n_k \cdot a_k\}$ in the two "extreme" cases:

(i) $n_1 = n_2 = \cdots = n_k = 1$; (i.e., T is a set) and

(ii) $n_1 = n_2 = \cdots = n_k = r$.

We shall illustrate how the inclusion–exclusion principle can be applied to obtain solutions for the remaining cases. Although we shall take a specific example, it should be clear that the method works in general.

Example. Determine the number of 10-combinations of the multiset $T = \{3 \cdot a, 4 \cdot b, 5 \cdot c\}$.

We shall apply the inclusion–exclusion principle to the set S of all 10-combinations of the multiset $T^* = \{\infty \cdot a, \infty \cdot b, \infty \cdot c\}$. Let P_1 be the property that a 10-combination of T^* has more than 3 a's. Let P_2 be the property that a 10-combination of T^* has more than 4 b's. Finally, let P_3 be the property that a 10-combination of T^* has more than 5 c's. The number of 10-combinations of T is then the number of 10-combinations of T^* that have none of the properties P_1, P_2, and P_3. As usual, let A_i consist of those 10-combinations of T^* which have property P_i, $(i = 1, 2, 3)$. We wish to determine the size of the set $\overline{A_1} \cap \overline{A_2} \cap \overline{A_3}$. By the inclusion–exclusion principle,

$$
\begin{aligned}
|\overline{A_1} \cap \overline{A_2} \cap \overline{A_3}| = \ & |S| - (|A_1| + |A_2| + |A_3|) \\
& + (|A_1 \cap A_2| + |A_1 \cap A_3| + |A_2 \cap A_3|) \\
& - |A_1 \cap A_2 \cap A_3|.
\end{aligned}
$$

By Theorem 3.5.1,

$$
|S| = \binom{10 + 3 - 1}{10} = \binom{12}{10} = 66.
$$

The set A_1 consists of all 10-combinations of T^* in which a occurs at least 4 times. If we take any one of these 10-combinations in A_1 and remove 4 a's, we are left with a 6-combination of T^*. Conversely, if we take a 6-combination of T^* and add 4 a's to it, we get a 10-combination of T^* in which a occurs at least 4 times. Thus, the number of 10-combinations in A_1 equals the number of 6-combinations of T^*. Hence,

$$|A_1| = \binom{6+3-1}{6} = \binom{8}{6} = 28.$$

In a similar way we see that the number of 10-combinations in A_2 equals the number of 5-combinations of T^*, and the number of 10-combinations in A_3 equals the number of 4-combinations of T^*. Consequently,

$$|A_2| = \binom{5+3-1}{5} = \binom{7}{5} = 21 \text{ and } |A_3| = \binom{4+3-1}{4} = \binom{6}{4} = 15.$$

The set $A_1 \cap A_2$ consists of all 10-combinations of T^* in which a occurs at least 4 times and b occurs at least 5 times. If, from any of these 10-combinations, we remove 4 a's and 5 b's, we are left with a 1-combination of T^*. Conversely, if to a 1-combination of T^* we add 4 a's and 5 b's we obtain a 10-combination in which a occurs at least 4 times and b occurs at least 5 times. Thus, the number of 10-combinations in $A_1 \cap A_2$ equals the number of 1-combinations of T^*, so that

$$|A_1 \cap A_2| = \binom{1+3-1}{1} = \binom{3}{1} = 3.$$

We can deduce in a similar way that the number of 10-combinations in $A_1 \cap A_3$ equals the number of 0-combinations in T^*, and that there are no 10-combinations in $A_2 \cap A_3$. Therefore,

$$|A_1 \cap A_3| = \binom{0+3-1}{0} = \binom{2}{0} = 1$$

and

$$|A_2 \cap A_3| = 0.$$

Also,

$$|A_1 \cap A_2 \cap A_3| = 0.$$

Putting all these results into the inclusion–exclusion principle, we obtain

$$\begin{aligned}|\overline{A}_1 \cap \overline{A}_2 \cap \overline{A}_3| &= 66 - (28 + 21 + 15) + (3 + 1 + 0) - 0 \\ &= 6.\end{aligned}$$

Can you list the six 10-combinations? □

    In the proof of Theorem 3.5.1, we have already pointed out the connection between r-combinations and solutions of equations in integers. The number of r-combinations of the multiset $\{n_1 \cdot a_1, n_2 \cdot a_2, \cdots, n_k \cdot a_k\}$ equals the number of integral solutions of the equation

$$x_1 + x_2 + \cdots + x_k = r$$

that satisfy

$$0 \leq x_i \leq n_i \qquad (i = 1, 2, \ldots, k).$$

Thus, the number of these solutions can be calculated by the method just illustrated.

Example. What is the number of integral solutions of the equation

$$x_1 + x_2 + x_3 + x_4 = 18$$

that satisfy

$$1 \leq x_1 \leq 5, \quad -2 \leq x_2 \leq 4, \quad 0 \leq x_3 \leq 5, \quad 3 \leq x_4 \leq 9?$$

We introduce new variables

$$y_1 = x_1 - 1, \ y_2 = x_2 + 2, \ y_3 = x_3, \ \text{and} \ y_4 = x_4 - 3,$$

and our equation becomes

$$y_1 + y_2 + y_3 + y_4 = 16. \tag{6.4}$$

The inequalities on the x_i's are satisfied if and only if

$$0 \leq y_1 \leq 4, \quad 0 \leq y_2 \leq 6, \quad 0 \leq y_3 \leq 5, \quad 0 \leq y_4 \leq 6.$$

    Let S be the set of all nonnegative integral solutions of equation (6.4). The size of S is

$$|S| = \binom{16 + 4 - 1}{16} = \binom{19}{16} = 969.$$

Let P_1 be the property that $y_1 \geq 5$, P_2 the property that $y_2 \geq 7$, P_3 the property that $y_3 \geq 6$, and P_4 the property that $y_4 \geq 7$. Let A_i denote the subset of S consisting of the solutions satisfying property P_i, $(i = 1, 2, 3, 4)$. We wish to evaluate the size of the set $\overline{A}_1 \cap \overline{A}_2 \cap \overline{A}_3 \cap \overline{A}_4$, and we do so by applying the inclusion–exclusion principle. The set

A_1 consists of all those solutions in S for which $y_1 \geq 5$. Performing a change in variable ($z_1 = y_1 - 5, z_2 = y_2, z_3 = y_3, z_4 = y_4$), we see that the number of solutions in A_1 is the same as the number of nonnegative integral solutions of

$$z_1 + z_2 + z_3 + z_4 = 11.$$

Hence,

$$|A_1| = \binom{14}{11} = 364.$$

In a similar way, we obtain

$$|A_2| = \binom{12}{9} = 220, \; |A_3| = \binom{13}{10} = 286, \; |A_4| = \binom{12}{9} = 220.$$

The set $A_1 \cap A_2$ consists of all those solutions in S for which $y_1 \geq 5$ and $y_2 \geq 7$. Performing a change in variable ($u_1 = y_1 - 5, u_2 = y_2 - 7, u_3 = y_3, u_4 = y_4$), we see that the number of solutions in $A_1 \cap A_2$ is the same as the number of nonnegative integral solutions of

$$u_1 + u_2 + u_3 + u_4 = 4.$$

Hence,

$$|A_1 \cap A_2| = \binom{7}{4} = 35.$$

Similarly, we get

$$|A_1 \cap A_3| = \binom{8}{5} = 56, \; |A_1 \cap A_4| = \binom{7}{4} = 35,$$

$$|A_2 \cap A_3| = \binom{6}{3} = 20, \; |A_2 \cap A_4| = \binom{5}{2} = 10,$$

$$\text{and } |A_3 \cap A_4| = \binom{6}{3} = 20.$$

The intersection of any three of the sets A_1, A_2, A_3, A_4 is empty. We now apply the inclusion–exclusion principle to obtain

$$
\begin{aligned}
|\overline{A}_1 \cap \overline{A}_2 \cap \overline{A}_3 \cap \overline{A}_4| &= 969 - (364 + 220 + 286 + 220) \\
&\quad + (35 + 56 + 35 + 20 + 10 + 20) \\
&= 55.
\end{aligned}
$$

□

6.3 Derangements

At a party, 10 gentlemen check their hats. In how many ways can their hats be returned so that no gentleman gets the hat with which he arrived? The 8 spark plugs of a V-8 engine are removed from their cylinders for cleaning. In how many ways can they be returned to the cylinders so that no spark plug goes into the cylinder whence it came? In how many ways can the letters M,A,D,I,S,O,N be written down so that the "word" spelled disagrees completely with the spelling of the word MADISON in the sense that no letter occupies the same position as it does in the word MADISON? Each of these questions is an instance of the following general problem.

We are given an n-element set X in which each element has a specified location, and we are asked to find the number of permutations of the set X in which no element is in its specified location. In the first question, the set X is the set of 10 hats, and the specified location of a hat is (the head of) the gentleman to which it belongs. In the second question, X is the set of spark plugs, and the location of a spark plug is the cylinder which contained it. In the third question, $X = \{M,A,D,I,S,O,N\}$, and the location of a letter is that specified by the word MADISON.

Since the actual nature of the objects is irrelevant, we may take X to be the set $\{1, 2, \ldots, n\}$ in which the location of each of the integers is that specified by its position in the sequence $1, 2, \ldots, n$. A *derangement* of $\{1, 2, \ldots, n\}$ is a permutation $i_1 i_2 \ldots i_n$ of $\{1, 2, \ldots, n\}$ such that $i_1 \neq 1$, $i_2 \neq 2, \ldots, i_n \neq n$. Thus, a derangement of $\{1, 2, \ldots, n\}$ is a permutation $i_1 i_2 \cdots i_n$ of $\{1, 2, \ldots, n\}$ in which no integer is in its natural position:

$$i_1 \neq 1 \quad i_2 \neq 2 \quad \cdots \quad i_n \neq n.$$

We denote by D_n the number of derangements of $\{1, 2, \ldots, n\}$. The preceding questions ask us to evaluate, respectively, D_{10}, D_8, and D_7. For $n = 1$, there are no derangements. The only derangement for $n = 2$ is 2 1. For $n = 3$, there are two derangements, namely, 2 3 1 and 3 1 2. The derangements for $n = 4$ are as follows:

2 1 4 3	3 1 4 2	4 1 2 3
2 3 4 1	3 4 1 2	4 3 1 2
2 4 1 3	3 4 2 1	4 3 2 1.

Thus, we have $D_1 = 0$, $D_2 = 1$, $D_3 = 2$, and $D_4 = 9$.

Theorem 6.3.1 *For $n \geq 1$,*

$$D_n = n! \left(1 - \frac{1}{1!} + \frac{1}{2!} - \frac{1}{3!} + \cdots + (-1)^n \frac{1}{n!} \right).$$

Proof. Let S be the set of all $n!$ permutations of $\{1, 2, \ldots, n\}$. For $j = 1, 2, \ldots, n$, let P_j be the property that, in a permutation, j is in its natural position. Thus, the permutation $i_1 i_2 \cdots i_n$ of $\{1, 2, \ldots, n\}$ has property P_j provided $i_j = j$. A permutation of $\{1, 2, \ldots, n\}$ is a derangement if and only if it has none of the properties $P_1, P_2, \ldots, P_n$. Let A_j denote the set of permutations of $\{1, 2, \ldots, n\}$ with property P_j, $(j = 1, 2, \ldots, n)$. The derangements of $\{1, 2, \ldots, n\}$ are precisely those permutations in $\overline{A}_1 \cap \overline{A}_2 \cap \cdots \cap \overline{A}_n$. Hence,

$$D_n = |\overline{A}_1 \cap \overline{A}_2 \cap \cdots \cap \overline{A}_n|,$$

and we use the inclusion–exclusion principle to evaluate D_n. The permutations in A_1 are of the form $1 i_2 \cdots i_n$, where $i_2 \cdots i_n$ is a permutation of $\{2, \ldots, n\}$. Thus, $|A_1| = (n-1)!$, and more generally we have $|A_j| = (n-1)!$ for $j = 1, 2, \ldots, n$. The permutations in $A_1 \cap A_2$ are of the form $1 \ 2 \ i_3 \cdots i_n$, where $i_3 \cdots i_n$ is a permutation of $\{3, \ldots, n\}$. Therefore, $|A_1 \cap A_2| = (n-2)!$, and more generally we have $|A_i \cap A_j| = (n-2)!$ for any 2-combination $\{i, j\}$ of $\{1, 2, \ldots, n\}$. For any integer k with $1 \leq k \leq n$, the permutations in $A_1 \cap A_2 \cap \cdots \cap A_k$ are of the form $1 \ 2 \cdots k i_{k+1} \cdots i_n$, where $i_{k+1} \cdots i_n$ is a permutation of $\{k+1, \ldots, n\}$. Consequently, $|A_1 \cap A_2 \cap \cdots \cap A_k| = (n-k)!$; more generally,

$$|A_{i_1} \cap A_{i_2} \cap \cdots \cap A_{i_k}| = (n-k)!$$

for any k-combination $\{i_1, i_2, \ldots, i_k\}$ of $\{1, 2, \ldots, n\}$. Since there are $\binom{n}{k}$ k-combinations of $\{1, 2, \ldots, n\}$, applying the inclusion–exclusion principle (see (6.3) at the end of section 6.1), we obtain

$$
\begin{aligned}
D_n &= n! - \binom{n}{1}(n-1)! + \binom{n}{2}(n-2)! - \binom{n}{3}(n-3)! \\
&\quad + \cdots + (-1)^n \binom{n}{n} 0! \\
&= n! - \frac{n!}{1!} + \frac{n!}{2!} - \frac{n!}{3!} + \cdots + (-1)^n \frac{n!}{n!} \\
&= n! \left(1 - \frac{1}{1!} + \frac{1}{2!} - \frac{1}{3!} + \cdots + (-1)^n \frac{1}{n!} \right).
\end{aligned}
$$

Thus, the theorem is proved.   □

We can use the formula obtained to calculate that

$$D_5 = 5!\left(1 - \frac{1}{1!} + \frac{1}{2!} - \frac{1}{3!} + \frac{1}{4!} - \frac{1}{5!}\right) = 44.$$

In a similar way, one can calculate that

$$D_6 = 265, \quad D_7 = 1854, \quad \text{and } D_8 = 14,833.$$

Recalling the series expansion

$$e^{-1} = 1 - \frac{1}{1!} + \frac{1}{2!} - \frac{1}{3!} + \frac{1}{4!} - \cdots,$$

we may write

$$e^{-1} = \frac{D_n}{n!} + (-1)^{n+1}\frac{1}{(n+1)!} + (-1)^{n+2}\frac{1}{(n+2)!} + \cdots.$$

From elementary facts about alternating infinite series, we conclude that e^{-1} and $D_n/n!$ differ by less than $1/(n+1)!$; in fact, D_n is the integer closest to $n!/e$. A calculation shows that, for $n \geq 7$, e^{-1} and $D_n/n!$ agree to at least three decimal places. Thus, from a practical point of view, e^{-1} and $D_n/n!$ are the same for $n \geq 7$. The number $D_n/n!$ is the ratio of the number of derangements of $\{1, 2, \ldots, n\}$ to the total number of permutations of $\{1, 2, \ldots, n\}$. Thus, $D_n/n!$ represents the probability, if we select a permutation of $\{1, 2, \ldots, n\}$ at random, that it is a derangement. In terms of the hat question posed at the beginning of this section, if the hats are returned to the gentlemen at random, the probability that no gentleman receives his own hat is $D_{10}/10!$, and this is effectively e^{-1}. From the preceding remarks, it is apparent that the probability that no gentleman receives his own hat would be essentially the same if the number of gentlemen were 1,000,000.

The derangement numbers D_n satisfy other relations that facilitate their evaluation. The first of these that we discuss is

$$D_n = (n-1)(D_{n-2} + D_{n-1}), \quad (n = 3, 4, 5, \ldots). \tag{6.5}$$

This formula is an example of a linear recurrence relation.[1] Starting with the initial information $D_1 = 0$, $D_2 = 1$, we can use (6.5) to calculate D_n for any positive integer n. For instance,

$$D_3 = 2(D_1 + D_2) = 2(0 + 1) = 2,$$
$$D_4 = 3(D_2 + D_3) = 3(1 + 2) = 9,$$
$$D_5 = 4(D_3 + D_4) = 4(2 + 9) = 44, \text{ and}$$
$$D_6 = 5(D_4 + D_5) = 5(9 + 44) = 265.$$

---

[1]Recurrence relations are taken up in Chapter 7.

In the next chapter we show how to solve linear recurrence relations with constant coefficients. The techniques introduced there will not apply here, however, since the formula (6.5) has a variable coefficient $n - 1$.

We can verify the formula (6.5) combinatorially as follows: Let $n \geq 3$, and consider the D_n derangements of $\{1, 2, \ldots, n\}$. These derangements can be partitioned into $n-1$ parts according to which of the integers $2, 3, \ldots, n$ is in the first position of the permutation. It should be clear that each part contains the same number of derangements. Thus, D_n equals $(n - 1)d_n$, where d_n is the number of derangements in which 2 is in the first position. Such derangements are of the form

$$2i_2i_3 \cdots i_n, \qquad i_2 \neq 2, i_3 \neq 3, \ldots, i_n \neq n.$$

These d_n derangements can be partitioned further into two subparts according as to whether $i_2 = 1$ or $i_2 \neq 1$. Let d'_n be the number of derangements of the form

$$21i_3i_4 \cdots i_n, \qquad i_3 \neq 3, \ldots, i_n \neq n.$$

Let d''_n be the number of derangements of the form

$$2i_2i_3 \cdots i_n, \qquad i_2 \neq 1, i_3 \neq 3, \ldots, i_n \neq n.$$

Then $d_n = d'_n + d''_n$, and it follows that

$$D_n = (n - 1)d_n = (n - 1)(d'_n + d''_n).$$

We first observe that d'_n is the same as the number of permutations $i_3i_4 \cdots i_n$ of $\{3, 4, \ldots, n\}$ in which $i_3 \neq 3, i_4 \neq 4, \ldots, i_n \neq n$. In other words, d'_n is the number of permutations of $\{3, 4, \ldots, n\}$ in which 3 is not in the first position, 4 is not in the second position, and so on. Thus, $d'_n = D_{n-2}$. We next observe that d''_n equals the number of permutations $i_2i_3 \cdots i_n$ of $\{1, 3, \ldots, n\}$ in which 1 is not in the first position, 3 is not in the second position, $\ldots, n$ is not in the $(n - 1)$th position. Hence, $d''_n = D_{n-1}$, and we conclude that

$$D_n = (n - 1)(d'_n + d''_n) = (n - 1)(D_{n-2} + D_{n-1}),$$

which is (6.5).

We can rewrite the formula (6.5) as

$$D_n - nD_{n-1} = -[D_{n-1} - (n - 1)D_{n-2}], \qquad (n \geq 3). \qquad (6.6)$$

The expression in the brackets on the right side is the same as the expression on the left side with n replaced by $n - 1$. Thus we can apply (6.6) recursively[2] to get

$$
\begin{aligned}
D_n - nD_{n-1} &= -[D_{n-1} - (n-1)D_{n-2}] \\
&= (-1)^2[D_{n-2} - (n-2)D_{n-3}] \\
&= (-1)^3[D_{n-3} - (n-3)D_{n-4}] \\
&= \cdots \\
&= (-1)^{n-2}(D_2 - 2D_1).
\end{aligned}
$$

Since $D_2 = 1$ and $D_1 = 0$, we obtain the simpler recurrence relation numbers:

$$
D_n = nD_{n-1} + (-1)^{n-2}
$$

for the derangement numbers, or, equivalently,

$$
D_n = nD_{n-1} + (-1)^n \qquad \text{for} \quad n = 2, 3, 4, \ldots. \tag{6.7}
$$

(Strictly speaking, our verification applies only for $n = 3, 4, \ldots$, but it is simple to check that (6.7) holds also when $n = 2$.) Using (6.7) and the value $D_6 = 265$ previously computed, we see that

$$
D_7 = 7D_6 + (-1)^7 = 7 \times 265 - 1 = 1854.
$$

By repeated application of the formula (6.7), or using it and mathematical induction, we can obtain a different proof of Theorem 6.3.1. (See Exercise 20.) Since (6.7) follows from (6.5), which was given an independent combinatorial proof, this provides a proof of Theorem 6.3.1 that does not use the inclusion–exclusion principle.

The formulas (6.5) and (6.7) are similar to formulas that hold for factorials:

$$
\begin{aligned}
n! &= (n-1)\left((n-2)! + (n-1)!\right), && (n = 3, 4, 5, \ldots) \tag{6.8} \\
n! &= n(n-1)!, && (n = 2, 3, 4, \ldots). \tag{6.9}
\end{aligned}
$$

Example. At a party there are n men and n women. In how many ways can the n women choose male partners for the first dance? How many ways are there for the second dance if everyone has to change partners?

For the first dance there are $n!$ possibilities. For the second dance, each woman has to choose as a partner a man other than the one

---

[2]That is, over and over again, with smaller and smaller values of n.

with whom she first danced. The number of possibilities is the nth derangement number D_n. □

Example. Suppose the n men and the n women at the party check their hats before the dance. At the end of the party their hats are returned randomly. In how many ways can they be returned if each man gets a male hat and each woman gets a female hat, but no one gets the hat he or she checked?

With no restrictions, the hats can be returned in $(2n)!$ ways. With the restriction that each man gets a male hat and each women gets a female hat, there are $n! \times n!$ ways. With the additional restriction that no one gets the correct hat, there are $D_n \times D_n$ ways.

6.4   Permutations with Forbidden Positions

In this section we consider the general problem of counting permutations of $\{1, 2, \ldots, n\}$ with restrictions on which integers can occupy each place of the permutation.

Let

$$X_1, X_2, \ldots, X_n$$

be (possibly empty) subsets of $\{1, 2, \ldots, n\}$. We denote by

$$P(X_1, X_2, \ldots, X_n)$$

the set of all permutations $i_1 i_2 \cdots i_n$ of $\{1, 2, \ldots, n\}$ such that

$$i_1 \text{ is not in } X_1,$$
$$i_2 \text{ is not in } X_2,$$
$$\vdots$$
$$i_n \text{ is not in } X_n.$$

A permutation of $\{1, 2, \ldots, n\}$ belongs to the set $P(X_1, X_2, \ldots, X_n)$ provided that an element of X_1 does not occupy the first place (thus, the only elements that can be in the first place are those in the complement $\overline{X_1}$ of X_1), an element of X_2 does not occupy the second place, ..., and an element of X_n does not occupy the nth place. The number of permutations in $P(X_1, X_2, \ldots, X_n)$ is denoted by

$$p(X_1, X_2, \ldots, X_n) = |P(X_1, X_2, \ldots, X_n)|.$$

Example. Let $n = 4$ and let $X_1 = \{1,2\}$, $X_2 = \{2,3\}$, $X_3 = \{3,4\}$, $X_4 = \{1,4\}$. Then $P(X_1, X_2, X_3, X_4)$ consists of all permutations $i_1 i_2 i_3 i_4$ of $\{1, 2, 3, 4\}$ such that

$$i_1 \neq 1, 2; \; i_2 \neq 2, 3; \; i_3 \neq 3, 4; \; \text{and} \; i_4 \neq 1, 4.$$

The set $P(X_1, X_2, X_3, X_4)$ contains only the two permutations

$$3 \; 4 \; 1 \; 2 \quad \text{and} \quad 4 \; 1 \; 2 \; 3.$$

Thus, $p(X_1, X_2, X_3, X_4) = 2$. □

Example. Let $X_1 = \{1\}, X_2 = \{2\}, \ldots, X_n = \{n\}$. Then the set $P(X_1, X_2, \ldots, X_n)$ equals the set of all permutations $i_1 i_2 \cdots i_n$ of $\{1, 2, \ldots, n\}$ for which $i_1 \neq 1, i_2 \neq 2, \ldots, i_n \neq n$. We conclude that $P(X_1, X_2, \ldots, X_n)$ is the set of derangements of $\{1, 2, \ldots, n\}$, and we have $p(X_1, X_2, \ldots, X_n) = D_n$. □

As seen in Section 3.4 there is a one-to-one correspondence between permutations of $\{1, 2, \ldots, n\}$ and placements of n nonattacking, indistinguishable rooks on an n-by-n board. The permutation $i_1 i_2 \cdots i_n$ of $\{1, 2, \ldots, n\}$ corresponds to the placement of n rooks on the board in the squares with coordinates $(1, i_1), (2, i_2), \ldots, (n, i_n)$. (Recall that the square with coordinates (k, l) is the square occupying the kth row and the lth column of the board.) The permutations in $P(X_1, X_2, \ldots, X_n)$ correspond to placements of n nonattacking rooks on an n-by-n board in which there are certain squares in which it is forbidden to put a rook.

Example. Let $n = 5$ and let $X_1 = \{1, 4\}, X_2 = \{3\}, X_3 = \emptyset, X_4 = \{1, 5\}, X_5 = \{2, 5\}$. Then the permutations in $P(X_1, X_2, X_3, X_4, X_5)$ are in one-to-one correspondence with the placements of 5 nonattacking rooks on the board with forbidden positions as shown.

	1	2	3	4	5
1	×			×	
2			×		
3					
4	×				×
5		×			×

□

Generalizing the derivation of the formula for the number D_n of derangements of $\{1, 2, \ldots, n\}$, we apply the inclusion–exclusion principle to obtain a formula for $p(X_1, X_2, \ldots, X_n)$. However, as we will point out later, this formula is not always of computational value. For convenience, our argument will be couched in the language of nonattacking rooks on an n-by-n board.

Let S be the set of all $n!$ placements of n nonattacking rooks on an n-by-n board. We say that such a placement of n nonattacking rooks satisfies property P_j provided that the rook in the jth row is in a column belonging to X_j, $(j = 1, 2, \ldots, n)$. As usual, A_j denotes the set of rook placements satisfying property P_j, $(j = 1, 2, \ldots, n)$. The set $P(X_1, X_2, \ldots, X_n)$ consists of all the placements of n nonattacking rooks that satisfy none of the properties $P_1, P_2, \ldots, P_n$. Hence,

$$
\begin{aligned}
p(X_1, X_2, \ldots, X_n) &= |\overline{A}_1 \cap \overline{A}_2 \cap \cdots \cap \overline{A}_n| \\
&= n! - \Sigma|A_i| + \Sigma|A_i \cap A_j| \\
&\quad - \cdots + (-1)^k \Sigma|A_{i_1} \cap A_{i_2} \cap \cdots \cap A_{i_k}| \\
&\quad + \cdots + (-1)^n |A_1 \cap A_2 \cap \cdots \cap A_n| \quad (6.10)
\end{aligned}
$$

where the kth summation is over all k-combinations of $\{1, 2, \ldots, n\}$. We now evaluate the n sums in the preceding formula. What does, for instance, $|A_1|$ count? It counts the number of ways to place n nonattacking rooks on the board where the rook in row 1 is in one of the columns in X_1. We can choose the column of that rook in $|X_1|$ ways and then place the remaining $n - 1$ nonattacking rooks in $(n-1)!$ ways. Thus, $|A_1| = |X_1|(n - 1)!$ and, more generally,

$$
|A_i| = |X_i|(n - 1)!, \qquad (i = 1, 2, \ldots, n).
$$

Hence,

$$
\Sigma|A_i| = (|X_1| + |X_2| + \cdots + |X_n|)(n - 1)!
$$

We let $r_1 = |X_1| + |X_2| + \cdots + |X_n|$ and obtain

$$
\Sigma|A_i| = r_1(n - 1)!.
$$

The number r_1 equals the number of forbidden squares of the board. Equivalently, r_1 equals the number of ways to place one rook on the board in a forbidden square.

Now consider $|A_1 \cap A_2|$. This number counts the number of ways to place n nonattacking rooks on the board where the rooks in row 1 and row 2 are both in forbidden positions (in X_1 and X_2, respectively). Each placement of two nonattacking rooks in rows 1 and 2 in forbidden positions can be completed to n nonattacking rooks in $(n - 2)!$

ways. Similar considerations hold for any $|A_i \cap A_j|$, and we obtain the following: Let r_2 equal the number of ways to place two nonattacking rooks on the board in forbidden positions. Then

$$\Sigma|A_i \cap A_j| = r_2(n-2)!.$$

We may directly generalize the preceding argument and evaluate the kth sum in (6.10). We define r_k as follows:

> r_k is the number of ways to place k nonattacking rooks on the n-by-n board where each of the k rooks is in a forbidden position, $(k = 1, 2 \ldots, n)$.

Then

$$\Sigma|A_{i_1} \cap A_{i_2} \cap \cdots \cap A_{i_k}| = r_k(n-k)!, \quad (k = 1, 2, \ldots, n).$$

Substituting this formula into (6.10), we obtain the next theorem.

Theorem 6.4.1 *The number of ways to place n nonattacking, indistinguishable rooks on an n-by-n board with forbidden positions equals*

$$n! - r_1(n-1)! + r_2(n-2)! - \cdots + (-1)^k r_k(n-k)! + \cdots + (-1)^n r_n.$$

$\square$

Example. Determine the number of ways to place 6 nonattacking rooks on the following 6-by-6 board, with forbidden positions as shown.

×					
×	×				
		×	×		
		×	×		

Since r_1 equals the number of forbidden positions, we have $r_1 = 7$. Before evaluating $r_2, r_3, \ldots, r_6$, we note that the set of forbidden positions can be partitioned into two "independent" parts, one part F_1 containing three positions and the other part F_2 containing four. Here by "independent" we mean that squares in different parts do not belong to a common row or column. We now evaluate r_2, the number of ways to place 2 nonattacking rooks in forbidden positions. The rooks may be both in F_1, both in F_2, or one in F_1 and one in F_2. In the last case they are automatically nonattacking because F_1 and F_2 are independent. Counting in this way, we obtain

$$r_2 = 1 + 2 + 3 \times 4 = 15.$$

For r_3 we need two nonattacking rooks in F_1 and one rook in F_2, or one rook in F_1 and two nonattacking rooks in F_2. Thus,

$$r_3 = 1 \times 4 + 3 \times 2 = 10.$$

For r_4 we need two nonattacking rooks in F_1 and two nonattacking rooks in F_2; hence,

$$r_4 = 1 \times 2 = 2.$$

Clearly, $r_5 = r_6 = 0$, and, by Theorem 6.4.1, the number of ways to place six nonattacking rooks on the board so that no rook occupies a forbidden position equals

$$6! - 7 \times 5! + 15 \times 4! - 10 \times 3! + 2 \times 2! = 226.$$

$\square$

In conclusion, we note that the formula in Theorem 6.4.1 is of computational value only if it is easier to evaluate the numbers $r_1, r_2, \ldots, r_n$ than to evaluate directly the number of ways to place n nonattacking rooks on an n-by-n board with forbidden positions. Note that the number r_n equals the number of ways to place n nonattacking rooks on the n-by-n "complementary" board, obtained by interchanging the forbidden and nonforbidden positions. If there are a lot of forbidden squares, then it may be more difficult to evaluate r_n than it is to count directly the number of ways to place n nonattacking rooks on the board.

6.5   Another Forbidden Position Problem

In Sections 6.3 and 6.4 we counted permutations of $\{1, 2, \ldots, n\}$ in which there are certain absolute forbidden positions. We consider in this section a problem of counting permutations in which there are certain *relative* forbidden positions and show how the inclusion–exclusion principle can be used to count the number of these permutations.

We introduce the problem as follows: Suppose a class of 8 boys takes a walk every day. The students walk in a line of 8 so that every boy except the first is preceded by another. In order that a child not see the same person in front of him, on the second day they decide to switch positions so that no boy is preceded by the same boy who preceded him on the first day. In how many ways can they switch positions?

One possibility is to reverse the order of the boys so that the first is now last, and so on, but there are many other possibilities. If we assign to the boys the numbers $1, 2, \ldots, 8$, with the last boy in the column of the first day receiving the number 1, $\ldots$, and the first boy receiving the number 8, as in

$$1 \quad 2 \quad 3 \quad 4 \quad 5 \quad 6 \quad 7 \quad 8,$$

then we are asked to determine the number of permutations of the set $\{1, 2, \ldots, 8\}$ in which the patterns 12, 23, $\ldots$, 78 do not occur. Thus, 31542876 is an allowable permutation, but 84312657 is not. For each positive integer n, we let Q_n denote the number of permutations of $\{1, 2, \ldots, n\}$ in which none of the patterns 12, 23, $\ldots$, $(n-1)n$ occurs. We use the inclusion–exclusion principle to evaluate Q_n. If $n = 1$, 1 is an allowable permutation. If $n = 2$, 21 is an allowable permutation. If $n = 3$, the allowable permutations are 213, 321, and 132, while if $n = 4$, they are as follows:

4 1 3 2	4 3 2 1	4 2 1 3
3 2 1 4	3 2 4 1	2 1 4 3
2 4 3 1	2 4 1 3	3 1 4 2
1 3 2 4	1 4 3 2	

Hence, $Q_1 = 1$, $Q_2 = 1$, $Q_3 = 3$, and $Q_4 = 11$.

Theorem 6.5.1 *For $n \geq 1$,*

$$Q_n = n! - \binom{n-1}{1}(n-1)! + \binom{n-1}{2}(n-2)!$$
$$- \binom{n-1}{3}(n-3)! + \cdots + (-1)^{n-1}\binom{n-1}{n-1}1!.$$

Proof. Let S be the set of all $n!$ permutations of $\{1, 2, \ldots, n\}$. Let P_j be the property that, in a permutation, the pattern $j(j+1)$ does occur, $(j = 1, 2, \ldots, n-1)$. Thus, a permutation of $\{1, 2, \ldots, n\}$ is counted in the number Q_n if and only if it has none of the properties $P_1, P_2, \ldots, P_{n-1}$. As usual, let A_j denote the set of permutations of $\{1, 2, \ldots, n\}$ that satisfy property P_j, $(j = 1, 2, \ldots, n-1)$. Then

$$Q_n = |\bar{A}_1 \cap \bar{A}_2 \cap \cdots \cap \bar{A}_{n-1}|,$$

and we apply the inclusion–exclusion principle to evaluate Q_n. We first calculate the number of permutations in A_1. A permutation is in A_1 if and only if the pattern 12 occurs in it. Thus, a permutation in A_1 may be regarded as a permutation of the $n-1$ symbols $\{12, 3, 4, \ldots, n\}$. We conclude that $|A_1| = (n-1)!$, and in general we see that

$$|A_j| = (n-1)! \qquad (j = 1, 2, \ldots, n-1).$$

Permutations that are in two of the sets $A_1, A_2, \ldots, A_{n-1}$ contain two patterns. These patterns either share an element, such as the patterns 12 and 23, or have no element in common, such as the patterns 12 and 34 . A permutation which contains the two patterns 12 and 34 can be regarded as a permutation of the $n-2$ symbols $\{12, 34, 5, \ldots, n\}$. Thus, $|A_1 \cap A_3| = (n-2)!$. A permutation that contains the two patterns 12 and 23 contains the pattern 123 and thus can be regarded as a permutation of the $n-2$ symbols $\{123, 4, \ldots, n\}$. Hence, $|A_1 \cap A_2| = (n-2)!$. In general, we see that

$$|A_i \cap A_j| = (n-2)!$$

for each 2-combination $\{i, j\}$ of $\{1, 2, \ldots, n-1\}$. More generally, we see that a permutation which contains k specified patterns from the list $12, 23, \ldots, (n-1)n$ can be regarded as a permutation of $n-k$ symbols, and thus that

$$|A_{i_1} \cap A_{i_2} \cap \cdots \cap A_{i_k}| = (n-k)!$$

for each k-combination $\{i_1, i_2, \ldots, i_k\}$ of $\{1, 2, \ldots, n-1\}$. Since, for each $k = 1, 2, \ldots, n-1$, there are $\binom{n-1}{k}$ k-combinations of $\{1, 2, \ldots, n-1\}$, applying the inclusion–exclusion principle we obtain the formula in the theorem. $\qquad\square$

Using the formula of Theorem 6.5.1, we calculate that

$$Q_5 = 5! - \binom{4}{1}4! + \binom{4}{2}3! - \binom{4}{3}2! + \binom{4}{4}1! = 53.$$

The numbers $Q_1, Q_2, Q_3, \ldots$ are closely related to the derangement numbers. Indeed, we have $Q_n = D_n + D_{n-1}$, $(n \geq 2)$. (See Exercise 23.) Thus, knowing the derangement numbers, we can calculate all the numbers Q_n, $(n \geq 2)$. Since we have already seen in the preceding section that $D_5 = 44$, $D_6 = 265$, we conclude that $Q_6 = D_6 + D_5 = 265 + 44 = 309$.

6.6   Möbius Inversion

The inclusion–exclusion principle is an instance of Möbius inversion on a finite[3] partially ordered set. In order to set the stage for the generality of Möbius inversion, we first discuss a somewhat more general version of the inclusion–exclusion principle.

Let n be a positive integer and consider the set $X_n = \{1, 2, \ldots, n\}$ of n elements, and the partially ordered set $(\mathcal{P}(X_n), \subseteq)$ of all subsets of X_n partially ordered by containment. Let

$$F : \mathcal{P}(X_n) \to \Re$$

be a real-valued function defined on $\mathcal{P}(X_n)$. We use F to define a new function

$$G : \mathcal{P}(X_n) \to \Re$$

by

$$G(K) = \sum_{L \subseteq K} F(L), \quad (K \subseteq X_n). \tag{6.11}$$

Möbius inversion allows one to *invert* equation (6.11) and to recover F from G; specifically, we have

$$F(K) = \sum_{L \subseteq K} (-1)^{|K| - |L|} G(L), \quad (K \subseteq X_n). \tag{6.12}$$

Notice that F is obtained from G in (6.12) in a way similar to that in which G is obtained from F in (6.11); the only difference is that in (6.12) we insert in front of each term of the summation either a 1 or -1 depending on whether $|K| - |L|$ is even or odd.

Let $A_1, A_2, \ldots, A_n$ be subsets of a finite set S, and for $K \subseteq X_n$, define $F(K)$ to be the number of elements of S that belong only to those sets A_i with $i \notin K$. Thus, for $s \in S$, s is counted by $F(K)$ if and only if

$$s \notin A_i, \quad \text{for each } i \in K, \text{ and}$$
$$s \in A_j, \quad \text{for each } j \notin K.$$

Then

$$G(K) = \sum_{L \subseteq K} F(L)$$

counts the number of elements that belong to all of the sets A_j with j not in K and possibly other sets as well. Thus,

$$G(K) = |\cap_{i \notin K} A_i|.$$

---

[3] One can replace the property of being finite by a weaker property called locally finite, which asserts that, for all a and b with $a \leq b$, the *interval* $\{x : a \leq x \leq b\}$ is a finite set.

By (6.12),

$$F(K) = \sum_{L \subseteq K} (-1)^{|K|-|L|} G(L). \tag{6.13}$$

Taking $K = X_n$ in (6.13), we get

$$F(X_n) = \sum_{L \subseteq X_n} (-1)^{n-|L|} G(L). \tag{6.14}$$

Now, $F(X_n)$ counts the number of elements of S that belong only to those sets A_i with $i \notin X_n$; that is, $F(X_n)$ is the number of elements of S that belong to none of the sets $A_1, A_2, \ldots, A_n$ and thus equals the number of elements contained in $\overline{A}_1 \cap \overline{A}_2 \cap \cdots \cap \overline{A}_n$. Substituting into (6.14), we obtain

$$|\overline{A}_1 \cap \overline{A}_2 \cap \cdots \cap \overline{A}_n| = \sum_{L \subseteq X_n} (-1)^{n-|L|} |\cap_{i \notin L} A_i|,$$

or equivalently, by replacing L with its complement J in X_n,

$$|\overline{A}_1 \cap \overline{A}_2 \cap \cdots \cap \overline{A}_n| = \sum_{J \subseteq X_n} (-1)^{|J|} |\cap_{i \in J} A_i|. \tag{6.15}$$

Equation (6.15) is equivalent to the formula for the inclusion–exclusion principle as given in Theorem 6.1.1.

We now replace $(\mathcal{P}(X_n), \subseteq)$ with an arbitrary finite partially ordered set $(X, \leq)$. To derive the formula for Möbius inversion, we first consider functions of two variables.

Let $\mathcal{F}(X)$ be the collection of all real-valued functions

$$f : X \times X \to \Re,$$

with the property that $f(x, y) = 0$ whenever $x \not\leq y$. Thus, $f(x, y)$ can be different from 0 only when $x \leq y$. We define the *convolution product* $h = f * g$ of two functions f and g in $\mathcal{F}(X)$ by

$$h(x, y) = \begin{cases} \sum_{\{z : x \leq z \leq y\}} f(x, z) g(z, y), & \text{if } x \leq y, \\ 0, & \text{otherwise.} \end{cases}$$

Thus, in the convolution product, to compute $h(x, y)$ when $x \leq y$, we add up all products $f(x, z) g(z, y)$ as z varies over all elements z between x and y in the given partial order. We leave it as an exercise to verify that the convolution product satisfies the associative law:

$$f * (g * h) = (f * g) * h, \quad (f, g, h \text{ in } \mathcal{F}(X)).$$

There are three special functions in $\mathcal{F}(X)$ of interest to us. The first is the *Kronecker delta function* δ, given by

$$\delta(x, y) = \begin{cases} 1, & \text{if } x = y \\ 0, & \text{otherwise.} \end{cases}$$

Note that $\delta * f = f * \delta = f$ for all functions $f \in \mathcal{F}(X)$, and thus δ acts as an identity function with respect to convolution product. The second is the *zeta function*, ζ defined by

$$\zeta(x, y) = \begin{cases} 1, & \text{if } x \leq y \\ 0, & \text{otherwise.} \end{cases}$$

The zeta function is a representation of the poset $(X, \leq)$ in that it contains all the information about which pairs x, y of elements satisfy $x \leq y$.

Let f be a function in $\mathcal{F}(X)$ such that $f(y, y) \neq 0$ for all y in X. We can inductively define a function g in $\mathcal{F}(X)$ by first letting

$$g(y, y) = \frac{1}{f(y, y)}, \quad (y \in X), \tag{6.16}$$

and then letting

$$g(x, y) = - \sum_{\{z : x \leq z < y\}} g(x, z) \frac{f(z, y)}{f(y, y)}, \quad (x < y). \tag{6.17}$$

From (6.17), we get

$$\sum_{\{z : x \leq z \leq y\}} g(x, z) f(z, y) = \delta(x, y), \quad (x \leq y). \tag{6.18}$$

Equation (6.18) tells us that

$$g * f = \delta,$$

and therefore g is a *left-inverse function* of f with respect to the convolution product $*$. In a similar way, one can show that f has a *right-inverse function* h satisfying

$$f * h = \delta.$$

Using the associative law for convolution product, we get

$$g = g * \delta = g * (f * h) = (g * f) * h = \delta * h = h.$$

Thus, $g = h$ and g is an *inverse function* of f. In sum, every function $f \in \mathcal{F}(X)$ with $f(y, y) \neq 0$ for all y in X has an inverse function g, inductively defined by (6.16) and (6.17), satisfying

$$g * f = f * g = \delta.$$

The third special function we define is the *Möbius function* μ. Since $\zeta(y, y) = 1$ for all $y \in X$, ζ has an inverse, and we define μ to be its inverse. Therefore,

$$\mu * \zeta = \delta,$$

and so, applying (6.18) with $f = \zeta$ and $g = \mu$, we get

$$\sum_{\{z:x \leq z \leq y\}} \mu(x, z) \zeta(z, y) = \delta(x, y), \quad (x \leq y),$$

or equivalently,

$$\sum_{\{z:x \leq z \leq y\}} \mu(x, z) = \delta(x, y), \quad (x \leq y). \tag{6.19}$$

Equation (6.19) implies that

$$\mu(x, x) = 1 \text{ for all } x \tag{6.20}$$

and

$$\mu(x, y) = - \sum_{\{z:x \leq z < y\}} \mu(x, z), \quad (x < y). \tag{6.21}$$

Example. In this example, we compute the Möbius function of the partially ordered set $(\mathcal{P}(X_n), \subseteq)$, where $X_n = \{1, 2, \ldots, n\}$. Let A and B be subsets of X_n with $A \subseteq B$. We prove by induction on $|B| - |A|$ that

$$\mu(A, B) = (-1)^{|B| - |A|}. \tag{6.22}$$

We have from (6.20) that $\mu(A, A) = 1$ and hence (6.22) holds if $B = A$. Suppose that $B \neq A$, and let $p = |B \setminus A| = |B| - |A|$. Then, from (6.21) and the induction hypothesis, we get

$$\mu(A, B) = - \sum_{\{C:A \subseteq C \subset B\}} \mu(A, C) \tag{6.23}$$

$$= - \sum_{\{C:A \subseteq C \subset B\}} (-1)^{|C| - |A|}$$

$$= - \sum_{k=0}^{p-1} (-1)^k \binom{p}{k}. \tag{6.24}$$

The last equality is a consequence of the fact that, for each integer k with $0 \leq k \leq p - 1$, there are as many sets C with $A \subseteq C \subset B$ and $|C| - |A| = k$ as there are subsets of cardinality k contained in the set $B \setminus A$ of cardinality p. Since

$$0 = (1 - 1)^p = \sum_{k=0}^{p} (-1)^k \binom{p}{k},$$

equation (6.24) now implies that

$$\mu(A, B) = (-1)^p \binom{p}{p} = (-1)^p = (-1)^{|B|-|A|}. \tag{6.25}$$

$\square$

Example. In this example we compute the Möbius function of a linearly ordered set. Let $X_n = \{1, 2, \ldots, n\}$ and consider the linearly ordered set $(X_n, \leq)$, where $1 < 2 < \cdots < n$. We have $\mu(k, k) = 1$ for $k = 1, 2, \ldots, n$, and $\mu(k, l) = 0$ for $1 \leq l < k \leq n$. Suppose that $l = k + 1$, where $1 \leq k \leq n - 1$. Then

$$\sum_{\{j : k \leq j \leq k+1\}} \mu(k, j) = 0;$$

hence,

$$\mu(k, k) + \mu(k, k + 1) = 0,$$

and this implies that $\mu(k, k + 1) = -\mu(k, k) = -1$. Now assume that $1 \leq k \leq n - 2$. Then

$$\mu(k, k) + \mu(k, k + 1) + \mu(k, k + 2) = 0;$$

therefore,

$$\mu(k, k + 2) = -(\mu(k, k) + \mu(k, k + 1)) = -(1 + (-1)) = 0.$$

Continuing like this, or using induction, we see that the Möbius function of a linearly ordered set $1 < 2 < \cdots < n$ satisfies

$$\mu(k, l) = \begin{cases} 1, & \text{if } l = k, \\ -1, & \text{if } l = k + 1, \\ 0, & \text{otherwise.} \end{cases}$$

$\square$

We now state and prove the general *Möbius Inversion Formula* for functions defined on a finite partially ordered set. In this theorem, we assume that $(X, \leq)$ has a *smallest element*—that is, an element 0 such that $0 \leq x$ for all $x \in X$. This holds, for instance, for the partially ordered set $(\mathcal{P}(X_n), \subseteq)$, where the smallest element is the empty set $\emptyset$.

Theorem 6.6.1 *Let $(X, \leq)$ be a partially ordered set with a smallest element 0. Let μ be its Möbius function, and let $F : X \to \Re$ be a real-valued function defined on X. Let the function $G : X \to \Re$ be defined by*

$$G(x) = \sum_{\{z:z \leq x\}} F(z), \quad (x \in X).$$

Then

$$F(x) = \sum_{\{y:y \leq x\}} G(y)\mu(y, x), \quad (x \in X).$$

Proof. Let ζ be the zeta function of $(X, \leq)$. Using the properties of ζ and μ previously discussed, we calculate as follows for x an arbitrary element in X:

$$
\begin{aligned}
\sum_{\{y:y \leq x\}} G(y)\mu(y, x) &= \sum_{\{y:y \leq x\}} \sum_{\{z:z \leq y\}} F(z)\mu(y, x) \\
&= \sum_{\{y:y \leq x\}} \mu(y, x) \sum_{\{z:z \in X\}} \zeta(z, y)F(z) \\
&= \sum_{\{z:z \in X\}} \sum_{\{y:y \leq x\}} \zeta(z, y)\mu(y, x)F(z) \\
&= \sum_{\{z:z \in X\}} \left(\sum_{\{y:z \leq y \leq x\}} \zeta(z, y)\mu(y, x) \right) F(z) \\
&= \sum_{\{z:z \in X\}} \delta(z, x)F(z) \\
&= F(x).
\end{aligned}
$$

$\square$

As a corollary, we get the general inclusion–exclusion principle as formulated earlier.

Corollary 6.6.2 *Let $X_n = \{1, 2, \ldots, n\}$ and let $F : \mathcal{P}(X_n) \to \Re$ be a function defined on the subsets of X_n. Let $G : \mathcal{P}(X_n) \to \Re$ be the function defined by*

$$G(K) = \sum_{L \subseteq K} F(L), \quad (K \subseteq X_n).$$

Then

$$F(K) = \sum_{L \subseteq K} (-1)^{|K|-|L|} G(L), \quad (K \subseteq X_n).$$

Proof. The corollary follows from Theorem 6.6.1 and the evaluation of the Möbius function of $(\mathcal{P}(X_n), \subseteq)$ as given in (6.25). $\qquad\square$

Example. We use Möbius inversion to obtain a formula for the number of ways to place n nonattacking rooks on an n-by-n board with forbidden positions, which is different from that given in Theorem 6.4.1. To facilitate our discussion, we now model an n-by-n board as an n-by-n matrix

$$A = [a_{ij} : 1 \leq i, j \leq n]$$

of 0's and 1's. We put a 0 in each position that is forbidden and a 1 in each position that is not. For example, the board

$$(6.26)$$

corresponds to the matrix

$$A = \begin{bmatrix} 0 & 1 & 0 & 1 \\ 1 & 1 & 1 & 0 \\ 1 & 0 & 1 & 1 \\ 1 & 1 & 0 & 1 \end{bmatrix}. \qquad (6.27)$$

A collection of four nonattacking rooks on the board corresponds to a collection of four 1's in A with the property that each row and column contains exactly one of these 1's (equivalently, no repeated 1 in a row or a column). For example, the four 1's

$$a_{14} = 1, a_{23} = 1, a_{31} = 1, \text{ and } a_{42} = 1$$

correspond to four nonattacking rooks in positions

$$(1, 4), (2, 3), (3, 1), (4, 2).$$

These four 1's correspond to the permutation $4, 3, 1, 2$ of $\{1, 2, 3, 4\}$, or equivalently, to the bijection

$$f : \{1, 2, 3, 4\} \to \{1, 2, 3, 4\},$$

with

$$f(1) = 4, f(2) = 3, f(3) = 1, \text{ and } f(4) = 2.$$

Again, let $X_n = \{1, 2, \ldots, n\}$, and let $\mathcal{P}_n$ denote the set of all $n!$ bijections $f : X_n \to X_n$. In general, n nonattacking rooks on an n-by-n

board correspond to n 1's in the matrix with exactly one 1 in each row and in each column. This, in turn, corresponds to a bijection

$$f : \{1, 2, \ldots, n\} \rightarrow \{1, 2, \ldots, n\}$$

in $\mathcal{P}_n$ with $a_{if(i)} = 1$ for $i = 1, 2, \ldots, n$, or equivalently, with

$$\prod_{i=1}^{n} a_{if(i)} = a_{1f(1)}a_{2f(2)} \cdots a_{nf(n)} = 1.$$

Since the only other values of such a product equal 0, we can say that the number of ways to place n nonattacking rooks on an n-by-n board with the associated n-by-n matrix $A = [a_{ij}]$ of 0's and 1's equals

$$\sum_{f \in \mathcal{P}_n} \prod_{i=1}^{n} a_{if(i)}. \tag{6.28}$$

(The expression in (6.28) is an important combinatorial function of a matrix A; it's called the *permanent* of A.)

Consider the partially ordered set $(\mathcal{P}(X_n), \subseteq)$. Each subset S of cardinality k of X_n picks out a set of k columns of A, and we denote the n-by-k submatrix formed by these columns by $A[S]$. Let $\mathcal{F}_n(S)$ denote the set of all functions $f : \{1, 2, \ldots, n\} \rightarrow S$, and let $\mathcal{G}_n(S)$ denote the subset of surjective ones. We then have

$$\mathcal{F}_n(S) = \cup_{T \subseteq S} \mathcal{G}_n(T).$$

Define the function $F : \mathcal{P}(X_n) \rightarrow \Re$ by

$$F(S) = \sum_{f \in \mathcal{G}_n(S)} \prod_{i=1}^{n} a_{if(i)}, \quad (S \subseteq X_n). \tag{6.29}$$

(Here, if $S = \emptyset$, then $F(S) = 0$.) Notice that $F(X_n)$ is equal to (6.28), since a surjective function $f : X_n \rightarrow X_n$ is a bijection. Thus, our goal is to calculate $F(X_n)$.

Let

$$G(S) = \sum_{T \subseteq S} F(T), \quad (S \subseteq X_n).$$

Then

$$G(S) = \sum_{g \in \mathcal{F}_n(S)} \prod_{i=1}^{n} a_{ig(i)}, \quad (S \subseteq X_n).$$

From Corollary 6.6.2, we get

$$F(X_n) = \sum_{S \subseteq X_n} (-1)^{n-|S|} G(S). \tag{6.30}$$

$G(S)$, being the summation of $a_{1g(1)}a_{2g(2)} \cdots a_{ng(n)}$ over *all* functions $g : X_n \to S$, is just the product

$$\prod_{i=1}^{n} \left(\sum_{j \in S} a_{ij} \right) ;$$

that is, $G(S)$ is the product of the sums of the elements in each row of $A[S]$. Thus, (6.30) becomes

$$F(X_n) = \sum_{S \subseteq X_n} (-1)^{n-|S|} \prod_{i=1}^{n} \left(\sum_{j \in S} a_{ij} \right) , \qquad (6.31)$$

and this gives a way to calculate the number of ways to place n nonattacking rooks on an n-by-n board: We pick a set of columns, evaluate the sum of the elements of each row in those columns, multiply these sums together, affix the appropriate sign, and add the results over all choices. The number of summands equals the number of subsets of a set of size n and hence equals 2^n.

Applying formula (6.30) to the board in (6.26) with associated 4-by-4 matrix (6.27), we we get by a tedious calculation that the number of ways to place 4 nonattacking rooks on the board (6.26) equals 6. In this case, with a small $n = 4$, it would be easier to arrive at this number 6 directly, but that's not the point. The point is that we have a way to count that depends only on simple, arithmetical calculations. □

In the next example we will make use of the direct product construction for partially ordered sets (see Exercise 38 of Chapter 4), which we review here. Let $(X, \leq_1)$ and $(Y, \leq_2)$ be partially ordered sets. Define the relation $\leq$ on the set

$$X \times Y = \{(x, y) : x \text{ in } X, y \text{ in } Y\}$$

by

$$(x, y) \leq (x', y') \text{ if and only if } x \leq_1 x' \text{ and } y \leq_2 y'.$$

It is straightforward to check that $(X \times Y, \leq)$ is a partially ordered set, called the *direct product* of $(X, \leq_1)$ with $(Y, \leq_2)$. We may generalize this direct product construction to any number of partially ordered sets.

The next theorem shows how the Möbius function of a direct product is determined from the Möbius functions of its component partially ordered sets.

Theorem 6.6.3 *Let $(X, \leq_1)$ and $(Y, \leq_2)$ be two finite partially ordered sets with Möbius functions μ_1 and μ_2, respectively. Let μ be the Möbius function of the direct product of $(X, \leq_1)$ and $(Y, \leq_2)$. Then*

$$\mu((x,y),(x',y')) = \mu(x,x')\mu(y,y'), \quad ((x,y),(x',y') \text{ in } X \times Y). \quad (6.32)$$

Proof. If $(x,y) \not\leq (y,y')$, then $\mu((x,y),(x',y')) = 0$, and either $x \not\leq_1 y$ or $x' \not\leq y'$, implying that either $\mu_1(x,x') = 0$ or $\mu_2(y,y') = 0$. Hence, (6.32) holds in this case.

Now suppose that $(x,y) \leq (y,y')$. We prove that (6.32) holds by induction on the number of pairs (u,v) that lie between (x,y) and (x',y') in the partial order. We have $x \leq_1 x'$ and $y \leq_2 y'$. If $(x,y) = (x',y')$, then $x = x'$ and $y = y'$ and both sides of (6.32) have value equal to 1. We assume that $(x,y) \neq (x',y')$ and proceed by induction:

$$
\begin{aligned}
\mu((x,y),(x',y')) &= - \sum_{\{(u,v):(x,y)\leq(u,v)<(x',y')\}} \mu((u,v),(x',y')) \\[2mm]
&= - \sum_{\{(u,v):(x,y)\leq(u,v)<(x',y')\}} \mu_1(u,x')\mu_2(v,y') \\
&\qquad \text{(by the inductive assumption)} \\[2mm]
&= - \left(\sum_{\{u:x\leq_1 u\leq_1 x'\}} \mu_1(u,x') \right) \left(\sum_{\{v:y\leq_2 v\leq_2 v'\}} \mu_2(v,y') \right) \\
&\qquad + \mu_1(x,x')\mu_2(y,y') \\[2mm]
&= (0)(0) + \mu_1(x,x')\mu_2(y,y').
\end{aligned}
$$

Thus, the theorem holds by induction. $\qquad\qquad\qquad\qquad\qquad\qquad\square$

We can express Theorem 6.6.3 by saying the Möbius function of the direct product of two partially ordered sets is the product of their Möbius functions. More generally, the Möbius function of the direct product of a finite number of finite partially ordered sets is the product of their Möbius functions.

Example. Let n be a positive integer and again let $X_n = \{1, 2, \ldots, n\}$. We now consider the partially ordered set $D_n = (X_n, \mid)$, where the partial order is that given by divisibility: $a \mid b$ if and only if a is a factor of b. For clarity, we use the divisibility symbol "$\mid$" rather than the general symbol "$\leq$" for a partial order. Our goal is to compute $\mu(1, n)$ for this partially ordered set. From this, one can then compute

$\mu(a, b)$ for any integers a and b in X_n by $\mu(a, b) = \mu(1, \frac{b}{a})$ if $a \mid b$. (See the exercises.)

The integer n has a unique factorization into primes, and thus,

$$n = p_1^{\alpha_1} p_2^{\alpha_2} \cdots p_k^{\alpha_k}$$

where $p_1, p_2, \ldots, p_k$ are distinct primes and $\alpha_1, \alpha_2, \ldots, \alpha_k$ are positive integers.[4] Since $\mu(1, n)$ is given inductively by

$$\mu(1, n) = - \sum_{\{m \geq 1 : m \mid n, m \neq n\}} \mu(1, m),$$

we need consider only $(X_n^*, \mid)$, where X_n^* is the subset of X_n consisting of all positive integers k such that $k \mid n$. Let r and s be integers in X_n^*. We have

$$r = p_1^{\beta_1} p_2^{\beta_2} \cdots p_k^{\beta_k} \text{ and } s = p_1^{\gamma_1} p_2^{\gamma_2} \cdots p_k^{\gamma_k},$$

where $0 \leq \beta_i, \gamma_i \leq \alpha_i, (i = 1, 2, \ldots, k)$.[5] Then $r \mid s$ if and only if $\beta_i \leq \gamma_i, (i = 1, 2, \ldots, k)$. Thus, the partially ordered set $(X_n^*, \mid)$ is just the direct product of k linear orders of sizes $\alpha_1 + 1, \alpha_2 + 1, \ldots, \alpha_k + 1$, respectively. From Theorem 6.6.3, we get

$$\mu(1, n) = \prod_{i=1}^{k} \mu(1, p^{\alpha_i}).$$

From our evaluation of the Möbius function of a linear order, we see that

$$\mu(1, p_i^{\alpha_i}) = \begin{cases} 1, & \text{if } \alpha_i = 0, \\ -1, & \text{if } \alpha_i = 1, \\ 0, & \text{if } \alpha_i \geq 2. \end{cases}$$

Hence,

$$\mu(1, n) = \begin{cases} 1, & \text{if } n = 1, \\ (-1)^k, & \text{if } n \text{ is a product of distinct primes,} \\ 0, & \text{otherwise.} \end{cases} \quad (6.33)$$

$\square$

The next theorem gives the classical Möbius inversion formula.

---

[4]The factorization is unique apart from the order in which the primes are written down.

[5]In order to have the same primes in these factorizations of r and s, we allow some of the exponents to be 0.

Theorem 6.6.4 *Let F be a real-valued function defined on the set of positive integers. Define a real-valued function G on the positive integers by*

$$G(n) = \sum_{k:k\mid n} F(k),$$

Then, for each positive integer n, we have

$$F(n) = \sum_{k:k\mid n} \mu(n/k)G(k),$$

where we write $\mu(n/k)$ for $\mu(1, n/k)$.

Proof. Since, for any fixed n, the definition of $G(n)$ depends only on the values of F on the set $X_n = \{1, 2, \ldots, n\}$, we may confine our attention to the partially ordered set $(X_n, \mid)$. By Theorem 6.6.1, we have

$$F(n) = \sum_{\{k:k\mid n\}} \mu(k,n)G(k) = \sum_{\{k:k\mid n\}} \mu(1, n/k)G(k).$$

$\square$

In the next two examples we apply Theorem 6.6.4 to solve two counting problems.

Example. In this example, we compute the value of the *Euler ϕ function* defined for a positive integer n by $\phi(n) = |S_n|$, where

$$S_n = \{k : 1 \le k \le n, \text{GCD}(k, n) = 1\}.$$

Thus, $\phi(n)$ equals the number of positive integers not exceeding n that are relatively prime to n. For example, $\phi(1) = 1$,

$$\phi(9) = |\{1, 2, 4, 5, 7, 8\}| = 6,$$

and $\phi(13) = 1$ (the value of ϕ at a prime number is always 1). Let

$$S_n^d = \{k : 1 \le k \le n, \text{GCD}(k, n) = d\}, \quad (d \text{ a positive divisor of } n).$$

Then $S_n = S_n^1$. We also have $|S_n^d| = \phi(n/d)$, since any integer k with $\text{GCD}(k, n) = 1$ is of the form $k = dk'$, where $1 \le k' \le n/d$ and $\text{GCD}(k', n/d) = 1$. We take the function F in Möbius inversion to be the Euler ϕ function, and we define

$$G(n) = \sum_{\{d:d\mid n\}} \phi(d).$$

Since $\phi(d)$ equals the number of integers k between 1 and n such that $\mathrm{GCD}(k,n) = d$, and since, for each such integer k, $\mathrm{GCD}(k,n) = d$ for some integer d with $d \mid n$, we conclude that $G(n) = n$. Thus, we have

$$n = \sum_{\{d:d\mid n\}} \phi(d),$$

and inverting this equation, we get

$$\phi(n) = \sum_{\{d:d\mid n\}} \mu(n/d)d = \sum_{\{d:d\mid n\}} \mu(d)\, n/d. \qquad (6.34)$$

Now, $\mu(d)$ is nonzero if and only if $d = 1$ or d is a product of distinct primes; in the latter case, $\mu(d) = (-1)^r$, where r is the number of distinct primes in d. Let the distinct primes dividing n be $p_1, p_2, \ldots, p_r$. Then (6.34) implies that $\phi(n)$ equals

$$n - \left(\frac{n}{p_1} + \frac{n}{p_2} + \cdots\right) + \left(\frac{n}{p_1 p_2} + \frac{n}{p_1 p_3} + \cdots\right) + \cdots +$$
$$(-1)^r \frac{n}{p_1 p_2 \cdots p_r} (-1)^r \frac{n}{p_1 p_2 \cdots p_r},$$

and this is just the product expansion

$$n \prod_{i=1}^{r} \left(1 - \frac{1}{p_i}\right).$$

Thus,

$$\mu(n) = n \prod_{p\mid n} \left(1 - \frac{1}{p}\right),$$

where the product is over all distinct primes p dividing n. □

We conclude this section with an application of classical Möbius inversion.

Example. We count the number of circular n-permutations of k different symbols $a_1, a_2, \ldots, a_k$, where each symbol can be used any number of times, equivalently, the number of circular n-permutations of the multiset $\{n \cdot a_1, n \cdot a_2, \ldots, n \cdot a_k\}$. We define the *period* of such a circular permutation to be the smallest positive number d of clockwise, circular shifts by one position required to leave the circular word unchanged. For example,

$$a_1$$

$$a_2 \qquad\qquad a_2$$

$$a_1$$

has period 2, since

$$
\begin{array}{ccccc}
a_1 & & a_2 & & a_1 \\
a_2 & a_2 \;\rightarrow\; a_1 & & a_1 \;\rightarrow\; a_2 & a_2 \;. \\
a_1 & & a_2 & & a_1
\end{array}
$$

The circular permutation

$$
\begin{array}{ccc}
 & a_1 & \\
a_2 & & a_1 \\
 & a_3 &
\end{array}
$$

has period 4, since we don't return to it until we have made a complete revolution (four position shifts). The period d of a circular n-permutation satisfies $1 \le d \le n$ and $d \mid n$, since period d implies that a particular pattern is repeated n/d times. We can consider a circular permutation as a linear string of symbols in which the first symbol is regarded as following the last symbol. Thus, a_1, a_2, a_1, a_2 corresponds to the first circular permutation just considered. Shifting, we get the string a_2, a_1, a_2, a_1; one more shift gets us back to a_1, a_2, a_1, a_2. The string

$$
a_1, a_2, a_3, a_1, a_2, a_3
$$

corresponds to a circular 6-permutation of period 3. Shifting three times we get

$$
a_1, a_2, a_3, a_1, a_2, a_3 {\rightarrow} a_3, a_1, a_2, a_3, a_1, a_2 {\rightarrow} a_2, a_3, a_1, a_2, a_3, a_1 \rightarrow
$$

$$
a_1, a_2, a_3, a_1, a_2, a_3,
$$

and we are back to the original string for the first time. In general, a circular n-permutation of period d corresponds in this way to exactly d different linear strings, each of period d.

Let $h(n)$ be the number of circular n-words possible using the symbols $a_1, a_2, \ldots, a_k$.[6] For m a positive integer, let $f(m)$ equal the number of strings of length m possible using the symbols $a_1, a_2, \ldots, a_k$. Since each string has a period d, where $d \mid n$, it follows that

$$
h(n) = \sum_{\{d:d\mid n\}} f(d). \tag{6.35}
$$

---

[6] $h(n)$ depends on k but this is not reflected in our notation.

Therefore, if we can calculate the number of strings of length n of each possible period d, then we can calculate $h(n)$. Let

$$g(m) = \sum_{\{e:e|m\}} f(e). \tag{6.36}$$

Then $g(m)$ is the total number of strings of length m, and so $g(m) = k^m$. By classical Möbius inversion, Theorem 6.6.4, we get

$$f(m) = \sum_{\{e:e|m\}} \mu(m/e)g(e) = \sum_{\{e:e|m\}} \mu(m/e)k^e. \tag{6.37}$$

Using (6.37) in (6.35), we obtain

$$
\begin{aligned}
h(n) &= \sum_{\{d:d|n\}} f(d) \\
&= \sum_{\{d:d|n\}} \frac{1}{d} \sum_{\{e:e|d\}} \mu(d/e)k^e \\
&= \sum_{\{e:e|n\}} \left(\sum_{\{m:m|n/e\}} \frac{1}{me}\mu(m) \right) k^e
\end{aligned}
$$

(since $e \mid d$ and $d \mid n$, we have $d = me$
where $me \mid n$ and so $m \mid n/e$)

$$
\begin{aligned}
&= \sum_{\{e:e|n\}} \left(\sum_{\{r:r|n/e\}} \frac{r}{n}\mu((n/e)/r) \right) k^e \\
&= \sum_{\{e:e|n\}} \frac{\phi(n/e)}{n} k^e \\
&= \frac{1}{n} \sum_{\{e:e|n\}} \phi(n/e)k^e.
\end{aligned}
$$

Therefore, the number of circular n-words that can be made from an alphabet of size k equals

$$\frac{1}{n} \sum_{\{e:e|n\}} \phi(n/e)k^e.$$

$\square$

6.7   Exercises

1. Find the number of integers between 1 and 10,000 inclusive that are not divisible by 4, 5, or 6.

2. Find the number of integers between 1 and 10,000 inclusive that are not divisible by 4, 6, 7, or 10.

3. Find the number of integers between 1 and 10,000 that are neither perfect squares nor perfect cubes.

4. Determine the number of 12-combinations of the multiset

$$S = \{4 \cdot a, 3 \cdot b, 4 \cdot c, 5 \cdot d\}.$$

5. Determine the number of 10-combinations of the multiset

$$S = \{\infty \cdot a, 4 \cdot b, 5 \cdot c, 7 \cdot d\}.$$

6. A bakery sells chocolate, cinnamon, and plain doughnuts and at a particular time has 6 chocolate, 6 cinnamon, and 3 plain. If a box contains 12 doughnuts, how many different options are there for a box of doughnuts?

7. Determine the number of solutions of the equation $x_1 + x_2 + x_3 + x_4 = 14$ in nonnegative integers x_1, x_2, x_3, and x_4 not exceeding 8.

8. Determine the number of solutions of the equation $x_1 + x_2 + x_3 + x_4 = 14$ in positive integers x_1, x_2, x_3, and x_4 not exceeding 8.

9. Determine the number of integral solutions of the equation

$$x_1 + x_2 + x_3 + x_4 = 20$$

that satisfy

$$1 \le x_1 \le 6, \ 0 \le x_2 \le 7, \ 4 \le x_3 \le 8, \ 2 \le x_4 \le 6.$$

10. Let S be a multiset with k distinct objects whose repetition numbers are $n_1, n_2, \ldots, n_k$, respectively. Let r be a positive integer such that there is at least one r-combination of S. Show that, in applying the inclusion–exclusion principle to determine the number of r-combinations of S, one has $A_1 \cap A_2 \cap \cdots \cap A_k = \emptyset$.

11. Determine the number of permutations of $\{1, 2, \ldots, 8\}$ in which no even integer is in its natural position.

12. Determine the number of permutations of $\{1, 2, \ldots, 8\}$ in which exactly four integers are in their natural position.

13. Determine the number of permutations of $\{1, 2, \ldots, 9\}$ in which at least one odd integer is in its natural position.

14. Determine a general formula for the number of permutations of the set $\{1, 2, \ldots, n\}$ in which exactly k integers are in their natural positions.

15. At a party seven gentlemen check their hats. In how many ways can their hats be returned so that

    (a) no gentleman receives his own hat?
    (b) at least one of the gentlemen receives his own hat?
    (c) at least two of the gentlemen receive their own hats?

16. Use combinatorial reasoning to derive the identity

$$n! = \binom{n}{0}D_n + \binom{n}{1}D_{n-1} + \binom{n}{2}D_{n-2}$$

$$+ \cdots + \binom{n}{n-1}D_1 + \binom{n}{n}D_0.$$

(Here, D_0 is defined to be 1.)

17. Determine the number of permutations of the multiset

$$S = \{3 \cdot a, 4 \cdot b, 2 \cdot c\},$$

where, for each type of letter, the letters of the same type do not appear consecutively. (Thus *abbbbcaca* is not allowed, but *abbbacacb* is.)

18. Verify the factorial formula

$$n! = (n-1)((n-2)! + (n-1)!), \qquad (n = 2, 3, 4, \ldots).$$

19. Using the evaluation of the derangement numbers as given in Theorem 6.3.1, provide a proof of the relation

$$D_n = (n-1)(D_{n-2} + D_{n-1}), \qquad (n = 3, 4, 5, \ldots).$$

20. Starting from the formula $D_n = nD_{n-1} + (-1)^n$, $(n = 2, 3, 4, \ldots)$, give a proof of Theorem 6.3.1.

21. Prove that D_n is an even number if and only if n is an odd number.

22. Show that the numbers Q_n of Section 6.5 can be rewritten in the form

$$Q_n = (n-1)! \left(n - \frac{n-1}{1!} + \frac{n-2}{2!} - \frac{n-3}{3!} + \cdots + \frac{(-1)^{n-1}}{(n-1)!} \right).$$

23. (Continuation of Exercise 22.) Verify the identity

$$(-1)^k \frac{n-k}{k!} = (-1)^k \frac{n}{k!} + (-1)^{k-1} \frac{1}{(k-1)!},$$

and use it to prove that $Q_n = D_n + D_{n-1}$, $(n = 2, 3, \ldots)$.

24. What is the number of ways to place six nonattacking rooks on the 6-by-6 boards with forbidden positions as shown?

(a)

(b)

(c)

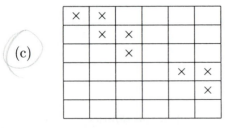

25. Count the permutations $i_1 i_2 i_3 i_4 i_5 i_6$ of $\{1, 2, 3, 4, 5, 6\}$, where $i_1 \neq 1, 5$; $i_3 \neq 2, 3, 5$; $i_4 \neq 4$ and $i_6 \neq 5, 6$.

26. Count the permutations $i_1 i_2 i_3 i_4 i_5 i_6$ of $\{1, 2, 3, 4, 5, 6\}$, where $i_1 \neq 1, 2, 3$; $i_2 \neq 1$; $i_3 \neq 1$; $i_5 \neq 5, 6$ and $i_6 \neq 5, 6$.

27. A carousel has eight seats, each representing a different animal. Eight girls are seated on the carousel facing forward (each girl looks at another girl's back). In how many ways can they change seats so that each has a different girl in front of her? How does the problem change if all the seats are identical?

28. A carousel has eight seats each representing a different animal. Eight boys are seated on the carousel but facing inward, so that each boy faces another (each boy looks at another boy's front). In how many ways can they change seats so that each faces a different boy? How does the problem change if all the seats are identical?

29. How many circular permutations are there of the multiset

$$\{3 \cdot a, 4 \cdot b, 2 \cdot c, 1 \cdot d\},$$

where, for each type of letter, all letters of that type do not appear consecutively?

30. How many circular permutations are there of the multiset

$$\{2 \cdot a, 3 \cdot b, 4 \cdot c, 5 \cdot d\},$$

where, for each type of letter, all letters of that type do not appear consecutively?

31. Let n be a positive integer and let $p_1, p_2, \ldots, p_k$ be all the different prime numbers that divide n. Consider the Euler function ϕ defined by

$$\phi(n) = |\{k : 1 \leq k \leq n, \mathrm{GCD}\{k, n\} = 1\}|.$$

Use the inclusion–exclusion principle to show that

$$\phi(n) = n \prod_{i=1}^{k} (1 - \frac{1}{p_i}).$$

32. * Let n and k be positive integers with $k \leq n$. Let $a(n, k)$ be the number of ways to place k nonattacking rooks on an n-by-n board in which the positions $(1, 1), (2, 2), \ldots, (n, n)$ and

$(1,2), (2,3), \ldots, (n-1, n), (n, 1)$ are forbidden. For example, if $n = 6$ the board is

		×	×	×	×
×			×	×	×
×	×			×	×
×	×	×			×
×	×	×	×		
	×	×	×	×	

Prove that

$$a(n, k) = \frac{2n}{2n - k} \binom{2n - k}{k}.$$

Note that $a(n, k)$ is the number of ways to choose k children from a group of $2n$ children arranged in a circle so that no two consecutive children are chosen.

33. Prove that the convolution product satisfies the associative law: $f * (g * h) = (f * g) * h$.

34. Consider the linearly ordered set $1 < 2 < \cdots < n$. Let $F : \{1, 2, \ldots, n\} \to \Re$ be a function and let $G : \{1, 2, \ldots, n\} \to \Re$ be defined by

$$G(m) = \sum_{k=1}^{m} F(k), \quad (1 \le k \le n).$$

Apply Möbius inversion to get F in terms of G.

35. Consider the board with forbidden positions as shown:

	×	×	
×			
			×
	×		

Use formula (6.31) to compute the number of ways to place 4 nonattacking rooks on this board.

36. Consider the partially ordered set $(\mathcal{P}(X_3), \subseteq)$ of subsets of $\{1, 2, 3\}$ partially ordered by containment. Let a function f in $\mathcal{F}(\mathcal{P}(X))$ be defined by

$$f(A, B) = \begin{cases} 1, & \text{if } A = B, \\ 2, & \text{if } A \subset B \text{ and } |B| - |A| = 1, \\ 1, & \text{if } A \subset B \text{ and } |B| - |A| = 2, \\ -1, & \text{if } A \subset B \text{ and } |B| - |A| = 3. \end{cases}$$

Find the inverse of f with respect to the convolution product.

37. Recall the partially ordered set Π_n of all partitions of $\{1, 2, \ldots n\}$, where the partial order is that of refinement (see Exercise 47 of Chapter 4). Determine the Möbius functions of Π_3 and Π_4.

38. Let n be a positive integer and consider the partially ordered set $(X_n, \mid)$. Let a and b be positive integers in X_n, where $a|b$. Prove that $\mu(a, b) = \mu(1, b/a)$.

39. Consider the multiset $X = \{n_1 \cdot a_1, n_2 \cdot a_2, \ldots, n_k \cdot a_k\}$ of k distinct elements with positive repetition numbers $n_1, n_2, \ldots, n_k$. We introduce a partial order on the submultisets of X by stating the following relationship: If $A = \{p_1 \cdot a_1, p_2 \cdot a_2, \ldots, p_k \cdot a_k\}$ and $B = \{q_1 \cdot a_1, q_2 \cdot a_2, \ldots, q_k \cdot a_k\}$ are submultisets of X, then $A \leq B$ provided that $p_i \leq q_i$ for $i = 1, 2, \ldots, k$. Prove that this statement defines a partial order on X and then compute its Möbius function.

Chapter 7

Recurrence Relations and
Generating Functions

Many combinatorial counting problems depend on an integer parameter n. This parameter n often denotes the size of some underlying set or multiset in the problem, the size of combinations, the number of positions in permutations, and so on. Thus, a counting problem is often not one individual problem but a sequence of individual problems. For example, let h_n denote the number of permutations of $\{1, 2, \ldots, n\}$. We know that $h_n = n!$, and hence we obtain a sequence of numbers

$$h_0, h_1, h_2, \ldots, h_n, \ldots$$

for which the general term h_n equals $n!$. An instance of this problem is obtained by choosing n to be a specific integer. If we take $n = 5$, then we obtain $h_5 = 5!$ as the answer to the problem of determining the number of permutations of $\{1, 2, 3, 4, 5\}$.

As another example, let g_n denote the number of nonnegative integral solutions of the equation

$$x_1 + x_2 + x_3 + x_4 = n.$$

From Chapter 3, we know that the general term of the sequence

$$g_0, g_1, g_2, \ldots, g_n, \ldots$$

satisfies

$$g_n = \binom{n+3}{n}.$$

In this chapter, we develop algebraic methods for solving some counting problems involving an integer parameter n. Our methods lead either to an explicit formula or to a function, a *generating function*, the coefficients of whose power series give the answers to the counting problem.

7.1   Some Number Sequences

Let
$$h_0, h_1, h_2, \ldots, h_n, \ldots \tag{7.1}$$
denote a sequence of numbers. We call h_n the *general term* or *generic term* of the sequence. Two familiar types of sequences are

> *arithmetic sequences*, in which each term is a constant q more than the previous term,

and

> *geometric sequences*, in which each term is a constant multiple q of the previous term.

In both instances, the sequence is uniquely determined once the initial term h_0 and the constant q are specified:

(arithmetic sequence)
$$h_0, h_0 + q, h_0 + 2q, \ldots, h_0 + nq, \ldots. \tag{7.2}$$

(geometric sequence)
$$h_0, qh_0, q^2 h_0, \ldots, q^n h_0, \ldots. \tag{7.3}$$

In the case of an arithmetic sequence, we have the rule
$$h_n = h_{n-1} + q, \qquad (n \geq 1) \tag{7.4}$$
and the general term is
$$h_n = h_0 + nq, \qquad (n \geq 0).$$

In the case of a geometric sequence, we have the rule
$$h_n = qh_{n-1}, \qquad (n \geq 1) \tag{7.5}$$
and the general term is
$$h_n = h_0 q^n, \qquad (n \geq 0).$$

Example. (Arithmetic sequences)

(a) $h_0 = 1$, $q = 2$: $1, 3, 5, \ldots, 1 + 2n, \ldots$.

This is the sequence of odd positive integers.

(b) $h_0 = 4$, $q = 0$: $4, 4, 4, \ldots, 4, \ldots$.

This is the constant sequence with each term equal to 4.

(c) $h_0 = 0$, $q = 1$: $0, 1, 2, \ldots, n, \ldots$.

This is the sequence of nonnegative integers (the counting numbers). □

Example. (Geometric sequences)

(a) $h_0 = 1$, $q = 2$: $1, 2, 2^2, \ldots, 2^n, \ldots$.

This is the sequence of nonnegative integral powers of 2. Its combinatorial significance is that it is the sequence for the counting problem that asks for the number of combinations of an n-element set. It is also the sequence used in order to determine the base 2 representation of a number.

(b) $h_0 = 5$, $q = 3$: $5, 3 \times 5, 3^2 \times 5, \ldots, 3^n \times 5, \ldots$.

This is the sequence for the counting problem that asks for the number of submultisets of the multiset consisting of $n + 1$ different objects whose repetition numbers are given by $4, 2, 2, \ldots, 2$ (n 2's), respectively. □

The *partial sums* for a sequence (7.1) are the sums

$$
\begin{aligned}
s_0 &= h_0 \\
s_1 &= h_0 + h_1 \\
s_2 &= h_0 + h_1 + h_2 \\
&\;\;\vdots \\
s_n &= h_0 + h_1 + h_2 + \cdots + h_n = \textstyle\sum_{k=0}^{n} h_k \\
&\;\;\vdots
\end{aligned}
$$

The partial sums form a new sequence $s_0, s_1, s_2, \ldots, s_n, \ldots$ with general term s_n.

The partial sums for an arithmetic sequence are

$$
s_n = \sum_{k=0}^{n} (h_0 + kq) = (n+1)h_0 + \frac{qn(n+1)}{2}.
$$

The partial sums for a geometric sequence are

$$
s_n = \sum_{k=0}^{n} q^k h_0 = \begin{cases} \frac{q^{n+1}-1}{q-1} h_0 & (q \neq 1) \\[2mm] (n+1)h_0 & (q = 1). \end{cases}
$$

The rules (7.4) and (7.5) for obtaining the next term in either an arithmetic sequence or geometric sequence are simple instances of recurrence relations. In our study of the derangement numbers in Chapter 6 we have obtained two recurrence relations for D_n. In (7.4) and (7.5) the nth term h_n of the sequence is obtained from the $(n-1)$th term h_{n-1} and a constant q. We defer the general definition of a recurrence relation until the next section. We next give an example of a recurrence relation that arises in a geometric counting problem.

Example. Determine the number h_n of regions that are created by n mutually overlapping circles in general position in the plane. By *mutually overlapping* we mean that each two circles intersect in two distinct points (thus nonintersecting or tangent circles are not allowed). By *general position*, we mean that there do not exist three circles with a common point.[1]

We have $h_0 = 1$ (one region that is the entire plane), $h_1 = 2$ (the inside and outside of a circle), $h_2 = 4$, and $h_3 = 8$. It is tempting now to think that $h_4 = 16$. However, a picture quickly reveals that $h_4 = 14$ (see Fig. 7.1).

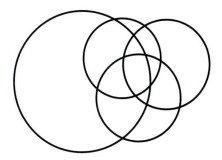

Figure 7.1: Four mutually overlapping circles

We obtain a recurrence relation as follows: Assume that $n \geq 2$ and that $n - 1$ mutually overlapping circles in general position have been drawn in the plane creating h_{n-1} regions. Now put in an nth circle so that there are now n mutually overlapping circles in general position. Each of the first $n - 1$ circles intersects the nth circle in two points, and since the circles are in general position we obtain $2(n - 1)$ distinct points $P_1, P_2, \ldots, P_{2(n-1)}$. These $2(n - 1)$ points divide the nth circle into $2(n - 1)$ arcs: those between P_1 and P_2, between P_2 and $P_3, \ldots$, between $P_{2(n-1)-1}$ and $P_{2(n-1)}$, and between $P_{2(n-1)}$ and P_1. Each of these $2(n-1)$ arcs divides a region formed by the first $n-1$ circles into

---

[1]It is not necessary that the circles be "round". Closed convex curves are sufficient.

two, creating $2(n - 1)$ more regions. Thus, h_n satisfies the recurrence relation

$$h_n = h_{n-1} + 2(n - 1), \qquad (n \geq 2). \tag{7.6}$$

We can use the recurrence relation (7.6) to obtain a formula for h_n in terms of the parameter n. By iterating (7.6),[2] we obtain

$$
\begin{aligned}
h_n &= h_{n-1} + 2(n - 1) \\
&= h_{n-2} + 2(n - 2) + 2(n - 1) \\
&= h_{n-3} + 2(n - 3) + 2(n - 2) + 2(n - 1) \\
&\vdots \\
h_n &= h_1 + 2(1) + 2(2) + \cdots + 2(n - 2) + 2(n - 1).
\end{aligned}
$$

Since $h_1 = 2$, and $1 + 2 + \cdots + (n - 1) = n(n - 1)/2$, we get

$$h_n = 2 + 2\frac{n(n - 1)}{2} = n^2 - n + 2, \qquad (n \geq 2).$$

This formula is also valid for $n = 1$ (since $h_1 = 2$), although it does not hold for $n = 0$ (since $h_0 = 1$). $\square$

The remainder of this section concerns a counting sequence called the *Fibonacci sequence*. In his book *Liber Abaci*,[3] published in 1202, Leonardo of Pisa[4] posed a problem of determining how many pairs of rabbits are born from one pair in a year.

The problem posed by Leonardo [Fibonacci] is the following:

A newly born pair of rabbits of opposite sexes is placed in an enclosure at the beginning of a year. Beginning with the second month, the female gives birth to a pair of rabbits of opposite sexes each month. Each new pair also gives birth to a pair of rabbits each month starting with their second month. Find the number of pairs of rabbits in the enclosure after one year.

In the beginning, there is one pair of rabbits who mature during the first month, so that at the beginning of the second month there is also only one pair of rabbits in the enclosure. During the second month the original pair gives birth to a pair of rabbits, so that there will be two pairs of rabbits at the beginning of the third month. During

---

[2] Applying (7.6) over and over again.

[3] Literally, a book about the abacus.

[4] Leonardo, who is better known by the name Fibonacci (meaning "son of Bonacci"), was largely responsible for the introduction of our present system of numeration in Western Europe.

the third month the newborn pair of rabbits is maturing and only the original pair gives birth. Therefore at the beginning of the fourth month there will be a $2 + 1 = 3$ pairs of rabbits in the enclosure. In general let f_n denote the number of pairs of rabbits in the enclosure at the beginning of month n (equivalently, at the end of month $n - 1$). We have calculated that $f_1 = 1, f_2 = 1, f_3 = 2$, and $f_4 = 3$, and we are asked to find f_{13}.

We derive a recurrence relation for f_n from which we can then easily calculate f_{13}. At the beginning of month n the pairs of rabbits in the enclosure can be partitioned into two parts: those present at the beginning of month $n - 1$ and those born during month $n - 1$. The number of pairs born during month $n - 1$ is, because of the 1 month maturation process, the number of pairs that there were at the beginning of month $n - 2$. Thus, at the beginning of month n, there are $f_{n-1} + f_{n-2}$ pairs of rabbits, giving us the recurrence relation

$$f_n = f_{n-1} + f_{n-2}, \qquad (n \geq 3).$$

Using this relation and the values for f_1, f_2, f_3, and f_4 already computed, we now see that

$$
\begin{aligned}
f_5 &= f_4 + f_3 &&= 3 + 2 &&= 5 \\
f_6 &= f_5 + f_4 &&= 5 + 3 &&= 8 \\
f_7 &= f_6 + f_5 &&= 8 + 5 &&= 13 \\
f_8 &= f_7 + f_6 &&= 13 + 8 &&= 21 \\
f_9 &= f_8 + f_7 &&= 21 + 13 &&= 34 \\
f_{10} &= f_9 + f_8 &&= 34 + 21 &&= 55 \\
f_{11} &= f_{10} + f_9 &&= 55 + 34 &&= 89 \\
f_{12} &= f_{11} + f_{10} &&= 89 + 55 &&= 144 \\
f_{13} &= f_{12} + f_{11} &&= 144 + 89 &&= 233.
\end{aligned}
$$

Consequently, after one year there are 233 pairs of rabbits in the enclosure. We define $f_0 = 0$ so that $f_2 = 1 = 1 + 0 = f_1 + f_0$. The sequence of numbers $f_0, f_1, f_2, f_3, \ldots$ satisfying the recurrence relation and initial conditions

$$f_n = f_{n-1} + f_{n-2} \qquad (n \geq 2)$$
$$f_0 = 0, \quad f_1 = 1 \tag{7.7}$$

is called the *Fibonacci sequence*, and the terms of the sequence are called *Fibonacci numbers*. The recurrence relation in (7.7) is also called the *Fibonacci recurrence*. From our calculations, the first few terms of the Fibonacci sequence are

$$0, 1, 1, 2, 3, 5, 8, 13, 21, 34, 55, 89, 144, 233, \ldots.$$

The Fibonacci sequence has many remarkable properties. We give two in the next two examples.

Example. The partial sums of the terms of the Fibonacci sequence are

$$s_n = f_0 + f_1 + f_2 + \cdots + f_n = f_{n+2} - 1. \tag{7.8}$$

We prove (7.8) by induction on n. For $n = 0$, (7.8) reduces to $f_0 = f_2 - 1$, which is certainly valid since $0 = 1 - 1$.

Now, let $n \geq 1$. We assume that (7.8) holds for n and then prove that it holds when n is replaced by $n + 1$:

$$
\begin{aligned}
f_0 + f_1 + f_2 + \cdots + f_{n+1} &= (f_0 + f_1 + f_2 \cdots + f_n) + f_{n+1} \\
&= (f_{n+2} - 1) + f_{n+1} \\
&\qquad \text{(by the induction assumption)} \\
&= f_{n+2} + f_{n+1} - 1 \\
&= f_{n+3} - 1 \\
&\qquad \text{(by the Fibonacci recurrence).}
\end{aligned}
$$

Thus, (7.8) holds by induction.   □

Example. The Fibonacci number f_n is even if and only if n is divisible by 3.

This certainly agrees with the values for the Fibonacci numbers f_0, f_1, f_2. It follows in general because, if we have

$$\text{even, odd, odd,}$$

then the next three numbers are, because of the Fibonacci recurrence,

$$\text{odd} + \text{odd} = \text{even,}$$

$$\text{odd} + \text{even} = \text{odd,}$$

and

$$\text{even} + \text{odd} = \text{odd.}$$

□

Several other properties of the Fibonacci numbers are given in the exercises.

Our goal now is to obtain a formula for the Fibonacci numbers, and in doing so, we illustrate a technique for solving recurrence relations that we develop in the next section.

Consider the Fibonacci recurrence relation in the form

$$f_n - f_{n-1} - f_{n-2} = 0, \qquad (n \geq 2), \qquad (7.9)$$

and for the moment, ignore any initial values for f_0 and f_1. One way to solve this recurrence relation is to look for a solution of the form

$$f_n = q^n,$$

where q is a nonzero number. Thus, we seek a solution among the familiar geometric sequences with first term equal to $q^0 = 1$. We observe that $f_n = q^n$ satisfies the Fibonacci recurrence relation if and only if

$$q^n - q^{n-1} - q^{n-2} = 0,$$

or, equivalently,

$$q^{n-2}(q^2 - q - 1) = 0, \qquad (n = 2, 3, 4, \ldots).$$

Since q is assumed to be different from zero, we conclude that $f_n = q^n$ is a solution of the Fibonacci recurrence relation if and only if $q^2 - q - 1 = 0$ or, equivalently, if and only if q is a root of the quadratic equation

$$x^2 - x - 1 = 0.$$

Using the quadratic formula, we find that the roots of this equation are

$$q_1 = \frac{1 + \sqrt{5}}{2}, \qquad q_2 = \frac{1 - \sqrt{5}}{2}.$$

Thus,

$$f_n = \left(\frac{1 + \sqrt{5}}{2} \right)^n \quad \text{and} \quad f_n = \left(\frac{1 - \sqrt{5}}{2} \right)$$

are both solutions of the Fibonacci recurrence relation. Since the Fibonacci recurrence relation is linear (there are no powers of f different from 1) and homogeneous (the right-hand side of (7.9) is 0), it follows by straightforward computation that

$$f_n = c_1 \left(\frac{1 + \sqrt{5}}{2} \right)^n + c_2 \left(\frac{1 - \sqrt{5}}{2} \right)^n \qquad (7.10)$$

is also a solution of the recurrence relation (7.9) for any choice of constants c_1 and c_2.

The Fibonacci sequence has the initial values $f_0 = 0$ and $f_1 = 1$. Can we choose c_1 and c_2 in (7.10) so that these initial values are attained? If so, then (7.10) will give a formula for the Fibonacci numbers. To satisfy these initial values, we must have

$$(n = 0) \qquad c_1 + c_2 = 0,$$

$$(n = 1) \qquad c_1 \left(\frac{1+\sqrt{5}}{2} \right) + c_2 \left(\frac{1-\sqrt{5}}{2} \right) = 1.$$

This is a simultaneous system of two linear equations in the unknowns c_1 and c_2, whose unique solution is computed to be

$$c_1 = \frac{1}{\sqrt{5}}, \qquad c_2 = \frac{-1}{\sqrt{5}}.$$

Substituting into (7.10), we obtain the next formula.

Theorem 7.1.1 *The Fibonacci numbers satisfy the formula*

$$f_n = \frac{1}{\sqrt{5}} \left(\frac{1+\sqrt{5}}{2} \right)^n - \frac{1}{\sqrt{5}} \left(\frac{1-\sqrt{5}}{2} \right)^n, \qquad (n \geq 0). \qquad (7.11)$$

Even though the Fibonacci numbers are whole numbers, an explicit formula for them involves the irrational number $\sqrt{5}$. When the binomial theorem is used to expand the nth powers in (7.11), all of the $\sqrt{5}$'s miraculously cancel out.

The solution (7.10) is the general solution of the Fibonacci recurrence relation (7.9) in the sense that no matter what the initial values $f_0 = a$ and $f_1 = b$, constants c_1 and c_2 can be determined so that the initial values hold. This is so because the matrix of coefficients of the linear system

$$c_1 + c_2 \qquad\qquad = a,$$

$$c_1 \left(\frac{1+\sqrt{5}}{2} \right) + c_2 \left(\frac{1-\sqrt{5}}{2} \right) = b$$

is invertible; its determinant,

$$\det \begin{bmatrix} 1 & 1 \\ \frac{1+\sqrt{5}}{2} & \frac{1-\sqrt{5}}{2} \end{bmatrix} = -\sqrt{5},$$

is different from zero. Thus, no matter what the values of a and b, the linear system can be solved uniquely for c_1 and c_2.

Example. Let $g_0, g_1, g_2, \ldots, g_n, \ldots$ be the sequence of numbers satisfying the Fibonacci recurrence relation and the initial conditions as follows:

$$g_n = g_{n-1} + g_{n-2} \quad (n \geq 2)$$
$$g_0 = 2, \quad g_1 = -1.$$

We would like to determine c_1 and c_2 that satisfy

$$
\begin{aligned}
c_1 + c_2 &= 2, \\
c_1 \left(\tfrac{1+\sqrt{5}}{2} \right) + c_2 \left(\tfrac{1-\sqrt{5}}{2} \right) &= -1.
\end{aligned}
$$

Solving this system, we obtain

$$c_1 = \frac{\sqrt{5} - 2}{\sqrt{5}}, \qquad c_2 = \frac{\sqrt{5} + 2}{\sqrt{5}}.$$

Thus, a formula for g_n is

$$g_n = \frac{\sqrt{5} - 2}{\sqrt{5}} \left(\frac{1 + \sqrt{5}}{2} \right)^n + \frac{\sqrt{5} + 2}{\sqrt{5}} \left(\frac{1 - \sqrt{5}}{2} \right)^n.$$

$\square$

The Fibonacci numbers also occur in other combinatorial problems.

Example. Determine the number h_n of ways to perfectly cover a 2-by-n board with dominoes. (See Chapter 1 for a definition of this.)

We define $h_0 = 1$.[5] We also compute that $h_1 = 1$, $h_2 = 2$, and $h_3 = 3$. Let $n \geq 2$. We partition the perfect covers of a 2-by-n board into two parts A and B. In A we put those perfect covers in which there is a vertical domino covering the square in the upper left hand corner. In B we put the other perfect covers, that is, the perfect covers in which there is a horizontal domino covering the square in the upper left hand corner and thus another horizontal domino covering the square in the lower left hand corner. The perfect covers in A are equinumerous with the perfect covers of a 2-by-$(n - 1)$ board. Thus, the number of perfect covers in A is

$$|A| = h_{n-1}.$$

The perfect covers in B are equinumerous with the perfect covers of a 2-by-$(n - 2)$ board, and hence the number of perfect covers in B is

$$|B| = h_{n-2}.$$

---

[5] A 2-by-0 board is empty and has exactly one perfect cover, namely the empty cover.

We conclude that

$$h_n = |A| + |B| = h_{n-1} + h_{n-2}, \qquad (n \geq 2).$$

Since $h_0 = h_1 = 1$ (the values of the Fibonacci numbers f_1 and f_2) and $h_n = h_{n-1} + h_{n-2}$ $(n \geq 2)$ (the Fibonacci recurrence relation), we conclude that $h_0, h_1, h_2, \ldots, h_n, \ldots$ is the Fibonacci sequence $f_1, f_2, \ldots, f_n, \ldots$ with f_0 deleted. $\qquad\square$

Example. Determine the number b_n of ways to perfectly cover a 1-by-n board with monominoes and dominoes.

If we take a perfect cover of a 2-by-n board with dominoes and look only at its first row, we see a perfect cover of a 1-by-n board with monominoes and dominoes. Conversely, each perfect cover of a 1-by-n board with monominoes and dominoes can be "extended" uniquely to a perfect cover of a 2-by-n board with dominoes. Thus, the number of perfect covers of a 1-by-n board with monominoes and dominoes equals the number of perfect covers of a 2-by-n board with dominoes. Therefore, $b_0, b_1, b_2, \ldots, b_n, \ldots$ is also the Fibonacci sequence with f_0 deleted. $\qquad\square$

In the next theorem we show how the Fibonacci numbers occur as sums of binomial coefficients.

Theorem 7.1.2 *The sums of the binomial coefficients along the diagonals of Pascal's triangle running upward from the left are Fibonacci numbers. More precisely, the nth Fibonacci number f_n satisfies*

$$f_n = \binom{n-1}{0} + \binom{n-2}{1} + \binom{n-3}{2} + \cdots + \binom{n-k}{k-1},$$

where $k = \lfloor \frac{n+1}{2} \rfloor$ is the floor of $\frac{n+1}{2}$.

Proof. Define

$$g_n = \binom{n-1}{0} + \binom{n-2}{1} + \cdots + \binom{n-k}{k-1}, \qquad (n \geq 0),$$

where $k = \lfloor \frac{n+1}{2} \rfloor$. Since $\binom{n}{p} = 0$ for each integer $p > n$, we can also write

$$g_n = \binom{n-1}{0} + \binom{n-2}{1} + \binom{n-3}{2} + \cdots + \binom{0}{n-1},$$

or, using summation notation,

$$g_n = \sum_{k=0}^{n-1} \binom{n-1-k}{k}.$$

To prove the theorem, it will suffice to show that g_n satisfies the Fibonacci recurrence relation and has the same initial values as the Fibonacci sequence. We have

$$g_0 = \binom{0}{-1} = 0$$

$$g_1 = \binom{0}{0} = 1.$$

Using Pascal's formula, we see that, for each $n \geq 2$,

$$g_{n-1} + g_{n-2} = \sum_{k=0}^{n-2} \binom{n-2-k}{k} + \sum_{j=0}^{n-3} \binom{n-3-j}{j}$$

$$= \binom{n-2}{0} + \sum_{k=1}^{n-2} \binom{n-2-k}{k} + \sum_{k=1}^{n-2} \binom{n-2-k}{k-1}$$

$$= \binom{n-2}{0} + \sum_{k=1}^{n-2} \left(\binom{n-2-k}{k} + \binom{n-2-k}{k-1} \right)$$

$$= \binom{n-2}{0} + \sum_{k=1}^{n-2} \binom{n-1-k}{k}$$

$$= \binom{n-1}{0} + \sum_{k=1}^{n-2} \binom{n-1-k}{k} + \binom{0}{n-1}$$

$$= \sum_{k=0}^{n-1} \binom{n-1-k}{k} = g_n.$$

Here, we have used the facts that

$$\binom{n-1}{0} = 1 = \binom{n-2}{0} \quad \text{and} \quad \binom{0}{n-1} = 0, \quad (n \geq 2).$$

We conclude that $g_0, g_1, g_2, \ldots, g_n, \ldots$ is the Fibonacci sequence and this proves the theorem. $\square$

7.2   Linear Homogeneous Recurrence Relations

Let
$$h_0, h_1, h_2, \ldots, h_n, \ldots \qquad (7.12)$$
be a sequence of numbers. This sequence is said to satisfy a *linear recurrence relation of order* k, provided that there exist quantities $a_1, a_2, \ldots, a_k$, with $a_k \neq 0$, and a quantity b_n (each of these quantities $a_1, a_2, \ldots, a_k, b_n$ may depend on n) such that
$$h_n = a_1 h_{n-1} + a_2 h_{n-2} + \cdots + a_k h_{n-k} + b_n, \quad (n \geq k). \qquad (7.13)$$

Example. The sequence of derangement numbers
$$D_0, D_1, D_2, \ldots, D_n, \ldots$$
satisfies the two recurrence relations
$$D_n = (n-1)D_{n-1} + (n-1)D_{n-2}, \qquad (n \geq 2)$$
$$D_n = nD_{n-1} + (-1)^n, \qquad (n \geq 1).$$
The first recurrence relation has order 2 and we have $a_1 = n - 1$, $a_2 = n - 1$ and $b_n = 0$. The second recurrence relation has order 1 and we have $a_1 = n$ and $b_n = (-1)^n$. $\square$

Example. The Fibonacci sequence $f_0, f_1, f_2, \ldots, f_n, \ldots$ satisfies the recurrence relation
$$f_n = f_{n-1} + f_{n-2} \qquad (n \geq 2)$$
of order 2 with $a_1 = 1, a_2 = 1$, and $b_n = 0$. $\square$

Example. The factorial sequence $h_0, h_1, h_2, \ldots, h_n, \ldots$, where $h_n = n!$, satisfies the recurrence relation
$$h_n = nh_{n-1} \qquad (n \geq 1)$$
of order 1 with $a_1 = n$ and $b_n = 0$. $\square$

Example. The geometric sequence $h_0, h_1, h_2, \ldots, h_n, \ldots$, where $h_n = q^n$, satisfies the recurrence relation
$$h_n = qh_{n-1} \qquad (n \geq 1)$$
of order 1 with $a_1 = q$ and $b_n = 0$. $\square$

As these examples indicate, the quantities $a_1, a_2, \ldots, a_k$ in (7.13) may be constant or may depend on n. Similarly, the quantity b_n in (7.13) may be a constant (possibly zero) or also may depend on n.

The linear recurrence relation (7.13) is called *homogeneous* provided that b_n is the zero constant and is said to have *constant coefficients* provided that $a_1, a_2, \ldots, a_k$ are constants. In this section, we discuss a special method for solving linear homogeneous recurrence relations with constant coefficients—that is, recurrence relations of the form

$$h_n = a_1 h_{n-1} + a_2 h_{n-2} + \cdots + a_k h_{n-k}, \quad (n \geq k), \tag{7.14}$$

where $a_1, a_2, \ldots, a_k$ are constants and $a_k \neq 0$.[6] The success of the method to be described depends on being able to find the roots of a certain polynomial equation associated with (7.14).

The recurrence relation (7.14) can be rewritten in the form

$$h_n - a_1 h_{n-1} - a_2 h_{n-2} - \cdots - a_k h_{n-k} = 0, \quad (n \geq k). \tag{7.15}$$

A sequence of numbers $h_0, h_1, h_2, \ldots, h_n, \ldots$ satisfying the recurrence relation (7.15) (or more generally, (7.13)) is uniquely determined once the values of $h_0, h_1, \ldots, h_{k-1}$, the so-called *initial values*, are prescribed. The recurrence relation (7.15) "kicks in" beginning with $n = k$. First, we ignore the initial values and look for solutions of (7.15) without prescribed initial values. It turns out that we can find "enough" solutions by considering solutions that form geometric sequences (and by suitably modifying them).

Example.[7] In this example we recall a method for solving linear homogeneous differential equations with constant coefficients. Consider the differential equation

$$y'' - 5y' + 6y = 0. \tag{7.16}$$

Here y is a function of a real variable x. We seek solutions of this equation among the basic exponential functions $y = e^{qx}$. Let q be a constant. Since $y' = qe^{qx}$ and $y'' = q^2 e^{qx}$, it follows that $y = e^{qx}$ is a solution of (7.16) if and only if

$$q^2 e^{qx} - 5qe^{qx} + 6e^{qx} = 0.$$

Since the exponential function e^{qx} is never zero, it may be cancelled and we obtain the following equation, which does not depend on x:

$$q^2 - 5q + 6 = 0.$$

---

[6]If a_k were 0, we would delete the term $a_k h_{n-k}$ from (7.14) and obtain a lower-order recurrence relation.

[7]For those who have not studied differential equations, this example can be omitted.

This equation has two roots, namely, $q = 2$ and $q = 3$. Hence

$$y = e^{2x} \quad \text{and} \quad y = e^{3x}$$

are both solutions of (7.16). Since the differential equation is linear and homogeneous,

$$y = c_1 e^{2x} + c_2 e^{3x} \tag{7.17}$$

is also a solution of (7.16) for any choice of the constants c_1 and c_2.[8] Now we bring in initial conditions for (7.16). These are conditions that prescribe both the value of y and its first derivative when $x = 0$, and with the differential equation (7.16) uniquely determine y. Suppose we prescribe the initial conditions

$$y(0) = a, \quad y'(0) = b \tag{7.18}$$

where a and b are fixed but unspecified numbers. Then, in order that the solution (7.17) of the differential equation (7.16) satisfy these initial conditions, we must have

$$\begin{aligned} y(0) = a : & \quad c_1 + \ c_2 = a \\ y'(0) = b : & \quad 2c_1 + 3c_2 = b. \end{aligned}$$

This system of equations has a unique solution for each choice of a and b, namely,

$$c_1 = 3a - b, \quad c_2 = b - 2a. \tag{7.19}$$

Thus, no matter what the initial conditions (7.18), we can choose c_1 and c_2 using (7.19) so that the function (7.17) is a solution of the differential equation (7.16). In this sense (7.17) is the *general solution* of the differential equation. Each solution of (7.16) with prescribed initial conditions can be written in the form (7.17) for suitable choice of the constants c_1 and c_2. □

The solution of linear homogeneous recurrence relations proceeds along similar lines with the role of the exponential function e^{qx} taken up by the discrete function q^n defined only for nonnegative integers n (the geometric sequences).

Theorem 7.2.1 *Let q be a nonzero number. Then $h_n = q^n$ is a solution of the linear homogeneous recurrence relation*

$$h_n - a_1 h_{n-1} - a_2 h_{n-2} - \cdots - a_k h_{n-k} = 0, \quad (a_k \neq 0, n \geq k) \tag{7.20}$$

---

[8]This can be verified by computing y' and y'' and substituting into (7.16).

with constant coefficients if and only if q is a root of the polynomial equation

$$x^k - a_1 x^{k-1} - a_2 x^{k-2} - \cdots - a_k = 0. \tag{7.21}$$

If the polynomial equation has k distinct roots $q_1, q_2, \ldots, q_k$, then

$$h_n = c_1 q_1^n + c_2 q_2^n + \cdots + c_k q_k^n \tag{7.22}$$

is the general solution of (7.20) in the following sense: No matter what initial values for $h_0, h_1, \ldots, h_{k-1}$ are given, there are constants $c_1, c_2, \ldots, c_k$ so that (7.22) is the unique sequence which satisfies both the recurrence relation (7.20) and the initial conditions.

Proof. We see that $h_n = q^n$ is a solution of (7.20) if and only if

$$q^n - a_1 q^{n-1} - a_2 q^{n-2} - \cdots - a_k q^{n-k} = 0$$

for all $n \geq k$. Since we assume $q \neq 0$, we may cancel q^{n-k}. Thus, these equations (there is one for each $n \geq k$) are equivalent to the *one* equation

$$q^k - a_1 q^{k-1} - a_2 q^{k-2} - \cdots - a_k = 0.$$

We conclude that $h_n = q^n$ is a solution of (7.20) if and only if q is a root of the polynomial equation (7.21).

Since a_k is assumed to be different from zero, 0 is not a root of (7.21). Hence, (7.21) has k roots, $q_1, q_2, \ldots, q_k$ all different from zero. These roots may be complex numbers. In general, $q_1, q_2, \ldots, q_k$ need not be distinct (the equation may have multiple roots), but we now assume that the roots $q_1, q_2, \ldots, q_k$ are distinct. Thus,

$$h_n = q_1^n, \quad h_n = q_2^n, \quad \ldots, \quad h_n = q_k^n$$

are k different solutions of (7.20). The linearity and the homogeneity of the recurrence relation (7.20) imply that, for any choice of constants $c_1, c_2, \ldots, c_k$,

$$h_n = c_1 q_1^n + c_2 q_2^n + \cdots + c_k q_k^n \tag{7.23}$$

is also a solution of (7.20).[9] We now show that (7.23) is the general solution of (7.20) in the sense given in the statement of the theorem.

Suppose we prescribe the initial values

$$h_0 = b_0, \quad h_1 = b_1, \quad \ldots, \quad \text{and } h_{k-1} = b_{k-1}.$$

---

[9]This can be verified by direct substitution.

Can we choose the constants $c_1, c_2, \ldots, c_k$ so that h_n as given in (7.23) satisfies these initial conditions? Equivalently, can we always solve the system of equations

$$
\begin{aligned}
(n = 0) \quad & c_1 + c_2 + \cdots + c_k = b_0 \\
(n = 1) \quad & c_1 q_1 + c_2 q_2 + \cdots + c_k q_k = b_1 \\
(n = 2) \quad & c_1 q_1^2 + c_2 q_2^2 + \cdots + c_k q_k^2 = b_2 \\
& \vdots \\
(n = k-1) \quad & c_1 q_1^{k-1} + c_2 q_2^{k-1} + \cdots + c_k q_k^{k-1} = b_{k-1},
\end{aligned}
\tag{7.24}
$$

no matter what the choice of $b_0, b_1, \ldots, b_{k-1}$?

Now we shall rely on a little bit of linear algebra. The coefficient matrix of this system of equations is

$$
\begin{bmatrix}
1 & 1 & \cdots & 1 \\
q_1 & q_2 & \cdots & q_k \\
q_1^2 & q_2^2 & \cdots & q_k^2 \\
\vdots & \vdots & \ddots & \vdots \\
q_1^{k-1} & q_2^{k-1} & \cdots & q_k^{k-1}
\end{bmatrix}.
\tag{7.25}
$$

The matrix in (7.25) is an important matrix called the *Vandermonde matrix*. The Vandermonde matrix is an invertible matrix if and only if $q_1, q_2, \ldots, q_k$ are distinct. Indeed, its determinant equals

$$
\prod_{1 \le i < j \le k} (q_j - q_i)
$$

and hence is nonzero exactly when $q_1, q_2, \ldots, q_k$ are distinct.[10] Thus, our assumption of the distinctness of $q_1, q_2, \ldots, q_k$ implies that the system (7.24) has a unique solution for each choice of $b_0, b_1, \ldots, b_{k-1}$. Therefore, (7.23) is the general solution of (7.20) and the proof of the theorem is complete. $\square$

The polynomial equation (7.21) is called the *characteristic equation* of the recurrence relation (7.20) and its k roots are the *characteristic roots*. By Theorem 7.2.1, if the characteristic roots are distinct, (7.22) is the general solution of (7.20).

Example. Solve the recurrence relation

$$
h_n = 2h_{n-1} + h_{n-2} - 2h_{n-3}, \quad (n \ge 3),
$$

subject to the initial values $h_0 = 1$, $h_1 = 2$, and $h_2 = 0$.

---

[10]The proof of this fact is nontrivial.

The characteristic equation of this recurrence relation is

$$x^3 - 2x^2 - x + 2 = 0,$$

and its three roots are $1, -1, 2$. By Theorem 7.2.1,

$$h_n = c_1 1^n + c_2(-1)^n + c_3 2^n = c_1 + c_2(-1)^n + c_3 2^n$$

is the general solution. We now want constants c_1, c_2, and c_3 so that

$$\begin{aligned}
(n = 0) \quad & c_1 + c_2 + c_3 = 1, \\
(n = 1) \quad & c_1 - c_2 + 2c_3 = 2, \\
(n = 2) \quad & c_1 + c_2 + 4c_3 = 0.
\end{aligned}$$

The unique solution of this system can be found to be $c_1 = 2, c_2 = -\frac{2}{3}, c_3 = -\frac{1}{3}$. Thus,

$$h_n = 2 - \frac{2}{3}(-1)^n - \frac{1}{3}2^n$$

is the solution of the given recurrence relation. $\qquad\square$

Example. Words of length n, using only the three letters a, b, c, are to be transmitted over a communication channel subject to the condition that no word in which two a's appear consecutively is to be transmitted. Determine the number of words allowed by the communication channel.

Let h_n denote the number of allowed words of length n. We have $h_0 = 1$ (the empty word) and $h_1 = 3$. Let $n \geq 2$. If the first letter of the word is b or c, then the word can be completed in h_{n-1} ways. If the first letter of the word is a, then the second letter is b or c. If the second letter is b, the word can be completed in h_{n-2} ways. If the second letter is c, the word can also be completed in h_{n-2} ways. Hence, h_n satisfies the recurrence relation

$$h_n = 2h_{n-1} + 2h_{n-2}, \qquad (n \geq 2).$$

The characteristic equation is

$$x^2 - 2x - 2 = 0,$$

and the characteristic roots are

$$q_1 = 1 + \sqrt{3}, \qquad q_2 = 1 - \sqrt{3}.$$

Therefore, the general solution is

$$h_n = c_1(1 + \sqrt{3})^n + c_2(1 - \sqrt{3})^n, \qquad (n \geq 3).$$

To determine h_n, we find c_1 and c_2 such that the initial values $h_0 = 1$ and $h_1 = 3$ hold. This leads to the system of equations

$$
\begin{aligned}
(n = 0) \quad & c_1 + c_2 && = 1 \\
(n = 1) \quad & c_1(1 + \sqrt{3}) + c_2(1 - \sqrt{3}) = 3,
\end{aligned}
$$

which has solution

$$
c_1 = \frac{2 + \sqrt{3}}{2\sqrt{3}}, \quad c_2 = \frac{-2 + \sqrt{3}}{2\sqrt{3}}.
$$

Therefore,

$$
h_n = \frac{2 + \sqrt{3}}{2\sqrt{3}}(1 + \sqrt{3})^n + \frac{-2 + \sqrt{3}}{2\sqrt{3}}(1 - \sqrt{3})^n, \quad (n \geq 0)
$$

is the desired solution.   □

If the roots $q_1, q_2, \ldots, q_k$ of the characteristic equation are not distinct, then

$$
h_n = c_1 q_1^n + c_2 q_2^n + \cdots + c_k q_k^n \tag{7.26}
$$

is not a general solution of the recurrence relation.

Example. The recurrence relation

$$
h_n = 4h_{n-1} - 4h_{n-2} \qquad (n \geq 2)
$$

has characteristic equation

$$
x^2 - 4x + 4 = (x - 2)^2 = 0.
$$

Thus, 2 is a twofold characteristic root. In this case, (7.26) becomes

$$
h_n = c_1 2^n + c_2 2^n = (c_1 + c_2)2^n = c2^n
$$

where $c = c_1 + c_2$ is a new constant. Consequently, we have only a single constant to choose in order to satisfy two initial conditions and it is not always possible to do so. For instance, suppose we prescribe the initial values $h_0 = 1$ and $h_1 = 3$. To satisfy these initial values, we must have

$$
\begin{aligned}
(n = 0) \quad & c = 1, \\
(n = 1) \quad & 2c = 3.
\end{aligned}
$$

But these equations are contradictory. Thus, $h_n = c2^n$ is not a general solution of the given recurrence relation.   □

If, as in the preceding example, some characteristic root is repeated, we would like to find another solution associated with that root. The situation is similar to that which occurs in differential equations.

Example. [For those who have studied differential equations.] Solve

$$y'' - 4y' + 4y = 0.$$

We have $y = e^{qx}$ is a solution if and only if

$$q^2 e^{qx} - 4q e^{qx} + 4e^{qx} = 0,$$

or, equivalently,

$$q^2 - 4q + 4 = 0.$$

The roots of this equation are $2, 2$ (2 is a double root) and lead directly to only one solution $y = e^{2x}$. But in this case, $y = xe^{2x}$ is also a solution:

$$y' = 2xe^{2x} + e^{2x}$$

$$y'' = 4xe^{2x} + 2e^{2x} + 2e^{2x} = 4xe^{2x} + 4e^{2x}$$

$$y'' - 4y' + 4y = (4xe^{2x} + 4e^{2x}) - 4(2xe^{2x} + e^{2x}) + 4xe^{2x} = 0.$$

Thus $y = e^{2x}$ and $y = xe^{2x}$ are both solutions of the differential equation, and hence, so is

$$y = c_1 e^{2x} + c_2 x e^{2x}. \tag{7.27}$$

We now verify that (7.27) is the general solution. Suppose we prescribe the initial conditions $y(0) = a$ and $y'(0) = b$. In order for (7.27) to satisfy these initial conditions, we must have

$$y(0) = a : \quad c_1 = a$$

$$y'(0) = b : \quad 2c_1 + c_2 = b.$$

These equations have the unique solution $c_1 = a$ and $c_2 = b - 2a$. Hence, constants c_1 and c_2 can be uniquely chosen to satisfy any given initial conditions, and (7.27) is the general solution. □

Example. Find the general solution of the recurrence relation

$$h_n - 4h_{n-1} + 4h_{n-2} = 0, \quad (n \geq 2).$$

The characteristic equation is

$$x^2 - 4x + 4 = (x - 2)^2 = 0$$

and has roots $2, 2$. We know that $h_n = 2^n$ is a solution of the recurrence relation. We show that $h_n = n2^n$ is also a solution. We have

$$h_n = n2^n, \; h_{n-1} = (n-1)2^{n-1}, \; h_{n-2} = (n-2)2^{n-2};$$

hence,

$$
\begin{aligned}
h_n - 4h_{n-1} + 4h_{n-2} &= n2^n - 4(n-1)2^{n-1} + 4(n-2)2^{n-2} \\
&= 2^{n-2}(4n - 8(n-1) + 4(n-2)) \\
&= 2^{n-2}(0) = 0.
\end{aligned}
$$

We now conclude that

$$h_n = c_1 2^n + c_2 n2^n \tag{7.28}$$

is a solution for each choice of constants c_1 and c_2. Now let us impose the initial conditions

$$h_0 = a \quad \text{and} \quad h_1 = b.$$

In order that these be satisfied, we must have

$$
\begin{aligned}
(n = 0) \quad & c_1 = a \\
(n = 1) \quad & 2c_1 + 2c_2 = b.
\end{aligned}
$$

These equations have the unique solution $c_1 = a$ and $c_2 = (b - 2a)/2$. Hence, constants c_1 and c_2 can be uniquely chosen to satisfy the initial conditions, and we conclude that (7.28) is the general solution of the given recurrence relation.   □

More generally, if a (possibly complex) number q is a root of multiplicity s of the characteristic equation of a linear homogeneous recurrence relation with constant coefficients, then it can be shown that each of

$$h_n = q^n, h_n = nq^n, h_n = n^2 q^n, \dots, h_n = n^{s-1} q^n$$

is a solution, and hence, so is

$$h_n = c_1 q^n + c_2 n q^n + c_2 n^2 q^n + \cdots + c_s n^{s-1} q^n,$$

for each choice of constants $c_1, c_2, \dots, c_s$.

The more general situation in which the characteristic equation has several roots of various multiplicities is treated in the next theorem, which we state without proof.

Theorem 7.2.2 *Let $q_1, q_2, \ldots, q_t$ be the distinct roots of the following characteristic equation of the linear homogeneous recurrence relation with constant coefficients:*

$$h_n = a_1 h_{n-1} + a_2 h_{n-2} + \cdots + a_k h_{n-k}, \quad a_k \neq 0, \quad (n \geq k). \quad (7.29)$$

Then, if q_i is an s_i-fold root of the characteristic equation of (7.29), then the part of the general solution of this recurrence relation corresponding to q_i is

$$
\begin{aligned}
H_n^{(i)} &= c_1 q_i^n + c_2 n q_i^n + \cdots + c_{s_i} n^{s_i-1} q_i^n \\
&= (c_1 + c_2 n + \cdots + c_{s_i} n^{s_i-1}) q_i^n.
\end{aligned}
$$

The general solution of the recurrence relation is

$$h_n = H_n^{(1)} + H_n^{(2)} + \cdots + H_n^{(t)}. \quad (7.30)$$

Example. Solve the recurrence relation

$$h_n = -h_{n-1} + 3h_{n-2} + 5h_{n-3} + 2h_{n-4}, \quad (n \geq 4)$$

subject to the initial values $h_0 = 1, h_1 = 0, h_2 = 1$, and $h_3 = 2$.

The characteristic equation of this recurrence relation is

$$x^4 + x^3 - 3x^2 - 5x - 2 = 0,$$

which has roots $-1, -1, -1, 2$. Thus, the part of the general solution corresponding to the root -1 is

$$H_n^{(1)} = c_1(-1)^n + c_2 n(-1)^n + c_3 n^2(-1)^n,$$

while the part of a general solution corresponding to the root 2 is

$$H_n^{(2)} = c_4 2^n.$$

The general solution is

$$h_n = H_n^{(1)} + H_n^{(2)} = c_1(-1)^n + c_2 n(-1)^n + c_3 n^2(-1)^n + c_4 2^n.$$

We want to determine c_1, c_2, c_3, and c_4 so that the initial conditions hold:

$(n=0)$	c_1			$+ \quad c_4$	$= 1,$
$(n=1)$	$-c_1$	$- \quad c_2$	$- \quad c_3$	$+ \quad 2c_4$	$= 0,$
$(n=2)$	c_1	$+ \quad 2c_2$	$+ \quad 4c_3$	$+ \quad 4c_4$	$= 1,$
$(n=3)$	$-c_1$	$- \quad 3c_2$	$- \quad 9c_3$	$+ \quad 8c_4$	$= 2.$

The unique solution of this system of equations is $c_1 = \frac{7}{9}$, $c_2 = -\frac{3}{9}$, $c_3 = 0$, $c_4 = \frac{2}{9}$. Thus, the solution is

$$h_n = \frac{7}{9}(-1)^n - \frac{3}{9}n(-1)^n + \frac{2}{9}2^n.$$

$\square$

The practical application of the method discussed in this section is limited by the difficulty in finding all the roots of a polynomial equation.

7.3   Nonhomogeneous Recurrence Relations

Recurrence relations that are not homogeneous are, in general, more difficult to solve and require special techniques, depending on the non-homogeneous part of the relation (the term b_n in (7.13)). In this section we consider several examples of linear nonhomogeneous recurrence relations with constant coefficients. Our first example is a famous puzzle.

Example. (*Towers of Hanoi puzzle*). There are three pegs and n circular disks of increasing size on one peg, with the largest disk on the bottom. These disks are to be transferred, one at a time, onto another of the pegs, with the provision that at no time is one allowed to place a larger disk on top of a smaller one. The problem is to determine the number of moves necessary for the transfer.

Let h_n be the number of moves required to transfer n disks. One verifies that $h_0 = 0$, $h_1 = 1$ and $h_2 = 3$. Can we find a recurrence relation that is satisfied by h_n? To transfer n disks to another peg we must first transfer the top $n-1$ disks to a peg, transfer the largest disk to the vacant peg, and then transfer the $n-1$ disks to the peg which now contains the largest disk. Thus, h_n satisfies

$$
\begin{aligned}
h_n &= 2h_{n-1} + 1, && (n \geq 1) \\
h_0 &= 0.
\end{aligned}
\tag{7.31}
$$

This is a linear recurrence relation of order 1 with constant coefficients, but it is not homogeneous because of the presence of the term 1. To find h_n we iterate (7.31):

$$
\begin{aligned}
h_n &= 2h_{n-1} + 1 \\
&= 2(2h_{n-2} + 1) + 1 = 2^2 h_{n-2} + 2 + 1
\end{aligned}
$$

$$= 2^2(2h_{n-3}+1)+2+1 = 2^3 h_{n-3} + 2^2 + 2 + 1$$

$$\vdots$$

$$= 2^{n-1}(h_0+1) + 2^{n-2} + \cdots + 2^2 + 2 + 1$$
$$= 2^{n-1} + \cdots + 2^2 + 2 + 1.$$

Therefore, the numbers h_n are the partial sums of the geometric sequence

$$1, 2, 2^2, \ldots, 2^n, \ldots$$

and hence satisfy

$$h_n = \frac{2^n - 1}{2 - 1} = 2^n - 1, \qquad (n \geq 0). \tag{7.32}$$

Now that we have a formula for h_n, it can easily be verified by mathematical induction and the recurrence relation (7.31). Here is how such a verification goes. Since $h_0 = 0$, (7.32) holds for $n = 0$. Assume that (7.32) holds for n. We then show that it holds with n replaced by $n+1$; that is,

$$h_{n+1} = 2h_n + 1 = 2(2^n - 1) + 1 = 2^{n+1} - 1,$$

proving the formula (7.32).

With only two pegs and $n > 1$ disks, it is impossible to transfer the disks on one peg to the other, subject to the rule that a smaller disk is never below a larger disk. As we have just seen, with three pegs the minimum number of moves is $2^n - 1$. In the case of $k \geq 4$ pegs, it is an unsolved problem to determine the minimum number of moves needed to transfer n disks of different sizes on one peg onto a different peg, again subject to the rule that a smaller disk is never below a larger disk. The case $k = 4$ is sometimes called the *Brahma* or *Reve's Puzzle*, and the puzzle is unsolved even in this case.[11]   □

Our success in the preceding example was made possible by the fact that, after we iterated the recurrence relation, we obtained a sum (in this case $2^{n-1} + \cdots + 2^2 + 2 + 1$) that we could evaluate. A similar situation occurred in Section 7.1 in our determination of the number of regions created by n mutually overlapping circles in general position. However, these are very special situations and iteration of a recurrence relation does not usually lead to a simple formula.

---

[11]There is an algorithm—the Frame-Stewart algorithm—to transfer the n disks whose number of moves is conjectured to be minimal in this case. One can find more information in "Variations on the Four-Post Tower of Hanoi Puzzle" by P.K. Stockmeyer, *Congressus Numerantium*, 102 (1994), 3-12.

We now illustrate a technique for solving linear recurrence relations of order 1 with constant coefficients—that is, recurrence relations of the form

$$h_n = ah_{n-1} + b_n, \qquad (n \geq 1). \qquad (7.33)$$

Example. Solve

$$h_n = 3h_{n-1} - 4n, \qquad (n \geq 1)$$
$$h_0 = 2.$$

We first consider the corresponding homogeneous recurrence relation

$$h_n = 3h_{n-1}, \qquad (n \geq 1).$$

Its characteristic equation is

$$x - 3 = 0,$$

and hence it has one characteristic root $q = 3$, giving the general solution

$$h_n = c3^n, \qquad (n \geq 1). \qquad (7.34)$$

We now seek a particular solution of the nonhomogeneous recurrence relation

$$h_n = 3h_{n-1} - 4n, \qquad (n \geq 1). \qquad (7.35)$$

We try to find a solution of the form

$$h_n = rn + s \qquad (7.36)$$

for appropriate numbers r and s. In order for (7.36) to satisfy (7.35), we must have

$$rn + s = 3(r(n-1) + s) - 4n$$

or, equivalently,

$$rn + s = (3r - 4)n + (-3r + 3s).$$

Equating the coefficients of n and the constant terms on both sides of this equation, we obtain

$$r = 3r - 4 \quad \text{or, equivalently,} \quad 2r = 4$$
$$s = -3r + 3s \quad \text{or, equivalently,} \quad 2s = 3r.$$

Hence, $r = 2$ and $s = 3$, and

$$h_n = 2n + 3 \qquad (7.37)$$

satisfies (7.35). We now combine the general solution (7.34) of the homogeneous relation with the particular solution (7.37) of the nonhomogeneous relation to obtain

$$h_n = c3^n + 2n + 3. \tag{7.38}$$

In (7.38) we have, for each choice of the constant c, a solution of (7.35). Now we try to choose c so that the initial condition $h_0 = 2$ is satisfied:

$$(n = 0) \qquad 2 = c \times 3^0 + 2 \times 0 + 3.$$

This gives $c = -1$, and hence

$$h_n = -3^n + 2n + 3 \qquad (n \geq 0)$$

is the solution of the original problem. □

The preceding technique is the discrete analogue of a technique used to solve nonhomogeneous differential equations. It can be summarized as follows:

(1) Find the general solution of the homogeneous relation.

(2) Find a particular solution of the nonhomogeneous relation.

(3) Combine the general solution and the particular solution, and determine values of the constants arising in the general solution so that the combined solution satisfies the initial conditions.

The main difficulty (besides the difficulty in finding the roots of the characteristic equation) is finding a particular solution in step (2). For some nonhomogeneous parts b_n in (7.33), there are certain types of particular solutions to try.[12] We mention only two:

(a) If b_n is a polynomial of degree k in n, then look for a particular solution h_n that is also a polynomial of degree k in n. Thus, try

$$
\begin{aligned}
&\text{(i)} \quad h_n = r \ (\text{a constant}) && \text{if } b_n = d \ (\text{a constant}), \\
&\text{(ii)} \quad h_n = rn + s && \text{if } b_n = dn + e, \\
&\text{(iii)} \quad h_n = rn^2 + sn + t && \text{if } b_n = fn^2 + dn + e.
\end{aligned}
$$

---

[12]These are solutions to *try*. Whether or not they work depends on the characteristic polynomial.

(b) If b_n is an exponential, then look for a particular solution that is also an exponential. Thus, try

$$h_n = pd^n \quad \text{if} \quad b_n = d^n.$$

The preceding example was of the type (a)(ii) above.

Example. Solve

$$h_n = 2h_{n-1} + 3^n, \qquad (n \geq 1)$$
$$h_0 = 2.$$

Since the homogeneous relation $h_n = 2h_{n-1}$ $(n \geq 1)$ has only one characteristic root $q = 2$, its general solution is

$$h_n = c2^n, \qquad (n \geq 1).$$

For a particular solution of $h_n = 2h_{n-1} + 3^n$ $(n \geq 1)$, we try

$$h_n = p3^n.$$

To be a solution, p must satisfy the equation

$$p3^n = 2p3^{n-1} + 3^n,$$

which, after cancellation, reduces to

$$3p = 2p + 3 \quad \text{or, equivalently,} \quad p = 3.$$

Hence

$$h_n = c2^n + 3^{n+1}$$

is a solution for each choice of the constant c. We now want to determine c so that the initial condition $h_0 = 2$ is satisfied:

$$(n = 0) \qquad c2^0 + 3 = 2.$$

This gives $c = -1$ and the solution of the problem is

$$h_n = -2^n + 3^{n+1}, \qquad (n \geq 0).$$

$\square$

The method just discussed for solving the recurrence relation

$$h_n = ah_{n-1} + b_n, \qquad (n \geq 1)$$

in (7.33) fails in general in case that $a = 1$. In this case, the recurrence relation becomes

$$h_n = h_{n-1} + b_n, \qquad (n \geq 1) \tag{7.39}$$

and iteration yields

$$h_n = h_0 + b_1 + b_2 + \cdots + b_n.$$

Thus, solving (7.39) is the same as summing the series

$$b_1 + b_2 + \cdots + b_n.$$

Example. Solve

$$h_n = h_{n-1} + n^3, \qquad (n \geq 1)$$
$$h_0 = 0.$$

We have, after iteration,

$$h_n = 0^3 + 1^3 + 2^3 + \cdots + n^3,$$

the sum of the cubes of the first n positive integers.[13] We calculate that

$$
\begin{array}{rclclclcl}
h_0 &=& 0^3 &=& 0 &=& 0^2 &=& 0^2 \\
h_1 &=& 0 + 1^3 &=& 1 &=& 1^2 &=& (0+1)^2 \\
h_2 &=& 1 + 2^3 &=& 9 &=& 3^2 &=& (0+1+2)^2 \\
h_3 &=& 9 + 3^3 &=& 36 &=& 6^2 &=& (0+1+2+3)^2 \\
h_4 &=& 36 + 4^3 &=& 100 &=& 10^2 &=& (0+1+2+3+4)^2.
\end{array}
$$

A reasonable conjecture is that

$$
\begin{aligned}
h_n &= (0+1+2+3+\cdots+n)^2 = \left(\frac{n(n+1)}{2}\right)^2 \\
&= \frac{n^2(n+1)^2}{4}.
\end{aligned}
$$

This formula can now be verified by induction on n as follows: Assuming that it holds for an integer n, we show that it also holds for $n + 1$:

$$
\begin{aligned}
h_{n+1} &= h_n + (n+1)^3 \\
&= \frac{n^2(n+1)^2}{4} + (n+1)^3 \\
&= \frac{(n+1)^2(n^2 + 4(n+1))}{4} \\
&= \frac{(n+1)^2(n+2)^2}{4}.
\end{aligned}
$$

---

[13] In the next chapter we shall see how to sum the kth powers of the first n positive integers for any k.

The latter is the formula with n replaced by $n + 1$. Therefore, by mathematical induction,

$$h_n = \frac{n^2(n+1)^2}{4}, \qquad (n \geq 0). \qquad \square$$

Example. Solve

$$h_n = 3h_{n-1} + 3^n, \qquad (n \geq 1)$$
$$h_0 = 2.$$

The general solution of the corresponding homogeneous relation is

$$h_n = c3^n.$$

We first try

$$h_n = p3^n$$

as a particular solution. Substituting, we get

$$p3^n = 3p3^{n-1} + 3^n,$$

which, after cancellation, gives

$$p = p + 1,$$

an impossibility. So instead we try, as a particular solution,

$$h_n = pn3^n.$$

Substituting, we now get

$$pn3^n = 3p(n-1)3^{n-1} + 3^n,$$

which, after cancellation, gives $p = 1$. Thus, $h_n = n3^n$ is a particular solution, and

$$h_n = c3^n + n3^n$$

is a solution for each choice of the constant c. To satisfy the initial condition $h_0 = 2$, we must choose c so that

$$(n = 0) \qquad c(3^0) + 0(3^0) = 2,$$

and this gives $c = 2$. Therefore,

$$h_n = 2 \times 3^n + n3^n = (2 + n)3^n$$

is the solution. $\qquad \square$

7.4   Generating Functions

In this section we discuss the method of generating functions as it pertains to solving counting problems. On one level, generating functions can be regarded as algebraic objects whose formal manipulation allows one to count the number of possibilities for a problem by means of algebra. On another level, generating functions are Taylor series (power series expansions) of infinitely differentiable functions. If we can find the function and its Taylor series, then the coefficients of the Taylor series give the solution to the problem. For the most part we keep questions of convergence in the background and manipulate power series on a formal basis.

Let

$$h_0, h_1, h_2, \ldots, h_n, \ldots \tag{7.40}$$

be an infinite sequence of numbers. Its *generating function* is defined to be the infinite series

$$g(x) = h_0 + h_1 x + h_2 x^2 + \cdots + h_n x^n + \cdots . \tag{7.41}$$

The coefficient of x^n in $g(x)$ is the nth term h_n of (7.40); thus, x^n acts as a placeholder for h_n. A finite sequence

$$h_0, h_1, h_2 \ldots, h_m$$

can be regarded as the infinite sequence

$$h_0, h_1, h_2, \ldots, h_m, 0, 0, \ldots$$

in which all but a finite number of terms equal 0. Hence, every finite sequence has a generating function

$$g(x) = h_0 + h_1 x + h_2 x^2 + \cdots + h_m x^m,$$

which is a polynomial.

Example. The generating function of the infinite sequence

$$1, 1, 1, \ldots, 1, \ldots,$$

each of whose terms equals 1, is

$$g(x) = 1 + x + x^2 + \cdots + x^n + \cdots .$$

This generating function $g(x)$ is the sum of a geometric series[14] with value

$$g(x) = \frac{1}{1 - x}. \tag{7.42}$$

---

[14]See Section 5.6.

The formula (7.42) holds the information about the infinite sequence of all 1's in exceedingly compact form!   □

Example. Let m be a positive integer. The generating function for the binomial coefficients

$$\binom{m}{0}, \binom{m}{1}, \binom{m}{2}, \ldots, \binom{m}{m}$$

is

$$g_m(x) = \binom{m}{0} + \binom{m}{1}x + \binom{m}{2}x^2 + \cdots + \binom{m}{m}x^m.$$

By the binomial theorem,

$$g_m(x) = (1+x)^m,$$

which also displays the information about the sequence of binomial coefficients in compact form.   □

Example. Let α be a real number. By Newton's binomial theorem of Section 5.6, the generating function for the infinite sequence of binomial coefficients

$$\binom{\alpha}{0}, \binom{\alpha}{1}, \binom{\alpha}{2}, \ldots, \binom{\alpha}{n}, \ldots$$

is

$$(1+x)^\alpha = \binom{\alpha}{0} + \binom{\alpha}{1}x + \binom{\alpha}{2}x^2 + \cdots + \binom{\alpha}{n}x^n + \cdots.$$

□

Example. Let k be an integer, and let the sequence

$$h_0, h_1, h_2, \ldots, h_n, \ldots$$

be defined by letting h_n equal the number of nonnegative integral solutions of

$$e_1 + e_2 + \cdots + e_k = n.$$

From Chapter 3, we know that

$$h_n = \binom{n+k-1}{n}, \qquad (n \geq 0).$$

The generating function (using summation notation now) is

$$g(x) = \sum_{n=0}^{\infty} \binom{n+k-1}{n}x^n.$$

From Chapter 5, we know that this generating function is

$$g(x) = \frac{1}{(1-x)^k}.$$

It is instructive to recall the derivation of this formula. We have

$$\frac{1}{(1-x)^k} = \frac{1}{1-x} \times \frac{1}{1-x} \times \cdots \times \frac{1}{1-x} \qquad (k \text{ factors})$$

$$= (1 + x + x^2 + \cdots)(1 + x + x^2 + \cdots) \cdots (1 + x + x^2 + \cdots)$$

$$= \left(\sum_{e_1=0}^{\infty} x^{e_1} \right) \left(\sum_{e_2=0}^{\infty} x^{e_2} \right) \cdots \left(\sum_{e_k=0}^{\infty} x^{e_k} \right). \qquad (7.43)$$

In the preceding notation, x^{e_1} is a typical term of the first factor, x^{e_2} is a typical term of the second factor, ... , x^{e_k} is a typical term of the kth factor. Multiplying these typical terms, we get

$$x^{e_1} x^{e_2} \cdots x^{e_k} = x^n, \text{ provided that}$$

$$e_1 + e_2 + \cdots + e_k = n. \qquad (7.44)$$

Thus, the coefficient of x^n in (7.43) equals the number of nonnegative integral solutions of (7.44), and this number we know to be

$$\binom{n+k-1}{n}.$$

□

The ideas used in the previous example apply to more general circumstances.

Example. For what sequence is

$$(1 + x + x^2 + x^3 + x^4 + x^5)(1 + x + x^2)(1 + x + x^2 + x^3 + x^4)$$

the generating function?

Let x^{e_1}, $(0 \le e_1 \le 5)$, x^{e_2}, $(0 \le e_2 \le 2)$, and x^{e_3}, $(0 \le e_3 \le 4)$ denote typical terms in the first, second, and third factors, respectively. Multiplying we obtain

$$x^{e_1} x^{e_2} x^{e_3} = x^n,$$

provided that

$$e_1 + e_2 + e_3 = n.$$

Thus, the coefficient of x^n in the product is the number h_n of integral solutions of $e_1 + e_2 + e_3 = n$ in which $0 \le e_1 \le 5$, $0 \le e_2 \le 2$, and $0 \le e_3 \le 4$. Note that $h_n = 0$ if $n > 5 + 2 + 4 = 11$.   □

Example. Determine the generating function for the number of n-combinations of apples, bananas, oranges, and pears where, in each n-combination, the number of apples is even, the number of bananas is odd, the number of oranges is between 0 and 4, and there is at least one pear.

First, we note that the problem is equivalent to finding the number h_n of nonnegative integral solutions of

$$e_1 + e_2 + e_3 + e_4 = n,$$

where e_1 is even (e_1 counts the number of apples), e_2 is odd (e_2 counts the number of bananas), $0 \le e_3 \le 4$ (e_3 counts the number of oranges), and $e_4 \ge 1$ (e_4 counts the number of pears). We create one factor for each type of fruit, where the exponents are the allowable numbers in the n-combinations for that type of fruit:

$$g(x) =$$

$$(1+x^2+x^4+\cdots)(x+x^3+x^5+\cdots)(1+x+x^2+x^3+x^4)(x+x^2+x^3++\cdots).$$

The first factor is the "apple factor," the second is the "banana factor," and so on. We now notice that

$$1 + x^2 + x^4 + \cdots = 1 + x^2 + (x^2)^2 + \cdots = \frac{1}{1-x^2}$$

$$x + x^3 + x^5 + \cdots = x(1 + x^2 + x^4 + \cdots) = \frac{x}{1-x^2}$$

$$1 + x + x^2 + x^3 + x^4 = \frac{1-x^5}{1-x}$$

$$x + x^2 + x^3 + \cdots = x(1 + x + x^2 + \cdots)$$
$$= \frac{x}{1-x}.$$

Thus,

$$g(x) = \frac{1}{1-x^2}\frac{x}{1-x^2}\frac{1-x^5}{1-x}\frac{x}{1-x}$$

$$= \frac{x^2(1-x^5)}{(1-x^2)^2(1-x)^2}.$$

Therefore, the coefficients in the Taylor series for this rational function count the number of combinations of the type considered! □

The next example shows how a counting problem can sometimes be explicitly solved by means of generating functions.

Example. Find the number h_n of bags of fruit that can be made out of apples, bananas, oranges, and pears where, in each bag, the number of apples is even, the number of bananas is a multiple of 5, the number of oranges is at most 4, and the number of pears is 0 or 1.

We are asked to count certain n-combinations of apples, bananas, oranges, and pears. We determine the generating function $g(x)$ for the sequence $h_0, h_1, h_2, \ldots, h_n, \ldots$. We introduce a factor for each type of fruit, and we find that

$$
\begin{aligned}
g(x) &= (1 + x^2 + x^4 + \cdots)(1 + x^5 + x^{10} + \cdots) \times \\
&\qquad (1 + x + x^2 + x^3 + x^4)(1 + x)
\end{aligned}
$$

$$
= \frac{1}{1 - x^2} \frac{1}{1 - x^5} \frac{1 - x^5}{1 - x}(1 + x)
$$

$$
= \frac{1}{(1 - x)^2} = \sum_{n=0}^{\infty} \binom{n + 1}{n} x^n
$$

$$
= \sum_{n=0}^{\infty} (n + 1) x^n.
$$

Thus, we see that $h_n = n + 1$. Notice how this formula for the counting number h_n was obtained merely by algebraic manipulation. □

Example. Determine the generating function for the number h_n of solutions of the equation

$$
e_1 + e_2 + \cdots + e_k = n
$$

in nonnegative *odd* integers $e_1, e_2, \ldots, e_k$.

We have

$$
\begin{aligned}
g(x) &= (x + x^3 + x^5 + \cdots) \cdots (x + x^3 + x^5 + \cdots) \quad (k \text{ factors}) \\
&= x(1 + x^2 + x^4 + \cdots) \cdots x(1 + x^2 + x^4 + \cdots) \\
&= \frac{x}{1 - x^2} \cdots \frac{x}{1 - x^2} \\
&= \frac{x^k}{(1 - x^2)^k}.
\end{aligned}
$$

□

We know that the number h_n of nonnegative integral solutions of the equation

$$e_1 + e_2 + \cdots + e_k = n \qquad (7.45)$$

is

$$h_n = \binom{n + k - 1}{n},$$

and we have determined that

$$g(x) = \frac{1}{(1 - x)^k}$$

is its generating function. It is much more difficult to determine an explicit formula for the number of nonnegative integral solutions of an equation obtained from (7.45) by putting arbitrary positive integral coefficients in front of the e_i. Nevertheless, the generating function for the number of solutions is readily obtained, using the ideas we have already discussed. We illustrate with the next example.

Example. Let h_n denote the number of nonnegative integral solutions of the equation

$$3e_1 + 4e_2 + 2e_3 + 5e_4 = n. \qquad (7.46)$$

Find the generating function $g(x)$ for $h_0, h_1, h_2, \ldots, h_n, \ldots$.

We introduce a change of variable by letting

$$f_1 = 3e_1, \quad f_2 = 4e_2, \quad f_3 = 2e_3, \quad \text{and} \quad f_4 = 5e_4.$$

Then h_n also equals the number of nonnegative integral solutions of

$$f_1 + f_2 + f_3 + f_4 = n,$$

where f_1 is a multiple of 3, f_2 is a multiple of 4, f_3 is even, and f_4 is a multiple of 5. Equivalently, h_n is the number of n-combinations of apples, bananas, oranges, and pears in which the number of apples is a multiple of 3, the number of bananas is a multiple of 4, the number of oranges is even, and the number of pears is a multiple of 5. Hence,

$$
\begin{aligned}
g(x) &= (1 + x^3 + x^6 + \cdots)(1 + x^4 + x^8 + \cdots) \times \\
&\qquad (1 + x^2 + x^4 + \cdots)(1 + x^5 + x^{10} + \cdots) \\[2mm]
&= \frac{1}{1 - x^3} \frac{1}{1 - x^4} \frac{1}{1 - x^2} \frac{1}{1 - x^5}.
\end{aligned}
$$

□

We conclude this section with the following change-making example:

Example. There is available an unlimited number of pennies, nickels, dimes, quarters, and half-dollar pieces. Determine the generating function $g(x)$ for the number h_n of ways of making n cents with these pieces.

The number h_n equals the number of nonnegative integral solutions of the equation

$$e_1 + 5e_2 + 10e_3 + 25e_4 + 50e_5 = n.$$

The generating function is

$$g(x) = \frac{1}{1-x} \frac{1}{1-x^5} \frac{1}{1-x^{10}} \frac{1}{1-x^{25}} \frac{1}{1-x^{50}}.$$

□

7.5 Recurrences and Generating Functions

In this section we show how to use generating functions in order to solve linear homogeneous recurrence relations with constant coefficients. This will provide an alternative means of solution for such recurrence relations to that given in Section 7.2. An important role in this method is played by Newton's binomial theorem. Specifically, the following case of Newton's binomial theorem will be used: If n is a positive integer and r is a nonzero real number, then

$$(1-rx)^{-n} = \sum_{k=0}^{\infty} \binom{-n}{k}(-rx)^k,$$

or, equivalently,

$$\frac{1}{(1-rx)^n} = \sum_{k=0}^{\infty} (-1)^k \binom{-n}{k} r^k x^k, \qquad \left(|x| < \frac{1}{|r|}\right).$$

We have seen in Section 5.6 that

$$\binom{-n}{k} = (-1)^k \binom{n+k-1}{k},$$

and hence we can write the preceding formula as

$$\frac{1}{(1-rx)^n} = \sum_{k=0}^{\infty} \binom{n+k-1}{k} r^k x^k, \qquad \left(|x| < \frac{1}{|r|}\right). \tag{7.47}$$

Example. Determine the generating function for the sequence of squares

$$0, 1, 4, \ldots, n^2, \ldots.$$

By (7.47), with $n = 2$ and $r = 1$,

$$\frac{1}{(1 - x)^2} = 1 + 2x + 3x^2 + \cdots + nx^{n-1} + \cdots,$$

and hence,

$$\frac{x}{(1 - x)^2} = x + 2x^2 + 3x^3 + \cdots + nx^n + \cdots.$$

Differentiating, we get

$$\frac{1 + x}{(1 - x)^3} = 1 + 2^2 x + 3^2 x^2 + \cdots + n^2 x^{n-1} + \cdots.$$

Multiplying by x, we obtain

$$\frac{x(1 + x)}{(1 - x)^3} = x + 2^2 x^2 + 3^2 x^3 + \cdots + n^2 x^n + \cdots.$$

Therefore, $x(1 + x)/(1 - x)^3$ is the desired generating function. □

Example. Solve the recurrence relation

$$h_n = 5h_{n-1} - 6h_{n-2}, \quad (n \geq 2),$$

subject to the initial values $h_0 = 1$ and $h_1 = -2$.

Let $g(x) = h_0 + h_1 x + h_2 x^2 + \cdots + h_n x^n + \cdots$ be the generating function for $h_0, h_1, h_2, \ldots, h_n, \ldots$. We then have the following equations:

$$
\begin{aligned}
g(x) &= h_0 + h_1 x + h_2 x^2 + \cdots + h_n x^n + \cdots, \\
-5xg(x) &= -5h_0 x - 5h_1 x^2 - \cdots - 5h_{n-1}x^n + \cdots, \\
6x^2 g(x) &= 6h_0 x^2 + \cdots + 6h_{n-2}x^n + \cdots.
\end{aligned}
$$

Adding these three equations, we obtain

$$
\begin{aligned}
(1 - 5x + 6x^2)g(x) = {}& h_0 + (h_1 - 5h_0)x + (h_2 - 5h_1 + 6h_0)x^2 + \cdots \\
& + (h_n - 5h_{n-1} + 6h_{n-2})x^n + \cdots.
\end{aligned}
$$

Since $h_n - 5h_{n-1} + 6h_{n-2} = 0$ $(n \geq 2)$, and since $h_0 = 1$ and $h_1 = -2$, we have

$$(1 - 5x + 6x^2)g(x) = h_0 + (h_1 - 5h_0)x = 1 - 7x.$$

Thus,

$$g(x) = \frac{1 - 7x}{1 - 5x + 6x^2}.$$

From this closed formula for the generating function $g(x)$ we would like to be able to determine a formula for h_n. To obtain such a formula, we use the method of partial fractions along with (7.47). We observe that

$$1 - 5x + 6x^2 = (1 - 2x)(1 - 3x),$$

and thus it is possible to write

$$\frac{1 - 7x}{1 - 5x + 6x^2} = \frac{c_1}{1 - 2x} + \frac{c_2}{1 - 3x}$$

for some constants c_1 and c_2. We can determine c_1 and c_2 by multiplying both sides of this equation by $1 - 5x + 6x^2$ to get

$$1 - 7x = (1 - 3x)c_1 + (1 - 2x)c_2,$$

or

$$1 - 7x = (c_1 + c_2) + (-3c_1 - 2c_2)x.$$

Hence,

$$c_1 + c_2 = 1,$$
$$-3c_1 - 2c_2 = -7.$$

Solving these equations simultaneously, we find that $c_1 = 5$ and $c_2 = -4$. Thus,

$$g(x) = \frac{1 - 7x}{1 - 5x + 6x^2} = \frac{5}{1 - 2x} - \frac{4}{1 - 3x}.$$

By (7.47),

$$\frac{1}{1 - 2x} = 1 + 2x + 2^2 x^2 + \cdots + 2^n x^n + \cdots,$$

and

$$\frac{1}{1 - 3x} = 1 + 3x + 3^2 x^2 + \cdots + 3^n x^n + \cdots.$$

Therefore,

$$
\begin{aligned}
g(x) &= 5(1 + 2x + 2^2 x^2 + \cdots + 2^n x^n + \cdots) \\
&\quad -4(1 + 3x + 3^2 x^2 + \cdots + 3^n x^n + \cdots) \\
&= 1 + (-2)x + (-15)x^2 + \cdots + (5 \times 2^n - 4 \times 3^n)x^n + \cdots.
\end{aligned}
$$

Since this is the generating function for $h_0, h_1, h_2, \ldots, h_n, \ldots$, we obtain
$$h_n = 5 \times 2^n - 4 \times 3^n \qquad (n = 0, 1, 2, \ldots).$$ □

The method used in the preceding example can be generalized to enable one to solve theoretically any linear homogeneous recurrence relation of order k with constant coefficients. The associated generating function will be of the form $p(x)/q(x)$, where $p(x)$ is a polynomial of degree less than k and where $q(x)$ is a polynomial of degree k having constant term equal to 1. To find a general formula for the terms of the sequence, we first use the method of partial fractions to express $p(x)/q(x)$ as a sum of algebraic fractions of the form

$$\frac{c}{(1 - rx)^t},$$

where t is a positive integer, r is a real number, and c is a constant. We then use (7.47) to find a power series for $1/(1 - rx)^t$. Combining like terms, we obtain a power series for the generating function, from which we can read off the terms of the sequence.

Example. Let $h_0, h_1, h_2, \ldots, h_n, \ldots$ be a sequence of numbers satisfying the recurrence relation

$$h_n + h_{n-1} - 16h_{n-2} + 20h_{n-3} = 0, \qquad (n \geq 3)$$

where $h_0 = 0$, $h_1 = 1$ and $h_2 = -1$. Find a general formula for h_n.

Let $g(x) = h_0 + h_1 x + h_2 x^2 + \cdots + h_n x^n + \cdots$ be the generating function for $h_0, h_1, h_2, \ldots, h_n, \ldots$. Adding the four equations,

$$
\begin{aligned}
g(x) &= h_0 + h_1 x + h_2 x^2 + h_3 x^3 + \cdots + h_n x^n + \cdots, \\
x g(x) &= \qquad h_0 x + h_1 x^2 + h_2 x^3 + \cdots + h_{n-1} x^n + \cdots, \\
-16x^2 g(x) &= \qquad\qquad - 16h_0 x^2 - 16h_1 x^3 - \cdots - 16h_{n-2} x^n - \cdots, \\
20x^3 g(x) &= \qquad\qquad\qquad 20h_0 x^3 + \cdots + 20h_{n-3} x^n + \cdots,
\end{aligned}
$$

we obtain

$$
\begin{aligned}
(1 + x - 16x^2 + 20x^3)g(x) &= h_0 + (h_1 + h_0)x + (h_2 + h_1 - 16h_0)x^2 \\
&\quad + (h_3 + h_2 - 16h_1 + 20h_0)x^3 + \cdots \\
&\quad + (h_n + h_{n-1} - 16h_{n-2} + 20h_{n-3})x^n + \cdots.
\end{aligned}
$$

Since $h_n + h_{n-1} - 16h_{n-2} + 20h_{n-3} = 0$, $(n \geq 3)$ and since $h_0 = 0$, $h_1 = 1$, and $h_2 = -1$, we get

$$(1 + x - 16x^2 + 20x^3)g(x) = x.$$

Hence,

$$g(x) = \frac{x}{1 + x - 16x^2 + 20x^3}.$$

We observe that $(1 + x - 16x^2 + 20x^3) = (1 - 2x)^2(1 + 5x)$. Thus, for some constants c_1, c_2, and c_3,

$$\frac{x}{1 + x - 16x^2 + 20x^3} = \frac{c_1}{1 - 2x} + \frac{c_2}{(1 - 2x)^2} + \frac{c_3}{1 + 5x}.$$

To determine the constants, we multiply both sides of this equation by $1 + x - 16x^2 + 20x^3$ to get

$$x = (1 - 2x)(1 + 5x)c_1 + (1 + 5x)c_2 + (1 - 2x)^2 c_3,$$

or, equivalently,

$$x = (c_1 + c_2 + c_3) + (3c_1 + 5c_2 - 4c_3)x + (-10c_1 + 4c_3)x^2.$$

Hence,

$$
\begin{array}{rrrrcl}
c_1 & + & c_2 & + & c_3 & = & 0, \\
3c_1 & + & 5c_2 & - & 4c_3 & = & 1, \\
-10c_1 & & & + & 4c_3 & = & 0.
\end{array}
$$

Solving these equations simultaneously, we find that

$$c_1 = -\frac{2}{49}, \quad c_2 = \frac{7}{49}, \quad \text{and } c_3 = -\frac{5}{49}.$$

Therefore,

$$g(x) = \frac{x}{1 + x - 16x^2 + 20x^3} = -\frac{2/49}{1 - 2x} + \frac{7/49}{(1 - 2x)^2} - \frac{5/49}{1 + 5x}.$$

By (7.47)

$$\frac{1}{1 - 2x} = \sum_{k=0}^{\infty} 2^k x^k,$$

$$\frac{1}{(1 - 2x)^2} = \sum_{k=0}^{\infty} \binom{k + 1}{k} 2^k x^k = \sum_{k=0}^{\infty} (k + 1)2^k x^k,$$

$$\frac{1}{1 + 5x} = \sum_{k=0}^{\infty} (-5)^k x^k.$$

Consequently,

$$g(x) = -\frac{2}{49}\left(\sum_{k=0}^{\infty} 2^k x^k\right) + \frac{7}{49}\left(\sum_{k=0}^{\infty}(k + 1)2^k x^k\right) - \frac{5}{49}\left(\sum_{k=0}^{\infty}(-5)^k x^k\right)$$

$$= \sum_{k=0}^{\infty} \left[-\frac{2}{49} 2^k + \frac{7}{49}(k+1)2^k - \frac{5}{49}(-5)^k \right] x^k.$$

Since $g(x)$ is the generating function for $h_0, h_1, h_2, \ldots, h_n, \ldots$, it follows that

$$h_n = -\frac{2}{49} 2^n + \frac{7}{49}(n+1)2^n - \frac{5}{49}(-5)^n, \qquad (n = 0, 1, 2, \ldots). \quad \square$$

The preceding formula for h_n should bring to mind the solution of recurrence relations, using the roots of the characteristic equation as described in Section 7.2. Indeed, the formula suggests that the roots of the characteristic equation for the given recurrence relation are 2, 2, and -5. The following discussion should clarify the relationship between the two methods.

In the foregoing example, we have expressed the generating function $g(x)$ in the form

$$g(x) = \frac{p(x)}{q(x)},$$

where

$$q(x) = 1 + x - 16x^2 + 20x^3.$$

Since the recurrence relation is

$$h_n + h_{n-1} - 16h_{n-2} + 20h_{n-3} = 0, \quad (n = 3, 4, 5, \ldots),$$

the associated characteristic equation is $r(x) = 0$, where

$$r(x) = x^3 + x^2 - 16x + 20.$$

If we replace x in $r(x)$ by $1/x$ (this amounts to the change in variable $y = 1/x$), we obtain

$$r(1/x) = \frac{1}{x^3} + \frac{1}{x^2} - 16\frac{1}{x} + 20,$$

or

$$x^3 r(1/x) = 1 + x - 16x^2 + 20x^3 = q(x).$$

The roots of the characteristic equation $r(x) = 0$ are $2, 2$, and -5. Since $r(x) = (x-2)^2(x+5)$, it follows that

$$q(x) = x^3 \left(\frac{1}{x} - 2 \right)^2 \left(\frac{1}{x} + 5 \right) = (1 - 2x)^2(1 + 5x),$$

which checks with our previous calculation.

The preceding relationships hold in general. Let the sequence of numbers $h_0, h_1, h_2, \ldots, h_n, \ldots$ be defined by the recurrence relation

$$h_n + a_1 h_{n-1} + \cdots + a_k h_{n-k} = 0, \ (n \geq k)$$

of order k and with initial values for $h_0, h_1, \ldots, h_{k-1}$. Recall that, since the recurrence relation has order k, a_k is assumed to be different from 0. Let $g(x)$ be the generating function for our sequence. Using the method given in the examples, we find that there are polynomials $p(x)$ and $q(x)$ such that

$$g(x) = \frac{p(x)}{q(x)},$$

where $q(x)$ has degree k and $p(x)$ has degree less than k. Indeed, we have

$$q(x) = 1 + a_1 x + a_2 x^2 + \cdots + a_k x^k$$

and

$$\begin{aligned}
p(x) =\ & h_0 + (h_1 + a_1 h_0)x + (h_2 + a_1 h_1 + a_2 h_0)x^2 \\
& + \cdots + (h_{k-1} + a_1 h_{k-2} + \cdots + a_{k-1} h_0)x^{k-1}.
\end{aligned}$$

The characteristic equation for this recurrence relation is $r(x) = 0$, where

$$r(x) = x^k + a_1 x^{k-1} + a_2 x^{k-2} + \cdots + a_k.$$

Hence,

$$q(x) = x^k r(1/x).$$

Thus, if the roots of $r(x) = 0$ are $q_1, q_2, \ldots, q_k$, then

$$r(x) = (x - q_1)(x - q_2) \cdots (x - q_k)$$

and

$$q(x) = (1 - q_1 x)(1 - q_2 x) \cdots (1 - q_k x).$$

Conversely, if we are given a polynomial

$$q(x) = b_0 + b_1 x + \cdots + b_k x^k$$

of degree k with $b_0 \neq 0$ and a polynomial

$$p(x) = d_0 + d_1 x + \cdots + d_{k-1} x^{k-1}$$

of degree less than k, then, using partial fractions and (7.47), we can find a power series[15] $h_0 + h_1 x + \cdots + h_n x^n + \cdots$ such that

$$\frac{p(x)}{q(x)} = h_0 + h_1 x + \cdots + h_n x^n + \cdots.$$

We can write the preceding equation in the form

$$
\begin{aligned}
d_0 + d_1 x + \cdots + d_{k-1} x^{k-1} = \ & (b_0 + b_1 x + \cdots + b_k x^k) \\
& \times (h_0 + h_1 x + \cdots + h_n x^n + \cdots).
\end{aligned}
$$

Multiplying out the right side and comparing coefficients, we obtain

$$
\begin{aligned}
b_0 h_0 &= d_0, \\
b_0 h_1 + b_1 h_0 &= d_1, \\
&\ \vdots \\
b_0 h_{k-1} + b_1 h_{k-2} + \cdots + b_{k-1} h_0 &= d_{k-1},
\end{aligned}
\tag{7.48}
$$

and

$$b_0 h_n + b_1 h_{n-1} + \cdots + b_k h_{n-k} = 0, \qquad (n \geq k). \tag{7.49}$$

Since $b_0 \neq 0$, equation (7.49) can be written in the form

$$h_n + \frac{b_1}{b_0} h_{n-1} + \cdots + \frac{b_k}{b_0} h_{n-k} = 0, \qquad (n \geq k).$$

This is a linear homogeneous recurrence relation with constant coefficients that is satisfied by $h_0, h_1, h_2, \ldots, h_n, \ldots$. The initial values $h_0, h_1, \ldots, h_{k-1}$ can be determined by solving the triangular system of equations (7.48), using the fact that $b_0 \neq 0$. We summarize in the next theorem.

Theorem 7.5.1 *Let*

$$h_0, h_1, h_2, \ldots, h_n, \ldots$$

be a sequence of numbers that satisfies the linear homogeneous recurrence relation

$$h_n + c_1 h_{n-1} + \cdots + c_k h_{n-k} = 0, \quad c_k \neq 0, \quad (n \geq k) \tag{7.50}$$

---

[15]This power series will converge to $p(x)/q(x)$ for all x with $|x| < t$, where t is the smallest absolute value of a root of $q(x) = 0$. Since we assume that $b_0 \neq 0$, 0 is not a root of $q(x) = 0$.

of order k with constant coefficients. Then its generating function g(x) is of the form

$$g(x) = \frac{p(x)}{q(x)}, \tag{7.51}$$

where $q(x)$ is a polynomial of degree k with a nonzero constant term and $p(x)$ is a polynomial of degree less than k. Conversely, given such polynomials $p(x)$ and $q(x)$, there is a sequence $h_0, h_1, h_2, \ldots, h_n, \ldots$ satisfying a linear homogeneous recurrence relation with constant coefficients of order k of the type (7.50) whose generating function is given by (7.51).

7.6   A Geometry Example

A set K of points in the plane or in space is said to be *convex*, provided that for any two points p and q in K, all the points on the line segment joining p and q are in K. Triangular regions, circular regions, and rectangular regions in the plane are all convex sets of points. On the other hand, the region on the left in Figure 7.2 is not convex since, for the two points p and q shown, the line segment joining p and q goes outside the region.

The regions in Figure 7.2 are examples of *polygonal regions*—that is, regions whose boundaries consist of a finite number of line segments, called their *sides*. Triangular regions and rectangular regions are polygonal, but circular regions are not. Any polygonal region must have at least three sides. The region on the right in Figure 7.2 is a convex polygonal region with six sides.

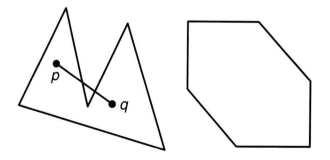

Figure 7.2

In a polygonal region, the points at which the sides meet are called *corners* (or *vertices*). A *diagonal* is a line segment joining two nonconsecutive corners.

Let K be a polygonal region with n sides. We can count the number of its diagonals as follows: Each corner is joined by a diagonal to $n-3$ other corners. Thus, counting the number of diagonals at each corner and summing, we get $n(n-3)$. Since each diagonal has two corners, each diagonal is counted twice in this sum. Hence, the number of diagonals is $n(n-3)/2$. We can arrive at this same number indirectly in the following way: There are

$$\binom{n}{2} = \frac{n(n-1)}{2}$$

line segments joining the n corners. Of these, n are sides of the polygonal region. The remaining ones are diagonals. Consequently, there are

$$\frac{n(n-1)}{2} - n = \frac{n(n-3)}{2}$$

diagonals.

Now assume that K is convex. Then each diagonal of K lies wholly within K. Thus, each diagonal of K divides K into one convex polygonal region with k sides and another with $n - k + 2$ sides for some $k = 3, 4, \ldots, n - 1$.

We can draw $n - 3$ diagonals meeting a particular corner of K, and in doing so divide K into $n - 2$ triangular regions. But, there are other ways of dividing the region into triangular regions by inserting $n - 3$ diagonals no two of which intersect in the interior of K, as the example in Figure 7.3 shows for $n = 8$.

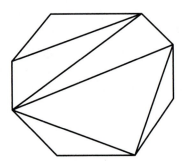

Figure 7.3

In the next theorem, we determine the number of different ways to divide a convex polygonal region into triangular regions by drawing diagonals that do not intersect in the interior. For notational convenience, we deal with a convex polygonal region of $n + 1$ sides which is then divided into $n - 1$ triangular regions by $n - 2$ diagonals.

Theorem 7.6.1 *Let h_n denote the number of ways of dividing a convex polygonal region with $n+1$ sides into triangular regions by inserting diagonals that do not intersect in the interior. Define $h_1 = 1$. Then h_n satisfies the recurrence relation*

$$\begin{aligned}
h_n &= h_1 h_{n-1} + h_2 h_{n-2} + \cdots + h_{n-1} h_1 \\
&= \sum_{k=1}^{n-1} h_k h_{n-k}, \quad (n \geq 2).
\end{aligned} \tag{7.52}$$

The solution of this recurrence relation is

$$h_n = \frac{1}{n}\binom{2n-2}{n-1}, \qquad (n = 1, 2, 3, \ldots). \tag{7.53}$$

Proof. We have defined $h_1 = 1$, and we think of a line segment as a polygonal region with two sides and no interior. We have $h_2 = 1$, since a triangular region has no diagonals, and it cannot be further subdivided. The recurrence relation (7.52) holds for $n = 2$,[16] since

$$\sum_{k=1}^{2-1} h_k h_{2-k} = \sum_{k=1}^{1} h_k h_{2-k} = h_1 h_1 = 1.$$

Now let $n \geq 3$. Consider a convex polygonal region K with $n + 1 \geq 4$ sides. We distinguish one side of K and call it the *base*. In each division of K into triangular regions, the base is a side of one of the triangular regions T, and this triangular region divides the remainder of K into a polygonal region K_1 with $k + 1$ sides and a polygonal region K_2 with $n - k + 1$ sides, for some $k = 1, 2, \ldots, n - 1$. (See Fig. 7.4.)

The further subdivision of K is accomplished by dividing K_1 and K_2 into triangular regions by inserting diagonals of K_1 and K_2, respectively, which do not intersect in the interior. Since K_1 has $k + 1$ sides, K_1 can be divided into triangular regions in h_k ways. Since K_2 has $n - k + 1$ sides, K_2 can be divided into triangular regions in h_{n-k} ways. Hence, for a particular choice of the triangular region T containing the base, there are $h_k h_{n-k}$ ways of dividing K into triangular regions by diagonals that do not intersect in the interior. Hence, there is a total of

$$h_n = \sum_{k=1}^{n-1} h_k h_{n-k}$$

ways to divide K into triangular regions in this way. This establishes the recurrence relation (7.52).

---

[16] This is why we defined $h_1 = 1$.

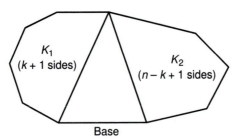

Base

Polygonal region with $n + 1$ sides

Figure 7.4

We now turn to the solution of (7.52) with the initial condition $h_1 = 1$. Ths recurrence relation is not linear. Moreover, h_n does not depend on a fixed number of values that come before it, but on all the values $h_1, h_2, \ldots, h_{n-1}$ that come before it. Thus, none of our methods for solving recurrence relations apply. Let

$$g(x) = h_1 x + h_2 x^2 + \cdots + h_n x^n + \cdots$$

be the generating function for the sequence $h_1, h_2, h_3, \ldots, h_n, \ldots$. Multiplying $g(x)$ by itself, we find that

$$
\begin{aligned}
(g(x))^2 &= h_1^2 x^2 + (h_1 h_2 + h_2 h_1)x^3 + (h_1 h_3 + h_2 h_2 + h_3 h_1)x^4 \\
&\quad + \cdots + (h_1 h_{n-1} + h_2 h_{n-2} + \cdots + h_{n-1} h_1)x^n + \cdots.
\end{aligned}
$$

Using (7.52) and the fact that $h_1 = h_2 = 1$, we obtain

$$
\begin{aligned}
(g(x))^2 &= h_1^2 x^2 + h_3 x^3 + h_4 x^4 + \cdots + h_n x^n + \cdots \\
&= h_2 x^2 + h_3 x^3 + h_4 x^4 + \cdots + h_n x^n + \cdots \\
&= g(x) - h_1 x = g(x) - x.
\end{aligned}
$$

Thus, $g(x)$ satisfies the equation

$$(g(x))^2 - g(x) + x = 0.$$

This is a quadratic equation for $g(x)$, so, by the quadratic formula,[17] $g(x) = g_1(x)$ or $g(x) = g_2(x)$ where

$$g_1(x) = \frac{1 + \sqrt{1 - 4x}}{2} \quad \text{and} \quad g_2(x) = \frac{1 - \sqrt{1 - 4x}}{2}.$$

From the definition of $g(x)$, it follows that $g(0) = 0$. Since $g_1(0) = 1$ and $g_2(0) = 0$, we conclude that

$$g(x) = g_2(x) = \frac{1 - \sqrt{1 - 4x}}{2} = \frac{1}{2} - \frac{1}{2}(1 - 4x)^{1/2}.$$

---

[17] But omitting some subtleties.

By Newton's binomial theorem (see, in particular, the calculation done at the end of Section 5.6),

$$(1+z)^{1/2} = 1 + \sum_{n=1}^{\infty} \frac{(-1)^{n-1}}{n \times 2^{2n-1}} \binom{2n-2}{n-1} z^n, \qquad (|z| < 1).$$

If we replace z by $-4x$, we get

$$\begin{aligned}
(1-4x)^{1/2} &= 1 + \sum_{n=1}^{\infty} \frac{(-1)^{n-1}}{n \times 2^{2n-1}} \binom{2n-2}{n-1} (-1)^n 4^n x^n \\
&= 1 + \sum_{n=1}^{\infty} (-1)^{2n-1} \frac{2}{n} \binom{2n-2}{n-1} x^n \\
&= 1 - 2 \sum_{n=1}^{\infty} \frac{1}{n} \binom{2n-2}{n-1} x^n, \qquad \left(|x| < \frac{1}{4}\right).
\end{aligned}$$

Thus,

$$g(x) = \frac{1}{2} - \frac{1}{2}(1-4x)^{1/2} = \sum_{n=1}^{\infty} \frac{1}{n} \binom{2n-2}{n-1} x^n, \qquad (7.54)$$

and hence,

$$h_n = \frac{1}{n} \binom{2n-2}{n-1}, \qquad (n \geq 1).$$

$\square$

The numbers

$$\frac{1}{n} \binom{2n-2}{n-1}$$

in the previous theorem are the Catalan numbers, and these will be investigated more throughly in Chapter 8.

7.7 Exponential Generating Functions

In Section 7.5, we defined the generating function for a sequence of numbers $h_0, h_1, h_2, \ldots, h_n, \ldots$ by using the set of monomials

$$\{1, x, x^2, \ldots, x^k, \ldots\}.$$

This is particularly suited to some counting sequences, especially those involving binomial coefficients, because of the form of Newton's binomial theorem. However, for sequences whose terms count permutations, it is more useful to consider a generating function with respect to the monomials

$$\left\{1, x, \frac{x^2}{2!}, \ldots, \frac{x^n}{n!}, \ldots\right\}. \qquad (7.55)$$

These monomials arise in the Taylor series

$$e^x = \sum_{n=0}^{\infty} \frac{x^n}{n!} = 1 + x + \frac{x^2}{2!} + \cdots + \frac{x^n}{n!} + \cdots. \tag{7.56}$$

Generating functions considered with respect to the monomials (7.55) are called *exponential generating functions.*[18] The *exponential generating function* for the sequence $h_0, h_1, h_2, \ldots, h_n, \ldots$ is defined to be

$$g^{(e)}(x) = \sum_{n=0}^{\infty} h_n \frac{x^n}{n!} = h_0 + h_1 x + h_2 \frac{x^2}{2!} + \cdots + h_n \frac{x^n}{n!} + \cdots.$$

Example. Let n be a positive integer. Determine the exponential generating function for the sequence of numbers

$$P(n,0), P(n,1), P(n,2), \ldots, P(n,n),$$

where $P(n,k)$ denotes the number of k-permutations of an n-element set, and thus has the value $n!/(n-k)!$ for $k = 0, 1, \ldots, n$. The exponential generating function is

$$
\begin{aligned}
g^{(e)}(x) &= P(n,0) + P(n,1)x + P(n,2)\frac{x^2}{2!} + \ldots + P(n,n)\frac{x^n}{n!} \\
&= 1 + nx + \frac{n!}{2!(n-2)!}x^2 + \cdots + \frac{n!}{n!0!}x^n \\
&= (1+x)^n.
\end{aligned}
$$

Thus, $(1+x)^n$ is both the exponential generating function for the sequence $P(n,0), P(n,1), \ldots, P(n,n)$ and, as we have seen in Section 7.5, the ordinary generating function for the sequence

$$\binom{n}{0}, \binom{n}{1}, \ldots, \binom{n}{n}.$$

$\square$

Example. The exponential generating function for the sequence

$$1, 1, 1, \ldots, 1, \ldots$$

is

$$g^{(e)}(x) = \sum_{n=0}^{\infty} \frac{x^n}{n!} = e^x.$$

---

[18]We reserve the phrase "generating function" or "ordinary generating function" for the case in which we use the monomials $\{1, x, x^2, \ldots, x^n, \ldots\}$.

More generally, if a is any real number, the exponential generating function for the sequence

$$a^0 = 1, a, a^2, \ldots, a^n, \ldots$$

is

$$g^{(e)}(x) = \sum_{n=0}^{\infty} a^n \frac{x^n}{n!} = \sum_{n=0}^{\infty} \frac{(ax)^n}{n!} = e^{ax}.$$

We recall from Section 3.4 that, for a positive integer k, k^n represents the number of n-permutations of a multiset with objects of k different types, each with an infinite repetition number. Thus, the exponential function for this sequence of counting numbers is e^{kx}. □

For a multiset S with objects of k different types, each with a finite repetition number, the next theorem determines the exponential generating function for the number of n-permutations of S. This is the solution in the form of an exponential generating function that was promised at the end of Section 3.4. We define the number of 0-permutations of a multiset to be equal to 1.

Theorem 7.7.1 *Let S be the multiset $\{n_1 \cdot a_1, n_2 \cdot a_2, \ldots, n_k \cdot a_k\}$, where $n_1, n_2, \ldots, n_k$ are nonnegative integers. Let h_n be the number of n-permutations of S. Then the exponential generating function $g^{(e)}(x)$ for the sequence $h_0, h_1, h_2, \ldots, h_n, \ldots$ is given by*

$$g^{(e)}(x) = f_{n_1}(x) f_{n_2}(x) \cdots f_{n_k}(x), \tag{7.57}$$

where, for $i = 1, 2, \ldots, k$,

$$f_{n_i}(x) = 1 + x + \frac{x^2}{2!} + \cdots + \frac{x^{n_i}}{n_i!}. \tag{7.58}$$

Proof. Let

$$g^{(e)}(x) = h_0 + h_1 x + h_2 \frac{x^2}{2!} + \cdots + h_n \frac{x^n}{n!} + \cdots$$

be the exponential generating function for $h_0, h_1, h_2, \ldots, h_n, \ldots$. Note that $h_n = 0$ for $n > n_1 + n_2 + \cdots + n_k$, so that $g^{(e)}(x)$ is a finite sum. From (7.58), we see that, when (7.57) is multiplied out, we get terms of the form

$$\frac{x^{m_1}}{m_1!} \frac{x^{m_2}}{m_2!} \cdots \frac{x^{m_k}}{m_k!} = \frac{x^{m_1+m_2+\cdots+m_k}}{m_1! m_2! \cdots m_k!}, \tag{7.59}$$

where

$$0 \le m_1 \le n_1, \ 0 \le m_2 \le n_2, \ldots, 0 \le m_k \le n_k.$$

Let $n = m_1 + m_2 + \cdots + m_k$. Then the expression in (7.59) can be written as

$$\frac{x^n}{m_1!m_2!\cdots m_k!} = \frac{n!}{m_1!m_2!\cdots m_k!}\frac{x^n}{n!}.$$

Thus, the coefficient of $x^n/n!$ in (7.57) is

$$\sum \frac{n!}{m_1!m_2!\cdots m_k!}, \tag{7.60}$$

where the summation extends over all integers $m_1, m_2, \ldots, m_k$, with $0 \le m_1 \le n_1, 0 \le m_2 \le n_2, \ldots, 0 \le m_k \le n_k$, and $m_1+m_2+\cdots+m_k = n$. But, from Section 3.4, we know that a term

$$\frac{n!}{m_1!m_2!\cdots m_k!}$$

in the sum (7.60) equals the number of n-permutations (or simply, permutations) of the submultiset $\{m_1{\cdot}e_1, m_2{\cdot}e_2, \ldots, m_k{\cdot}e_k\}$ of S. Since the number of n-permutations of S equals the number of permutations of all such submultisets with $m_1 + m_2 + \cdots + m_k = n$, the number h_n equals the number in (7.60). Since this is also the coefficient of $x^n/n!$ in (7.57), we conclude that

$$g^{(e)}(x) = f_{n_1}(x)f_{n_2}(x)\cdots f_{n_k}(x). \qquad \square$$

Using the same type of reasoning as used in the proof of the preceding theorem, we can calculate the exponential generating function for sequences of numbers that count n-permutations of a multiset with additional restrictions. Let us first observe that, if, in (7.58), we define

$$f_\infty(x) = 1 + x + \frac{x^2}{2!} + \cdots + \frac{x^k}{k!} + \cdots = e^x, \tag{7.61}$$

then the theorem continues to hold if some of the repetition numbers $n_1, n_2, \ldots, n_k$ are equal to ∞.

Example. Let h_n denote the number of n-digit numbers with digits $1, 2,$ or 3, where the number of 1's is even, the number of 2's is at least three, and the number of 3's is at most four. Determine the exponential generating function $g^{(e)}(x)$ for the resulting sequence of numbers $h_0, h_1, h_2, \ldots, h_n, \ldots$.

The function $g^{(e)}(x)$ has a factor for each of the three digits $1, 2,$ and 3. The restrictions on the digits are reflected in the factors as follows: The factor of $g^{(e)}(x)$ corresponding to the digit 1 is

$$h_1(x) = 1 + \frac{x^2}{2!} + \frac{x^4}{4!} + \cdots,$$

since the number of 1's is to be even. The factors of $g^{(e)}(x)$ corresponding to the digits 2 and 3 are, respectively,

$$h_2(x) = \frac{x^3}{3!} + \frac{x^4}{4!} + \frac{x^5}{5!} + \cdots,$$

and

$$h_3(x) = 1 + \frac{x}{1!} + \frac{x^2}{2!} + \frac{x^3}{3!} + \frac{x^4}{4!}.$$

The exponential generating function is the product of the preceding three factors:

$$g^{(e)}(x) = h_1(x)h_2(x)h_3(x).$$

$\square$

Exponential generating functions can sometimes be used to find explicit formulas for counting problems. We illustrate this with three examples.

Example. Determine the number of ways to color the squares of a 1-by-n chessboard, using the colors, red, white, and blue, if an even number of squares are to be colored red.

Let h_n denote the number of such colorings, where we define h_0 to be 1. Then h_n equals the number of n-permutations of a multiset of three colors (red, white, and blue), each with an infinite repetition number, in which red occurs an even number of times. Thus, the exponential generating function for $h_0, h_1, \ldots, h_n, \ldots$ is the product of red, white, and blue factors:

$$
\begin{aligned}
g^{(e)} &= \left(1 + \frac{x^2}{2!} + \frac{x^4}{4!} + \cdots\right)\left(1 + \frac{x}{1!} + \frac{x^2}{2!} + \cdots\right)\left(1 + \frac{x}{1!} + \frac{x^2}{2!} + \cdots\right) \\
&= \frac{1}{2}(e^x + e^{-x})e^x e^x = \frac{1}{2}(e^{3x} + e^x) \\
&= \frac{1}{2}\left(\sum_{n=0}^{\infty} 3^n \frac{x^n}{n!} + \sum_{n=0}^{\infty} \frac{x^n}{n!}\right) \\
&= \frac{1}{2}\sum_{n=0}^{\infty}(3^n + 1)\frac{x^n}{n!}.
\end{aligned}
$$

Hence, $h_n = (3^n + 1)/2$.

The form of this solution suggests that h_n may satisfy a linear homogeneous recurrence relation of order 2 whose characteristic equation has roots 3 and 1. Yet we show, as an alternative way to solve this problem, that h_n satisfies a nonhomogeneous recurrence relation. We have $h_1 = 2$. Let $n \geq 2$. If the first square is colored white or blue, there are h_{n-1} ways to complete the coloring. If the first square is colored red, then there must be an odd number of red squares among the remaining $n-1$ squares; hence, there are $3^{n-1} - h_{n-1}$ ways to complete the coloring in this case. Therefore, h_n satisfies the recurrence relation

$$h_n = 2h_{n-1} + (3^{n-1} - h_{n-1}) = h_{n-1} + 3^{n-1}, \quad (n \geq 2).$$

If we iterate this recurrence relation $h_n = h_{n-1} + 3^{n-1}$ and use $h_1 = 2$, we again find that $h_n = (3^n + 1)/2$.

$\square$

Example. Determine the number h_n of n-digit numbers with each digit odd, where the digits 1 and 3 occur an even number of times.

Let $h_0 = 1$. The number h_n equals the number of n-permutations of the multiset $S = \{\infty \cdot 1, \infty \cdot 3, \infty \cdot 5, \infty \cdot 7, \infty \cdot 9\}$, in which 1 and 3 occur an even number of times. The exponential generating function for $h_0, h_1, h_2, \ldots, h_n, \ldots$ is a product of five factors, one for each of the allowable digits:

$$
\begin{aligned}
g^{(e)}(x) &= \left(1 + \frac{x^2}{2!} + \frac{x^4}{4!} + \cdots\right)^2 \left(1 + x + \frac{x^2}{2!} + \cdots\right)^3 \\
&= \left(\frac{e^x + e^{-x}}{2}\right)^2 e^{3x} \\
&= \left(\frac{e^{2x} + 1}{2}\right)^2 e^x \\
&= \frac{1}{4}(e^{4x} + 2e^{2x} + 1)e^x \\
&= \frac{1}{4}(e^{5x} + 2e^{3x} + e^x) \\
&= \frac{1}{4}\left(\sum_{n=0}^{\infty} 5^n \frac{x^n}{n!} + 2\sum_{n=0}^{\infty} 3^n \frac{x^n}{n!} + \sum_{n=0}^{\infty} \frac{x^n}{n!}\right) \\
&= \sum_{n=0}^{\infty} \left(\frac{5^n + 2 \times 3^n + 1}{4}\right) \frac{x^n}{n!}.
\end{aligned}
$$

Hence,

$$h_n = \frac{5^n + 2 \times 3^n + 1}{4}, \quad (n \geq 0).$$

$\square$

Example. Determine the number h_n of ways to color the squares of a 1-by-n board with the colors red, white, and blue, where the number of red squares is even and there is at least one blue square.

The exponential generating function $g^{(e)}(x)$ is

$$g^{(e)}(x) = \left(1 + \frac{x^2}{2!} + \frac{x^4}{4!} + \cdots\right)\left(1 + \frac{x}{1!} + \frac{x^2}{2!} + \cdots\right)\left(\frac{x}{1!} + \frac{x^2}{2!} + \cdots\right)$$

$$= \frac{e^x + e^{-x}}{2}e^x(e^x - 1)$$

$$= \frac{e^{3x} - e^{2x} + e^x - 1}{2}$$

$$= -\frac{1}{2} + \sum_{n=0}^{\infty} \frac{3^n - 2^n + 1}{2}\frac{x^n}{n!}.$$

Thus,

$$h_n = \frac{3^n - 2^n + 1}{2}, \qquad (n = 1, 2, \ldots)$$

and

$$h_0 = 0.$$

Note that h_0 should be 0. A 1-by-0 board is empty, no squares get colored, and so we cannot satisfy the condition that the number of blue squares is at least 1. □

7.8 Exercises

1. Let $f_0, f_1, f_2, \ldots, f_n, \ldots$ denote the Fibonacci sequence. By evaluating each of the following expressions for small values of n, conjecture a general formula and then prove it, using mathematical induction and the Fibonacci recurrence:

   (a) $f_1 + f_3 + \cdots + f_{2n-1}$
   (b) $f_0 + f_2 + \cdots + f_{2n}$
   (c) $f_0 - f_1 + f_2 - \cdots + (-1)^n f_n$
   (d) $f_0^2 + f_1^2 + \cdots + f_n^2$

2. Prove that the nth Fibonacci number f_n is the integer that is closest to the number

$$\frac{1}{\sqrt{5}}\left(\frac{1 + \sqrt{5}}{2}\right)^n.$$

3. Prove the following about the Fibonacci numbers:

   (a) f_n is even if and only if n is divisible by 3.

   (b) f_n is divisible by 3 if and only if n is divisible by 4.

   (c) f_n is divisible by 4 if and only if n is divisible by 6.

4. Prove that the Fibonacci sequence is the solution of the recurrence relation

$$a_n = 5a_{n-4} + 3a_{n-5}, \quad (n \geq 5),$$

   where $a_0 = 0, a_1 = 1, a_2 = 1, a_3 = 2,$ and $a_4 = 3$. Then use this formula to show that the Fibonacci numbers satisfy the condition that f_n is divisible by 5 if and only if n is divisible by 5.

5. By examining the Fibonacci sequence, make a conjecture about when f_n is divisible by 7 and then prove your conjecture.

6. * Let m and n be positive integers. Prove that, if m is divisible by n, then f_m is divisible by f_n.

7. * Let m and n be positive integers whose greatest common divisor is d. Prove that the greatest common divisor of the Fibonacci numbers f_m and f_n is the Fibonacci number f_d.

8. Consider a 1-by-n chessboard. Suppose we color each square of the chessboard with one of the two colors red and blue. Let h_n be the number of colorings in which no two squares that are colored red are adjacent. Find and verify a recurrence relation that h_n satisfies. Then derive a formula for h_n.

9. Let h_n equal the number of different ways in which the squares of a 1-by-n chessboard can be colored, using the colors red, white, and blue so that no two squares that are colored red are adjacent. Find and verify a recurrence relation that h_n satisfies. Then find a formula for h_n.

10. Suppose that, in his problem, Fibonacci had placed two pairs of rabbits in the enclosure at the beginning of a year. Find the number of pairs of rabbits in the enclosure after one year. More generally, find the number of pairs of rabbits in the enclosure after n months.

11. The *Lucas numbers* $l_0, l_1, l_2, \ldots, l_n \ldots$ are defined on the basis of the same recurrence relation defining the Fibonacci numbers, but with different initial conditions:

$$l_n = l_{n-1} + l_{n-2}, \ (n \geq 2), l_0 = 2, l_1 = 1.$$

Prove that

(a) $l_n = f_{n-1} + f_{n+1}$ for $n \geq 1$.

(b) $l_0^2 + l_1^2 + \cdots + l_n^2 = l_n l_{n+1} + 2$ for $n \geq 0$.

12. Solve the recurrence relation $h_n = 4h_{n-2}, \ (n \geq 2)$ with initial values $h_0 = 0$ and $h_1 = 1$.

13. Solve the recurrence relation $h_n = (n + 2)h_{n-1}, \ (n \geq 1)$ with initial value $h_0 = 2$.

14. Solve the recurrence relation $h_n = h_{n-1} + 9h_{n-2} - 9h_{n-3}, \ (n \geq 3)$ with initial values $h_0 = 0$, $h_1 = 1$, and $h_2 = 2$.

15. Solve the recurrence relation $h_n = 8h_{n-1} - 16h_{n-2}, \ (n \geq 2)$ with initial values $h_0 = -1$ and $h_1 = 0$.

16. Solve the recurrence relation $h_n = 3h_{n-2} - 2h_{n-3}, \ (n \geq 3)$ with initial values $h_0 = 1$, $h_1 = 0$, and $h_2 = 0$.

17. Solve the recurrence relation $h_n = 5h_{n-1} - 6h_{n-2} - 4h_{n-3} + 8h_{n-4}$, $(n \geq 4)$ with initial values $h_0 = 0$, $h_1 = 1$, $h_2 = 1$, and $h_3 = 2$.

18. Determine a recurrence relation for the number a_n of ternary strings (made up of 0's, 1's, and 2's) of length n that do not contain two consecutive 0's or two consecutive 1's. Then, find a formula for a_n.

19. Solve the following recurrence relations by examining the first few values for a formula and then proving your conjectured formula by induction.

(a) $h_n = 3h_{n-1}, \ (n \geq 1); h_0 = 1$

(b) $h_n = h_{n-1} - n + 3, \ (n \geq 1); h_0 = 2$

(c) $h_n = -h_{n-1} + 1, \ (n \geq 1); h_0 = 0$

(d) $h_n = -h_{n-1} + 2, \ (n \geq 1); h_0 = 1$

(e) $h_n = 2h_{n-1} + 1, \ (n \geq 1); h_0 = 1$

20. Let h_n denote the number of ways to perfectly cover a 1-by-n board with monominoes and dominoes in such a way that no two dominoes are consecutive. Find, but do not solve, a recurrence relation and initial conditions satisfied by h_n.

21. Let a_n equal the number of ternary strings of length n made up of 0's, 1's, and 2's, such that the substrings 00, 01, 10, and 11 never occur. Prove that

$$a_n = a_{n-1} + 2a_{n-2}, \quad (n \geq 2),$$

with $a_0 = 1$ and $a_1 = 3$. Then find a formula for a_n.

22. * Let $2n$ equally spaced points be chosen on a circle. Let h_n denote the number of ways to join these points in pairs so that the resulting line segments do not intersect. Establish a recurrence relation for h_n.

23. Solve the nonhomogeneous recurrence relation

$$\begin{aligned} h_n &= 4h_{n-1} + 3 \times 2^n, \quad (n \geq 1) \\ h_0 &= 1. \end{aligned}$$

24. Solve the nonhomogeneous recurrence relation

$$\begin{aligned} h_n &= 3h_{n-1} - 2, \quad (n \geq 1) \\ h_0 &= 1. \end{aligned}$$

25. Solve the nonhomogeneous recurrence relation

$$\begin{aligned} h_n &= 2h_{n-1} + n, \quad (n \geq 1) \\ h_0 &= 1. \end{aligned}$$

26. Solve the nonhomogeneous recurrence relation

$$\begin{aligned} h_n &= 6h_{n-1} - 9h_{n-2} + 2n, \quad (n \geq 2) \\ h_0 &= 1 \\ h_1 &= 0. \end{aligned}$$

27. Solve the nonhomogeneous recurrence relation

$$\begin{aligned} h_n &= 4h_{n-1} - 4h_{n-2} + 3n + 1, \quad (n \geq 2) \\ h_0 &= 1 \\ h_1 &= 2. \end{aligned}$$

28. Determine the generating function for each of the following sequences:

   (a) $c^0 = 1, c, c^2, \ldots, c^n, \ldots$.
   (b) $1, -1, 1, -1, \ldots, (-1)^n, \ldots$.
   (c) $\begin{pmatrix} \alpha \\ 0 \end{pmatrix}, -\begin{pmatrix} \alpha \\ 1 \end{pmatrix}, \begin{pmatrix} \alpha \\ 2 \end{pmatrix}, \ldots, (-1)^n \begin{pmatrix} \alpha \\ n \end{pmatrix}, \ldots,$
       (α is a real number.)
   (d) $1, \frac{1}{1!}, \frac{1}{2!}, \ldots, \frac{1}{n!}, \ldots$.
   (e) $1, -\frac{1}{1!}, \frac{1}{2!}, \ldots, (-1)^n \frac{1}{n!}, \ldots$.

29. Let S be the multiset $\{\infty \cdot e_1, \infty \cdot e_2, \infty \cdot e_3, \infty \cdot e_4\}$. Determine the generating function for the sequence $h_0, h_1, h_2, \ldots, h_n, \ldots$ where h_n is the number of n-combinations of S with the following added restrictions:

   (a) Each e_i occurs an odd number of times.
   (b) Each e_i occurs a multiple-of-3 number of times.
   (c) The element e_1 does not occur, and e_2 occurs at most once.
   (d) The element e_1 occurs $1, 3$, or 11 times, and the element e_2 occurs $2, 4$, or 5 times.
   (e) Each e_i occurs at least 10 times.

30. Solve the following recurrence relations by using the method of generating functions as described in Section 7.5:

   (a) $h_n = 4h_{n-2}, (n \geq 2); h_0 = 0, h_1 = 1$
   (b) $h_n = h_{n-1} + h_{n-2}, (n \geq 2); h_0 = 1, h_1 = 3$
   (c) $h_n = h_{n-1} + 9h_{n-2} - 9h_{n-3}, (n \geq 3); h_0 = 0, h_1 = 1, h_2 = 2$
   (d) $h_n = 8h_{n-1} - 16h_{n-2}, (n \geq 2); h_0 = -1, h_1 = 0$
   (e) $h_n = 3h_{n-2} - 2h_{n-3}, (n \geq 3); h_0 = 1, h_1 = 0, h_2 = 0$
   (f) $h_n = 5h_{n-1} - 6h_{n-2} - 4h_{n-3} + 8h_{n-4}, (n \geq 4); h_0 = 0, h_1 = 1, h_2 = 1, h_3 = 2$

31. Solve the nonhomogeneous recurrence relation

$$h_n = 4h_{n-1} + 4^n, \quad (n \geq 1)$$
$$h_0 = 3.$$

32. Determine the generating function for the sequence of cubes

$$0, 1, 8, \ldots, n^3, \ldots.$$

33. Let $h_0, h_1, h_2, \ldots, h_n, \ldots$ be the sequence defined by

$$h_n = n^3, \quad (n \geq 0).$$

Show that $h_n = h_{n-1} + 3n^2 - 3n + 1$ is the recurrence relation for the sequence.

34. Formulate a combinatorial problem that leads to the following generating function:

$$(1+x+x^2)(1+x^2+x^4+x^6)(1+x^2+x^4+\cdots)(x+x^2+x^3+\cdots).$$

35. Determine the generating function for the number h_n of bags of fruit of apples, oranges, bananas, and pears in which there are an even number of apples, at most two oranges, a multiple of three number of bananas, and at most one pear. Then find a formula for h_n from the generating function.

36. Determine the generating function for the number h_n of nonnegative integral solutions of

$$2e_1 + 5e_2 + e_3 + 7e_4 = n.$$

37. Let $h_0, h_1, h_2, \ldots, h_n, \ldots$ be the sequence defined by $h_n = \binom{n}{2}$, $(n \geq 0)$. Determine the generating function for the sequence.

38. Let $h_0, h_1, h_2, \ldots, h_n, \ldots$ be the sequence defined by $h_n = \binom{n}{3}$, $(n \geq 0)$. Determine the generating function for the sequence.

39. * Let h_n denote the number of regions into which a convex polygonal region with $n+2$ sides is divided by its diagonals, assuming no three diagonals have a common point. Define $h_0 = 0$. Show that

$$h_n = h_{n-1} + \binom{n+1}{3} + n, \quad (n \geq 1).$$

Then determine the generating function and also obtain a formula for h_n.

40. Determine the exponential generating function for the sequence of factorials: $0!, 1!, 2!, 3!, \ldots, n!, \ldots$.

41. Let α be a real number. Let the sequence $h_0, h_1, h_2, \ldots, h_n, \ldots$ be defined by $h_0 = 1$, and $h_n = \alpha(\alpha - 1) \cdots (\alpha - n + 1)$, $(n \geq 1)$. Determine the exponential generating function for the sequence.

42. Let S denote the multiset $\{\infty \cdot e_1, \infty \cdot e_2, \ldots, \infty \cdot e_k\}$. Determine the exponential generating function for the sequence $h_0, h_1, h_2, \ldots, h_n, \ldots$ where $h_0 = 1$ and, for $n \geq 1$:

    (a) h_n equals the number of n-permutations of S in which each object occurs an odd number of times.

    (b) h_n equals the number of n-permutations of S in which each object occurs at least four times.

    (c) h_n equals the number of n-permutations of S in which e_1 occurs at least once, e_2 occurs at least twice, $\ldots$, e_k occurs at least k times.

    (d) h_n equals the number of n-permutations of S in which e_1 occurs at most once, e_2 occurs at most twice, $\ldots$, e_k occurs at most k times.

43. Let h_n denote the number of ways to color the squares of a 1-by-n board with the colors red, white, blue, and green in such a way that the number of squares colored red is even and the number of squares colored white is odd. Determine the exponential generating function for the sequence $h_0, h_1, \ldots, h_n, \ldots$, and then find a simple formula for h_n.

44. Determine the number of ways to color the squares of a 1-by-n chessboard, using the colors red, blue, green, and orange if an even number of squares is to be colored red and an even number is to be colored green.

45. Determine the number of n digit numbers with all digits odd, such that 1 and 3 each occur a nonzero, even number of times.

46. Determine the number of n digit numbers with all digits at least 4, such that 4 and 6 each occur an even number of times, and 5 and 7 each occur at least once, there being no restriction on the digits 8 and 9.

47. We have used exponential generating functions to show that the number h_n of n digit numbers with each digit odd, where the digits 1 and 3 occur an even number of times, satisfies the formula

$$h_n = \frac{5^n + 2 \times 3^n + 1}{4}, \quad (n \geq 0).$$

Obtain an alternative derivation of this formula.

48. We have used exponential generating functions to show that the number h_n of ways to color the squares of a 1-by-n board with the colors red, white , and blue, where the number of red squares is even and there is at least one blue square, satisfies the formula

$$h_n = \frac{3^n - 2^n + 1}{2}, \quad (n \geq 1)$$

with $h_0 = 0$. Obtain an alternative derivation of this formula by finding a recurrence relation satisfied by h_n and then solving the recurrence relation.

Chapter 8

Special Counting Sequences

We have already considered several special counting sequences in the previous chapters. The counting sequence for permutations of a set of n elements is

$$0!, 1!, 2!, \ldots, n!, \ldots.$$

The counting sequence for derangements of a set of n elements is

$$D_0, D_1, D_2, \ldots, D_n, \ldots,$$

where D_n has been evaluated in Theorem 6.3.1. In addition, we have investigated the Fibonacci sequence

$$f_0, f_1, f_2, \ldots, f_n, \ldots,$$

and a formula for f_n has been given in Theorem 7.1.1. In this chapter, we study primarily four famous and important counting sequences, the sequence of Catalan numbers, the sequences of the Stirling numbers of the first and second kind, and the sequence of the number of partitions of a positive integer n.

8.1 Catalan Numbers

The *Catalan sequence*[1] is the sequence

$$C_0, C_1, C_2, \ldots, C_n, \ldots,$$

---

[1] After Eugène Catalan (1814–1894).

where

$$C_n = \frac{1}{n+1}\binom{2n}{n}, \qquad (n = 0, 1, 2, \ldots)$$

is the nth *Catalan number*. The first few Catalan numbers are

$$
\begin{array}{ll}
C_0 = 1 & C_5 = 42 \\
C_1 = 1 & C_6 = 132 \\
C_2 = 2 & C_7 = 429 \\
C_3 = 5 & C_8 = 1430 \\
C_4 = 14 & C_9 = 4862.
\end{array}
$$

The Catalan number

$$C_{n-1} = \frac{1}{n}\binom{2n-2}{n-1}$$

arose in Section 7.6 as the number of ways to divide a convex polygonal region with $n + 1$ sides into triangles by inserting diagonals that do not intersect in the interior. The Catalan numbers occur in several seemingly unrelated counting problems and we discuss some of them in this section.

Theorem 8.1.1 *The number of sequences*

$$a_1, a_2, \ldots, a_{2n} \qquad (8.1)$$

of $2n$ terms that can be formed by using n $+1$'s and n -1's whose partial sums satisfy

$$a_1 + a_2 + \cdots + a_k \geq 0, \qquad (k = 1, 2, \ldots, 2n) \qquad (8.2)$$

equals the nth Catalan number

$$C_n = \frac{1}{n+1}\binom{2n}{n}, \qquad (n \geq 0).$$

Proof. We call a sequence (8.1) of n $+1$'s and n -1's *acceptable* if it satisfies (8.2) and *unacceptable* otherwise. Let A_n denote the number of acceptable sequences of n $+1$'s and n -1's, and let U_n denote the number of unacceptable ones. The total number of sequences of n $+1$'s and n -1's is

$$\binom{2n}{n} = \frac{(2n)!}{n!n!},$$

since such sequences can be regarded as the permutations of objects of two different types with n objects of one type (the $+1$'s) and n of the other (the -1's). Hence,

$$A_n + U_n = \binom{2n}{n},$$

and we evaluate A_n by first evaluating U_n and then subtracting from $\binom{2n}{n}$.

Consider an unacceptable sequence (8.1) of n $+1$'s and n -1's. Because the sequence is unacceptable, there is a *smallest* k such that the partial sum

$$a_1 + a_2 + \cdots + a_k$$

is negative. Because k is smallest, there is an equal number of $+1$'s and -1's preceding a_k, and we have

$$a_1 + a_2 + \cdots + a_{k-1} = 0$$

and

$$a_k = -1.$$

In particular, k is an odd integer. We now reverse the signs of each of the first k terms; that is, we replace a_i by $-a_i$ for each $i = 1, 2, \ldots, k$ and leave unchanged the remaining terms. The resulting sequence

$$a_1', a_2', \ldots, a_{2n}'$$

is a sequence of $(n+1)$ $+1$'s and $(n-1)$ -1's. This process is reversible: Given a sequence of $(n+1)$ $+1$'s and $(n-1)$ -1's, there is a first instance when the number of $+1$'s exceeds the number of -1's (since there are more $+1$'s than -1's). Reversing the signs of the $+1$'s and -1's up to that point results in an unacceptable sequence of n $+1$'s and n -1's. Thus, there are as many unacceptable sequences as there are sequences of $(n+1)$ $+1$'s and $(n-1)$ -1's. The number of sequences of $(n+1)$ $+1$'s and $(n+1)$ -1's is the number

$$\frac{(2n)!}{(n+1)!(n-1)!}$$

of permutations of objects of two types, with $n+1$ objects of one type and $n-1$ of the other. Hence,

$$U_n = \frac{(2n)!}{(n+1)!(n-1)!},$$

and therefore,

$$
\begin{aligned}
A_n &= \frac{(2n)!}{n!n!} - \frac{(2n)!}{(n+1)!(n-1)!} \\
&= \frac{(2n)!}{n!(n-1)!}\left(\frac{1}{n} - \frac{1}{n+1}\right) \\
&= \frac{(2n)!}{n!(n-1)!}\left(\frac{1}{n(n+1)}\right) \\
&= \frac{1}{n+1}\binom{2n}{n}.
\end{aligned}
$$

□

There are many different interpretations of Theorem 8.1.1. We discuss two of them in the next examples.

Example. There are $2n$ people in line to get into a theatre. Admission is 50 cents.[2] Of the $2n$ people, n have a 50-cent piece and n have a \$1 dollar bill.[3] The box office at the theatre rather foolishly begins with an empty cash register. In how many ways can the people line up so that whenever a person with a \$1 dollar bill buys a ticket, the box office has a 50-cent piece in order to make change? (After every one is admitted, there will be n \$1 dollar bills in the cash register.)

First, suppose that the people are regarded as "indistinguishable;" that is, we simply have a sequence of n 50-cent pieces and n dollar bills, and it doesn't matter who holds which and where they are in the line. If we identify a 50-cent piece with a $+1$ and a dollar bill with a -1, then the answer is the number

$$
C_n = \frac{1}{n+1}\binom{2n}{n}
$$

of acceptable sequences as defined in Theorem 8.1.1. Now suppose that the people are regarded as "distinguishable;" that is, we take into account who is who in the line. So we have n people holding 50-cent pieces and n holding dollar bills. The answer is now

$$
(n)!(n!)\frac{1}{n+1}\binom{2n}{n} = \frac{(2n)!}{n+1},
$$

---

[2]This problem shows its age!

[3]To have this problem more accurately reflect reality, perhaps we should have the theater charge \$5, and have n people with \$5 dollar bills and n with \$10 bills.

since, with each sequence of n 50-cent pieces and n dollar bills, there are $n!$ orders for the people with 50-cent pieces and $n!$ orders for the people with dollar bills. □

Example. A big city lawyer works n blocks north and n blocks east of her place of residence. Every day she walks $2n$ blocks to work. (See the map below for $n = 4$.) How many routes are possible if she never crosses (but may touch) the diagonal line from home to office?

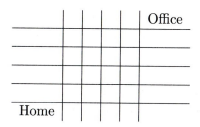

Each acceptable route is either above the diagonal or below the diagonal. We find the number of acceptable routes above the diagonal, and multiply by 2. Each route is a sequence of n northerly blocks and n easterly blocks. We identify north with $+1$ and east with -1. Thus, each route corresponds to a sequence

$$a_1, a_2, \ldots, a_{2n}$$

of n $+1$'s and n -1's, and in order that the route not dip below the diagonal, we must have

$$\sum_{i=1}^{k} a_i \geq 0, \qquad (k = 1, \ldots, 2n).$$

Hence, by Theorem 8.1.1, the number of acceptable routes above the diagonal equals the nth Catalan number, and the total number of acceptable routes is

$$2C_n = \frac{2}{n+1}\binom{2n}{n}.$$

□

We next show that the Catalan numbers satisfy a homogeneous recurrence relation of order 1 (but with a nonconstant coefficient). We have

$$C_n = \frac{1}{n+1}\binom{2n}{n} = \frac{1}{n+1}\frac{(2n)!}{n!n!}$$

and

$$C_{n-1} = \frac{1}{n}\binom{2n-2}{n-1} = \frac{1}{n}\frac{(2n-2)!}{(n-1)!(n-1)!}.$$

Dividing, we obtain

$$\frac{C_n}{C_{n-1}} = \frac{4n-2}{n+1}.$$

Therefore, the Catalan sequence is determined by the following recurrence relation and initial condition:

$$
\begin{aligned}
C_n &= \frac{4n-2}{n+1}C_{n-1}, \quad (n \geq 1)\\
C_0 &= 1.
\end{aligned}
\tag{8.3}
$$

Previously we noted that $C_9 = 4862$. It follows from the recurrence relation (8.3) that

$$C_{10} = \frac{38}{11}\,C_9 = \frac{38}{11}(4862) = 16,796.$$

We now define a new sequence of numbers

$$C_1^*, C_2^*, \ldots, C_n^*, \ldots,$$

which, in order to refer to them by name, we call the *pseudo-Catalan numbers*. We let

$$C_n^* = n!C_{n-1}, \qquad (n = 1, 2, 3, \ldots).$$

We have

$$C_1^* = 1!(1) = 1,$$

and using (8.3) with n replaced by $n-1$, we obtain

$$
\begin{aligned}
C_n^* &= n!C_{n-1}\\
&= n!\frac{4n-6}{n}C_{n-2}\\
&= (4n-6)(n-1)!C_{n-2}\\
&= (4n-6)C_{n-1}^*.
\end{aligned}
$$

Thus, the pseudo-Catalan numbers are determined by the following recurrence relation and initial condition:

$$
\begin{aligned}
C_n^* &= (4n-6)C_{n-1}^*, \quad (n \geq 2)\\
C_1^* &= 1.
\end{aligned}
\tag{8.4}
$$

Using this recurrence relation, we calculate the first few pseudo-Catalan numbers:

$$
\begin{array}{ll}
C_1^* = 1 & C_4^* = 120 \\
C_2^* = 2 & C_5^* = 1680 \\
C_3^* = 12 & C_6^* = 30240.
\end{array}
$$

The defining formula for the Catalan numbers and the definition of the pseudo-Catalan numbers imply the formula

$$
C_n^* = (n-1)! \binom{2n-2}{n-1} = \frac{(2n-2)!}{(n-1)!}, \qquad (n \geq 1)
$$

for the pseudo-Catalan numbers. This formula can also be derived from the recurrence relation (8.4).

Example. Let $a_1, a_2, \ldots, a_n$ be n numbers. By a *multiplication scheme* for these numbers we mean a scheme for carrying out the multiplication of $a_1, a_2, \ldots, a_n$. A multiplication scheme requires a total of $n-1$ multiplications between two numbers, each of which is either one of $a_1, a_2, \ldots, a_n$ or a partial product of them. Let h_n denote the number of multiplication schemes for n numbers. We have $h_1 = 1$ (this can be taken as the definition of h_1) and $h_2 = 2$, since

$$
(a_1 \times a_2) \quad \text{and} \quad (a_2 \times a_1)
$$

are two possible schemes. This example serves to show that the order of the numbers in the multiplication scheme is taken into consideration.[4] If $n = 3$, there are 12 schemes:

$$
\begin{array}{lll}
(a_1 \times (a_2 \times a_3)) & (a_2 \times (a_1 \times a_3)) & (a_3 \times (a_1 \times a_2)) \\
((a_2 \times a_3) \times a_1) & ((a_1 \times a_3) \times a_2) & ((a_1 \times a_2) \times a_3) \\
(a_1 \times (a_3 \times a_2)) & (a_2 \times (a_3 \times a_1)) & (a_3 \times (a_2 \times a_1)) \\
((a_3 \times a_2) \times a_1) & ((a_3 \times a_1) \times a_2) & ((a_2 \times a_1) \times a_3).
\end{array}
$$

Thus, $h_3 = 12$. Each multiplication scheme for three numbers requires two multiplications and each multiplication corresponds to a set of parentheses. The outside parentheses allow us to identify each multiplication $\times$ with a set of parentheses. In general, each multiplication scheme can be obtained by listing $a_1, a_2, \ldots, a_n$ in some order and then inserting $n-1$ pairs of parentheses so that each pair of parentheses designates a multiplication of two factors. But in order to derive a recurrence relation for h_n, we look at it in an inductive way. Each scheme for $a_1, a_2, \ldots, a_n$ can be gotten from a scheme for $a_1, a_2, \ldots, a_{n-1}$ in exactly one of the following ways:

---

[4]In more algebraic language, we are not assuming that the commutative law holds.

(i) Take a multiplication scheme for $a_1, a_2, \ldots, a_{n-1}$ (which has $n-2$ multiplications and $n - 2$ sets of parentheses) and insert a_n on either side of either factor in one of the $n - 2$ multiplications. Thus, each scheme for $n-1$ numbers gives $2 \times 2 \times (n-2) = 4(n-2)$ schemes for n numbers in this way.

(ii) Take a multiplication scheme for $a_1, a_2, \ldots, a_{n-1}$ and multiply it on the left or right by a_n. Thus, each scheme for $n - 1$ numbers gives two schemes for n numbers in this way.

To illustrate, let $n = 6$ and consider the multiplication scheme

$$((a_1 \times a_2) \times ((a_3 \times a_4) \times a_5))$$

for a_1, a_2, a_3, a_4, a_5.[5] There are 4 multiplications in this scheme. We take any one of them, say, the multiplication of $(a_3 \times a_4)$ and a_5, and insert a_6 on either side of either of these two factors to get

$$((a_1 \times a_2) \times (((a_6 \times (a_3 \times a_4)) \times a_5))$$
$$((a_1 \times a_2) \times (((a_3 \times a_4) \times a_6) \times a_5))$$
$$((a_1 \times a_2) \times ((a_3 \times a_4) \times (a_6 \times a_5)))$$
$$((a_1 \times a_2) \times ((a_3 \times a_4) \times (a_5 \times a_6))).$$

There are $4 \times 4 = 16$ schemes for $a_1, a_2, a_3, a_4, a_5, a_6$ obtained in this way. Besides these, we have two additional schemes in which a_6 enters into the final multiplication, namely,

$$(a_6 \times ((a_1 \times a_2) \times ((a_3 \times a_4) \times a_5)))), \quad ((((a_1 \times a_2) \times ((a_3 \times a_4) \times a_5)) \times a_6).$$

Thus, each multiplication scheme for five numbers gives 18 schemes for six numbers, and we have $h_6 = 18h_5$.

Let $n \geq 2$. Then, generalizing the foregoing analysis, we see that each of the h_{n-1} multiplication schemes for $n - 1$ numbers gives

$$4(n - 2) + 2 = 4n - 6$$

schemes for n numbers We thus obtain the recurrence relation

$$h_n = (4n - 6)h_{n-1}, \quad (n \geq 2),$$

which, together with the initial value $h_1 = 1$, determines the entire sequence $h_1, h_2, \ldots, h_n, \ldots$. This is the same type of recurrence relation

---

[5]Which multiplication $\times$ corresponds to each set of parentheses in the preceding scheme?

with the same initial value satisfied by the pseudo-Catalan numbers (8.4). Hence,

$$h_n = C_n^* = (n-1)! \binom{2n-2}{n-1}, \quad (n \geq 1).$$

$\square$

In the preceding example, suppose that we count only those multiplication schemes in which the n numbers are listed in the order $a_1, a_2, \ldots, a_n$. Thus, for instance, $((a_2 \times a_1) \times a_3)$ is no longer counted. Let g_n denote the number of multiplication schemes with this additional restriction. Then, since we consider only one of the $n!$ possible orderings, $h_n = n! g_n$, and hence,

$$g_n = \frac{h_n}{n!} = \frac{C_n^*}{n!} = \frac{1}{n}\binom{2n-2}{n-1} = C_{n-1}, \quad (n \geq 1), \qquad (8.5)$$

showing that g_n is the $(n-1)$st Catalan number.

We can also derive a recurrence relation for g_n, using its definition as follows: In each scheme for $a_1, a_2, \ldots, a_n$ there is a final multiplication $\times$, and it corresponds to the outer parentheses. We thus have

$$((\text{scheme for} \quad a_1, \ldots, a_k) \times (\text{scheme for} \quad a_{k+1}, \ldots, a_n)).$$

The multiplication scheme for $a_1, \ldots, a_k$ can be chosen in g_k ways, and the multiplication scheme for $a_{k+1}, \ldots, a_n$ can be chosen in g_{n-k} ways. Since k can be any of the numbers $1, 2, \ldots, n-1$, we have

$$g_n = g_1 g_{n-1} + g_2 g_{n-2} + \cdots + g_{n-1} g_1, \quad (n \geq 2). \qquad (8.6)$$

This recurrence relation, along with the initial condition $g_1 = 1$, uniquely determines the counting sequence

$$g_1, g_2, g_3, \ldots, g_n, \ldots .$$

From our calculations in (8.5), the solution of the recurrence relation (8.6) that satisfies the initial condition $g_1 = 1$ is

$$g_n = C_{n-1} = \frac{1}{n}\binom{2n-2}{n-1}, \quad (n \geq 1).$$

The recurrence relation (8.6) is the same recurrence relation that occurred in Section 7.6 in connection with the problem of dividing a convex polygonal region into triangles by means of its diagonals, where we showed by analytic means that its solution is C_{n-1}. Thus, we have a purely combinatorial derivation of the formula obtained in Section 7.6, and we conclude that the number of ways to divide a convex polygonal region with $n + 1$ sides into triangular regions by inserting diagonals that do not intersect in the interior is the same as the number of multiplication schemes for n numbers given in a specified order with the common value equal to the $(n - 1)$st Catalan number.

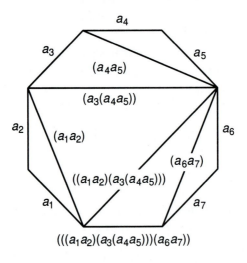

Figure 8.1

The correspondence between the multiplication schemes for the n numbers $a_1, a_2, \ldots, a_n$ and triangularizations of convex polygonal regions of $n + 1$ sides is indicated in Figure 8.1 for $n = 7$. Each diagonal corresponds to one of the multiplications other than the last, with the base of the polygon corresponding to the last multiplication.

8.2   Difference Sequences and Stirling Numbers

Let

$$h_0, h_1, h_2, \ldots, h_n, \ldots \tag{8.7}$$

be a sequence of numbers. We define a new sequence

$$\Delta h_0, \Delta h_1, \Delta h_2, \ldots, \Delta h_n, \ldots, \tag{8.8}$$

called the *(first-order) difference sequence* of (8.7), by

$$\Delta h_n = h_{n+1} - h_n, \qquad (n \geq 0).$$

The terms of the difference sequence (8.8) are the differences of consecutive terms of the sequence (8.7). We can form the difference sequence of (8.8) and obtain the *second-order difference sequence*

$$\Delta^2 h_0, \Delta^2 h_1, \Delta^2 h_2, \ldots, \Delta^2 h_n, \ldots.$$

Here,

$$
\begin{aligned}
\Delta^2 h_n &= \Delta(\Delta h_n) \\
&= \Delta h_{n+1} - \Delta h_n \\
&= (h_{n+2} - h_{n+1}) - (h_{n+1} - h_n) \\
&= h_{n+2} - 2h_{n+1} + h_n, \qquad (n \geq 0).
\end{aligned}
$$

More generally, we can inductively define the *pth-order difference* sequence of (8.7) by

$$\Delta^p h_0, \Delta^p h_1, \Delta^p h_2, \ldots, \Delta^p h_n, \ldots \qquad (p \geq 1),$$

where

$$\Delta^p h_n = \Delta(\Delta^{p-1} h_n).$$

Thus, the pth-order difference sequence is the first-order difference sequence of the $(p-1)$st-order difference sequence. We define the 0th-*order difference sequence* of a sequence to be itself; that is,

$$\Delta^0 h_n = h_n, \qquad (n \geq 0).$$

The *difference table* for the sequence (8.7) is obtained by listing the pth-order difference sequences in a row for each $p = 0, 1, 2, \ldots$:

$$h_0 \quad h_1 \quad h_2 \quad h_3 \quad h_4 \quad \cdots$$

$$\Delta h_0 \quad \Delta h_1 \quad \Delta h_2 \quad \Delta h_3 \quad \cdots$$

$$\Delta^2 h_0 \quad \Delta^2 h_1 \quad \Delta^2 h_2 \quad \cdots$$

$$\Delta^3 h_0 \quad \Delta^3 h_1 \quad \cdots$$

$$\cdots$$

The pth-order differences are in row p, with the sequence itself in row 0. (Thus, we start counting the rows with 0.)

Example. Let a sequence $h_0, h_1, h_2, \ldots, h_n, \ldots$ be defined by

$$h_n = 2n^2 + 3n + 1, \qquad (n \geq 0).$$

The difference table for this sequence is

$$1 \quad 6 \quad 15 \quad 28 \quad 45 \quad 66 \quad 91 \; \cdots$$
$$5 \quad 9 \quad 13 \quad 17 \quad 21 \quad 25 \; \cdots$$
$$4 \quad 4 \quad 4 \quad 4 \quad 4 \; \cdots$$
$$0 \quad 0 \quad 0 \quad 0 \; \cdots$$
$$\cdots$$

The third-order difference sequence in this case consists of all 0's and hence so do all higher-order differences sequences. $\quad\square$

We now show that if a sequence has the property that its general term is a polynomial of degree p in n, then the $(p+1)$th-order differences are all 0. When this happens, we may suppress all the rows of 0's after the first row of 0's.

Theorem 8.2.1 *Let the general term of a sequence be a polynomial of degree p in n:*

$$h_n = a_p n^p + a_{p-1} n^{p-1} + \cdots + a_1 n + a_0, \qquad (n \geq 0).$$

Then $\Delta^{p+1} h_n = 0$ for all $n \geq 0$.

Proof. We prove the theorem by induction on p. If $p = 0$, then we have

$$h_n = a_0, \text{a constant, for all } n \geq 0;$$

and hence,

$$\Delta h_n = h_{n+1} - h_n = a_0 - a_0 = 0, \qquad (n \geq 0).$$

We now suppose that $p \geq 1$ and assume that the theorem holds when the general term is a polynomial of degree at most $p - 1$ in n. We have

$$\Delta h_n = (a_p(n+1)^p + a_{p-1}(n+1)^{p-1} + \cdots + a_1 n + a_0)$$
$$- (a_p n^p + a_{p-1} n^{p-1} + \cdots + a_1 n + a_0).$$

By the binomial theorem,

$$a_p(n+1)^p - a_p n^p = a_p \left(n^p + \binom{p}{1} n^{p-1} + \cdots + 1 \right) - a_p n^p$$

$$= a_p \binom{p}{1} n^{p-1} + \cdots + a_p.$$

From this calculation, we conclude that the pth powers of n cancel in Δh_n and that Δh_n is a polynomial in n of degree at most $p - 1$. By the induction assumption,

$$\Delta^p(\Delta h_n) = 0, \qquad (n \geq 0).$$

Since $\Delta^{p+1} h_n = \Delta^p(\Delta h_n)$, it now follows that

$$\Delta^{p+1} h_n = 0, \qquad (n \geq 0).$$

Hence, the theorem holds by induction. $\square$

Now suppose that g_n and f_n are the general terms of two sequences, and another sequence is defined by

$$h_n = g_n + f_n, \qquad (n \geq 0).$$

Then

$$
\begin{aligned}
\Delta h_n &= h_{n+1} - h_n \\
&= (g_{n+1} + f_{n+1}) - (g_n + f_n) \\
&= (g_{n+1} - g_n) + (f_{n+1} - f_n) \\
&= \Delta g_n + \Delta f_n.
\end{aligned}
$$

More generally, it follows inductively that

$$\Delta^p h_n = \Delta^p g_n + \Delta^p f_n, \qquad (p \geq 0)$$

and, indeed, if c and d are constants, it also follows that

$$\Delta^p(cg_n + df_n) = c\Delta^p g_n + d\Delta^p f_n, \qquad (n \geq 0) \qquad (8.9)$$

for each integer $p \geq 0$. We refer to the property in (8.9) as the *linearity property* of differences.[6] From (8.9) we see that the difference table for the sequence of h_n's can be obtained by multiplying the entries of the difference table for the g_n's by c and multiplying the entries of the difference table for the f_n's by d, and then adding corresponding entries.

Example. Let $g_n = n^2 + n + 1$ and let $f_n = n^2 - n - 2$, $(n \geq 0)$. The difference table for the g_n's is

---

[6]In the language of linear algebra, the set of sequences forms a vector space, and Δ is a linear transformation on this vector space.

$$
\begin{array}{cccccc}
1 & 3 & 7 & 13 & 21 & \cdots \\
 & 2 & 4 & 6 & 8 & \cdots \\
 & & 2 & 2 & 2 & \cdots \\
 & & & 0 & 0 & \cdots
\end{array}
$$

The difference table for the f_n's is

$$
\begin{array}{cccccc}
-2 & -2 & 0 & 4 & 10 & \cdots \\
 & 0 & 2 & 4 & 6 & \cdots \\
 & & 2 & 2 & 2 & \cdots \\
 & & & 0 & 0 & \cdots
\end{array}
$$

Let

$$
\begin{aligned}
h_n = 2g_n + 3f_n &= 2(n^2 + n + 1) + 3(n^2 - n - 2) \\
&= 5n^2 - n - 4.
\end{aligned}
$$

The difference table for the h_n's is obtained by multiplying the entries of the first difference table by 2 and the entries of the second difference table by 3 and then adding corresponding entries. The result is

$$
\begin{array}{cccccc}
-4 & 0 & 14 & 38 & 72 & \cdots \\
 & 4 & 14 & 24 & 34 & \cdots \\
 & & 10 & 10 & 10 & \cdots \\
 & & & 0 & 0 & \cdots
\end{array}
$$

$\square$

By its very definition, the difference table for a sequence $h_0, h_1, h_2, \ldots, h_n, \ldots$ is determined by the entries in row number 0. We next observe that the difference table is also determined by the entries along the left edge, the *0th diagonal*—that is, by the numbers

$$
h_0 = \Delta^0 h_0, \Delta^1 h_0, \Delta^2 h_0, \Delta^3 h_0, \ldots
$$

along the leftmost diagonal of the difference table.[7] This property is a consequence of the fact that the entries on a diagonal (running from left to right) of the difference table are determined from those on the previous diagonal. For instance, the entries on the 1st diagonal are

$$
\begin{aligned}
h_1 = \Delta^0 h_1 &= \Delta^1 h_0 + \Delta^0 h_0 = \Delta h_0 + h_0 \\
\Delta h_1 &= \Delta^2 h_0 + \Delta h_0 \\
\Delta^2 h_1 &= \Delta^3 h_0 + \Delta^2 h_0
\end{aligned}
$$

$$
\cdots \qquad \cdots
$$

---

[7] This property is the discrete analogue of the fact that an analytic function $f(x)$ is determined (via its Taylor expansion) by the value of the function and all its derivatives at $x = 0$: $f(0), f'(0), f''(0), \ldots$.

If the 0th diagonal of a difference table contains only 0's, then the entire difference table contains only 0's. The next simplest 0th diagonal is one that contains only 0's except for one 1, say, in row p. (Thus there are p 0's preceding the 1.) From the fact that the entries on the 0th diagonal in rows $p + 1, p + 2, \ldots$ are all 0, it is apparent that all the entries in rows $p + 1, p + 2, \ldots$ equal 0.

Suppose, for instance, $p = 4$. Thus, rows 5 and greater contain only 0's. Can we find the general term of a sequence such that the 0th diagonal of its difference table is

$$0, 0, 0, 0, 1, 0, 0, \ldots? \qquad (8.10)$$

We use these entries on the left edge to determine a triangular portion of the difference table and obtain

$$
\begin{array}{ccccc}
0 & 0 & 0 & 0 & 1 \\
 & 0 & 0 & 0 & 1 \\
 & & 0 & 0 & 1 \\
 & & & 0 & 1 \\
 & & & & 1
\end{array}
$$

Since row number 5 consists of all 0's, we look for a sequence whose nth term h_n is a polynomial in n of degree 4. From the portion of the difference table just computed, we see that

$$h_0 = 0, \ h_1 = 0, \ h_2 = 0, \ h_3 = 0, \ \text{and} \ h_4 = 1.$$

Thus, if h_n is a polynomial of degree 4, it has roots $0, 1, 2, 3$, and hence,

$$h_n = cn(n-1)(n-2)(n-3)$$

for some constant c. Since $h_4 = 1$, we must have

$$1 = c(4)(3)(2)(1) \ \text{or, equivalently,} \ c = \frac{1}{4!}.$$

Accordingly, the sequence with general term

$$h_n = \frac{n(n-1)(n-2)(n-3)}{4!} = \binom{n}{4}, \qquad (n \geq 0)$$

has a difference table with 0th diagonal given by (8.10).

The same kind of argument shows that, more generally,

$$h_n = \frac{n(n-1)(n-2)\cdots(n-(p-1))}{p!} = \binom{n}{p}$$

is a polynomial in n of degree p whose difference table has its 0th diagonal equal to

$$\overbrace{0, 0, \ldots, 0}^{p}, 1, 0, 0, \ldots.$$

Using the linearity property of differences and the fact that the 0th diagonal of a difference table determines the entire difference table, and hence the sequence itself, we obtain the next theorem.

Theorem 8.2.2 *The general term of the sequence whose difference table has its 0th diagonal equal to*

$$c_0, c_1, c_2, \ldots, c_p, 0, 0, 0, \ldots \qquad where \ c_p \neq 0$$

is a polynomial in n of degree p satisfying

$$h_n = c_0 \binom{n}{0} + c_1 \binom{n}{1} + c_2 \binom{n}{2} + \cdots + c_p \binom{n}{p}. \qquad (8.11)$$

$\square$

Combining Theorems 8.2.1 and 8.2.2, we see that every polynomial in n of degree p can be expressed in the form (8.11) for some choice of constants $c_0, c_1, \ldots, c_p$. These constants are uniquely determined. (See Exercise 10.)

Example. Consider the sequence with general term

$$h_n = n^3 + 3n^2 - 2n + 1, \qquad (n \geq 0).$$

Computing differences, we obtain

$$
\begin{array}{cccc}
1 & 3 & 17 & 49 \\
 & 2 & 14 & 32 \\
 & & 12 & 18 \\
 & & & 6
\end{array}
$$

Since h_n is a polynomial in n of degree 3, the 0th diagonal of the difference table is

$$1, 2, 12, 6, 0, 0, \ \ldots.$$

Hence, by Theorem 8.2.2, another way to write h_n is

$$h_n = 1 \binom{n}{0} + 2 \binom{n}{1} + 12 \binom{n}{2} + 6 \binom{n}{3}. \qquad (8.12)$$

Why would we want to write h_n in this way? Here's one reason. Suppose we want to find the partial sums

$$\sum_{k=0}^{n} h_k = h_0 + h_1 + \cdots + h_n.$$

Using (8.12), we see that

$$\sum_{k=0}^{n} h_k = 1 \sum_{k=0}^{n} \binom{k}{0} + 2 \sum_{k=0}^{n} \binom{k}{1} + 12 \sum_{k=0}^{n} \binom{k}{2} + 6 \sum_{k=0}^{n} \binom{k}{3}.$$

From (5.14) we know that

$$\sum_{k=0}^{n} \binom{k}{p} = \binom{n+1}{p+1}. \tag{8.13}$$

Hence,

$$\sum_{k=0}^{n} h_k = 1 \binom{n+1}{1} + 2 \binom{n+1}{2} + 12 \binom{n+1}{3} + 6 \binom{n+1}{4},$$

a very simple formula for the partial sums. □

The foregoing procedure can be used to find the partial sums of any sequence whose general term is a polynomial in n.

Theorem 8.2.3 *Assume that the sequence* $h_0, h_1, h_2, \ldots, h_n, \ldots$ *has a difference table whose 0th diagonal equals*

$$c_0, c_1, c_2, \ldots, c_p, 0, 0, \ldots.$$

Then

$$\sum_{k=0}^{n} h_k = c_0 \binom{n+1}{1} + c_1 \binom{n+1}{2} + \cdots + c_p \binom{n+1}{p+1}.$$

Proof. By Theorem 8.2.2, we have

$$h_n = c_0 \binom{n}{0} + c_1 \binom{n}{1} + \cdots + c_p \binom{n}{p}.$$

Using formula (8.13), we obtain

$$\begin{aligned}
\sum_{k=0}^{n} h_k &= c_0 \sum_{k=0}^{n} \binom{k}{0} + c_1 \sum_{k=0}^{n} \binom{k}{1} + \cdots + c_k \sum_{k=0}^{n} \binom{k}{p} \\
&= c_0 \binom{n+1}{1} + c_1 \binom{n+1}{2} + \cdots + c_p \binom{n+1}{p+1}. \quad \square
\end{aligned}$$

Example. Find the sum of the fourth powers of the first n positive integers.

Let $h_n = n^4$. Computing differences, we obtain

$$
\begin{array}{ccccc}
0 & 1 & 16 & 81 & 256 \\
 & 1 & 15 & 65 & 175 \\
 & & 14 & 50 & 110 \\
 & & & 36 & 60 \\
 & & & & 24
\end{array}
$$

Because h_n is a polynomial of degree 4, the 0th diagonal of the difference table equals

$$0, 1, 14, 36, 24, 0, 0, \ldots .$$

Hence,

$$
\begin{aligned}
1^4 + 2^4 + \cdots + n^4 &= \sum_{k=0}^{n} k^4 \\
&= 0\binom{n+1}{1} + 1\binom{n+1}{2} + 14\binom{n+1}{3} \\
&\quad + 36\binom{n+1}{4} + 24\binom{n+1}{5}. \qquad \square
\end{aligned}
$$

In a similar way, we can evaluate the sum of the pth powers of the first n positive integers by considering the sequence whose general term is $h_n = n^p$. The preceding example treated the case $p = 4$.

The numbers that occur in the 0th diagonal of the difference tables are of combinatorial significance, and we now discuss them.

Let

$$h_n = n^p.$$

By Theorems 8.2.1 and 8.2.2, the 0th diagonal of the difference table for h_n has the form

$$c(p, 0), c(p, 1), c(p, 2), \ldots, c(p, p), 0, 0, \ldots ,$$

and it follows that

$$n^p = c(p, 0)\binom{n}{0} + c(p, 1)\binom{n}{1} + \cdots + c(p, p)\binom{n}{p}. \qquad (8.14)$$

If $p = 0$, then $h_n = 1$, a constant, and (8.14) reduces to

$$n^0 = 1 = 1\binom{n}{0} = 1;$$

in particular,

$$c(0,0) = 1.$$

Since n^p, as a polynomial in n, has a constant term equal to 0, if $p \geq 1$, then we also have

$$c(p,0) = 0, \quad (p \geq 1).$$

We rewrite (8.14) by introducing a new expression. Let

$$[n]_k = \begin{cases} n(n-1)\cdots(n-k+1) & \text{if } k \geq 1 \\ 1 & \text{if } k = 0. \end{cases}$$

We note that $[n]_k$ is the same as $P(n,k)$, the number of k-permutations of n distinct objects (see Section 3.2), but we wish now to use the less cumbersome notation $[n]_k$. We also note that

$$[n]_{k+1} = (n-k)[n]_k.$$

Since

$$\binom{n}{k} = \frac{n(n-1)\cdots(n-k+1)}{k!} = \frac{[n]_k}{k!},$$

we obtain

$$[n]_k = k!\binom{n}{k}.$$

Hence, (8.14) can be rewritten as

$$\begin{aligned} n^p &= c(p,0)\frac{[n]_0}{0!} + c(p,1)\frac{[n]_1}{1!} + \cdots + c(p,p)\frac{[n]_p}{p!} \\ &= \sum_{k=0}^{p} c(p,k)\frac{[n]_k}{k!} \\ &= \sum_{k=0}^{p} \frac{c(p,k)}{k!}[n]_k. \end{aligned}$$

Now we introduce the numbers

$$S(p,k) = \frac{c(p,k)}{k!}, \quad (0 \leq k \leq p)$$

and in terms of them, (8.14) becomes

$$\begin{aligned} n^p &= S(p,0)[n]_0 + S(p,1)[n]_1 + \cdots + S(p,p)[n]_p \\ &= \sum_{k=0}^{p} S(p,k)[n]_k. \end{aligned} \tag{8.15}$$

The numbers $S(p, k)$ just introduced are called the *Stirling numbers*[8] *of the second kind.*[9] Since

$$S(p, 0) = \frac{c(p, 0)}{0!} = c(p, 0),$$

we have

$$S(p, 0) = \begin{cases} 1 & \text{if } p = 0 \\ 0 & \text{if } p \geq 1. \end{cases} \tag{8.16}$$

In (8.14), the coefficient of n^p on the left-hand side is 1, and on the right-hand side it is

$$\frac{c(p, p)}{p!}.$$

(Only the last term on the right side of (8.14) contributes to the coefficient of n^p, since the other terms are polynomials in n of degree less than p.) Thus, we have

$$S(p, p) = \frac{c(p, p)}{p!} = 1, \qquad (p \geq 0). \tag{8.17}$$

We now show that the Stirling numbers of the second kind satisfy a Pascal-like recurrence relation.

Theorem 8.2.4 *If* $1 \leq k \leq p - 1$, *then*

$$S(p, k) = kS(p - 1, k) + S(p - 1, k - 1).$$

Proof. We first observe that, were it not for the factor k in front of $S(p - 1, k)$, we would have the Pascal recurrence. We have

$$n^p = \sum_{k=0}^{p} S(p, k)[n]_k \tag{8.18}$$

and

$$n^{p-1} = \sum_{k=0}^{p-1} S(p - 1, k)[n]_k.$$

---

[8] After James Stirling (1692-1770).

[9] So there must be Stirling numbers of the first kind! We discuss them later in this section.

Thus,

$$n^p = n \times n^{p-1} \;\; = \;\; n \sum_{k=0}^{p-1} S(p-1,k)[n]_k$$

$$= \;\; \sum_{k=0}^{p-1} S(p-1,k)n[n]_k$$

$$= \;\; \sum_{k=0}^{p-1} S(p-1,k)(n-k+k)[n]_k$$

$$= \;\; \sum_{k=0}^{p-1} S(p-1,k)(n-k)[n]_k + \sum_{k=0}^{p-1} kS(p-1,k)[n]_k$$

$$= \;\; \sum_{k=0}^{p-1} S(p-1,k)[n]_{k+1} + \sum_{k=1}^{p-1} kS(p-1,k)[n]_k.$$

We replace k by $k-1$ in the left summation in the line directly above and obtain

$$n^p \;\; = \;\; \sum_{k=1}^{p} S(p-1,k-1)[n]_k + \sum_{k=1}^{p-1} kS(p-1,k)[n]_k$$

$$= \;\; S(p-1,p-1)[n]_p + \sum_{k=1}^{p-1} \left(S(p-1,k-1) + kS(p-1,k) \right)[n]_k.$$

For each k with $1 \leq k \leq p-1$, comparing the coefficient of $[n]_k$ in this expression for n^p with the coefficient of $[n]_k$ in the expression (8.18), we obtain

$$S(p,k) = S(p-1,k-1) + kS(p-1,k).$$

$\square$

The recurrence relation given in Theorem 8.2.4 and the initial values

$$S(p,0) = 0, \quad (p \geq 1) \text{ and } S(p,p) = 1, \quad (p \geq 0)$$

from (8.16) and (8.17) determine the sequence of Stirling numbers of the second kind $S(p,k)$. As for the binomial coefficients we have a Pascal-like triangle for these Stirling numbers. (See Fig. 8.2.)

$p\backslash k$	0	1	2	3	4	5	6	7	$\cdots$
0	1								
1	0	1							
2	0	1	1						
3	0	1	3	1					
4	0	1	7	6	1				
5	0	1	15	25	10	1			
6	0	1	31	90	65	15	1		
7	0	1	63	301	350	140	21	1	
$\vdots$	$\vdots$	$\vdots$	$\vdots$	$\vdots$	$\vdots$	$\vdots$	$\vdots$	$\vdots$	$\ddots$

Figure 8.2: The triangle of $S(p,k)$

Each entry $S(p,k)$ in the triangle, other than those on the two sides of the triangle (these are the entries given by the initial values), is obtained by multiplying the entry in the row directly above it by k and adding the result to the entry immediately to its left in the row directly above it.

From the triangle of the Stirling numbers of the second kind, it appears that

$$S(p,1) \quad = \quad 1, \ (p \geq 1)$$

$$S(p,2) \quad = \quad 2^{p-1} - 1, \ (p \geq 2)$$

$$S(p,p-1) \quad = \quad \binom{p}{2}, \ (p \geq 1).$$

We leave the verification of these formulas as exercises. They are also readily verified, using the combinatorial interpretation of the Stirling numbers of the second kind given in the next theorem.

Theorem 8.2.5 *The Stirling number of the second kind $S(p,k)$ counts the number of partitions of a set of p elements into k indistinguishable boxes in which no box is empty.*

Proof. First, we give an explanation of what "indistinguishable" means in this case. To say that the boxes are indistinguishable means that we can't tell one box from another. They all look the same. If, for instance, the contents of some box are the elements a, b, and c, then it doesn't matter which box it is. The only thing that matters is what the contents of the various boxes are, not *which* box holds what.

Let $S^*(p,k)$ denote the number of partitions of a set of p elements into k indistinguishable boxes in which no box is empty. We easily see that

$$S^*(p, p) = 1, \qquad (p \geq 0)$$

because, if there are the same number of boxes as elements, each box contains exactly one element (and remember we can't tell one box from another), and

$$S^*(p, 0) = 0, \qquad (p \geq 1)$$

because if there is at least one element and no boxes, there can be no partitions. If we can show that the numbers $S^*(p, k)$ satisfy the same recurrence relation as the Stirling numbers of the second kind; that is, if we can show that

$$S^*(p, k) = kS^*(p - 1, k) + S^*(p - 1, k - 1), \qquad (1 \leq k \leq p - 1)$$

then we will be able to conclude that $S^*(p, k) = S(p, k)$ for all k and p with $0 \leq k \leq p$.

We argue as follows: Consider the set of the first p positive integers $1, 2, \ldots, p$ as the set to be partitioned. The partitions of $\{1, 2, \ldots, p\}$ into k nonempty, indistinguishable boxes are of two types:

(i) those in which p is all alone in a box; and

(ii) those in which p is not in a box by itself. Thus, the box containing p contains at least one more element.

In the case of type (i), if we remove p from the box that contains it, we are left with a partition of $\{1, 2, \ldots, p - 1\}$ into $k - 1$ nonempty, indistinguishable boxes. Hence, there are $S^*(p - 1, k - 1)$ partitions of $\{1, 2, \ldots, p\}$ of type (i).

Now consider a partition of type (ii). Suppose we remove p from the box that contains it. Since p was not all alone in its box, we are left with a partition $A_1, A_2, \ldots, A_k$ of $\{1, 2, \ldots, p-1\}$ into k nonempty, indistinguishable boxes. One might now want to conclude that there are $S^*(p - 1, k)$ partitions of type (ii), but this is not so. The reason is that the partition $A_1, A_2, \ldots, A_k$ of $\{1, 2, \ldots, p - 1\}$ which results upon the removal of p arises from k different partitions of $\{1, 2, \ldots, p\}$, namely, from

$$A_1 \cup \{p\}, A_2, \ldots, A_k,$$
$$A_1, A_2 \cup \{p\}, \ldots, A_k,$$
$$\vdots$$
$$A_1, A_2, \ldots, A_k \cup \{p\}.$$

Put another way, after we delete p, we can't tell which box it came from; it could have been any one of the k boxes, since all boxes remain nonempty upon the removal of p. It follows that there are $kS^*(p-1,k)$ partitions of $\{1, 2, \ldots, k\}$ of type (ii). Hence,

$$S^*(p, k) = kS^*(p-1, k) + S^*(p-1, k-1),$$

and the proof is complete.   □

Now that we know that $S(p, k)$ counts the number of partitions of a set of p elements into k nonempty, indistinguishable boxes, we have no use for the notation $S^*(p, k)$ introduced in the proof of Theorem 8.2.5. It is now redundant.

We now use our combinatorial interpretation of the Stirling numbers of the second kind to obtain a formula for them. In doing so, we shall first determine the number $S^{\#}(p, k)$[10] of partitions of $\{1, 2, \ldots, k\}$ into k nonempty, *distinguishable* boxes.[11] Think of one box as colored red, one colored blue, one green, and so on. Now it not only matters which elements are together in a box, but which box it is. (Is it the red box, the blue box, the green one, . . . ?) Once the contents of the k boxes are known, we can color the k boxes in $k!$ ways. Thus,

$$S^{\#}(p, k) = k!S(p, k), \tag{8.19}$$

and it follows that

$$S(p, k) = \frac{1}{k!}S^{\#}(p, k).$$

(Note that (8.19) implies that the numbers $S^{\#}(p, k)$ are the same as the numbers $c(p, k)$ introduced earlier.) Thus, it suffices to find a formula for $S^{\#}(p, k)$, and this we do by applying the inclusion–exclusion principle of Chapter 6. Before doing so, we remark that the validity of (8.19) rests on the fact that each box is nonempty. If boxes were allowed to be empty, we could not multiply $S(p, k)$ by $k!$ to get $S^{\#}(p, k)$. If r of the boxes of a partition were empty, then it would give rise to only $\frac{k!}{r!}$ partitions into distinguishable boxes, because permuting empty boxes amongst themselves doesn't change anything.[12]

Theorem 8.2.6 *For each integer k with $0 \le k \le p$, we have*

$$S^{\#}(p, k) = \sum_{t=0}^{k} (-1)^t \binom{k}{t} (k-t)^p;$$

---

[10]We abandoned one notation and almost immediately introduce another! In mathematics, notation is important. Properly used it adds clarity; briefness is not its only virtue.

[11]Just when you're starting to feel comfortable with indistinguishable boxes, we change the rules and distinguish them!

[12]What we really have is a multiset with r objects of the same type (the empty set) and $k - r$ other different objects (the contents of the nonempty boxes).

hence,

$$S(p, k) = \frac{1}{k!} \sum_{t=0}^{k} (-1)^t \binom{k}{t} (k - t)^p.$$

Proof. Let U be the set of *all* partitions of $\{1, 2, \ldots, p\}$ into k distinguishable boxes $\mathcal{B}_1, \mathcal{B}_2, \ldots, \mathcal{B}_k$. We define k properties $P_1, P_2, \ldots, P_k$, where P_i is the property that the ith box $\mathcal{B}_i$ is empty. Let A_i denote the subset of U consisting of those partitions for which box $\mathcal{B}_i$ is empty. Then

$$S^{\#}(p, k) = |\overline{A}_1 \cap \overline{A}_2 \cap \cdots \cap \overline{A}_k|.$$

We have

$$|U| = k^p$$

since each of the p elements can be put into any one of the k distinguishable boxes. Let t be an integer with $1 \leq t \leq k$. How many partitions of U belong to the intersection $A_1 \cap A_2 \cap \cdots \cap A_t$? For these partitions, boxes $\mathcal{B}_1, \mathcal{B}_2, \ldots, \mathcal{B}_t$ are empty and the remaining boxes $\mathcal{B}_{t+1}, \ldots, \mathcal{B}_k$ may or may not be empty. Thus, $|A_1 \cap A_2 \cap \cdots \cap A_t|$ counts the number of partitions of $\{1, 2, \ldots, p\}$ into $k - t$ distinguishable boxes and hence equals $(k - t)^p$. The same conclusion holds no matter which t boxes are assumed empty; that is,

$$|A_{i_1} \cap A_{i_2} \cap \cdots \cap A_{i_t}| = (k - t)^p$$

for each t-combination $\{i_1, i_2, \ldots, i_t\}$ of $\{1, 2, \ldots, k\}$. Thus, by the inclusion–exclusion principle (see formula (6.3)), we have

$$S^{\#}(p, k) = \sum_{t=0}^{k} (-1)^t \binom{k}{t} (k - t)^p.$$

$\square$

The *Bell number*[13] B_p is the number of partitions of a set of p elements into nonempty, indistinguishable boxes. Now we do not specify the number of boxes, but since no box is to be empty, the number of boxes cannot exceed p. The Bell numbers are just the sum of the entries in a row of the triangle of Stirling numbers of the second kind (see Figure 8.2); that is,

$$B_p = S(p, 0) + S(p, 1) + \cdots + S(p, p).$$

---

[13] After E.T. Bell (1883-1960).

We therefore have

$$
\begin{array}{ll}
B_0 = 1 & B_4 = 15 \\
B_1 = 1 & B_5 = 52 \\
B_2 = 2 & B_6 = 203 \\
B_3 = 5 & B_7 = 877.
\end{array}
$$

The Bell numbers satisfy a recurrence relation, but not one of constant order.

Theorem 8.2.7 *If $p \geq 1$, then*

$$
B_p = \binom{p-1}{0} B_0 + \binom{p-1}{1} B_1 + \cdots + \binom{p-1}{p-1} B_{p-1}.
$$

Proof. We partition the set $\{1, 2, \ldots, p\}$ into nonempty, indistinguishable boxes. The box containing p also contains a subset X (possibly empty) of $\{1, 2, \ldots, p-1\}$. The set X has t elements, where t is some integer between 0 and $p-1$. We can choose a set X of size t in $\binom{p-1}{t}$ ways and partition the $p-1-t$ elements of $\{1, 2, \ldots, p-1\}$ that don't belong to X into nonempty, indistinguishable boxes in B_{p-1-t} ways. Hence,

$$
B_p = \sum_{t=0}^{p-1} \binom{p-1}{t} B_{p-1-t}.
$$

As t takes on the values $0, 1, \ldots, p-1$, so does $(p-1)-t$. Hence, we obtain

$$
\begin{aligned}
B_p &= \sum_{t=0}^{p-1} \binom{p-1}{(p-1)-t} B_t \\
&= \sum_{t=0}^{p-1} \binom{p-1}{t} B_t.
\end{aligned}
$$

$\square$

The Stirling numbers of the second kind show us how to write n^p in terms of $[n]_0, [n]_1, \ldots, [n]_p$. The Stirling numbers of the first kind play the inverse role. They show us how to write $[n]_p$ in terms of $n^0, n^1, \ldots, n^p$.[14] By definition,

$$
\begin{aligned}
[n]_p &= n(n-1)(n-2) \cdots (n-p+1) \\
&= (n-0)(n-1)(n-2) \cdots (n-(p-1)). \qquad (8.20)
\end{aligned}
$$

---

[14]For those familiar with linear algebra: The polynomials of degree at most p with, say, real coefficients form a vector space of dimension $p + 1$. Both $1, n, n^2, \ldots, n^p$ and $[n]_0 = 1, [n]_1, \ldots, [n]_p$ are a basis for this vector space. The Stirling numbers of the first and second kind show us how to express one basis in terms of the other.

Thus,

(i) $[n]_0 = 1$,

(ii) $[n]_1 = n$,

(iii) $[n]_2 = n(n-1) = n^2 - n$,

(iv) $[n]_3 = n(n-1)(n-2) = n^3 - 3n^2 + 2n$,

(v) $[n]_4 = n(n-1)(n-2)(n-3) = n^4 - 6n^3 + 11n^2 - 6n$.

In general, the product on the right in (8.20) has p factors. If we multiply it out, we obtain a polynomial involving the powers

$$n^p, n^{p-1}, \ldots, n^1, n^0 = 1$$

of n in which the coefficients alternate in sign; that is, we obtain an expression of the form

$$
\begin{aligned}
[n]_p &= s(p,p)n^p - s(p,p-1)n^{p-1} + \cdots + \\
&\quad (-1)^{p-1}s(p,1)n^1 + (-1)^p s(p,0)n^0 \\
&= \sum_{k=0}^{p} (-1)^{p-k} s(p,k) n^k.
\end{aligned}
\tag{8.21}
$$

The *Stirling numbers of the first kind* are the coefficients

$$s(p,k), \qquad (0 \le k \le p).$$

that occur in (8.21). It follows readily from (8.20) and (8.21) that

$$s(p,0) = 0, \qquad (p \ge 1)$$

and

$$s(p,p) = 1, \qquad (p \ge 0).$$

Thus, the Stirling numbers of the first kind satisfy the same initial conditions as the Stirling numbers of the second kind. But they satisfy a different recurrence relation, whose proof follows the same basic outline as that of Theorem 8.2.4.

Theorem 8.2.8 *If $1 \le k \le p-1$, then*

$$s(p,k) = (p-1)s(p-1,k) + s(p-1,k-1).$$

Proof. By (8.21), we have

$$[n]_p = \sum_{k=0}^{p}(-1)^{p-k}s(p,k)n^k. \qquad (8.22)$$

Replacing p by $p-1$ in this equation, we also have

$$[n]_{p-1} = \sum_{k=0}^{p-1}(-1)^{p-1-k}s(p-1,k)n^k.$$

Next, we observe that

$$[n]_p = [n]_{p-1}(n-(p-1)).$$

Hence,

$$[n]_p = (n-(p-1))\sum_{k=0}^{p-1}(-1)^{p-1-k}s(p-1,k)n^k,$$

which, after rewriting, becomes

$$\sum_{k=0}^{p-1}(-1)^{p-1-k}s(p-1,k)n^{k+1} + \sum_{k=0}^{p-1}(-1)^{p-k}(p-1)s(p-1,k)n^k.$$

We replace k by $k-1$ in the first summation and obtain

$$[n]_p = \sum_{k=1}^{p}(-1)^{p-k}s(p-1,k-1)n^k + \sum_{k=0}^{p-1}(-1)^{p-k}(p-1)s(p-1,k)n^k.$$

Comparing the coefficient of n^k in this expression with the coefficient of n^k in the expression (8.22), we get

$$s(p,k) = s(p-1,k-1) + (p-1)s(p-1,k)$$

for each integer k with $1 \le k \le p-1$. $\square$

Like the Stirling numbers of the second kind, the Stirling numbers of the first kind also count something, and this is explained in the next theorem. Its proof is similar in structure to the proof of Theorem 8.2.5.

Theorem 8.2.9 *The Stirling number $s(p,k)$ of the first kind counts the number of arrangements of p objects into k nonempty circular permutations.*

Proof. We refer to the circular permutations in the statement of the theorem as circles. Let $s^{\#}(p, k)$ denote the number of ways to arrange p people in k nonempty circles. We have

$$s^{\#}(p, p) = 1, \qquad (p \geq 0)$$

because, if there are p people and p circles, then each circle contains one person.[15] We also have

$$s^{\#}(p, 0) = 0, \qquad (p \geq 1)$$

because, if there is at least one person, any arrangement contains at least one circle. Thus, the numbers $s^{\#}(p, k)$ satisfy the same initial conditions as the Stirling numbers of the first kind. We now show that they satisfy the same recurrence relation; that is,

$$s^{\#}(p, k) = (p - 1)s^{\#}(p - 1, k) + s^{\#}(p - 1, k - 1).$$

Let the people be labeled $1, 2, \ldots, p$. The arrangements of $1, 2, \ldots, p$ into k circles are of two types. Those of the first type have person p in a circle by himself; there are $s^{\#}(p - 1, k - 1)$ of these. In the second type, p is in a circle with at least one other person. These can be obtained from the arrangements of $1, 2, \ldots, p - 1$ into k circles by putting person p on the left of any one of $1, 2, \ldots, p - 1$. Thus, each arrangement of $1, 2, \ldots, p - 1$ gives $p - 1$ arrangements of $1, 2, \ldots, p$ in this way, and hence there is a total of $(p - 1)s^{\#}(p - 1, k)$ arrangements of the second type. Hence, the number of arrangements of p people into k circles is

$$s^{\#}(p, k) = s^{\#}(p - 1, k - 1) + (p - 1)s^{\#}(p - 1, k).$$

It now follows that $s(p, k) = s^{\#}(p, k)$. □

For emphasis, we note that what we have done in the proof of Theorem 8.2.9 is to partition the set $\{1, 2, \ldots, p\}$ into k nonempty, *indistinguishable* boxes and then arrange the elements in each of the boxes into a circular permutation.

8.3 Partition Numbers

A *partition of a positive integer n* is a representation of n as an unordered sum of one or more positive integers, called *parts*. Since the

---

[15]The right hand of each person holds the left hand of the same person!

order of the parts is unimportant, we can always arrange the parts so that they are ordered from largest to smallest. The partitions of 1, 2, 3, 4, and 5 are, respectively, 1; $2, 1 + 1$; $3, 2 + 1, 1 + 1 + 1$; $4, 3 + 1, 2 + 2, 2 + 1 + 1, 1 + 1 + 1 + 1$; and $5, 4 + 1, 3 + 2, 3 + 1 + 1, 2 + 2 + 1, 2 + 1 + 1 + 1, 1 + 1 + 1 + 1 + 1$.

A partition of n is sometimes written as

$$\lambda = n^{a_n} \ldots 2^{a_2} \ldots 1^{a_1}, \tag{8.23}$$

where a_i is a nonnegative integer equal to the number of parts equal to i. (This expression is purely symbolic; its terms are not exponentials nor is the expression a product.) When written in the form (8.23), the term i^{a_i} is usually omitted if $a_i = 0$. In this notation, the partitions of 5 are:

$$5^1, 4^1 1^1, 3^1 2^1, 3^1 1^2, 2^2 1^1, 2^1 1^3, 1^5.$$

Let p_n denote the number of different partitions of the positive integer n, and for convenience, let $p_0 = 1$. The *partition sequence* is the sequence of numbers

$$p_0, p_1, \ldots, p_n, \ldots.$$

By the preceding discussion, we have $p_0 = 1, p_1 = 1, p_2 = 2, p_3 = 3, p_4 = 5$, and $p_5 = 7$. It is a simple observation (cf. (8.23)) that p_n equals the number of solutions in nonnegative integers $a_n, \ldots, a_2, a_1$ of the equation

$$n a_n + \cdots 2 a_2 + 1 a_1 = n.$$

Let λ be the partition $n = n_1 + n_2 + \cdots + n_k$ of n, where $n_1 \geq n_2 \geq \cdots \geq n_k > 0$. The *Ferrers diagram*, or simply *diagram*, of λ is a left-justified array of dots that has k rows with n_i dots in row i. For example, the diagram of the partition $10 = 4 + 2 + 2 + 1 + 1$ is

$$
\begin{array}{cccc}
\bullet & \bullet & \bullet & \bullet \\
\bullet & \bullet & & \\
\bullet & \bullet & & \\
\bullet & & & \\
\bullet & & &
\end{array}
$$

The *conjugate partition* of the partition λ of n is the partition λ^* whose diagram is obtained from the diagram of λ by interchanging rows with columns (flipping the diagram over the diagonal running from the upper left to the lower right). For example, the diagram of the conjugate of the partition $10 = 4 + 2 + 2 + 1 + 1$ is

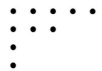

and is thus the partition $10 = 5 + 3 + 1 + 1$. The number of parts of the conjugate of a partition λ equals the largest part of λ.

Let λ be the partition $n = n_1 + n_2 + \cdots + n_k$ of n. More formally, the conjugate partition λ^* of λ is the partition $n = n_1^* + n_2^* + \cdots + n_l^*$ of n ($l = n_1$), where n_i^* is the number of parts of λ that are at least equal to i:

$$n_i^* = |\{j : n \geq i\}| \quad (i = 1, 2, \ldots, l).$$

Example. Let λ be the partition $12 = 4 + 4 + 2 + 2$ of 12, whose diagram is

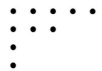

The conjugate λ^* is also the partition $12 = 4 + 4 + 2 + 2$. Thus, $\lambda^* = \lambda$ and λ is a *self-conjugate partition*. $\square$

We now obtain an expression for the generating function of the sequence of partition numbers.

Theorem 8.3.1

$$\sum_{n=0}^{\infty} p_n x^n = \prod_{k=1}^{\infty} (1 - x^k)^{-1}.$$

Proof. The expression on the right equals the product

$$(1 + x + \cdots + x^{1a_1} + \cdots)(1 + x^2 + \cdots + x^{2a_2} + \cdots)(1 + x^3 + \cdots + x^{3a_3} + \cdots) \cdots.$$

A term x^n arises in this product by choosing a term x^{1a_1} from the first factor, x^{2a_2} from the second, x^{3a_3} from the third, and so on, with $1a_1 + 2a_2 + 3a_3 + \cdots = n$. Thus, each partition of n contributes 1 to the coefficient of x^n, and the coefficient of x^n equals the number p_n of partitions of n. $\square$

Let $\mathcal{P}_n$ denote the set of all partitions of the positive integer n. There is a natural way to partially order the partitions in $\mathcal{P}_n$. (For this definition, it is notationally convenient to allow zero parts in order that when we compare two partitions they have the same number of parts.) Let

$$\lambda : n = n_1 + n_2 + \cdots + n_k \quad (n_1 \geq n_2 \geq \cdots \geq n_k)$$

and

$$\mu : n = m_1 + m_2 + \cdots + m_k \quad (m_1 \geq m_2 \geq \cdots \geq m_k)$$

be two partitions of n. Then we say that λ *is majorized by* μ (or that μ *majorizes* λ) and write

$$\lambda \leq \mu,$$

provided that the partial sums for λ are at most equal to the corresponding partial sums for μ:

$$n_1 + \cdots + n_i \leq m_1 + \cdots + m_i \quad (i = 1, 2, \ldots, k).$$

It is straightforward to check that the relation of *majorization* is reflexive, antisymmetric, and transitive, and hence is a partial order on $\mathcal{P}_n$.

Example. Consider the three partitions of 9:

$$\lambda : 9 = 5 + 1 + 1 + 1 + 1; \mu : 9 = 4 + 2 + 2 + 1; \nu : 9 = 4 + 4 + 1.$$

For the purpose of comparing all three of these partitions, we add trailing 0s to μ and ν, and think of μ as $9 = 4 + 2 + 2 + 1 + 0$ and ν as $9 = 4 + 4 + 1 + 0 + 0$. We have $\mu \leq \nu$ as

$$4 \leq 4,$$
$$4 + 2 \leq 4 + 4,$$
$$4 + 2 + 2 \leq 4 + 4 + 1,$$
$$4 + 2 + 2 + 1 \leq 4 + 4 + 1 + 0.$$

On the other hand, λ and μ are incomparable as $4 < 5$ but $4+2+2 > 5+1+1$. Similarly, λ and ν are incomparable.   □

In Section 4.3 we discussed the lexicographic order for n-tuples of 0s and 1s. The lexicographic order can also be used on partitions to produce a total order on $\mathcal{P}_n$ that turns out to be a linear extension of the partial order of majorization. Let $\lambda : n = n_1 + n_2 + \cdots + n_k$ $(n_1 \geq n_2 \geq \cdots \geq n_k)$, and $\mu : n = m_1 + m_2 + \cdots + m_k$ $(m_1 \geq m_2 \geq \cdots \geq m_k)$ be two partitions of n. Then we say that λ *precedes* μ in the *lexicographic order*,[16] provided that there is an integer i such

---

[16] The alphabet is the integers, with small integers preceding large integers in the alphabet. Also, just as in the lexicographic order of n-tuples of 0's and 1's, we read "words" from left to right.

that $n_j = m_j$ for $j < i$ and $n_i < m_i$. For instance, the partition $12 = 4 + 3 + 2 + 2 + 1$ precedes the partition $12 = 4 + 3 + 3 + 1 + 1$ since, reading from left to right, $4 = 4$, $3 = 3$, but $2 < 3$. It is simple to verify that lexicographic order is a partial order on $\mathcal{P}_n$.

Theorem 8.3.2 *Lexicographic order is a linear extension of the partial order of majorization on the set $\mathcal{P}_n$ of partitions of a positive integer n.*

Proof. The fact that lexicographic order is a total order (each two partitions of n are comparable) follows almost immediately from its definition. We continue with the notation preceding the statement of the theorem. Let λ and μ be different partitions of n, with λ majorized by μ. Choose the first integer i such that $n_j = m_j$ for $j < i$ but $n_i \neq m_i$. Since

$$n_1 + \cdots + n_{i-1} + n_i \leq m_1 + \cdots m_{i-1} + m_i,$$

we conclude that $n_i < m_i$, and hence λ precedes μ in the lexicographic order. $\qquad\square$

8.4   A Geometric Problem

In this section we shall obtain a combinatorial geometric interpretation of the sum

$$h_n^{(k)} = \binom{n}{0} + \binom{n}{1} + \cdots + \binom{n}{k} \qquad (0 \leq k \leq n) \qquad (8.24)$$

of the first $k + 1$ binomial coefficients with upper argument equal to n—that is, the sum of the first $k + 1$ numbers in row n of Pascal's triangle. For each fixed k, we obtain a sequence

$$h_0^{(k)}, h_1^{(k)}, h_2^{(k)}, \ldots, h_n^{(k)}, \ldots . \qquad (8.25)$$

If $k = 0$, we have

$$h_n^{(0)} = \binom{n}{0} = 1,$$

and (8.25) is the sequence of all 1's. If $k = 1$, we obtain

$$h_n^{(1)} = \binom{n}{0} + \binom{n}{1} = n + 1.$$

If $k = 2$, we have

$$
\begin{aligned}
h_n^{(2)} &= \binom{n}{0} + \binom{n}{1} + \binom{n}{2} \\
&= 1 + n + \frac{n(n-1)}{2} \\
&= \frac{n^2 + n + 2}{2}.
\end{aligned}
$$

We also note that $h_0^{(k)} = 1$ for all k. We use Pascal's formula to determine the differences of (8.25):

$$
\begin{aligned}
\Delta h_n^{(k)} &= h_{n+1}^{(k)} - h_n^{(k)} \\
&= \binom{n+1}{0} + \binom{n+1}{1} + \cdots + \binom{n+1}{k} - \binom{n}{0} - \binom{n}{1} - \cdots - \binom{n}{k} \\
&= \left[\binom{n+1}{1} - \binom{n}{1} \right] + \cdots + \left[\binom{n+1}{k} - \binom{n}{k} \right] \\
&= \binom{n}{0} + \cdots + \binom{n}{k-1}.
\end{aligned}
$$

Hence,

$$
\Delta h_n^{(k)} = h_n^{(k-1)}. \tag{8.26}
$$

It is a consequence of (8.26) that the difference table for the sequence

$$
h_0^{(k)}, h_1^{(k)}, h_2^{(k)}, h_2^{(k)}, \ldots, h_n^{(k)}, \ldots \tag{8.27}
$$

can be obtained from the difference table for

$$
h_0^{(k-1)}, h_1^{(k-1)}, h_2^{(k-1)}, \ldots, h_n^{(k-1)}, \ldots
$$

by inserting (8.27) on top as a new row.

The number $h_n^{(k)}$ counts the number of combinations with at most k elements of a set with n elements. We now show that $h_n^{(k)}$ also has an interpretation as a counting function for a geometrical problem:

$h_n^{(k)}$ counts the number of regions into which k-dimensional space is divided by n $(k-1)$-dimensional hyperplanes in general position.

We need to explain some of the terms in this assertion.

We start with $k = 1$. Consider a 1-dimensional space, that is, a line. A 0-dimensional space is a point and n points in general position means simply that the points are distinct. If we insert n distinct points on the line, then the line gets divided into $n + 1$ parts called regions. (See Fig. 8.3 in which 4 points divide the line into 5 regions.)

Figure 8.3

This result agrees with the definition of $h_n^{(1)}$ given in (8.24).

Now let $k = 2$, and consider n lines in a plane in general position. In this case, "general position" means that the lines are distinct and not parallel (so that each pair of lines intersects in exactly one point) and the points of intersection are all different—that is, no three of the lines meet in the same point. For n lines in general position in a plane, the number of points of intersection is $\binom{n}{2}$, since each pair of lines gives a different point. The number of regions into which a plane is divided by n lines in general position is given in the table below for $n = 0$ to 5.

Lines	Regions
0	1
1	2
2	4
3	7
4	11
5	16

This table is readily verified.

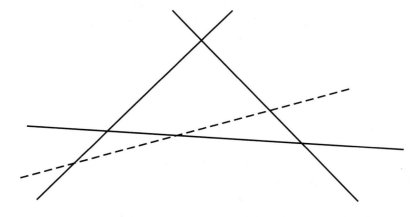

Figure 8.4

We now reason inductively. Suppose we have n lines in general position and we then insert a new line so that the resulting set of $n+1$ lines is in general position. The first n lines intersect the new line in n different points. The n points, as we have already verified, divide the new line into

$$h_n^{(1)} = n + 1$$

parts. Each of these $h_n^{(1)} = n + 1$ parts divides a region formed by the first n lines into two regions. (See Fig. 8.4 for the case $n = 3$ in which the new line is the dashed line.) Hence, the number of regions is increased by $h_n^{(1)} = n + 1$ in going from n lines to $n + 1$ lines. But this is exactly the relation expressed by (8.26) for the case $k = 2$:

$$\Delta h_n^{(2)} = h_{n+1}^{(2)} - h_n^{(2)} = h_n^{(1)} = n + 1.$$

Since $h_0^{(2)} = 1$, we conclude that

$$h_n^{(2)} = \binom{n}{0} + \binom{n}{1} + \binom{n}{2}$$

is the number of regions formed by n lines in general position in a plane.

The case $k = 3$ is similar. Consider n planes in 3-space in general position. General position now means that each pair of planes, but no three planes, meet in a line, and every three planes, but no four planes, meet in a point. We now insert a new plane so that the resulting set of $n + 1$ planes is also in general position. The first n planes intersect the new plane in n lines in general position (because the planes are in general position). These n lines divide the new plane into $h_n^{(2)}$ planar regions, as determined previously for $k = 2$. Each of these $h_n^{(2)}$ planar regions divides a space region formed by the first n planes into two. Hence, the number of space regions is increased by $h_n^{(2)}$ in going from n planes to $n + 1$ planes. This is exactly the relation expressed by (8.26) for the case $k = 3$:

$$\Delta h_n^{(3)} = h_{n+1}^{(3)} - h_n^{(3)} = h_n^{(2)}.$$

Since $h_0^{(3)} = 1$ (zero planes divide space into 1 region, namely, all of space), we conclude that

$$h_n^{(3)} = \binom{n}{0} + \binom{n}{1} + \binom{n}{2} + \binom{n}{3}$$

is the number of regions into which space is divided by n planes in general position in 3-space.

The same type of reasoning applies to higher dimensional space. The number of regions into which k-dimensional space is divided by n $(k-1)$-dimensional hyperplanes in general position equals

$$h_n^{(k)} = \binom{n}{0} + \binom{n}{1} + \cdots + \binom{n}{k}. \qquad (8.28)$$

We conclude by considering the case $k = n$. From our definition (8.24), we obtain

$$h_n^{(n)} = \binom{n}{0} + \binom{n}{1} + \cdots + \binom{n}{n} = 2^n.$$

Our geometrical assertion in this case is that n hyperplanes in general position in n-dimensional space divide n-dimensional space into 2^n regions. General position now means, since there are only n $(n-1)$-dimensional hyperplanes, that the n hyperplanes have exactly one point in common. This fact is familiar to all, at least for the cases $k = 1, 2$, and 3. Consider the case $k = 3$ of 3-dimensional space. We can coordinatize the space by associating with each point a triple of numbers (x_1, x_2, x_3). The three coordinate planes $x_1 = 0, x_2 = 0$, and $x_3 = 0$ divide the space into $2^3 = 8$ quadrants. (Each quadrant is determined by prescribing signs to each of x_1, x_2, x_3.) More generally, n-dimensional space is coordinatized by associating an n-tuple of numbers $(x_1, x_2, \ldots, x_n)$ with each point. There are n coordinate planes, namely, those determined by $x_1 = 0, x_2 = 0, \ldots$, and $x_n = 0$. These planes divide n-dimensional space into the 2^n "quadrants" determined by prescribing signs to each of $x_1, x_2, \ldots, x_n$. One such quadrant is the so-called nonnegative quadrant $x_1 \geq 0, x_2 \geq 0, \ldots, x_n \geq 0$.

8.5 Lattice Paths and Schröder Numbers

In this section, we formalize the notion of a lattice path, which we have experienced in the exercises in Chapter 3 and in the examples in Section 8.1.

We consider the *integral lattice* of points in the coordinate plane with integer coordinates. Given two such points (p, q) and (r, s), with $p \geq r$ and $q \geq s$, a *rectangular lattice path* from (r, s) to (p, q) is a path from (r, s) to (p, q) that is made up of *horizontal steps* $H = (1, 0)$ and *vertical steps* $V = (0, 1)$. Thus, a rectangular lattice path from

(r, s) to (p, q) starts at (r, s) and gets to (p, q) using unit horizontal and vertical segments.

Example. In Figure 8.5 we show a rectangular lattice path from $(0, 0)$ to $(7, 5)$, consisting of 7 horizontal steps and 5 vertical steps. Given that the path starts at $(0, 0)$, it is uniquely determined by the sequence

$$H, V, V, H, H, H, V, V, H, V, H, H$$

of 7 H's and 5 V's.

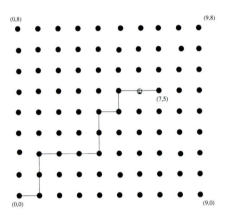

Figure 8.5

$\square$

Theorem 8.5.1 *The number of rectangular lattice paths from (r, s) to (p, q) equals the binomial coefficient*

$$\binom{p - r + q - s}{p - r} = \binom{p - r + q - s}{q - s}.$$

Proof. The two binomial coefficients in the statement of the theorem are equal. A rectangular lattice path from (r, s) to (p, q) is uniquely determined by its sequence of $p - r$ horizontal steps H and $q - s$ vertical steps V, and every such sequence determines a rectangular lattice path from (r, s) to (p, q). Hence, the number of paths equals the number of permutations of $p - r + q - s$ objects of which $p - r$ are H's and $q - s$ are V's. From Section 3.4 we know this number to be the binomial coefficient

$$\binom{p - r + q - s}{p - r}.$$

$\square$

Consider a rectangular lattice path from (r, s) to (p, q), where $p \geq r$ and $q \geq s$. Such a path uses exactly $(p - r) + (q - s)$ steps, and there is no loss in generality in assuming that $(r, q) = (0, 0)$. This is because we may simply translate (r, s) back to $(0, 0)$ and (p, q) back to $(p - r, q - s)$ and obtain a one-to-one correspondence between rectangular lattice paths from (r, s) to (p, q) and those from $(0, 0)$ to $(p - r, q - s)$. By Theorem 8.5.1, if $p \geq 0$ and $q \geq 0$, the number of rectangular lattice paths from $(0, 0)$ to (p, q) equals

$$\binom{p + q}{p} = \binom{p + q}{q}.$$

We now consider rectangular lattice paths from $(0, 0)$ to (p, q) that are restricted to lie on or below the line $y = x$ in the coordinate plane. We call such paths *subdiagonal rectangular lattice paths*. A subdiagonal path from $(0, 0)$ to $(9, 9)$ is shown in Figure 8.6.

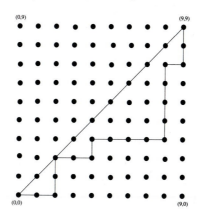

Figure 8.6

In Section 8.1 we already proved the next theorem.

Theorem 8.5.2 *Let n be a nonnegative integer. Then the number of subdiagonal rectangular lattice paths from $(0, 0)$ to (n, n) equals the nth Catalan number*

$$C_n = \frac{1}{n + 1}\binom{2n}{n}.$$

□

More generally, we can count the number of subdiagonal rectangular lattice paths from $(0, 0)$ to (p, q) whenever $p \geq q$. Of course, if $q > p$, there can be no subdiagonal rectangular lattice paths from $(0, 0)$ to (p, q).

Theorem 8.5.3 *Let p and q be positive integers with $p \geq q$. Then the number of subdiagonal rectangular lattice paths from $(0,0)$ to (p,q) equals*

$$\frac{q-p+1}{q+1}\binom{p+q}{q}.$$

Proof. For the proof, we generalize the proof given in Section 8.1, which showed that the Catalan number C_n counts the number of subdiagonal rectangular lattice paths from $(0,0)$ to (n,n), and, in particular, the proof of Theorem 8.1.1. To obtain our answer, we determine the number $l(p,q)$ of rectangular lattice paths γ from $(0,0)$ to (p,q) that cross the diagonal, and then subtract $l(p,q)$ from the total number $\binom{p+q}{q}$ of rectangular lattice paths from $(0,0)$ to (p,q). The number $l(p,q)$ is the same as the number of rectangular lattice paths γ' from $(0,-1)$ to $(p,q-1)$ that touch (possibly cross) the diagonal line $y = x$. This follows by shifting paths down one unit, thereby shifting a path γ into a path γ', and this establishes a one-to-one correspondence between the two kinds of paths.

Consider a path γ' from $(0,-1)$ to $(p,q-1)$ that touches the diagonal line $y = x$. Let γ'_1 be the subpath of γ' from $(0,-1)$ to the first diagonal point (d,d) touched by γ'. Let γ'_2 be the subpath of γ' from (d,d) to $(p,q-1)$. We reflect γ'_1 about the line $y = x$ and obtain a path γ_1^* from $(-1,0)$ to (d,d). Following γ_1^* with γ_2, we get a path γ^* from $(-1,0)$ to $(p,q-1)$. This construction is illustrated in Figure 8.7.

Now every rectangular lattice path θ from $(-1,0)$ to $(p,q-1)$ must cross the diagonal line $y = x$, since $(-1,0)$ is above the line and $(p,q-1)$ is below. If we reflect the part of θ that goes from $(-1,0)$ to the first crossing point, we get a path from $(0,-1)$ to $(p,q-1)$ that touches the line $y = x$. This shows that the correspondence γ' to γ^* is a one-to-one correspondence, and hence that $l(p,q)$ equals the number of rectangular lattice paths from $(-1,0)$ to $(p,q-1)$. By Theorem 8.5.1, we have

$$l(p,q) = \binom{p+1+q-1}{q-1} = \binom{p+q}{q-1}.$$

Therefore, the number of subdiagonal rectangular lattice paths from $(0,0)$ to (p,q) equals

$$\binom{p+q}{q} - l(p,q) = \binom{p+q}{q} - \binom{p+q}{q-1} = \frac{(p+q)!}{p!q!} - \frac{(p+q)!}{(q-1)!(p+1)!},$$

which simplifies to

$$\frac{p-q+1}{p+1}\binom{p+q}{q}.$$

□

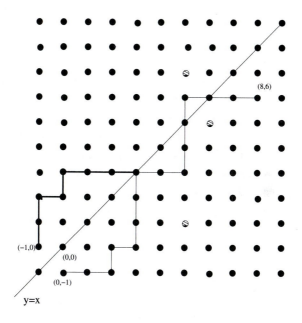

Figure 8.7

We now consider lattice paths for which, in addition to horizontal steps $H = (1,0)$ and vertical steps $V = (0,1)$, we allow *diagonal steps* $D = (1,1)$. Let p and q be nonnegative integers, and let $K(p,q)$ be the number of lattice paths from $(0,0)$ to (p,q) made up of horizontal, vertical, and diagonal steps, and let $K(p,q : rD)$ be the number of such paths that use exactly r diagonal steps D. We have $K(p,q : 0D)$ equal to the number of rectangular lattice paths from $(0,0)$ to (p,q); thus, by Theorem 8.5.1,

$$K(p,q : 0D) = \binom{p+q}{p}.$$

We also have $K(p,q : rD) = 0$ if $r > \min\{p,q\}$.

Theorem 8.5.4 *Let $r \le \min\{p,q\}$. Then*

$$K(p,q : rD) = \binom{p+q-r}{p-r \quad q-r \quad r} = \frac{(p+q-r)!}{(p-r)!(q-r)!r!},$$

and

$$K(p,q) = \sum_{r=0}^{\min\{p,q\}} \frac{(p+q-r)!}{(p-r)!(q-r)!r!}.$$

Proof. A lattice path from $(0,0)$ to (p,q) that uses r diagonal steps D must use $p - r$ horizontal steps H and $q - r$ vertical steps V, and is uniquely determined by its sequence of $p - r$ H's, $q - r$ V's, and r D's. Thus, the number of such paths is the number of permutations of the multiset

$$\{(p-r) \cdot H, (q-r) \cdot V, r \cdot D\}.$$

From Chapter 3, we know the number of such permutations to be the multinomial number in the statement of the theorem. If we do not specify the number r of diagonal steps, then by summing $K(p, q : rD)$ from $r = 0$ to $r = \min\{p, q\}$, we obtain $K(p, q)$ as given in the theorem. $\square$

Now let $p \geq q$ and let $R(p, q)$ be the number of subdiagonal lattice paths from $(0,0)$ to (p,q). Also, let $R(p, q : rD)$ be the number of subdiagonal lattice paths from $(0,0)$ to (p,q) that use exactly r diagonal steps D. We have

$$R(p,q) = \sum_{r=0}^{q} R(p,q : rD).$$

Theorem 8.5.5 *Let p and q be positive integers with $p \geq q$, and let r be a nonnegative integer with $r \leq q$. Then*

$$R(p,q:rD) = \frac{q-p+1}{q-r+1}\frac{(p+q-r)!}{r!(p-r)!(q-r)!}$$

$$= \frac{q-p+1}{q-r+1}\left(\begin{array}{ccc} & p+q-r & \\ r & (p-r) & (q-r) \end{array}\right),$$

and

$$R(p,q) = \sum_{r=0}^{q} \frac{q-p+1}{q-r+1}\frac{(p+q-r)!}{r!(p-r)!(q-r)!}.$$

Proof. A subdiagonal lattice path γ from $(0,0)$ to (p,q) with steps H, V, and D, with r diagonal steps D, becomes a subdiagonal rectangular lattice path π from $(0,0)$ to $(p-r, q-r)$ after removing the r diagonal steps D. Conversely, a subdiagonal rectangular lattice path π from $(0,0)$ to $(p-r, q-r)$ becomes a subdiagonal lattice path, with r diagonal steps, from $(0,0)$ to (p,q) by inserting r diagonal steps in any of the $p + q - 2r + 1$ places before, between, and after the horizontal and vertical steps. The number of ways to insert the diagonal steps D in π equals the number of solutions in nonnegative integers of the equation

$$x_1 + x_2 + \cdots + x_{p+q-2r+1} = r,$$

and from Section 3.5, we know this number to be

$$\binom{p + q - 2r + 1 + r - 1}{r} = \binom{p + q - r}{r}. \tag{8.29}$$

Thus, there correspond to each subdiagonal rectangular lattice path from $(0,0)$ to $(p - r, q - r)$ a number of subdiagonal lattice paths from $(0,0)$ to (p, q) with r diagonal steps, and this number is given by (8.29). Therefore,

$$R(p, q : rD) = \binom{p + q - r}{r} R(p - r, q - r : 0D).$$

Using Theorem 8.5.3, we get

$$R(p, q : rD) = \binom{p + q - r}{r} \frac{q - p + 1}{q - r + 1} \binom{p + q - 2r}{q - r},$$

which simplifies to

$$\frac{q - p + 1}{q - r + 1} \frac{(p + q - r)!}{r!(p - r)!(q - r)!} = \frac{q - p + 1}{q - r + 1} \binom{p + q - r}{r \quad (p - r) \quad (q - r)}.$$

Summing $R(p, q : rD)$ from $r = 0$ to q we get the formula for $R(p, q)$ given in the theorem. $\qquad\square$

Notice that, by taking $r = 0$ in Theorem 8.5.5, we get Theorem 8.5.3.

We now suppose that $p = q = n$. The subdiagonal lattice paths from $(0,0)$ to (n, n) are called *Schröder paths*.[17] The *large Schröder number* R_n is the number of Schröder paths from $(0,0)$ to (n, n). Thus, by Theorem 8.5.5,

$$R_n = R(n, n) = \sum_{r=0}^{n} \frac{1}{n - r + 1} \frac{(2n - r)!}{r!((n - r)!)^2}. \tag{8.30}$$

---

[17]After Friedrich Wilhelm Karl Ernst Schröder (1841-1902). See R.P. Stanley: Hipparchus, Plutarch, Schröder, and Hough, *American Mathematical Monthly*, 104 (1997), 344-350. Also see L.W. Shapiro and R.A. Sulanke: Bijections for Schröder numbers, *Mathematics Magazine*, 73 (2000), 369-376. We rely heavily on both of these articles for this section.

The sequence $R_0, R_1, R_2, \ldots, R_n, \ldots$ of large Schröder numbers begins as

$$1, 2, 6, 22, 90, 394, 1806, \ldots.$$

We now turn to the small Schröder numbers, which are defined in terms of constructs called bracketings. Let $n \geq 1$, and let $a_1, a_2, \ldots, a_n$ be a sequence of n symbols. We generalize the idea of a multiplication scheme for $a_1, a_2, \ldots, a_n$ described in Section 8.2 to that of a bracketing of the sequence $a_1, a_2, \ldots, a_n$. For our multiplication schemes, we had a binary operation $\times$ that combined two quantities, and a multiplication scheme was a way to put $n - 1$ sets of parentheses on the sequence $a_1, a_2, \ldots, a_n$, with each set of parentheses corresponding to a multiplication of two quantities. In a bracketing, a set of parentheses can enclose any number of symbols. For clarity, we shall now drop the symbol $\times$, since it's use now introduces some ambiguity. Before giving the formal definition of bracketing, we list the bracketings for $n = 1, 2, 3$, and 4 and, at the same time, introduce some of the simplifications we adopt for purposes of clarity.

Example. If $n = 1$, then there is only one bracketing, namely, a_1. To be precise, we should write this as (a_1) but, also for clarity, we shall remove parentheses around single elements and let the parantheses be implicit. For $n = 2$, there is also only one bracketing, namely, $(a_1 a_2)$, or, for more clarity, $a_1 a_2$. In general, we omit the last set of parentheses corresponding to the final bracketing of the remaining symbols. For $n = 3$, we have 3 bracketings:

$$a_1 a_2 a_3, (a_1 a_2) a_3, \text{ and } a_1(a_2 a_3).^{18}$$

For $n = 4$, we have 11 bracketings:

$$a_1 a_2 a_3 a_4, (a_1 a_2) a_3 a_4, (a_1 a_2 a_3) a_4, a_1(a_2 a_3) a_4, a_1(a_2 a_3 a_4), a_1 a_2(a_3 a_4),$$

and

$$((a_1 a_2) a_3) a_4, (a_1(a_2 a_3)) a_4, a_1((a_2 a_3) a_4), a_1(a_2(a_3 a_4)), (a_1 a_2)(a_3 a_4).^{19}$$

□

We now give the formal recursive definition of a *bracketing* of a sequence $a_1, a_2, \ldots, a_n$. Each symbol a_i is itself a bracketing; and

---

[18]Without any of our simplifications, these would be written as $(a_1 \times a_2 \times a_3), ((a_1 \times a_2) \times a_3)$, and $(a_1 \times (a_2 \times a_3))$. The last two are multiplication schemes, since each pair of parentheses in them corresponds to a multiplication of two quantities. The first is not, since a set of parentheses brackets three symbols.

[19]Only the last 5 are multiplication schemes.

any consecutive sequence of two or more bracketings enclosed by a set of parentheses is a bracketing. Thus, in contrast to multiplication schemes in Section 8.2, a pair of parentheses need not correspond to a multiplication of two symbols. Using this definition, we can construct all bracketings of the sequence $a_1, a_2, \ldots, a_n$ by carrying out the following recursive algorithm in all possible ways.

Algorithm to Construct Bracketings

Start with a sequence $a_1, a_2, \ldots, a_n$.

1. Let γ equal $a_1 a_2 \ldots a_n$.

2. While γ has at least 3 symbols, do the following:

    (a) Put a set of parentheses around any number $k \geq 2$ of consecutive symbols, say, $a_i a_{i+1} \cdots a_{i+k-1}$, to form a new symbol $(a_i a_{i+1} \cdots a_{i+k-1})$.

    (b) Replace γ with the expression in which $(a_i a_{i+1} \cdots a_{i+k-1})$ is now one symbol.[20]

3. Output the current expression.

A multiplication scheme for $a_1, a_2, \ldots, a_n$ is a *binary bracketing*–that is, a bracketing in which each set of parentheses encloses two symbols.

Example. We give an example of an application of the algorithm. Let $n = 9$ so that we start with $a_1 a_2 a_3 a_4 a_5 a_6 a_7 a_8 a_9$. We arrive at a bracketing by making the following choices:

$$
\begin{aligned}
a_1 a_2 a_3 a_4 a_5 a_6 a_7 a_8 a_9 \quad &\rightarrow \quad a_1 a_2 a_3 (a_4 a_5 a_6) a_7 a_8 a_9 \\
&\rightarrow \quad (a_1 a_2) a_3 (a_4 a_5 a_6) a_7 a_8 a_9 \\
&\rightarrow \quad (a_1 a_2) a_3 ((a_4 a_5 a_6) a_7 a_8) a_9 \\
&\rightarrow \quad (a_1 a_2)(a_3 ((a_4 a_5 a_6) a_7 a_8) a_9).
\end{aligned}
$$

This bracketing is not a binary bracketing, since there are sets of parenthesis which enclose more than two symbols; for instance, $(a_4 a_5 a_6)$ do, and so do $((a_4 a_5 a_6) a_7 a_8)$ (which encloses the three symbols $(a_4 a_5 a_6)$, a_7, and a_8), and $(a_3 ((a_4 a_5 a_6) a_7 a_8) a_9)$ (which encloses the three symbols a_3, $((a_4 a_5 a_6) a_7 a_8)$, and a_9) . □

---

[20]But recall that, if we choose the entire sequence of symbols, we don't put in parentheses. Since $k \geq 2$, we don't put a set of parentheses around one symbol.

For $n \geq 1$, the *small Schröder number* s_n is defined to be the number of bracketings of a sequence $a_1, a_2, \ldots, a_n$ of n symbols. We have seen that $s_1 = 1$, $s_2 = 1$, $s_3 = 3$ and $s_4 = 11$. In fact, the sequence $(s_n : n = 1, 2, 3, \ldots)$ begins as

$$1, 1, 3, 11, 45, 197, 903, \ldots .$$

Comparing this with the initial part of the sequence of large Schröder numbers leads one to the tentative conclusion that $R_n = 2s_{n+1}$ for $n \geq 1$ with $R_0 = 1$. We give a proof of this by computing the generating functions for both the small and large Schröder numbers.

Theorem 8.5.6 *The generating function for the sequence* $(s_n : n \geq 1)$ *of small Schröder numbers is*

$$\sum_{n=1}^{\infty} s_n x^n = \frac{1}{4} \left(1 + x - \sqrt{x^2 - 6x + 1} \; \right).$$

Proof. Let $g(x) = \sum_{n=1}^{\infty} s_n x^n$ be the generating function of the small Schröder numbers. The recursive definition of bracketing implies that

$$
\begin{aligned}
g(x) &= x + g(x)^2 + g(x)^3 + g(x)^4 + \cdots \\
&= x + g(x)^2 (1 + g(x) + g(x)^2 + \cdots) \\
&= x + \frac{g(x)^2}{1 - g(x)}.
\end{aligned}
$$

This gives

$$(1 - g(x))g(x) = (1 - g(x))x + g(x)^2;$$

hence,

$$2g(x)^2 - (1 + x)g(x) + x = 0.$$

Therefore, $g(x)$ is a solution of the quadratic equation

$$2y^2 - (1 + x)y + x = 0.$$

The two solutions of this quadratic equation are

$$y_1(x) = \frac{(1 + x) + \sqrt{(1 + x)^2 - 8x}}{4}$$

and

$$y_2(x) = \frac{(1 + x) - \sqrt{(1 + x)^2 - 8x}}{4}.$$

Since $g(0) = 0$, and $y_1(0) = 1/2$ and $y_2(0) = 0$, we have

$$g(x) = y_2(x) = \frac{1 + x - \sqrt{x^2 - 6x + 1}}{4}.$$

$\square$

The generating function $g(x) = \sum_{n=1}^{\infty} s_n x^n$, as evaluated in Theorem 8.5.6, can be used to obtain a recurrence relation for the small Schröder numbers that is useful for computation. Now recall the quadratic

$$2y^2 - (1 + x)y + x = 0$$

that arose in the proof of Theorem 8.5.6. If we differentiate this quadratic with respect to x, we get

$$4y\frac{dy}{dx} - y - (1 + x)\frac{dy}{dx} + 1 = 0;$$

hence,

$$\frac{dy}{dx} = \frac{y - 1}{4y - 1 - x}$$

$$= \frac{(x - 3)y - x + 1}{x^2 - 6x + 1}.$$

The last equality can be routinely verified by cross multiplying and then making use of the quadratic equation $2y^2 - (1 + x)y + x = 0$. We now have

$$(x^2 - 6x + 1)\frac{dy}{dx} - (x - 3)y + x - 1 = 0. \qquad (8.31)$$

Substituting $y = g(x) = \sum_{n=1}^{\infty} s_n x^n$ in (8.31), we get, after some simplification,

$$\sum_{n=1}^{\infty} (n - 1)s_n x^{n+1} - 3\sum_{n=1}^{\infty} (2n - 1)s_n x^n + \sum_{n=1}^{\infty} n s_n x^{n-1} + x - 1 = 0,$$

which can be written as

$$\sum_{n=1}^{\infty} (n - 1)s_n x^{n+1} - 3\sum_{n=0}^{\infty} (2n + 1)s_{n+1} x^{n+1} +$$

$$\sum_{n=-1}^{\infty} (n + 2)s_{n+2} x^{n+1} + x - 1 = 0.$$

Setting the coefficients of x^{n+1} equal to 0 for $n \geq 1$, we obtain

$$(n + 2)s_{n+2} - 3(2n + 1)s_{n+1} + (n - 1)s_n = 0, \quad (n \geq 1). \qquad (8.32)$$

The recurrence relation (8.32) is a homogeneous linear recurrence relation of order 2 with nonconstant coefficients.

We now compute the generating function for the large Schröder numbers.

Theorem 8.5.7 *The generating function for the sequence* $(R_n : n \geq 0)$ *of large Schröder numbers is*

$$\sum_{n=0}^{\infty} R_n x^n = \frac{1}{2x} \left(-(x-1) - \sqrt{x^2 - 6x + 1} \right).$$

Proof. Let $h(x) = \sum_{n=0}^{\infty} R_n x^n$ be the generating function for the large Schröder numbers. A subdiagonal lattice path from $(0,0)$ to (n,n) (i) is the empty path (if $n = 0$), (ii) starts with a diagonal step D, or (iii) starts with a horizontal step H. The number of paths of type (ii) equals the number of subdiagonal paths from $(1,1)$ to (n,n) and thus equals R_{n-1}. The paths of type (iii) begin with a horizontal step H and then follow a path γ from $(1,0)$ to (n,n) without going above the diagonal line joining $(1,1)$ and (n,n). Since γ ends on the diagonal at the point (n,n), there is a first point (k,k) of γ on the diagonal, where $1 \leq k \leq n$. Since (k,k) is the first point of γ on the diagonal, γ arrives at (k,k) by a vertical step V from the point $(k,k-1)$. The part of γ from $(1,0)$ to $(k,k-1)$ is a lattice path γ_1 that does not go above the diagonal line joining $(1,0)$ to $(k,k-1)$. The part of γ from (k,k) to (n,n) is a lattice path γ_2 that does not go above the diagonal line joining (k,k) to (n,n). There are R_{k-1} choices for γ_1 and R_{n-k} choices for γ_2, and hence the number of lattice paths of type (iii) equals $R_{k-1} R_{n-k}$. Summarizing, we get the recurrence relation

$$R_n = R_{n-1} + \sum_{k=1}^{n} R_{k-1} R_{n-k}, \quad (n \geq 1), \qquad (8.33)$$

or equivalently,

$$R_n = R_{n-1} + \sum_{k=0}^{n-1} R_k R_{n-1-k}, \quad (n \geq 1), \qquad (8.34)$$

where $R_0 = 1$. Thus,

$$x^n R_n = x(x^{n-1} R_{n-1}) + x \left(\sum_{k=0}^{n-1} x^k R_k x^{n-1-k} R_{n-1-k} \right), \quad (n \geq 1).$$

Since $R_0 = 1$, the preceding equation implies that the generating function $h(x)$ of the large Schröder numbers satisfies

$$h(x) = 1 + xh(x) + xh(x)^2. \qquad (8.35)$$

Therefore, $h(x)$ is a solution of the quadratic equation

$$xy^2 + (x - 1)y + 1 = 0.$$

The two solutions of this quadratic equation are

$$y_1(x) = \frac{-(x - 1) + \sqrt{x^2 - 6x + 1}}{2x}$$

and

$$y_2(x) = \frac{-(x - 1) - \sqrt{x^2 - 6x + 1}}{2x}.$$

The first of these cannot be the generating function of the large Schröder numbers as it does not give nonnegative integers. Hence,

$$h(x) = y_2(x) = \frac{1 - x - \sqrt{x^2 - 6x + 1}}{2x}.$$

$\square$

Comparing the generating functions for the large and small Schröder numbers, we obtain the following corollary.

Corollary 8.5.8 *The large and small Schröder numbers are related by*

$$R_n = 2s_{n+1}, \quad (n \geq 1).$$

$\square$

In Sections 7.6 and 8.1 we considered triangulating a convex polygonal region by means of its diagonals which do not intersect in the interior of the region. We showed that the number of such triangularizations of a convex polygonal region with $n + 1$ sides equals the number of multiplication schemes for n numbers given in a particular order, with the common value equal to the Catalan number

$$C_{n-1} = \frac{1}{n}\binom{2n - 2}{n - 1}.$$

Thus, the nth Catalan number C_n equals the number of triangularizations of a convex polygonal region with $n + 2$ sides. We now show that bracketings can be given a combinatorial geometric interpretation.

Consider a convex polygonal region Π_{n+1} with $n + 1$ sides, and the sequence $a_1, a_2, \ldots, a_n$. The base of Π_{n+1} is labeled as base, and the remaining n sides are labeled with $a_1, a_2, \ldots, a_n$, beginning with the side immediately to the left of the base being labeled a_1 and proceeding in order in a clockwise fashion. Bracketings of $a_1, a_2, \ldots, a_n$ are in one-to-one correspondence with *dissections* of Π_{n+1}, where, by a dissection of Π_{n+1}, we mean a partition of Π_{n+1} into regions obtained by inserting diagonals that do not intersect in the interior. In contrast to triangularizations, the regions in the partition of Π_{n+1} are not restricted to be triangles.

We illustrate the correspondence in Figure 8.8, using the example of a bracketing that we constructed with our algorithm:

$$
\begin{aligned}
a_1 a_2 a_3 a_4 a_5 a_6 a_7 a_8 a_9 \;\; &\to \;\; a_1 a_2 a_3 (a_4 a_5 a_6) a_7 a_8 a_9 \\
&\to \;\; (a_1 a_2) a_3 (a_4 a_5 a_6) a_7 a_8 a_9 \\
&\to \;\; (a_1 a_2) a_3 ((a_4 a_5 a_6) a_7 a_8) a_9 \\
&\to \;\; (a_1 a_2)(a_3 ((a_4 a_5 a_6) a_7 a_8) a_9).
\end{aligned}
$$

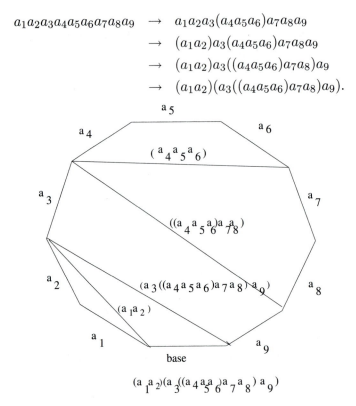

Figure 8.8

This correspondence works in general and establishes a one-to-one correspondence between bracketings and dissections, and also proves the next theorem. We adopt the convention that a polygonal region with 2 sides is a line segment and that it has exactly one dissection (the empty dissection).

Theorem 8.5.9 *Let n be a positive integer. Then the number of dissections of a convex polygonal region of $n + 1$ sides equals the small Schröder number s_n.* □

In terms of the polygonal region Π_{n+1}, our algorithm for constructing a bracketing of n is both natural and obvious.

Algorithm to Construct Dissections of Π_{n+1}

Start with the convex polygonal region Π_{n+1}, with the sides labeled $base, a_1, a_2, \ldots, a_n$ in a clockwise fashion.

1. Let $\Gamma = \Pi_{n+1}$.

   (a) While Γ has more than three sides, insert a diagonal of Γ, thereby partitioning Γ into two parts.

   (b) Replace Γ with the part containing the base. (This part will have at least one fewer side.)

3. Output the full dissected polygonal region Π_{n+1}.

8.6 Exercises

1. Let $2n$ (equally spaced) points on a circle be chosen. Show that the number of ways to join these points in pairs, so that the resulting n line segments do not intersect, equals the nth Catalan number C_n.

2. Prove that the number of 2-by-n arrays

$$\begin{bmatrix} x_{11} & x_{12} & \cdots & x_{1n} \\ x_{21} & x_{22} & \cdots & x_{2n} \end{bmatrix}$$

that can be made from the numbers $1, 2 \ldots, 2n$ such that

$$x_{11} < x_{12} < \cdots < x_{1n},$$

$$x_{21} < x_{22} < \cdots < x_{2n}$$

$$x_{11} < x_{21}, x_{12} < x_{22}, \ldots, x_{1n} < x_{2n},$$

equals the nth Catalan number, C_n.

3. Write out all of the multiplication schemes for four numbers and the triangularization of a convex polygonal region of five sides corresponding to them.

4. Determine the triangularization of a convex polygonal region corresponding to the following multiplication schemes:

   (a) $(a_1 \times (((a_2 \times a_3) \times (a_4 \times a_5)) \times a_6))$

   (b) $((((a_1 \times a_2) \times (a_3 \times (a_4 \times a_5)) \times ((a_6 \times a_7) \times a_8)))$

5. * Let m and n be nonnegative integers with $n \geq m$. There are $m + n$ people in line to get into a theatre for which admission is 50 cents. Of the $m + n$ people, n have a 50-cent piece and m have a \$1 dollar bill. The box office opens with an empty cash register. Show that the number of ways the people can line up so that change is available when needed is

$$\frac{n - m + 1}{n + 1} \binom{m + n}{m}.$$

   (The case $m = n$ is the case treated in Section 8.1.)

6. Let the sequence $h_0, h_1, \ldots, h_n, \ldots$ be defined by $h_n = 2n^2 - n + 3$, ($n \geq 0$). Determine the difference table, and find a formula for $\sum_{k=0}^{n} h_k$.

7. The general term h_n of a sequence is a polynomial in n of degree 3. If the first four entries of the 0th row of its difference table are $1, -1, 3, 10$, determine h_n and a formula for $\sum_{k=0}^{n} h_k$.

8. Find the sum of the fifth powers of the first n positive integers.

9. Prove the following formula for the kth-order differences of a sequence $h_0, h_1, \ldots, h_n, \ldots$:

$$\Delta^k h_n = \sum_{j=0}^{k} (-1)^{k-j} \binom{k}{j} h_{n+j}.$$

10. If h_n is a polynomial in n of degree m, prove that the constants $c_0, c_1, \ldots, c_m$ such that

$$h_n = c_0 \binom{n}{0} + c_1 \binom{n}{1} + \cdots + c_m \binom{n}{m}$$

   are uniquely determined. (Cf. Theorem 8.2.2.)

11. Compute the Stirling numbers of the second kind $S(8, k)$, ($k = 0, 1, \ldots, 8$).

12. Prove that the Stirling numbers of the second kind satisfy the following relations:

    (a) $S(n, 1) = 1, \quad (n \geq 1)$
    (b) $S(n, 2) = 2^{n-1} - 1, \quad (n \geq 2)$
    (c) $S(n, n-1) = \binom{n}{2}, \quad (n \geq 1)$
    (d) $S(n, n-2) = \binom{n}{3} + 3\binom{n}{4} \quad (n \geq 2).$

13. Let X be a p-element set and let Y be a k-element set. Prove that the number of functions $f : X \rightarrow Y$ which map X *onto* Y equals
$$k!S(p, k) = S^{\#}(p, k).$$

14. * Find and verify a general formula for
$$\sum_{k=0}^{n} k^p$$
involving Stirling numbers of the second kind.

15. The number of partitions of a set of n elements into k distinguishable boxes (some of which may be empty) is k^n. By counting in a different way, prove that
$$k^n = \binom{k}{1} 1!S(n, 1) + \binom{k}{2} 2!S(n, 2) + \cdots + \binom{k}{n} n!S(n, n).$$

(If $k > n$, define $S(n, k)$ to be 0.)

16. Compute the Bell number B_8. (Cf. Exercise 11.)

17. Compute the triangle of Stirling numbers of the first kind $s(n, k)$ up to $n = 7$.

18. Write $[n]_k$ as a polynomial in n for $k = 1, 2, \ldots, 7$.

19. Prove that the Stirling numbers of the first kind satisfy the following formulas:

    (a) $s(n, 1) = (n-1)!, \quad (n \geq 1)$
    (b) $s(n, n-1) = \binom{n}{2}, \quad (n \geq 1).$

20. Verify that $[n]_n = n!$, and write $n!$ as a polynomial in n using the Stirling numbers of the first kind. Do this explicitly for $n = 6$.

21. For each integer $n = 1, 2, 3, 4, 5$, construct the diagram of the set $\mathcal{P}_n$ of partitions of n partially ordered by majorization.

22. (a) Calculate p_6 and construct the diagram of the set $\mathcal{P}_6$ partially ordered by majorization.

    (b) Calculate p_7 and construct the diagram of the set $\mathcal{P}_7$ partially ordered by majorization.

23. A total order on a finite set has a unique maximal element (a largest element) and a unique minimal element (a smallest element). What are the largest partition and smallest partition in the lexicographic order on $\mathcal{P}(n)$?

24. A partial order on a finite set may have many maximal elements and minimal elements. In the set $\mathcal{P}_n$ of partitions of n partially ordered by majorization, prove that there is a unique maximal element and a unique minimal element.

25. Let $t_1, t_2, \ldots, t_m$ be distinct positive integers, and let

$$q_n = q_n(t_1, t_2, \ldots, t_n)$$

equal the number of partitions of n in which all parts are taken from $t_1, t_2, \ldots, t_m$. Define $q_0 = 1$. Show that the generating function for $q_0, q_1, \ldots, q_n, \ldots$ is

$$\prod_{k=1}^{m} (1 - x^{t_k})^{-1}.$$

26. Determine the conjugate of each of the following partitions:

    (a) $12 = 5 + 4 + 2 + 1$

    (b) $15 = 6 + 4 + 3 + 1 + 1$

    (c) $20 = 6 + 6 + 4 + 4$

    (d) $21 = 6 + 5 + 4 + 3 + 2 + 1$

    (e) $29 = 8 + 6 + 6 + 4 + 3 + 2$

27. For each integer $n > 2$, determine a self-conjugate partition of n that has at least two parts.

28. Prove that conjugation reverses the order of majorization; that is, if λ and μ are partitions of n and λ is majorized by μ, then μ^* is majorized by λ^*.

29. Evaluate $h^{(k)}_{k-1}$, the number of regions into which k-dimensional space is partitioned by $k-1$ hyperplanes in general position.

30. Use the recurrence relation (8.32) to compute the small Schröder numbers s_8 and s_9.

31. Use the recurrence relation (8.34) to compute the large Schröder numbers R_7 and R_8. Verify that $R_7 = 2s_8$ and $R_8 = 2s_9$, as stated in Corollary 8.5.8.

32. Use the generating function for the large Schröder numbers to compute the first few large Schröder numbers.

33. Use the generating function for the small Schröder numbers to compute the first few small Schröder numbers.

34. Prove that the large Schröder number R_n equals the number of lattice paths from $(0,0)$ to $(2n,0)$ with steps $(1,1)$ and $(1,-1)$ that never go above the horizontal axis. (These are sometimes called *Dyck paths*.)

35. * The large Schröder number R_n counts the number of subdiagonal lattice paths from $(0,0)$ to (n,n). The small Schröder number counts the number of dissections of a convex polygonal region of $n+1$. Since $R_n = 2s_{n+1}$ for $n \geq 1$, there are as many subdiagonal lattice paths from $(0,0)$ to (n,n) as there are dissections of a convex polygonal region of $n+1$ sides. Find a one-to-one correspondence between these lattice paths and these dissections.

Chapter 9

Matchings in Bipartite Graphs

We begin with the following three problems:

Problem 1. Consider an m-by-n chessboard in which certain squares are forbidden. What is the largest number of nonattacking rooks that can be placed on the board?

In previous sections we considered the problem of counting the number of ways to place n nonattacking rooks on an n-by-n board. It was presumed that this number was positive. Now we are concerned not only with whether or not it is possible to place n nonattacking rooks on the board but, more generally, with the question of the largest number of nonattacking rooks that can be placed on the board.

Problem 2. Consider again an m-by-n chessboard where certain squares are forbidden. What is the largest number of dominoes that can be placed on the board so that each domino covers two allowed squares and no two dominoes overlap (cover the same square)?

In Chapter 1 we considered the special case of this problem concerning when a board with forbidden squares has a perfect cover. For a perfect cover, we must have, in addition, every allowed square covered by a domino. If p is the total number of allowed squares, then there is a perfect cover if and only if both p is even and the answer to Problem 2 is $p/2$.

Problem 3. A company has n jobs available, with each job demanding certain qualifications. There are m people who apply for the n jobs. What is the largest number of jobs that can be filled from the available

m applicants if a job can be filled only by a person who meets its qualifications?

The first two problems are of a seemingly recreational nature. The third problem, however, is clearly of a more serious and applied nature. As a matter of fact, Problems 1 and 3 are different formulations of the same abstract problem, and Problem 2 is a special case. In this chapter we solve the abstract problem and thereby solve each of Problems 1, 2, and 3. Of course, in Problem 3, we want to know not only the largest number of jobs that can be filled, but also how to assign the applicants to the jobs. (A similar remark applies to Problems 1 and 2.) Thus, we shall also discuss how to find an assignment in which the largest number of jobs is filled.

9.1 General Problem Formulation

Each of Problems 1, 2, and 3 fits into the following framework: Let

$$X = \{x_1, x_2, \ldots, x_m\}$$

and

$$Y = \{y_1, y_2, \ldots, y_n\}$$

be two finite sets with m elements and n elements, respectively. We assume that the sets X and Y have no elements in common, that is,

$$X \cap Y = \emptyset.$$

Let Δ be a collection of pairs

$$e = \{x, y\},$$

where x is an element of X and y is an element of Y. The triple

$$G = (X, \Delta, Y)$$

is called a *bipartite graph*.[1] The elements of $X \cup Y$ are called the *vertices* of G, and X, Y is called a *bipartition* (partition into two parts) of the vertices of G. We regard X, Y and Y, X as the same bipartition and thus do not distinguish between (X, Δ, Y) and (Y, Δ, X), although we usually write the vertices of X on the left and the vertices of Y on the right. The pairs $e = \{x, y\}$ in Δ are called the *edges* of G. Note that

---

[1]As the name suggests bipartite graphs are instances of more general objects called graphs. Graphs in general are discussed in later chapters.

each edge $e = \{x, y\}$ is a set of two vertices, one of which, x, comes
from X and the other of which, y, comes from Y. We say that the
edge e *joins* the vertices x and y, and that the vertices x and y *meet*
the edge e. Thus, a bipartite graph is prescribed by

  (i) a set of vertices;

  (ii) a partition of that set of vertices into two parts; and

  (iii) a set of edges joining a vertex in one part to a vertex in the other
      part.

Example. Let $X = \{x_1, x_2, x_3, x_4\}$ and $Y = \{y_1, y_2, y_3\}$, and let

$$\Delta = \{\{x_1, y_1\}, \{x_1, y_3\}, \{x_2, y_1\}, \{x_3, y_2\}, \{x_3, y_3\}, \{x_4, y_3\}\}.$$

The vertex x_1 meets two edges, namely the edges $\{x_1, y_1\}$ and $\{x_1, y_3\}$.
The vertex y_3 meets three edges. We can picture the bipartite graph
$G = (X, \Delta, Y)$ as shown in Figure 9.1.

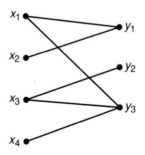

Figure 9.1

In this figure each vertex is represented by a dot, with the vertices
of X on the left and those of Y on the right. Each edge is represented
by a line segment joining the two vertices it contains, but one needs to
keep in mind that an edge consists of just two vertices.   □

Every bipartite graph can be pictured in a way similar to that in
the preceding example. As a result, we now speak of the vertices of
X as the *left vertices* and the vertices of Y as the *right vertices*. Each
edge joins a left vertex to a right vertex. According to our convention
of not distinguishing between (X, Δ, Y) and (Y, Δ, X), we could call
X the set of right vertices and Y the set of left vertices.

Let $G = (X, \Delta, Y)$ be a bipartite graph. A *matching* of G is defined
to be a subset M of the set Δ of edges, with the property that no two of

the edges of M have a common vertex. Thus, if M is a matching, then each left vertex meets at most one edge of M, and similarly each right vertex meets at most one edge of M. In the bipartite graph pictured in Figure 9.1, the three edges

$$\{x_1, y_3\}, \quad \{x_2, y_1\}, \quad \{x_3, y_2\}$$

form a matching of size three. (Note that the fact that the edges $\{x_1, y_3\}$ and $\{x_2, y_1\}$ cross in the figure is of no concern, and indeed this crossing is not part of the abstract definition of the bipartite graph determined by Figure 9.1. What is of concern is the fact that no two of the three edges meet at a vertex.)

Example. Consider the 4-by-5 board with forbidden positions pictured in Figure 9.2.

	y_1	y_2	y_3	y_4	y_5
x_1		×			
x_2			×		×
x_3	×		×		×
x_4	×				

Figure 9.2

We wish to associate a bipartite graph with this board. Corresponding to each row of the board, there is a left vertex:

x_i is the left vertex corresponding to row i, $(i = 1, 2, 3, 4)$.

Corresponding to each column of the board, there is a right vertex:

y_j is the right vertex corresponding to column j, $(j = 1, 2, 3, 4, 5)$.

The sets of left and right vertices are, respectively,

$$X = \{x_1, x_2, x_3, x_4\} \text{ and } Y = \{y_1, y_2, y_3, y_4, y_5\}.$$

In G we join vertex x_i and vertex y_j by an edge if and only if the square at the intersection of row i and column j is allowed. We let Δ be the set of edges obtained in this way and then define a bipartite graph G by

$$G = (X, \Delta, Y).$$

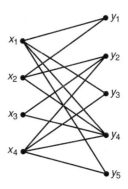

Figure 9.3

The set Δ of edges of G is in one-to-one correspondence with the allowed squares of the board. The graph G corresponding to the board in Figure 9.2 is pictured in Figure 9.3.

Consider the matching

$$M = \{\{x_1, y_1\}, \{x_2, y_4\}, \{x_4, y_2\}\}$$

of G. Each of the three edges of M corresponds to an allowed square of the board. Since no two of the edges of M have a common left vertex, no two of these squares are in the same row. Since no two of the edges of M have a common right vertex, no two of the squares are in the same column. Thus, if we put rooks on the three squares corresponding to the edges of M, we have three nonattacking rooks on the board. Conversely, a collection of nonattacking rooks on the board gives a matching of the bipartite graph. Therefore, there is a one-to-one correspondence between sets of nonattacking rooks on the board and matchings in the associated bipartite graph. In this one-to-one correspondence, the number of nonattacking rooks equals the number of edges of the matching. □

The discussion in the previous example applies in general. We can associate a bipartite graph $G = (X, \Delta, Y)$, called a *rook-bipartite graph*, with any m-by-n board B with forbidden positions. This graph has vertices $X = \{x_1, x_2, \ldots, x_m\}$ corresponding to the rows of the board, and vertices $Y = \{y_1, y_2, \ldots, y_n\}$ corresponding to the columns. The pair $\{x_i, y_j\}$ is an edge of Δ if and only if the square at the intersection of row i and column j is allowed. Nonattacking rooks on the board B correspond to matchings in the bipartite graph G. Moreover, the largest number of nonattacking rooks that can be placed on the board B equals the largest number of edges in a matching of the bipartite graph G. Each bipartite graph is the rook-bipartite graph of some board with forbidden positions. (See Exercise 3.)

If G is any bipartite graph, we now define

$$\rho(G) = \max\{|M| : M \text{ a matching}\}$$

to be the size of the largest matching M of G. Problem 1 is equivalent to determining $\rho(G)$ for the rook-bipartite graph corresponding to a board with forbidden positions.

Example. Consider a 4-by-5 board whose squares are alternately colored black and white, and then forbid the same squares as in the previous example. (See Fig. 9.2.) For identification we label the nonforbidden white squares $w_1, w_2, \ldots, w_7$ and the nonforbidden black squares $b_1, b_2, \ldots, b_6$, as shown in Figure 9.4.

w_1	×	w_2	b_1	w_3
b_2	w_4	×	w_5	×
×	b_3	×	b_4	×
×	w_6	b_5	w_7	b_6

Figure 9.4

Except for the black-white labeling, this is the same board as in Figure 9.2. We associate with this board a different bipartite graph $G = (X, \Delta, Y)$. This time, we let

$$X = \{w_1, w_2, \ldots, w_7\}$$

be the set of white squares, and we let

$$Y = \{b_1, b_2, \ldots, b_6\}$$

be the set of black squares. Thus, the left vertices of G are the white squares, and the right vertices are the black squares. There is an edge $\{w_i, b_j\}$ in Δ joining a white square w_i and black square b_j if and only if the two squares have a common side. Hence, two squares are joined by an edge in G if and only if one domino can simultaneously cover both squares. Each edge of G thus corresponds to a possible domino on the board, and each possible domino on the board corresponds to an edge. The bipartite graph G just defined is shown in Figure 9.5.

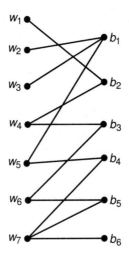

Figure 9.5

Consider the matching

$$M = \{\{w_1, b_2\}, \{w_3, b_1\}, \{w_6, b_3\}, \{w_7, b_4\}\}$$

of G. Each of the edges of M corresponds to a domino, and because no two edges in M have a common vertex, no two of these dominoes overlap. Conversely, from a set of nonoverlapping dominoes on the board, we obtain a matching of G. Therefore, there is a one-to-one correspondence between sets of nonoverlapping dominoes on the board and matchings in the bipartite graph associated with the board, in the manner indicated. In this one-to-one correspondence, the number of dominoes equals the number of edges of the matching.   □

The discussion in the previous example applies generally to the problem of determining the largest number of nonoverlapping dominoes that can be placed on a board B with forbidden positions. Given an m-by-n board B with forbidden positions, we associate a *domino-bipartite graph* $G = (X, \Delta, Y)$ where

$$X = \{w_1, w_2, \ldots, w_p\}$$

is the set of white squares,

$$Y = \{b_1, b_2, \ldots, b_q\}$$

is the set of black squares, and there is an edge $\{w_i, b_j\}$ in Δ if and only if one domino can simultaneously cover both w_i and b_j. Non-overlapping dominoes on the board B correspond to matching edges in the bipartite graph G. The largest number of nonoverlapping dominoes that can be placed on the board equals the largest number $\rho(G)$ of edges in a matching of G. Problem 2 is therefore equivalent to determining $\rho(G)$ for the domino-bipartite graph G corresponding to a board with forbidden positions. In contrast with rook-bipartite graphs, not every bipartite graph is the domino-bipartite graph of a board with forbidden positions. This is because, in a board, a square has a common side with at most four other squares. This implies that, in the corresponding domino-bipartite graph, each vertex can meet at most four edges. (See Exercise 4.)

Example. Four people x_1, x_2, x_3, x_4 apply for five jobs y_1, y_2, y_3, y_4, y_5. Suppose that

x_1 is qualified for y_1, y_3, y_4, y_5;

x_2 is qualified for y_1, y_2, y_4;

x_3 is qualified for y_2, y_4; and

x_4 is qualified for y_2, y_3, y_4, y_5.

We construct a bipartite graph $G = (X, \Delta, Y)$ in such a way that should seem quite natural. We let $X = \{x_1, x_2, x_3, x_4\}$ be the set of people (applying for a job) and $Y = \{y_1, y_2, y_3, y_4, y_5\}$ be the set of available jobs, and we put an edge $\{x_i, y_j\}$ in Δ if and only if x_i is qualified for job y_j. The resulting bipartite graph is the same bipartite graph pictured in Figure 9.3 (that is, the rook-bipartite graph of the board in Figure 9.2). There is a one-to-one correspondence between matchings in the bipartite graph G and assignments of qualified persons to the jobs. For example, the matching

$$M = \{\{x_1, y_1\}, \{x_2, y_4\}, \{x_4, y_2\}\}$$

corresponds to the assignment

person	assigned to	job
x_1	$\longrightarrow$	y_1
x_2	$\longrightarrow$	y_4
x_4	$\longrightarrow$	y_2

In the problem of placing nonattacking rooks on the board in Figure 9.2, the preceding matching corresponds to placing rooks at the intersections of row 1 and column 1, row 2 and column 4, and row 4 and column 2. Thus, we see that *assigning people to jobs for which they qualify is really the same abstract mathematical problem as putting rooks on a board so that no rook can attack another!*   □

As should be clear from the foregoing discussion, with any group $X = \{x_1, x_2, \ldots, x_m\}$ of people and any group $Y = \{y_1, y_2, \ldots, y_n\}$ of jobs, we can associate a bipartite graph (X, Δ, Y), where there is an edge $\{x_i, y_j\}$ joining x_i and y_j if and only if person x_i qualifies for job y_j. There is a one-to-one correspondence between matchings in G and possible assignments of qualified people to jobs. The fact that no two edges in a matching M meet at the same vertex in X means that each person is assigned, at most, one job. The fact that no two edges in M meet at the same vertex in Y means that two different people are not assigned the same job. This assignment problem can also be regarded as a nonattacking rooks problem: the square at the intersection of row i and column j is forbidden if person x_i does not qualify for job y_j and is allowed otherwise.

In sum, all three of the introductory problems are concerned with the problem of determining the largest number $\rho(G)$ of edges of a matching in a bipartite graph (X, Δ, Y). In the next section we show how to find $\rho(G)$.

9.2   Matchings

We consider a bipartite graph

$$G = (X, \Delta, Y),$$

where

$$X = \{x_1, x_2, \ldots, x_m\} \text{ and } Y = \{y_1, y_2, \ldots, y_n\}.$$

Recall that the largest number of edges in a matching is denoted by $\rho(G)$. Our goal is not only to determine $\rho(G)$, but also to determine a matching M^* with

$$|M^*| = \rho(G). \tag{9.1}$$

By the pigeonhole principle, a matching can have, at most, m edges, because if there were more than m edges, two edges would have to meet at the same left vertex. Similarly, a matching can have, at most, n edges. Thus, we have the simple inequality

$$\rho(G) \leq \min\{m, n\}. \tag{9.2}$$

Each matching M satisfies $|M| \le \rho(G)$. A matching M^* satisfying (9.1)—that is, a matching M with the largest possible number of edges among all matchings in G—is called a *max-matching*. If we know $\rho(G)$, we can determine whether any matching M is a max-matching by counting the number $|M|$ of edges in M and checking whether $|M| = \rho(G)$.

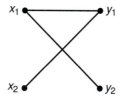

Figure 9.6

Example. Consider the bipartite graph G in Figure 9.6. The edges $\{x_1, y_2\}$ and $\{x_2, y_1\}$ form a matching of size two and hence, since $\rho(G)$ cannot be more than 2, we have $\rho(G) = 2$. Notice that the edge $\{x_1, y_1\}$ determines a matching M with one edge. There is no way to add another edge to this matching M in order to obtain a matching with two edges. Thus, one cannot draw the conclusion that a matching has the largest number of edges just from knowing that it is impossible to enlarge the matching by including more edges. □

We now discuss how to recognize whether a matching is a max-matching without knowing the value of $\rho(G)$. Of course, once we are able to conclude that a certain matching M is a max-matching, we then know that $\rho(G) = |M|$ and we have determined $\rho(G)$.

Let u and v be two vertices in the bipartite graph $G = (X, \Delta, Y)$. A *path* joining u and v is a sequence of distinct vertices (except that u may equal v)

$$\gamma : u = u_0, u_1, u_2, \ldots, u_{p-1}, u_p = v \qquad (9.3)$$

such that any two consecutive vertices are joined by an edge. Thus, in order for (9.3) to be a path,

$$\{u_0, u_1\}, \{u_1, u_2\}, \ldots, \{u_{p-1}, u_p\} \qquad (9.4)$$

must all be edges in Δ. The edges in (9.4) are called the *edges of the path* γ. The *length* of the path γ is the number p of its edges. The vertices u and v are called the *end-vertices* of the path γ. The vertices in a path must be alternately left and right vertices. The end-vertices can be either both left vertices, both right vertices, or one of each type. If $u = v$ in the path (9.3), then the path is called a *cycle*. A cycle in a bipartite graph necessarily has even length.

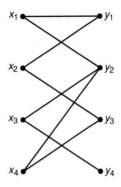

Figure 9.7

Example. In the bipartite graph G pictured in Figure 9.7,

$$x_1, y_2, x_3, y_4$$

is a path of length 3 joining x_1 and y_4;

$$y_1, x_2, y_3, x_4$$

is a path joining y_1 and x_4; and

$$x_3, y_2, x_4, y_3, x_2$$

is a path joining x_3 and x_2. Also

$$x_1, y_1, x_2, y_3, x_4, y_2, x_1$$

is a cycle of length 6.   □

Now let M be a matching in the bipartite graph $G = (X, \Delta, Y)$. Let $\overline{M}$ be the complement of M, that is, the set of edges of G that do not belong to M. Let u and v be vertices, where one of u and v is a left vertex and one is a right vertex. A path γ joining u and v is an *alternating path with respect to the matching M* (for brevity, an *M-alternating path*) provided that the following properties hold:

(1) The first, third, fifth, ... edges of γ do not belong to the matching M (thus they belong to $\overline{M}$).

(2) The second, fourth, sixth, ... edges of γ belong to the matching M.

(3) Neither u nor v meets an edge of the matching M.

Notice that the length of an M-alternating path γ is an odd number $2k+1$ with $k \geq 0$, and that $k+1$ of the edges of γ are edges of $\overline{M}$ while k of the edges of γ are edges of M. We introduce further notation as follows:

M_γ denotes those edges of γ that belong to M;

and

$\overline{M}_\gamma$ denotes those edges of γ that do not belong to M.

It follows from our discussion that

$$|\overline{M}_\gamma| = |M_\gamma| + 1.$$

Example. Consider the bipartite graph G pictured in Figure 9.7. The set

$$M = \{\{x_1, y_1\}, \{x_2, y_3\}, \{x_3, y_4\}\}$$

is a matching of three edges. The path

$$\gamma : u = x_4, y_3, x_2, y_1, x_1, y_2 = v$$

is an M-alternating path. We have

$$M_\gamma = \{\{x_2, y_3\}, \{x_1, y_1\}\},$$

and

$$\overline{M}_\gamma = \{\{x_4, y_3\}, \{x_2, y_1\}, \{x_1, y_2\}\}.$$

If we remove the edges of M_γ from M and replace them with the edges of $\overline{M}_\gamma$, we obtain a matching

$$
\begin{aligned}
M' &= (M - M_\gamma) \cup \overline{M}_\gamma \\
&= \{\{x_3, y_4\}, \{x_4, y_3\}, \{x_2, y_1\}, \{x_1, y_2\}\}
\end{aligned}
$$

of four edges.   □

As illustrated in the previous example, if M is a matching and there is an M-alternating path γ, then

$$(M - M_\gamma) \cup \overline{M}_\gamma$$

is a matching with one more edge than M, and hence M is not a max-matching. We now show that the converse holds as well; that is, the only way a matching M cannot be a max-matching is for there to exist an M-alternating path γ.

Theorem 9.2.1 *Let M be a matching in the bipartite graph $G = (X, \Delta, Y)$. Then M is a max-matching if and only if there does not exist an M-alternating path.*

Proof. As previously observed, if M is a max-matching, then there does not exist an M-alternating path.

To establish the converse, we now assume that M is not a max-matching and prove that there exists an M-alternating path. Let M' be a matching satisfying

$$|M'| > |M|.$$

We consider the bipartite graph

$$G^* = (X, \Delta^*, Y)$$

where

$$\Delta^* = (M - M') \cup (M' - M).$$

The bipartite graph G^* has the same left and right vertices as G. The edges of G^* are those edges of G which either belong to M, but not to M' (the edges of $M - M'$), or belong to M', but not to M (the edges of $M' - M$). Thus, to get G^*, we remove from G all edges that belong neither to M nor to M' and also those edges that belong to both M and M'. Since $|M'| > |M|$, we have

$$|M' - M| > |M - M'|. \tag{9.5}$$

The bipartite graph G^* has the property that each of its vertices meets, at most, two edges: Each vertex meets, at most, one edge of $M - M'$, since M is a matching, and meets, at most, one edge of $M' - M$, since M' is a matching. This property of G^* implies that the set of edges of G^* can be partitioned into paths and cycles. In each of the paths and cycles of this partition, the edges alternate between $M - M'$ and $M' - M$. A path in the partition has the property that both its first and last vertices meet only one edge of G^*. These paths and cycles are of the following four types:

Type 1. A path whose first and last edges are both in $M' - M$. (See Fig. 9.8. In this and the other figures the bold lines denote the edges of M.) These paths have odd length and contain one more edge of M' than they do of M. Included among the Type 1 paths are paths of length 1, where the first edge is the same as the last edge.

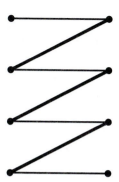

Figure 9.8. Type 1 path

Type 2. A path whose first and last edges are both in $M - M'$. (See Fig. 9.9). These paths also have odd length, but they contain one more edge of M than they do of M'.

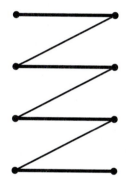

Figure 9.9. Type 2 path

Type 3. A path whose first edge is in $M - M'$ and whose last edge is in $M' - M$ (or vice-versa). (See Fig. 9.10.) These paths have even length and contain as many edges of M as they do of M'.

Figure 9.10 Type 3 path

Type 4. A cycle. (See Fig. 9.11.) These cycles have even length and contain as many edges of M as they do of M'.

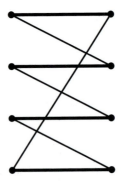

Figure 9.11 Type 4 cycle

There are more edges of $M - M'$ than of $M' - M$ in a path of Type 2, and the same number of edges of $M - M'$ as of $M - M'$ in a path of Type 3 and in a cycle of Type 4. In a path of Type 1, there are more edges of $M' - M$ than of $M - M'$. By (9.5), $M' - M$ has more edges than $M - M'$ does. Hence, in G^*, there is a path of Type 1. A path of Type 1 is by definition an M-alternating path. Thus, if a matching M is not a max-matching, there is an M-alternating path. □

Theorem 9.2.1 characterizes max-matchings among all the matchings in a bipartite graph. Its strength lies in the fact that, given a matching M, in order to determine whether M is a max-matching, we need search only for an M-alternating path γ. If we find such a path γ, then, by removing from M those edges of γ that belong to M and replacing them with the edges of γ that do not belong to M, we obtain a matching M' that has more edges than M. If we cannot find an M-alternating path γ, then by Theorem 9.2.1, M is a max-matching.

The weakness of Theorem 9.2.1 lies in the last statement of the previous paragraph. After searching for an M-alternating path and not finding one, we need to know that we didn't find one because there wasn't any to be found, not because we didn't look hard enough! We cannot expect to examine all possible paths in order to determine whether among them there is an M-alternating path. Such a task would require, in general, too much time and effort. What we seek is some way of establishing that a matching is a max-matching that is easy to check. In other words, we seek an easily verifiable *certification* that a matching is a max-matching. We now discuss such a certification.

Let $G = (X, \Delta, Y)$ be a bipartite graph. A subset S of the set $X \cup Y$ of vertices of G is called a *cover*, provided that each edge of G has at least one of its two vertices in S:

$$\{x, y\} \cap S \neq \emptyset \quad \text{for all } \{x, y\} \text{ in } \Delta.$$

The set X of left vertices of G is a cover, since each edge has a left vertex. The set Y of right vertices is also a cover. Indeed, the set $X \cup Y$ of all vertices of G is a cover. However, our interest lies in small covers.

Example. Let G be the bipartite graph pictured in Figure 9.12. In addition to the covers $\{x_1, x_2, x_3, x_4\}$ and $\{y_1, y_2, y_3, y_4\}$, we have the cover $S = \{x_1, x_3, y_2\}$ with only three vertices. The fact that S is a cover means that there is no edge whose left vertex is one of $\{x_2, x_4\}$ *and* whose right vertex is one of $\{y_1, y_3, y_4\}$, and this is readily checked by inspection. □

We define the *cover number* of G to be

$$c(G) = \min\{|S| : S \text{ a cover of } G\},$$

the smallest number of vertices in a cover of G. Every cover S satisfies

$$|S| \geq c(G).$$

We call a cover S of G that satisfies

$$|S| = c(G)$$

(i.e., a cover with the smallest number of vertices) a *min-cover*.

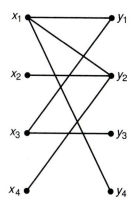

Figure 9.12

Example. Let G be the bipartite graph pictured in Figure 9.12. As already observed,

$$S = \{x_1, x_3, y_2\}$$

is a cover of G. It is straightforward to check that $c(G) = 3$ and S is a min-cover: If we do not include x_1 in a cover, then each of the three

vertices y_1, y_2, and y_4 (the vertices joined by an edge to x_1) would have
to be in the cover. If we do not include y_2 in the cover, we would have
to include each of the three vertices x_1, x_2 and x_4 in the cover. Since
$\{x_1, y_2\}$ is not a cover, each cover contains at least three vertices. $\quad\square$

In the next lemma we show that the matching number cannot exceed the cover number.

Lemma 9.2.2 *If G is a bipartite graph, then*

$$\rho(G) \leq c(G); \tag{9.6}$$

*that is, the largest number of edges in a matching of G does not exceed
the smallest number of vertices in a cover of G.*

Proof. Let G be the bipartite graph (X, Δ, Y) and let S^* be a cover
satisfying $|S^*| = c(G)$. Let M be a matching. Since S^* is a cover,
each edge of M has at least one of its vertices in S^*. Suppose that
$|M| > |S^*|$. Then, by the pigeonhole principle, two different edges in
M contain the same vertex of S^*. But this contradicts the fact that
M is a matching. Hence,

$$|M| \leq |S^*| = c(G).$$

$\quad\square$

Lemma 9.2.2 has the following consequence: Suppose that, in some
way or other, we have found a matching M in a bipartite graph G
which we think might be a max-matching. If we can find a cover S
such that
$$|M| = |S|,$$

then M is a max-matching (and S is a min-cover). This fact is a
consequence of the inequalities

$$c(G) \leq |S| = |M| \leq \rho(G), \tag{9.7}$$

which together imply that $c(G) \leq \rho(G)$. Applying (9.6), we conclude
that $c(G) = \rho(G)$. Now, (9.7) implies that

$$|M| = \rho(G) \quad \text{and} \quad |S| = c(G);$$

that is, M is a max-matching and S is a min-cover. Thus, in this case,
S acts as a certification that there is no matching with a larger number
of edges than M.

Example. Continuing with the bipartite graph in Figure 9.12, we see that

$$M = \{\{x_1, y_1\}, \{x_2, y_2\}, \{x_3, y_3\}\}$$

is a matching of three edges. As already observed, $S = \{x_1, x_3, y_2\}$ is a cover of three vertices. Hence,

$$3 = |M| \leq \rho(G) \leq c(G) \leq |S| = 3.$$

We have equality throughout, and hence M is a max-matching, S is a min-cover, and $\rho(G) = c(G) = 3$. □

We now turn to showing that we can always find a matching M and a cover S satisfying

$$|M| = |S|, \tag{9.8}$$

from which we will be able to conclude, as before, that $\rho(G) = c(G)$, M is a max-matching, and S is a min-cover. Thus, our sought-after certification is a cover S with the same size as the matching M.

Let $G = (X, \Delta, Y)$ be a bipartite graph and let M be a matching in G. We describe an algorithm that is a systematic search for an M-alternating path. Either the algorithm produces an M-alternating path (and we use the proof of Theorem 9.2.1 to obtain a matching with one more edge than M), or it fails to produce an M-alternating path, but does produce a min-cover S with $|M| = |S|$ (and we thus conclude that M is a max-matching and S is a certification for M; the algorithm didn't produce an M-alternating path, because no such alternating path exists). The next algorithm is a special instance of a more general network algorithm of Ford and Fulkerson.[2]

Matching Algorithm

Let $G = (X, \Delta, Y)$ be a bipartite graph where $X = \{x_1, x_2, \ldots, x_m\}$ and $Y = \{y_1, y_2, \ldots, y_n\}$. Let M be any matching in G.

(0) Begin by labeling with (*) all vertices in X that do not meet any edge in M and call all vertices *unscanned*. Go to (1).

(1) If in the previous step, no new label has been given to a vertex of X, then stop. Otherwise go to (2).

---

[2]L.R. Ford, Jr. and D.R. Fulkerson: *Flows in Networks*, Princeton University Press, Princeton (1962). Also, see Section 12.2.

(2) While there exists a labeled, but unscanned, vertex of X, select such a vertex, say, x_i, and label with (x_i) all vertices in Y joined to x_i by an edge *not belonging to M not previously labeled*. The vertex x_i is now scanned. If there are no labeled but unscanned vertices go to (3).

(3) If, in step (2), no new label has been given to a vertex of Y, then stop. Otherwise go to (4).

(4) While there exists a labeled, but unscanned vertex, of Y, select such a vertex, say, y_j, and label with (y_j) any vertex of X joined to y_j by an edge *belonging to M not previously labeled*. The vertex y_j is now scanned. If there are no labeled but unscanned vertices, go to (1).

Since each vertex receives, at most, one label, and since each vertex is scanned, at most, once, the Matching Algorithm halts after a finite number of steps. There are two possibilities to consider:

Breakthrough: There is a labeled vertex of Y that does not meet an edge of M.

Nonbreakthrough: The algorithm has come to a halt and Breakthrough has not occurred; that is, each vertex of Y that is labeled also meets some edge of M.

In the case of Breakthrough, the Matching Algorithm has succeeded in finding an M-alternating path γ. One end vertex of γ is the vertex v of Y, which is labeled, but does not meet any edge of M. The other end vertex of γ is a vertex u of X with label (*) (and which therefore does not meet any edge of M). The M-alternating path γ can be constructed by starting at v and working backwards through the labels until a vertex u with label (*) is found. With Breakthrough and the M-alternating path γ, we can obtain (as in the proof of Theorem 9.2.1) a matching with one more edge than M.

If Nonbreakthrough occurs, we show that it is because M is a max-matching; that is, according to Theorem 9.2.1, there isn't any M-alternating path. Thus, Breakthrough occurs exactly when M is not a max-matching, and when Breakthrough occurs, we have a way of obtaining an M-alternating path and hence a matching with one more edge than M.

Theorem 9.2.3 *Assume Nonbreakthrough occurs in the Matching Algorithm. Let X^{un} consist of all the unlabeled vertices of X and let Y^{lab} consist of all the labeled vertices of Y. Then both of the following hold:*

(i) $S = X^{un} \cup Y^{lab}$ is a min-cover of the bipartite graph G;

(ii) $|M| = |S|$ and M is a max-matching.

Proof. We show that S is a cover by assuming that there is an edge $e = \{x, y\}$, neither of whose vertices belongs to S, and obtaining a contradiction. Thus, assume that x is in $X - X^{un}$ and y is in $Y - Y^{lab}$ and $e = \{x, y\}$ is an edge. Since x is not in X^{un}, x is labeled; since y is not in Y^{lab}, y is unlabeled. Either e belongs to M or it does not. If e does not belong to M, then in applying step (2) of the algorithm, y receives the label (x), a contradiction. We now assume that e belongs to M. Since x meets the edge e of M, it follows from step (0) that the label of x is not (*). But then it follows from the algorithm that x has label (y). (See step (4).) By the algorithm again, vertex y can give label (y) to a vertex of X only if y is already labeled. Since y is not labeled, we have a contradiction again. Since both possibilities lead to a contradiction, we conclude that S is a cover.

We complete the proof of the theorem by showing that $|M| = |S|$. As we have already demonstrated, this equality also implies that S is a min-cover and M is a max-matching. We establish a one-to-one correspondence between the vertices in S and the edges in M, thereby proving $|M| = |S|$. Let y be a vertex in Y^{lab} so that y is labeled. Since Breakthrough has not occurred, y meets an edge of M, and hence exactly one edge, say, the edge $\{x, y\}$ of M. By step (4) of the algorithm, x gets the label (y) and hence x is not in X^{un}. Thus, each vertex of Y^{lab} meets an edge of M whose other vertex belongs to $X - X^{un}$. Now consider a vertex x' in X^{un}. Since x' is not labeled, it follows from step (0) that x' meets an edge of M (otherwise x' would have the label (*)), and hence exactly one edge, say $\{x', y'\}$, of M. The vertex y' cannot be in Y^{lab} since we previously showed that the unique edge of M meeting a vertex in Y^{lab} has its other vertex in $X - X^{un}$. Thus, we have shown that, for each vertex of $X^{un} \cup Y^{lab}$, there is a unique edge of M containing it and all these edges are distinct. Hence,

$$|S| = |X^{un} \cup Y^{lab}| \leq |M|,$$

and we conclude that $|S| = |M|$. $\qquad\square$

The following important corollary is known as König's theorem.

Corollary 9.2.4 *Let* $G = (X, \Delta, Y)$ *be a bipartite graph. Then*

$$\rho(G) = c(G);$$

that is, the largest number of edges in a matching equals the smallest number of vertices in a cover.

Proof. By Lemma 9.2.2,

$$\rho(G) \leq c(G).$$

By Theorem 9.2.3, there is a matching M and a cover S such that $|M| = |S|$. Hence,

$$\rho(G) \geq |M| = |S| \geq c(G),$$

and we conclude that $\rho(G) = c(G)$. $\qquad\qquad\qquad\qquad\qquad\square$

The Matching Algorithm can be applied to obtain a max-matching in a bipartite graph (X, Δ, Y) as follows: We first choose a matching in a greedy fashion—we pick any edge e_1, then any edge e_2 that does not meet e_1, then any edge e_3 that does not meet e_1 or e_2, and continue like this until we run out of choices.[3] We call the resulting matching M^1 and apply the Matching Algorithm to it. If Nonbreakthrough occurs, then, by Theorem 9.2.3, M^1 is a max-matching. If Breakthrough occurs, then we obtain a matching M^2 with more edges than M^1. We now apply the matching algorithm to M^2. In this way, we obtain a sequence of matchings $M^1, M^2, M^3, \dots$, each with more edges than the preceding one. After a finite number of applications of the matching algorithm, we obtain a matching M^k for which the matching algorithm results in Nonbreakthrough, and hence M^k is a max-matching.

Example. We determine a max-matching in the bipartite graph $G = (X, \Delta, Y)$ in Figure 9.13. We choose the edges $\{x_2, y_2\}, \{x_3, y_3\}$, and $\{x_4, y_4\}$ and obtain a matching M^1 of size 3. The edges of M^1 are in boldface in Figure 9.13. We now apply the Matching Algorithm to M^1, and the results as shown in Figure 9.13 are as follows:

(i) Step (0): The vertices x_1, x_5, and x_6, which do not meet an edge of M^1, are labeled (*).

(ii) Step (2): We scan x_1, x_5 and x_6, in turn, and label y_3 with (x_1) and y_4 with (x_5). Since all vertices joined to x_6 already have a label, no vertex of Y gets labeled (x_6).

---

[3] Or perhaps we stop because there are no more obvious choices.

(iii) Step (4): We scan the vertices y_3 and y_4, labeled in (ii), and label x_3 with (y_3) and x_4 with (y_4).

(iv) Step (2): We scan the vertices x_3 and x_4, labeled in (iii), and label y_2 with (x_3).

(v) Step (4): We scan the vertex y_2, labeled in (iv), and label x_2 with (y_2).

(vi) Step (2): We scan the vertex x_2, labeled in (v), and label y_1, y_5 and y_6 with (x_2).

(vii) Step (4): We scan the vertices y_1, y_5, and y_6, labeled in (vi), and find that no new labels are possible.

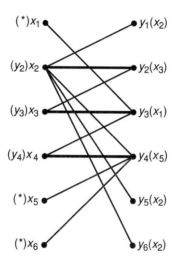

Figure 9.13

The algorithm has now come to an end, and since we have labeled a vertex of Y that does not meet an edge of M^1 (in fact, the three vertices y_1, y_5, and y_6 have this property), we have achieved Breakthrough.[4] If we trace backwards from y_1, using the labels as a guide, we find the M^1-alternating path

$$\gamma : y_1, x_2, y_2, x_3, y_3, x_1.$$

We have

$$M_\gamma^1 = \{\{x_2, y_2\}, \{x_3, y_3\}\}$$

<hr>

[4]The algorithm can be halted as soon as Breakthrough is achieved.

and

$$\overline{M_\gamma^1} = \{\{y_1, x_2\}, \{y_2, x_3\}, \{y_3, x_1\}\}.$$

Then

$$
\begin{aligned}
M^2 &= (M^1 - M_\gamma^1) \cup (\overline{M_\gamma^1}) \\
&= \{\{x_4, y_4\}, \{y_1, x_2\}, \{y_2, x_3\}, \{y_3, x_1\}\}
\end{aligned}
$$

is a matching of four edges.

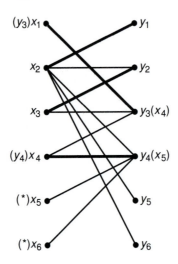

Figure 9.14

We now apply the matching algorithm to M^2. The resulting label-ing of the vertices is shown in Figure 9.14. In this case, Breakthrough has not occurred. By Theorem 9.2.3, M^2 is a max-matching of size 4, and the set

$$S = \{x_2, x_3, y_3, y_4\},$$

of size 4, consisting of the unlabeled vertices of X and the labeled vertices of Y, is a min-cover.   □

The theory of matchings, as developed in this section, solves each of Problems 1, 2, and 3, discussed at the beginning of this chapter.

Let $G = (X, \Delta, Y)$ be a bipartite graph such that the set X of left vertices and the set Y of right vertices have the same size n. A matching in G can contain at most n edges. A matching M in G with n edges is called a *perfect matching*. Each vertex in X and each vertex in Y meets exactly one edge of a perfect matching M. Hence, a perfect matching M establishes a one-to-one correspondence

$$f: \quad X \to Y$$

between the vertices in X and the vertices in Y, where

$$f(x) = y \text{ if } \{x, y\} \text{ is an edge of } M.$$

It follows from Corollary 9.2.4 that G has a perfect matching if and only if no set of fewer than n vertices covers all the edges of G. In some instances, this condition is not difficult to check.

A bipartite graph $G = (X, \Delta, Y)$ is called *regular of degree p*, provided that each of its vertices meets exactly p edges. If G is regular of degree $p \geq 1$, then X and Y must have the same size n. This is because, counting the number of edges by looking at the left vertices, we see that the total number of edges of G is $p|X|$, while counting by looking at right vertices, we see that the total number is $p|Y|$. Equating these two counts, we get

$$p|X| = p|Y|,$$

and since $p \neq 0$, we obtain $|X| = |Y|$.

Theorem 9.2.5 *A bipartite graph* $G = (X, \Delta, Y)$ *that is regular of degree* $p \geq 1$ *always has a perfect matching.*

Proof. Let X and Y each contain n vertices. Let S be any cover of G. Because S is a cover, every edge of G meets at least one vertex of S. Since G is regular of degree p, each vertex of S meets exactly p edges. Hence, the total number of edges of G is at most $p|S|$.[5] But the total number of edges of G is pn. Thus,

$$p|S| \geq pn,$$

and it follows that

$$|S| \geq n.$$

Therefore every cover of G has at least n vertices and by Corollary 9.2.4, G has a perfect matching. $\square$

Example. At a party there are n boys and n girls. Suppose that there exists a positive integer p such that each boy has been previously introduced to p girls, and each girl has been previously introduced to p boys. Show that the boys and girls can be paired up so that the boy and girl of each pair are acquainted.

We construct a bipartite graph $G = (X, \Delta, Y)$ as follows: The set X of left vertices consists of the n boys, and the set of the right vertices

---

[5]The reason that the number of edges is *at most* $p|S|$ is that some edges may have both of their vertices in S, and these edges are counted twice.

consists of the n girls. We join a boy and a girl by an edge if and only if they are acquainted. The assumptions imply that G is regular of positive degree p. By Theorem 9.2.5, G has a perfect matching M, and this matching M describes the required pairing.   □

9.3   Systems of Distinct Representatives

Let Y be a finite set, and let $\mathcal{A} = (A_1, A_2, \ldots, A_n)$ be a family[6] of n subsets of Y. A family $(e_1, e_2, \ldots, e_n)$ of elements of Y is called a *system of representatives* of $\mathcal{A}$, provided that

$$e_1 \text{ is in } A_1, e_2 \text{ is in } A_2, \ldots, e_n \text{ is in } A_n.$$

In a system of representatives, the element e_i belongs to A_i and thus "represents" the set A_i. If, in a system of representatives, the elements $e_1, e_2, \ldots, e_n$ are all different, then $(e_1, e_2, \ldots, e_n)$ is called a *system of distinct representatives*, abbreviated SDR.

Example.   Let (A_1, A_2, A_3, A_4) be the family of subsets of the set $Y = \{a, b, c, d, e\}$, defined by

$$A_1 = \{a, b, c\}, A_2 = \{b, d\}, A_3 = \{a, b, d\}, A_4 = \{b, d\}.$$

Then (a, b, b, d) is a system of representatives, and (c, b, a, d) is an SDR.
                        □

A family $\mathcal{A} = (A_1, A_2, \ldots, A_n)$ of *nonempty* sets always has a system of representatives. We need pick only one element from each of the sets to obtain a system of representatives. However, the family $\mathcal{A}$ need not have an SDR even though all the sets in the family are nonempty. For instance, if there are two sets in the family, say, A_1 and A_2, each containing only one element, and the element in A_1 is the same as the element in A_2, that is,

$$A_1 = \{x\}, \quad A_2 = \{x\},$$

then the family A does not have an SDR. This is because, in any system of representatives, x has to represent both A_1 and A_2, and thus no SDR exists (no matter what $A_3, \ldots, A_n$ equal). However, a family A can fail to have an SDR for somewhat more complicated reasons.

---

[6]A family is really the same as a sequence. We have here a sequence whose terms are sets. As in sequences of numbers, different terms can be equal; that is, the sets need not be different.

Example. Let the family $\mathcal{A} = (A_1, A_2, A_3, A_4)$ be defined by

$$A_1 = \{a, b\},\ A_2 = \{a, b\},\ A_3 = \{a, b\},\ A_4 = \{a, b, c, d\}.$$

Then $\mathcal{A}$ does not have an SDR because in any system of representatives, A_1 has to be represented by a or b, A_2 has to be represented by a or b, and A_3 has to be represented by a or b. So we have two elements, namely, a and b, from which the representatives of three sets, namely, A_1, A_2, and A_3, have to be drawn. By the pigeonhole principle, two of the three sets A_1, A_2 and A_3 have to be represented by the same element. $\qquad\square$

We can obtain a general necessary condition for the existence of an SDR by generalizing the argument in the preceding example. Let $\mathcal{A} = (A_1, A_2, \ldots, A_n)$ be a family of sets. Let k be an integer with $1 \le k \le n$. In order for $\mathcal{A}$ to have an SDR, it is necessary that the union of every k sets of the family A contain at least k elements. Suppose, to the contrary, that there are k sets, say, $A_1, A_2, \ldots, A_k$, which together contain fewer than k elements; that is, $A_1 \cup A_2 \cup \cdots \cup A_k = F$, where

$$|F| < k.$$

Then the representatives of each of the k sets $A_1, A_2, \ldots, A_k$ have to be drawn from the elements of the set F. Since F has fewer than k elements, it follows from the pigeonhole principle that two of the k sets $A_1, A_2, \ldots, A_k$ have to be represented by the same element. Hence, there can be no SDR. We formulate this necessary condition as the next lemma.

Lemma 9.3.1 *In order for the family $\mathcal{A} = (A_1, A_2, \ldots, A_n)$ of sets to have an SDR, it is necessary that the following condition hold:*

(MC): *For each $k = 1, 2, \ldots, n$ and each choice of k distinct indices $i_1, i_2, \ldots, i_k$ from $\{1, 2, \ldots, n\}$,*

$$|A_{i_1} \cup A_{i_2} \cup \cdots \cup A_{i_k}| \ge k. \tag{9.9}$$

Condition (MC) in Lemma 9.3.1 is often called the *Marriage Condition*. The reason stems from the following amusing and classical formulation of the problem of systems of distinct representatives.

Example. (*The Marriage Problem*). There are n men and m women, and all the men are eager to marry. If there were no restrictions on who marries whom, then we need only require that the number m

of women be at least as large as the number n of men, in order to marry off all the men. But we would expect that each man would insist on some compatibility with a spouse and would thereby eliminate some of the women as potential spouses. Thus, each man would arrive at a certain set of acceptable spouses from the available women. (A woman could also eliminate herself from the set of potential spouses of any man whom she finds unacceptable.) Let $(A_1, A_2, \ldots, A_n)$ be the family of subsets of the women, where A_i denotes the set of spouses acceptable to the ith man $(i = 1, \ldots, n)$. Then a *complete marriage* of the men corresponds to an SDR $(w_1, w_2, \ldots, w_n)$ of $(A_1, A_2, \ldots, A_n)$. The correspondence is that the ith man marries the woman w_i, $(i = 1, 2, \ldots, n)$. Since w_i is in A_i, w_i is an acceptable spouse for the ith man. Since $(w_1, w_2, \ldots, w_n)$ is a system of *distinct* representatives, no two men are claiming the same woman.[7] In the context of this example, the Marriage Condition asserts that the combined lists of any set of k men contain at least k women, and thus this is a necessary condition for a complete marriage of the men.

The Marriage Condition (9.9) is not only a necessary condition for the existence of SDR, but a sufficient condition as well. It thus provides a characterization for the existence of an SDR. We obtain the sufficiency of the Marriage Condition for an SDR from Corollary 9.2.4.

We associate a bipartite graph $G = (X, \Delta, Y)$ to each family $\mathcal{A} = (A_1, A_2, \ldots, A_n)$ of subsets of a set $Y = \{y_1, y_2, \ldots, y_m\}$. We take X equal to the set $\{1, 2, \ldots, n\}$–the set indexing the members of the family $\mathcal{A}$–and define the set of edges Δ by

$$\Delta = \{\{i, y_j\} : y_j \text{ is in } A_i\}.$$

Thus, the vertices i and y_j are joined by an edge in G if and only if y_j is an element of the set A_i. Put another way, vertex i is joined by an edge to those elements of Y which can serve as representatives of A_i. A system of representatives of $\mathcal{A}$ corresponds to a set of n edges, one meeting each vertex of X, but there may be more than one edge meeting a vertex of Y, since, in a system of representatives, the same element may represent two different sets. An SDR corresponds to a set of n edges, one meeting each vertex of X and, at most, one meeting each vertex of Y. We thus conclude that *the family of sets $\mathcal{A} = (A_1, A_2, \ldots, A_n)$ has an SDR if an only if the associated bipartite graph G has a matching M of n edges.* Since X has only n vertices, G cannot have a matching of more than n edges. Thus, $\mathcal{A}$ has an SDR if and only if $\rho(G) = n$.

---

[7]We forgot to say that no woman is allowed two spouses!

Theorem 9.3.2 *The family* $\mathcal{A} = (A_1, A_2, \ldots, A_n)$ *of sets has an SDR if and only if the Marriage Condition* (MC) *holds.*

Proof. By Lemma 9.3.1 we know that the Marriage Condition holds if $\mathcal{A}$ has an SDR. We now assume that the Marriage Condition holds and show that $\mathcal{A}$ has an SDR. Let $G = (X, \Delta, Y)$ be the bipartite graph associated with the family $\mathcal{A}$, as in the paragraph preceding the theorem. We need to show that $\rho(G) = n$. By Corollary 9.2.4, we can conclude that $\rho(G) = n$ if we show that $c(G) = n$, that is, if we show that there is no cover of G consisting of fewer than n vertices. Suppose, to the contrary, that there is a cover S of G with $|S| < n$. Let

$$S = S_1 \cup S_2,$$

where $S_1 = S \cap X$ are the left vertices in S and $S_2 = S \cap Y$ are the right vertices in S. Then, since $|S| < n$, we have

$$|S_1| + |S_2| < n. \tag{9.10}$$

Because S is a cover, there is no edge joining a vertex in $X - S_1$ to a vertex in $Y - S_2$. Let

$$k = |X - S_1| = n - |S_1|$$

and let

$$X - S_1 = \{i_1, i_2, \ldots, i_k\}.$$

Then, since there is no edge joining a vertex in $X - S_1$ to a vertex in $Y - S_2$, $A_{i_1}, A_{i_2}, \ldots$, and A_{i_k} are all subsets of S_2. Hence,

$$A_{i_1} \cup A_{i_2} \cup \cdots \cup A_{i_k} \subseteq S_2,$$

and thus

$$|A_{i_1} \cup A_{i_2} \cup \cdots \cup A_{i_k}| \leq |S_2|.$$

By (9.10),

$$|S_2| < n - |S_1| = k;$$

therefore,

$$|A_{i_1} \cup A_{i_2} \cup \cdots \cup A_{i_k}| < k,$$

contradicting the Marriage Condition. Consequently, there is no cover S of G with fewer than n vertices, $\rho(G) = n$, and $\mathcal{A}$ has an SDR. $\square$

In much the same way, Corollary 9.2.4 can be used to obtain the next characterization of the largest number of sets in a family which has an SDR.

Theorem 9.3.3 *Let $\mathcal{A} = (A_1, A_2, \ldots, A_n)$ be a family of subsets of a set Y. Then the largest number ρ of sets of $\mathcal{A}$ which can be chosen so that they have an SDR equals the smallest value taken on by the expression*

$$|A_{i_1} \cup A_{i_2} \cup \cdots \cup A_{i_k}| + n - k$$

over all choices of $k = 1, 2, \ldots, n$ and all choices of k distinct indices $i_1, i_2, \ldots, i_k$ from $\{1, 2, \ldots, n\}$.

The number ρ defined in Theorem 9.3.3 is the matching number $\rho(G)$ of the bipartite graph G that we have associated with the family $\mathcal{A}$.

Example. We define a family $\mathcal{A} = (A_1, A_2, A_3, A_4, A_5, A_6)$ of subsets of $\{a, b, c, d, e, f\}$ by

$$A_1 = \{a, b, c\}, \quad A_2 = \{b, c\}, \quad A_3 = \{b, c\},$$
$$A_4 = \{b, c\}, \quad A_5 = \{c\}, \quad A_6 = \{a, b, c, d\}.$$

We have

$$|A_2 \cup A_3 \cup A_4 \cup A_5| = |\{b, c\}| = 2;$$

hence,

$$|A_2 \cup A_3 \cup A_4 \cup A_5| + 6 - 4 = 4.$$

Thus, with $n = 6$ and $k = 4$, we see by Theorem 9.3.3 that, at most, four of the sets $\mathcal{A}$ can be chosen so that they have an SDR. Since (A_1, A_2, A_5, A_6) has (a, b, c, d) as an SDR, it follows that 4 is the largest number of sets with an SDR. In terms of marriage, 4 is the largest number of gentlemen that can marry if each gentleman is to marry an acceptable woman.   □

9.4   Stable Marriages

In this section[8] we consider a variation of the marriage problem discussed in the previous section.

There are n women and n men in a community. Each woman ranks each man in accordance with her preference for that man as a spouse. No ties are allowed, so that if a woman is indifferent between two men, we nonetheless require that she express some preference. The

---

[8]This section is partly based on the article "College admissions and the stability of marriage" by D. Gale and L.S. Shapely, *American Mathematical Monthly*, 69(1962), 9-15. A comprehensive treatment of the questions considered here can be found in the book *The stable marriage problem: Structure and algorithms* by D. Gusfield and R.W. Irving, The MIT Press, Cambridge (1989).

preferences are to be purely ordinal, and thus each woman ranks the men in the order $1, 2, \ldots, n$. Similarly, each man ranks the women in the order $1, 2, \ldots, n$. There are $n!$ ways in which the women and men can be paired so that a *complete marriage* takes place. We say that a complete marriage is *unstable*, provided that there exist two women A and B and two men a and b such that

(i) A and a get married;

(ii) B and b get married;

(iii) A prefers (i.e., ranks higher) b to a;

(iv) b prefers A to B.

Thus, in an unstable complete marriage, A and b could act independently of the others and run off with each other, since both would regard their new partner as more preferable than their current spouse. Thus, the complete marriage is "unstable" in the sense that it can be upset by a man and a woman acting together in a manner that is beneficial to both. A complete marriage is called *stable*, provided it is not unstable. The question that arises first is *Does there always exist a stable, complete marriage?*

We set up a mathematical model for this problem by using a bipartite graph again. Let $G = (X, \Delta, Y)$ be a bipartite graph in which

$$X = \{w_1, w_2, \ldots, w_n\}$$

is the set of n women and

$$Y = \{m_1, m_2, \ldots, m_n\}$$

is the set of n men. We join each woman-vertex (left is now woman) to each man-vertex (right is now man). The resulting bipartite graph is *complete* in the sense that it contains all possible edges between its two sets of vertices.[9] Corresponding to each edge $\{w_i, m_j\}$, there is a pair p, q of numbers where p denotes the position of m_j in w_i's ranking of the men, and q denotes the position of w_i in m_j's ranking of the women. A complete marriage of the women and men corresponds to a perfect matching (of n edges) in this bipartite graph G.

It is more convenient, for notational purposes, to use the model afforded by the *preferential ranking matrix*. This matrix is an n-by-n array of n rows, one for each of the women $w_1, w_2, \ldots, w_n$, and n

---

[9]In Chapter 11, this graph is called the complete bipartite graph $K_{n,n}$.

columns, one for each of the n men $m_1, m_2, \ldots, m_n$. In the position at the intersection of row i and column j, we place the pair p, q of numbers representing, respectively, the ranking of m_j by w_i and the ranking of w_i by m_j. A complete marriage corresponds to a set of n positions of the matrix that includes exactly one position from each row and one position from each column.[10]

Example. Let $n = 2$, and let the preferential ranking matrix be

$$
\begin{array}{c}
 \quad\ m_1\ \ m_2 \\
\begin{array}{c} w_1 \\ w_2 \end{array}
\left[\begin{array}{cc} 1,2 & 2,2 \\ 2,1 & 1,1 \end{array} \right].
\end{array}
$$

Thus, for instance, the entry $1, 2$ in the first row and first column means that w_1 has put m_1 first on her list and m_1 has put w_1 second on his list. There are two possible complete marriages:

(1) $w_1 \leftrightarrow m_1$, $w_2 \leftrightarrow m_2$,

(2) $w_1 \leftrightarrow m_2$, $w_2 \leftrightarrow m_1$.

The first is readily seen to be stable. The second is unstable since w_2 prefers m_2 to her spouse m_1, and similarly m_2 prefers w_2 to his spouse w_1. □

Example. Let $n = 3$, and let the preferential ranking matrix be

$$
\left[\begin{array}{ccc} 1,3 & 2,2 & 3,1 \\ 3,1 & 1,3 & 2,2 \\ 2,2 & 3,1 & 1,3 \end{array} \right]. \tag{9.11}
$$

There are $3! = 6$ possible complete marriages. One is

$$ w_1 \leftrightarrow m_1, \ w_2 \leftrightarrow m_2, \ w_3 \leftrightarrow m_3. $$

Since each woman gets her first choice, the complete marriage is stable, even though each man gets his last choice! Another stable complete marriage is obtained by giving each man his first choice. But note that, in general, there may not be a complete marriage in which every man (or every woman) gets first choice. For example, this happens when all the women have the same first choice and all the men have the same first choice. □

We now show that a stable complete marriage always exists and, in doing so, obtain an algorithm for determining a stable complete marriage. Thus, complete chaos can be avoided!

---

[10]The astute reader has no doubt noticed that a complete marriage corresponds to n nonattacking rooks, where we treat the n-by-n matrix as an n-by-n board.

Theorem 9.4.1 *For each preferential ranking matrix, there is a stable complete marriage.*

Proof. We define an algorithm, the *deferred acceptance algorithm*,[11] for determining a complete marriage:

Deferred Acceptance Algorithm

Begin with every woman marked as rejected.

While there exists a rejected woman, do the following:

(1) Each woman marked as rejected chooses the man whom she ranks highest among all those men who have not yet rejected her.

(2) Each man picks out the woman he ranks highest among all those women who have chosen him and whom he has not yet rejected, defers decision on her, and now rejects the others.

Thus, during the execution of the algorithm,[12] the women propose to the men, and some men and some women become *engaged*, but the men are able to break engagements if they receive a better offer. Once a man becomes engaged, he remains engaged throughout the execution of the algorithm, but his fiancée may change; in his eyes, a change is always an improvement. A woman, however, may be engaged and disengaged several times during the execution of the algorithm; however, each new engagement results in a less desirable partner for her. It follows from the description of the algorithm that, as soon as there are no rejected women, then each man is engaged to exactly one woman, and since there are as many men as women, each woman is engaged to exactly one man. We now pair each man with the woman to whom he is engaged and obtain a complete marriage. We now show that this marriage is stable.

Consider women A and B and men a and b such that A is paired with a and B is paired with b, but A prefers b to a. We show that b cannot prefer A to B. Since A prefers b to a, during some stage of the algorithm A chose b, but A was rejected by b for some woman he ranked higher. But the woman b eventually gets paired with is at least as high on his list as any woman that he rejected during the course of the algorithm. Since A was rejected by b, b must prefer B to A. Thus, there is no unstable pair, and this complete marriage is stable. $\square$

---

[11]Also called the Gale–Shapley algorithm.

[12]Note that we have reversed the traditional roles of men and women in which men are the suitors.

Example. We apply the deferred acceptance algorithm to the preferential ranking matrix in (9.11), designating the women as A, B, C, respectively, and the men as a, b, c, respectively.[13] In (1), A chooses a, B chooses b, and C chooses c. There are no rejections, the algorithm halts, and A marries a, B marries b, C marries c, and they live happily ever after.   □

Example. We apply the deferred acceptance algorithm to the preferential ranking matrix

$$
\begin{array}{c@{\quad}cccc}
 & a & b & c & d \\
A & \left[\begin{array}{cccc} 1,2 & 2,1 & 3,2 & 4,1 \end{array}\right. \\
B & 2,4 & 1,2 & 3,1 & 4,2 \\
C & 2,1 & 3,3 & 4,3 & 1,4 \\
D & \left. 1,3 & 4,4 & 3,4 & 2,3 \end{array}\right]
\end{array}
\qquad (9.12)
$$

The results of the algorithm are as follows:

(i) A chooses a, B chooses b, C chooses d, D chooses a; a rejects D.

(ii) D chooses d; d rejects C.

(iii) C chooses a; a rejects A.

(iv) A chooses b; b rejects B.

(v) B chooses a; a rejects B.

(vi) B chooses c.

In (vi), there are no rejections, and

$$A \leftrightarrow b, \ B \leftrightarrow c, \ C \leftrightarrow a, \ D \leftrightarrow d$$

is a stable complete marriage.   □

    If, in the deferred acceptance algorithm, we interchange the roles of the women and men and have the men choose women according to their rank preferences, we obtain a stable complete marriage which may, but need not, differ from the one obtained by having the women choose men.

Example. We apply the deferred acceptance algorithm to the preferential ranking matrix in (9.12), where the men choose the women. The results are as follows:

---

[13] The BIG guys versus the little guys.

(i) a chooses C, b chooses A, c chooses B, d chooses A; A rejects d.

(ii) d chooses B; B rejects d.

(iii) d chooses D.

The complete marriage

$$a \leftrightarrow C, \; b \leftrightarrow A, \; c \leftrightarrow B, \; d \leftrightarrow D$$

is stable. This is the same complete marriage obtained by applying the algorithm the other way around. □

Example. We apply the deferred acceptance algorithm to the preferential ranking matrix in (9.11), where the men choose the women. The results are as follows:

(i) a chooses B, b chooses C, c chooses A.

Since there are no rejections, the stable complete marriage obtained is

$$a \leftrightarrow B, \; b \leftrightarrow C, \; c \leftrightarrow A.$$

This is different from the complete marriage obtained by applying the algorithm the other way around. □

A stable complete marriage is called *optimal for a woman*, provided that a woman gets as a spouse a man whom she ranks at least as high as the spouse she obtains in every other stable complete marriage. In other words, there is no stable complete marriage in which the woman gets a spouse who is higher on her list. A stable complete marriage is called *women-optimal* provided that it is optimal for each woman. In a similar way, we define a *men-optimal* stable complete marriage. It is not obvious that there exist women-optimal and men-optimal stable complete marriages. In fact, it is not even obvious that, if each woman is independently given the best partner that she has in all the stable complete marriages, then this results in a pairing of the women and the men (it is conceivable that two women might end up with the same man in this way). Clearly, there can be only one women-optimal complete marriage and only one men-optimal complete marriage.

Theorem 9.4.2 *The stable complete marriage obtained from the deferred acceptance algorithm, with the women choosing the men, is women-optimal. If the men choose the women in the deferred acceptance algorithm, the resulting complete marriage is men-optimal.*

Proof. A man M is called *feasible* for a woman W, provided that there is some stable complete marriage in which M is W's spouse. We shall prove by induction that the complete marriage obtained by applying the deferred acceptance algorithm has the property that the men who reject a particular woman are not feasible for that woman. Because of the nature of the algorithm, this implies that each woman obtains as a spouse the man she ranks highest among all the men that are feasible for her, and hence the complete marriage is women-optimal.

The induction is on the number of rounds of the algorithm. To start the induction, we show that, at the end of the first round, no woman has been rejected by a man that is feasible for her. Suppose that both woman A and woman B choose man a, and a rejects A in favor of B. Then any complete marriage in which A is paired with a is not stable because a prefers B and B prefers a to whichever man she is eventually paired with.

We now proceed by induction and assume that at the end of some round $k \geq 1$, no woman has been rejected by a man who is feasible for her. Suppose that at the end of the $(k+1)$st round, woman A is rejected by man a in favor of woman B. Then B prefers a over all those men that have not yet rejected her. By the induction assumption, none of the men who have rejected B in the first k rounds is feasible for B, and so there is no stable complete marriage in which B is paired with one of them. Thus, in any stable marriage, B is paired with a man who is no higher on her list than a.

Now suppose that there is a stable complete marriage in which A is paired with a. Then a prefers B to A and, by the last remark, B prefers a to whomever she is paired with. This contradicts the fact that the complete marriage is stable. The inductive step is now complete and we conclude that the stable complete marriage obtained from the deferred acceptance algorithm is optimal for the women. $\square$

We now show that in the women-optimal complete marriage, each man has the *worst* partner he can have in any stable complete marriage.

Corollary 9.4.3 *In the women-optimal stable complete marriage, each man is paired with the woman he ranks lowest among all the partners that are possible for him in a stable complete marriage.*

Proof. Let man a be paired with woman A in the women-optimal stable complete marriage. By Theorem 9.4.2 A prefers a to all other men that are possible for her in a stable complete marriage. Suppose there is a stable complete marriage in which a is paired with woman B, where a ranks B lower than A. In this stable marriage, A is paired with some man b different from a whom she therefore ranks lower than a. But then A prefers a, and a prefers A, and this complete marriage is not stable contrary to assumption. Hence, there is no stable complete marriage in which a gets a worse partner than A. □

Suppose the men-optimal and women-optimal stable complete marriages are identical. Then, by Corollary 9.4.3, in the woman-optimal complete marriage, each man gets both his best and worst partner taken over all stable complete marriages. (A similar conclusion holds for the women.) It thus follows in this case that there is exactly one stable complete marriage. Of course, the converse holds as well: if there is only one stable complete marriage, then the men-optimal and women-optimal stable complete marriages are identical.

The deferred acceptance algorithm has been in use since 1952 to match medical residents in the United States to hospitals.[14] We can think of the hospitals as being the women and the residents as being the men. But now, since a hospital generally has places for several residents, polyandrous marriages in which a woman can have several spouses are allowed.

We conclude this section with a discussion of a similar problem for which the existence of a stable marriage is no longer guaranteed.

Example. Suppose an even number $2n$ of girls wish to pair up as roommates. Each girl ranks the other girls in the order $1, 2, \ldots, 2n - 1$ of preference. A *complete marriage* in this situation is a pairing of the girls into n pairs. A complete marriage is *unstable*, provided there exist two girls who are not roommates such that each of the girls prefers the other to her current roommate. A complete marriage is *stable*, provided it is not unstable. Does there always exist a stable complete marriage?

Consider the case of four girls, A, B, C, D, where A ranks B first, B ranks C first, C ranks A first, and each of $A, B,$ and C ranks D last. Then, irrespective of the other rankings, there is no stable complete marriage as the following argument shows. Suppose A and D are roommates. Then B and C are also roommates. But C prefers A to B, and since A ranks D last, A prefers C to D. Thus, this complete marriage is not stable. A similar conclusion holds if B and D are roommates or if C and D are roommates. Since D has a roommate, there is no stable complete marriage. □

---

[14]It can also be used to match students to colleges, etc.

9.5   Exercises

1. Consider the chessboard B with forbidden positions shown in Figure 9.15. Construct the rook-bipartite graph G associated with B. Find 6 positions for 6 nonattacking rooks on B, and determine the corresponding matching in G.

Figure 9.15

2. Construct the domino-bipartite graph G associated with the board B in Figure 9.15. Determine a matching of 10 edges in G and the associated perfect cover of the board by dominoes.

3. Show that every bipartite graph is the rook-bipartite graph of some board.

4. Give an example of a bipartite graph that is not the domino-bipartite graph of any board.

5. Consider an m-by-n chessboard in which both m and n are odd. The board has one more square of one color, say, black, than of white. Show that, if exactly one black square is forbidden on the board, the resulting board has a perfect cover with dominoes.

6. Consider an m-by-n chessboard, where at least one of m and n is even. The board has an equal number of white and black squares. Show that if m and n are at least 2 and if exactly one white and exactly one black square are forbidden, the resulting board has a perfect cover with dominoes.

7. Let $G = (X, \Delta, Y)$ be a bipartite graph. Suppose that there is a positive integer p such that each vertex in X meets at least p edges, and each vertex in Y meets at most p edges. By counting the total number of edges in G, prove that Y has at least as many vertices as X.

8. Let $G = (X, \Delta, Y)$ be a bipartite graph that is regular of degree $p \geq 1$. Use Theorem 9.2.5 and induction to show that the edges of G can be partitioned into p perfect matchings.

9. Consider an n-by-n chessboard with forbidden positions for which there exists a positive integer p such that each row and each column contains exactly p allowed squares. Prove that it is possible to place n nonattacking rooks on the board.

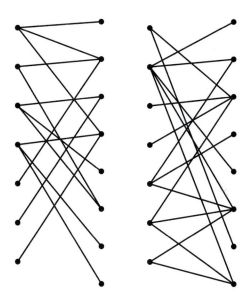

Figure 9.16

10. Use the matching algorithm to determine the largest number of edges in a matching M of the bipartite graphs in Figure 9.16. In each case, find a cover S with $|S| = |M|$.

11. A corporation has 7 available positions $y_1, y_2, \ldots, y_7$ and 10 applicants $x_1, x_2, \ldots, x_{10}$. The set of positions each applicant is qualified for is given, respectively, by $\{y_1, y_2, y_6\}$, $\{y_2, y_6, y_7\}$, $\{y_3, y_4\}$, $\{y_1, y_5\}$, $\{y_6, y_7\}$, $\{y_3\}$, $\{y_2, y_3\}$, $\{y_1, y_3\}$, $\{y_1\}$, $\{y_5\}$. Determine the largest number of positions that can be filled by the qualified applicants and justify your answer.

12. Let $\mathcal{A} = (A_1, A_2, A_3, A_4, A_5, A_6)$, where

$$A_1 = \{a, b, c\}, \quad A_2 = \{a, b, c, d, e\}, \quad A_3 = \{a, b\},$$
$$A_4 = \{b, c\}, \quad A_5 = \{a\}, \quad A_6 = \{a, c, e\}.$$

Does the family $\mathcal{A}$ have an SDR? If not, what is the largest number of sets in the family with an SDR?

13. Let $\mathcal{A} = (A_1, A_2, A_3, A_4, A_5, A_6)$, where

$$A_1 = \{1,2\}, \ A_2 = \{2,3\}, \ A_3 = \{3,4\},$$
$$A_4 = \{4,5\}, \ A_5 = \{5,6\}, \ A_6 = \{6,1\}.$$

Determine the number of different SDR's that $\mathcal{A}$ has. Generalize to n sets.

14. Let $\mathcal{A} = (A_1, A_2, \ldots, A_n)$ be a family of sets with an SDR. Let x be an element of A_1. Prove that there is an SDR containing x, but show by example that it may not be possible to find an SDR in which x represents A_1.

15. Suppose $\mathcal{A} = (A_1, A_2, \ldots, A_n)$ is a family of sets that "more than satisfies" the Marriage Condition. More precisely, suppose that

$$|A_{i_1} \cup A_{i_2} \cup \cdots \cup A_{i_k}| \geq k+1$$

for each $k = 1, 2, \ldots, n$ and each choice of k distinct indices $i_1, i_2, \ldots, i_k$. Let x be an element of A_1. Prove that $\mathcal{A}$ has an SDR in which x represents A_1.

16. Let $n > 1$, and let $\mathcal{A} = (A_1, A_2, \ldots, A_n)$ be the family of subsets of $\{1, 2, \ldots, n\}$, where

$$A_i = \{1, 2, \ldots, n\} - \{i\}, \qquad (i = 1, 2, \ldots, n).$$

Prove that $\mathcal{A}$ has an SDR and that the number of SDR's is the nth derangement number D_n.

17. Consider a chessboard with forbidden positions which has the property that, if a square is forbidden, so is every square to its right and every square below it. Prove that the chessboard has a perfect cover by dominoes if and only if the number of allowable white squares equals the number of allowable black squares.

18. * Let A be a matrix with n columns, with integer entries taken from the set $S = \{1, 2, \ldots, k\}$. Assume that each integer i in S occurs exactly nr_i times in A, where r_i is an integer. Prove that it is possible to permute the entries in each row of A to obtain a matrix B in which each integer i in S appears r_i times in each column.[15]

---

[15]E. Kramer, S. Magliveras, T. van Trung, and Q. Wu: "Some perpendicular arrays for arbitrary large t," *Discrete Math.*, 96 (1991), 101-110.

19. Find a 2-by-2 preferential ranking matrix for which both complete marriages are stable.

20. Consider a preferential ranking matrix in which woman A ranks man a first, and man a ranks A first. Show that, in every stable marriage, A is paired with a.

21. Consider the preferential ranking matrix

$$
\begin{bmatrix}
1, n & 2, n-1 & 3, n-2 & \cdots & n, 1 \\
n, 1 & 1, n & 2, n-1 & \cdots & n-1, 2 \\
n-1, 2 & n, 1 & 1, n & \cdots & n-2, 3 \\
\vdots & \vdots & \vdots & \vdots & \vdots \\
3, n-3 & 4, n-3 & 5, n-4 & \cdots & 2, n-1 \\
2, n-2 & 3, n-2 & 4, n-3 & \cdots & 1, n
\end{bmatrix}
$$

Prove that, for each $k = 1, 2, \ldots, n$, the complete marriage in which each woman gets her kth choice is stable.

22. Use the deferred acceptance algorithm to obtain both the women-optimal and men-optimal stable complete marriages for the preferential ranking matrix

$$
\begin{array}{c c}
& \begin{array}{cccc} a & b & c & d \end{array} \\
\begin{array}{c} A \\ B \\ C \\ D \end{array} &
\begin{bmatrix}
1,3 & 2,3 & 3,2 & 4,3 \\
1,4 & 4,1 & 3,3 & 2,2 \\
2,2 & 1,4 & 3,4 & 4,1 \\
4,1 & 2,2 & 3,1 & 1,4
\end{bmatrix}
\end{array} .
$$

Conclude that, for the given preferential ranking matrix, there is only one stable complete marriage.

23. Prove that in every application of the deferred acceptance algorithm with n women and n men, there are at most $n^2 - n + 1$ proposals.

24. * Extend the deferred acceptance algorithm to the case in which there are more men than women. In such a case, not all of the men will get partners.

25. Show, by using Exercise 22, that it is possible that in no complete marriage does any person get his or her first choice.

26. Apply the deferred acceptance algorithm to obtain a stable complete marriage for the preferential ranking matrix

$$
\begin{array}{c}
\quad\quad a \quad\ \ b \quad\ \ c \quad\ \ d \\
\begin{array}{c} A \\ B \\ C \\ D \end{array}
\left[
\begin{array}{cccc}
1,3 & 2,2 & 3,1 & 4,3 \\
1,4 & 2,3 & 3,2 & 4,4 \\
3,1 & 1,4 & 2,3 & 4,2 \\
2,2 & 3,1 & 1,4 & 4,1
\end{array}
\right]
\end{array}.
$$

27. Consider an n-by-n board in which there is a nonnegative number a_{ij} in the square in row i and column j, $(1 \le i, j \le n)$. Assume that the sum of the numbers in each row and in each column equals 1. Prove that it is possible to place n nonattacking rooks on the board at positions occupied by positive numbers.

Chapter 10

Combinatorial Designs

A *combinatorial design*, or simply a *design*, is an arrangement of the objects of a set into subsets satisfying certain prescribed properties. This is a very general definition and includes a vast amount of combinatorial theory. Many of the examples introduced in Chapter 1 can be viewed as designs: (i) perfect covers by dominoes of boards with forbidden positions, where we arrange the allowed squares into pairs so that each pair can be covered by one domino; (ii) magic squares, where we arrange the integers from 1 to n^2 in an n-by-n array so that certain sums are identical; (iii) Latin squares, where we arrange the integers from 1 to n in an n-by-n array so that each integer occurs once in each row and once in each column. We shall treat Latin squares and the notion of orthogonality, briefly introduced in Chapter 1, more thoroughly in this chapter.

The area of combinatorial designs is highly developed, yet many interesting and fundamental questions remain unanswered. Many of the methods for constructing designs rely on the algebraic structure called a finite field and more general systems of arithmetic. In the first section we give a brief introduction to these "finite arithmetics," concentrating mainly on modular arithmetic. Our discussion will not be comprehensive, but should be sufficient to enable us to do arithmetic comfortably in these systems.

10.1 Modular Arithmetic

Let Z denote the set of integers

$$\{\ldots, -2, -1, 0, 1, 2, \ldots\},$$

and let $+$ and $\times$ denote ordinary addition and multiplication of integers. The reason for being so cautious in pointing out the usual notations for addition and multiplication is that we are going to introduce new additions and new multiplications on certain subsets of the set Z of integers, and we don't want the reader to confuse them with ordinary addition and multiplication.

Let n be a positive integer with $n \geq 2$, and let

$$Z_n = \{0, 1, \ldots, n-1\}$$

be the set of nonnegative integers that are less than n. We can think of the integers in Z_n as the possible remainders when *any* integer is divided by n:

If m is an integer, then there exist unique integers q (the quotient) and r (the remainder) such that

$$m = q \times n + r, \quad 0 \leq r \leq n-1.$$

With this in mind, we define an addition, denoted $\oplus$, and a multiplication, denoted $\otimes$, on Z_n as follows:

For any two integers a and b in Z_n, $a \oplus b$ is the (unique) remainder when the ordinary sum $a + b$ is divided by n, and $a \otimes b$ is the (unique) remainder when the ordinary product $a \times b$ is divided by n.

This addition and multiplication depend on the chosen integer n, and we should be writing something like $\oplus_n$ and $\otimes_n$, but such notation gets a little cumbersome.[1] So we just caution the reader that $\oplus$ and $\otimes$ depend on n, and we call them *addition mod n* and *multiplication mod n*, and with this addition and multiplication we get the *system of integers mod n*.[2] We usually denote the arithmetic system of the integers mod n with the same symbol Z_n that we use for its set of elements.

---

[1]Shortly, after the reader has gotten familiar with these new additions and multiplications, we shall replace the notations $\oplus$ and $\otimes$ by the ordinary notations $+$ and $\times$ and preface our calculations with the statement that they are being done mod n.

[2]*Mod* is short for *modulo*, which means *with respect to a modulus* (a quantity, which in our case, is the quantity n). To compute, for instance, $a \otimes b$, we perform the usual multiplication $a \times b$, and then subtract enough multiples of n from $a \times b$ in order to get an integer in Z_n. The latter is sometimes referred to as "modding out" n.

Example. The simplest case is $n = 2$. We have $Z_2 = \{0, 1\}$, and addition and multiplication mod 2 are given in the following tables:

$\oplus$	0	1
0	0	1
1	1	0

$\otimes$	0	1
0	0	0
1	0	1

Notice that mod 2 arithmetic is just like ordinary arithmetic except that $1 \oplus 1 = 0$. This is because $1 + 1 = 2$ and subtracting 2 lands us back at 0 in Z_2. □

Example. The addition and multiplication tables for the integers mod 3 are as follows:

$\oplus$	0	1	2
0	0	1	2
1	1	2	0
2	2	0	1

$\otimes$	0	1	2
0	0	0	0
1	0	1	2
2	0	2	1

In particular, $2 \otimes 2 = 1$ since $2 \times 2 = 4$ and $4 = 1 \times 3 + 1$. □

Example. Some instances of addition and multiplication in the system of integers modulo 6 are

$$4 \oplus 5 = 3,$$
$$2 \oplus 3 = 5,$$
$$2 \otimes 2 = 4,$$
$$3 \otimes 5 = 3,$$
$$3 \otimes 2 = 0,$$
$$5 \otimes 5 = 1.$$

□

As these examples indicate, sometimes addition or multiplication mod n is like ordinary addition or multiplication (this happens when the ordinary result is an integer in Z_n). Other times, addition or multiplication modulo n is quite different from ordinary addition and multiplication, and the results can seem quite odd. For instance, as displayed in the preceding example, in the integers mod 6 we have $5 \otimes 5 = 1$, which is suggesting that the reciprocal of 5 is itself; that is, the number which, when multiplied by 5, gives 1, is 5 itself! We also have $3 \otimes 2 = 0$ in the integers mod 6, which should at least suggest caution, since, in ordinary multiplication, nonzero numbers never multiply to 0.

Before proceeding, we recall some basic notions of arithmetic and algebra as they relate to the integers mod n. First, we observe[3] that addition and multiplication mod n satisfy the usual laws of commutativity, associativity, and distributivity. An *additive inverse* of an integer a in Z_n is an integer b in Z_n such that $a \oplus b = 0$. There is an obvious candidate for the additive inverse for a: If $a = 0$, then it's 0; if $a \neq 0$, then $n - a$ is between 1 and $n - 1$, and $n - a$ is an additive inverse of a, since

$$a + (n - a) = n = 1 \times n + 0 \text{ implying } a \oplus (n - a) = 0.$$

In all cases, the additive inverse is uniquely determined. Following usual conventions, the additive inverse of a is denoted by $-a$, but keep in mind that $-a$ *denotes*[4] one of the integers in $\{0, 1, 2 \ldots, n-1\}$. The fact that all integers in Z_n have additive inverses means that we can always subtract in Z_n, since subtracting b from a is the same as adding $-b$ to a: $a \ominus b = a \oplus (-b)$.

A *multiplicative inverse* of an integer a in Z_n is an integer b in Z_n such that $a \otimes b = 1$. In contrast to additive inverses, there is no obvious candidate for the multiplicative inverse of a. In fact, it should come as no surprise that some nonzero a's may not have multiplicative inverses. In the system Z of integers, the integer 2 does not have a multiplicative inverse since there is no integer b such that $2 \times b = 1$.[5] Indeed, in Z the only numbers that have multiplicative inverses are 1 and -1. Following usual conventions, we denote a multiplicative inverse of an integer a in Z_n by a^{-1}, *if there is one*.

Example. In the integers modulo 10, the additive inverses are as follows:

$$-0 = 0 \quad -1 = 9 \quad -2 = 8 \quad -3 = 7 \quad -4 = 6$$
$$-5 = 5 \quad -9 = 1 \quad -8 = 2 \quad -7 = 3 \quad -6 = 4$$

Note that we have the unusual circumstance whereby $-5 = 5$, but remember that -5 denotes the integer in Z_{10} which, when added (mod 10) to 5, gives 0, and 5 does have this property: $5 \oplus 5 = 0$. Notice also that, if $-a = b$, then $-b = a$; put another way $-(-a) = a$.

---

[3]Actually, it's more than an observation, but it is elementary, if not tedious, to check that these properties hold. What is implicit in the word *observation* is that we don't want to bother to check these properties. A student who has never done this before probably should check at least some of them.

[4]If we were to follow our defined notation, we should probably be denoting the additive inverse of a by $\ominus a$

[5]Of course, 2 has a multiplicative inverse in the system of rational numbers, namely $1/2$, but $1/2$ is not an integer.

By simply checking all possibilities, we can see that the situation with multiplicative inverses in Z_{10} is the following:

$$1^{-1} = 1 \quad \text{(the multiplicative inverse of 1 is always 1)}$$
$$3^{-1} = 7 \quad (3 \otimes 7 = 1)$$
$$7^{-1} = 3 \quad (7 \otimes 3 = 1)$$
$$9^{-1} = 9 \quad (9 \otimes 9 = 1).$$

None of $0, 2, 4, 5, 6,$ and 8 has a multiplicative inverse in Z_{10}. We thus see that four of the integers in Z_{10} have multiplicative inverses and six do not. $\qquad\square$

In general, integers in Z_n may or may not have multiplicative inverses. Of course, 0 never has a multiplicative inverse since $0 \times b = 0$ for all b in Z_n. Our first theorem characterizes those integers in Z_n which have multiplicative inverses and, when this characterizing condition is satisfied, its proof points to a method for finding a multiplicative inverse. This method relies on the next simple algorithm for computing the greatest common divisor (GCD) of two positive integers a and b.

Algorithm to compute the GCD of a and b

Set $A = a$ and $B = b$.
While $A \times B \neq 0$, do the following:
    If $A \geq B$, then replace A by $A - B$.
    Else, replace B by $B - A$.
Set GCD $= B$.

In words, the algorithm says to subtract the smaller of the current A and B from the larger. Continue until one of A and B is 0 (it will be A because, in the case of a tie, we subtract B from A). We then let GCD equal the terminal value of B.

We prove in the next lemma that the algorithm terminates and computes the GCD of a and b correctly.

Lemma 10.1.1 *The preceding algorithm terminates and computes the GCD of a and b correctly.*

Proof. We first observe that the algorithm does terminate with the value of A equal to 0. This is so since A and B are always nonnegative integers and at each step one of them decreases. Since we subtract B from A when $A = B$, A achieves the value 0 before B does. We next observe that, given two positive integers m and n with $m \geq n$, we have

$$\text{GCD}\{m, n\} = \text{GCD}\{m - n, n\}.$$

This is because any common divisor of m and n is also a common divisor of $m - n$ and n (if p divides both m and n, then p divides their difference $m - n$); and, conversely, any common divisor of $m - n$ and n is also a common divisor of m and n (if p divides both $m - n$ and n, then p divides their sum $(m - n) + n = m$). Hence, it follows that throughout the algorithm, even though the values of A and B are changing, their GCD is a constant d. Since initially $A = a$ and $B = b$, we see that d is the GCD of a and b. At the termination of the algorithm, we have $A = 0$ and $B > 0$. Since the GCD of two integers, one of which is 0 and one of which is positive, is the positive one, it follows that upon termination the GCD of a and b is the value of B.
□

The GCD algorithm is a remarkably simple algorithm for computing the GCD of two nonnegative integers a and b, and entails nothing more than repeated subtraction. As illustrated in the next example, it is a consequence of this algorithm that the GCD, d, of a and b can be written as a linear combination of a and b with integral coefficients: there are integers x and y such that

$$d = a \times x + b \times y.$$

Example. Compute the GCD of 48 and 126.
    We apply the algorithm and display the results in tabular form:

A	B
48	126
48	78
48	30
18	30
18	12
6	12
6	6
0	6

We conclude that the GCD of 48 and 126 is the terminal value $d = 6$ of B.
    If, in applying the algorithm to compute the GCD of two positive integers a and b, we subtract A several times consecutively from B or B several times consecutively from A, as just occurred, then we can combine these consecutive steps and treat them as a division.[6] When

---

[6] Division of one positive integer by another is, after all, just successive subtraction. For example, when we divide 23 by 5 we get a quotient of 4 and a remainder of 3. This can be displayed as $23 = 4 \times 5 + 3$, which means we can subtract four (and no more) 5's from 23 without getting a negative number.

using the algorithm to compute the GCD by hand, it is generally more efficient to apply the algorithm in this way. The results for computing the GCD of 48 and 126 are displayed in the following table.

A	B	
48	126	$126 = 2 \times 48 + 30$
48	30	$48 = 1 \times 30 + 18$
30	18	$30 = 1 \times 18 + 12$
12	18	$18 = 1 \times 12 + 6$
12	6	$12 = 2 \times 6 + 0$
0	6	$d = 6$

The last nonzero remainder in these divisions is the GCD $d = 6$ of 48 and 126.

We now use the equations in the preceding table in order to write 6 as a linear combination of 48 and 126:

$$6 = 18 - 1 \times 12$$
$$6 = 18 - 1 \times (30 - 1 \times 18) = 2 \times 18 - 1 \times 30$$
$$6 = 2 \times (48 - 1 \times 30) - 1 \times 30 = 2 \times 48 - 3 \times 30$$
$$6 = 2 \times 48 - 3 \times (126 - 2 \times 48) = 8 \times 48 - 3 \times 126.$$

The final equation, $6 = 8 \times 48 - 3 \times 126$ expresses 6 as an integral linear combination of 48 and 126. $\square$

We next show how to determine which integers in Z_n have multiplicative inverses.

Theorem 10.1.2 *Let n be an integer with $n \geq 2$ and let a be a nonzero integer in $Z_n = \{0, 1, \ldots, n-1\}$. Then a has a multiplicative inverse in Z_n if and only if the greatest common divisor (GCD) of a and n is 1. If a has a multiplicative inverse, then it is unique.*

Proof. We first show that there can be, at most, one multiplicative inverse for an integer a in Z_n. We shall make use of the rules for addition and multiplication mod n that we have already pointed out, namely, commutativity and associativity. We let b and c be multiplicative inverses of a, and show that $b = c$. Thus, suppose that $a \otimes b = 1$ and $a \otimes c = 1$. Then

$$c \otimes (a \otimes b) \;=\; c \otimes 1 \;=\; c$$
$$c \otimes (a \otimes b) \;=\; (c \otimes a) \otimes b \;=\; 1 \otimes b \;=\; b.$$

We thus conclude that $b = c$, and each integer a in Z_n has, at most, one multiplicative inverse.

We next show that, if the GCD of a and n is not 1, then a does not have a multiplicative inverse. Let $m > 1$ be the GCD of a and n. Then n/m is a nonzero integer in Z_n, and since $a \times (n/m)$ is a multiple of n (because there is a factor of m in a), we have

$$a \otimes (n/m) = 0.$$

Suppose there is a multiplicative inverse a^{-1}. Then, using the associative law again,[7] we see that

$$a^{-1} \otimes (a \otimes (n/m)) = a^{-1} \otimes 0 = 0$$
$$a^{-1} \otimes (a \otimes (n/m)) = (a^{-1} \otimes a) \otimes (n/m) = 1 \otimes n/m = n/m.$$

Hence, we have $n/m = 0$, which is a contradiction since $1 \leq n/m < n$. Therefore, a does not have a multiplicative inverse.

We lastly suppose that the GCD of a and n is 1 and show that a has a multiplicative inverse. It is a consequence of the GCD algorithm, that there exist integers x and y in Z such that

$$a \times x + n \times y = 1. \tag{10.1}$$

The integer x cannot be a multiple of n, for otherwise equation (10.1) would imply that 1 is a multiple of n, contradicting our assumption that $n \geq 2$. Therefore, x has a nonzero remainder when divided by n. That is, there exist integers q and r with $1 \leq r \leq n - 1$ such that

$$x = q \times n + r.$$

Substituting into (10.1), we get

$$a \times (q \times n + r) + n \times y = 1,$$

which, upon rewriting, becomes

$$a \times r = 1 - (a \times q + y) \times n.$$

Thus, $a \times r$ differs from 1 by a multiple of n, and it follows that

$$a \otimes r = 1,$$

so r is a (and therefore the unique, by what we have already proved) multiplicative inverse of a in Z_n. $\square$

---

[7]For those students who might have thought that the associative law of arithmetic was not of much consequence and maybe even a nuisance, we now have seen two important applications of it. And there are more to come!

Corollary 10.1.3 *Let n be a prime number. Then each nonzero integer in Z_n has a multiplicative inverse.*

Proof. Since n is a prime number, the GCD of n and any integer a between 1 and $n - 1$, inclusively, is 1. We now apply Theorem 10.1.2 to complete the proof. $\qquad\square$

It is common to call two integers whose GCD is 1 *relatively prime*. Thus, by Theorem 10.1.2, the number of integers in Z_n that have multiplicative inverses equals the number of integers between 1 and $n - 1$ that are relatively prime to n.

Applying the algorithm for computing the GCD of two numbers to the nonzero number a in Z_n and n, we obtain an algorithm for determining whether a has a multiplicative inverse in Z_n. By Theorem 10.1.2, a has a multiplicative inverse if and only if this GCD equals 1. As in the proof of Theorem 10.1.2, we can use the results of this algorithm to determine the multiplicative inverse of a when it exists. We illustrate this technique in the next example.

Example. Determine whether 11 has a multiplicative inverse in Z_{30}, and if so, calculate the multiplicative inverse.

We apply the algorithm for computing the GCD to 11 and $n = 30$ and display the results in the following table.

A	B	
30	11	$30 = 2 \times 11 + 8$
8	11	$11 = 1 \times 8 + 3$
8	3	$8 = 2 \times 3 + 2$
2	3	$3 = 1 \times 2 + 1$
2	1	$2 = 2 \times 1 + 0$
0	1	$d = 1$

Thus, the GCD of 11 and 30 is $d = 1$, and by Theorem 10.1.2, 11 has a multiplicative inverse in Z_{30}. We use the equations in the preceding table in order to obtain an equation of the form (10.1) in the proof of Theorem 10.1.2:

$$1 = 3 - 1 \times 2$$
$$1 = 3 - 1 \times (8 - 2 \times 3) = 3 \times 3 - 1 \times 8$$
$$1 = 3 \times (11 - 1 \times 8) - 1 \times 8 = 3 \times 11 - 4 \times 8$$
$$1 = 3 \times 11 - 4 \times (30 - 2 \times 11) = 11 \times 11 - 4 \times 30.$$

The final equation expressing the GCD 1 as a linear combination of 11 and 30, namely,

$$1 = 11 \times 11 - 4 \times 30,$$

tells us that, in Z_{30},

$$1 = 11 \otimes 11.$$

Hence,

$$11^{-1} = 11.$$

Of course, now that we know this fact we can check: $11 \times 11 = 121$, and 121 has remainder 1 when divided by 30. □

Example. Find the multiplicative inverse of 16 in Z_{45}.

We display our calculations in the following table:

A	B	
45	16	$45 = 2 \times 16 + 13$
13	16	$16 = 1 \times 13 + 3$
13	3	$13 = 4 \times 3 + 1$
1	3	$3 = 3 \times 1 + 0$
1	0	$d = 1$

Note that, contrary to the rules for our algorithm to compute GCDs, we made B equal to 0. The reason we set up the algorithm the way we did is in order (for a computer program) to know where to look for the GCD. But if we are doing the calculations by hand, we can make either A or B equal to 0 (and then choose the other as the GCD).

Since the GCD is 1, we conclude that 16 has a multiplicative inverse in Z_{45}. The resulting equations yield

$$1 = 13 - 4 \times 3$$
$$1 = 13 - 4 \times (16 - 1 \times 13) = 5 \times 13 - 4 \times 16$$
$$1 = 5 \times (45 - 2 \times 16) - 4 \times 16 = 5 \times 45 - 14 \times 16.$$

We conclude that $16^{-1} = -14 = 31$ in Z_{45}. □

Let n be a prime number. By Corollary 10.1.3, each nonzero integer in Z_n has a multiplicative inverse. This implies that, not only can we add, subtract, and multiply in Z_n, but we can also divide by any nonzero integer in Z_n:

$$a \div b = a \times b^{-1}, \quad (b \neq 0).$$

In addition, multiplicative inverses imply that the following properties hold in Z_n if n is a prime:

(i) (Cancellation rule 1) $a \otimes b = 0$ implies $a = 0$ or $b = 0$.

[If $a \neq 0$, then multiplying by a^{-1}, we obtain

$$0 = a^{-1} \otimes (a \otimes b) = (a^{-1} \otimes a) \otimes b = 1 \otimes b = b.]$$

(ii) (Cancellation rule 2) $a \otimes b = a \otimes c$, $a \neq 0$ implies $b = c$.

[We apply Cancellation rule 1 to $a \otimes (b - c) = 0$.]

(iii) (Solutions of linear equations) If $a \neq 0$, the equation

$$a \otimes x = b$$

has the unique solution $x = a^{-1} \otimes b$.

[Multiplying the equation by a^{-1} and using the associative law once again shows that the only possible solution is $x = a^{-1} \otimes b$. Then, substituting $x = a^{-1} \otimes b$ into the equation, we see that

$$a \otimes (a^{-1} \otimes b) = (a^{-1} \otimes a) \otimes b = 1 \otimes b = b.]$$

The conclusion that we draw from this discussion is that the usual laws of arithmetic that we are accustomed to taking for granted in the arithmetic systems of real numbers or rational numbers also hold for Z_n, *provided n is a prime number.* If n is not a prime, then as we have seen, many but not all of the usual laws of arithmetic hold in Z_n. For example, if n has the nontrivial factorization $n = a \times b$, $(1 < a, b < n)$, then, in Z_n, $a \otimes b = 0$, and neither a nor b has a multiplicative inverse. What is unusual about these arithmetical systems is that they have only a finite number of elements (in contrast to the infinite number of rational, real, and complex numbers).

At this point, we stop using the more cumbersome notation $\oplus$ and $\otimes$ for addition and multiplication mod n and use instead $+$ and $\times$, respectively.

There are other methods, however, to obtain finite arithmetical systems which satisfy the laws of arithmetic that we are accustomed to. The name given to such an arithmetical system is a *field*.[8] The method is a generalization of that used to obtain the complex numbers from the real numbers and can be summarized as follows:

Recall that the polynomial $x^2 + 1$ (with real coefficients) has no root in the system of real numbers.[9] The complex numbers are obtained from the real numbers by "adjoining" a root, usually denoted by i, of

---

[8]The properties that an arithmetical system must satisfy in order to be labeled a field can be found in most books on abstract algebra.

[9]Because the square of a real number can never be the negative number -1. We hasten to point out that this is *not* one of the usual laws of arithmetic to which we have referred. For example, in Z_5 we have $2^2 = 4 = -1$; in fact, the notion of *negative* number has no significance here because $-1 = 4$, $-2 = 3$, $-3 = 2$, and $-4 = 1$. One should not think of the additive inverse as a negative number!

$x^2 + 1 = 0$. The system of complex numbers consists of all numbers of the form $a + bi$, where a and b are real numbers, for which the usual laws of arithmetic hold and where $i^2 + 1 = 0$ (i.e., $i^2 = -1$). For instance,

$$(2 + 3i) \times (4 + i) = 8 + 2i + 12i + 3i^2 = 8 + 14i - 3 = 5 + 14i.$$

This method can be used to construct fields with p^k elements for every prime p and integer $k \geq 2$, starting from the field Z_p. We illustrate the method by constructing fields with 4 and 27 elements, respectively.

Example. *Construction of a field of 4 elements.* We start with Z_2 and the polynomial $x^2 + x + 1$ with coefficients in Z_2. This polynomial has no root in Z_2, since the only possibilities are 0 and 1 and $0^2 + 0 + 1 = 1$ and $1^2 + 1 + 1 = 1$. Because this polynomial has degree 2, we conclude it cannot be factored in any nontrivial way. We adjoin a root i of this polynomial[10] to Z_2, getting $i^2 + i + 1 = 0$, or, equivalently,

$$i^2 = -i - 1 = i + 1.$$

(Recall that in Z_2, we have $-1 = 1$.) The elements of the resulting field are the 4 elements

$$\{0, 1, i, 1 + i\},$$

with addition table and multiplication tables as follows:

+	0	1	i	$1 + i$
0	0	1	i	$1 + i$
1	1	0	$1 + i$	i
i	i	$1 + i$	0	1
$1 + i$	$1 + i$	i	1	0

×	0	1	i	$1 + i$
0	0	0	0	0
1	0	1	i	$1 + i$
i	0	i	$1 + i$	1
$1 + i$	0	$1 + i$	1	i

Thus, $i^{-1} = 1 + i$, since $i \times (1 + i) = i + i^2 = i + (1 + i) = 1$. □

Example. *Construction of a field of $3^3 = 27$ elements.* We start with $Z_3 = \{0, 1, 2\}$, the integers mod 3. We look for a polynomial of

---

[10]We use i as a symbol for the root in order to stress the *analogy* with the complex numbers. It is not true that $i^2 = -1$.

degree 3 with coefficients in Z_3 that cannot be factored in a nontrivial way. A polynomial of degree 3 will have this property if and only if it has no root in Z_3.[11] The polynomial $x^3 + 2x + 1$ with coefficients in Z_3 does not have a root in Z_3 (one need only test the three elements $0, 1$, and 2 of Z_3). Thus, we adjoin a root i of this polynomial, getting $i^3 + 2i + 1 = 0$ or, equivalently,

$$i^3 = -1 - 2i = 2 + i.$$

(Recall that, in Z_3, we have $-1 = 2$ and $-2 = 1$.) Now, use the usual rules of arithmetic, but whenever an i^3 appears, replace it by $2 + i$. The elements of the resulting field are the 27 elements

$$\{a + bi + ci^2 : a, b \text{ and } c \text{ in } Z_3\}.$$

Since there are 27 elements, it is no longer practical to write out the addition and multiplication tables. But we illustrate some of the arithmetic in this system as follows:

$$(2+i+2i^2)+(1+i+i^2) = (2+1)+(1+1)i+(2+1)i^2 = 0+2i+0i^2 = 2i;$$

$$
\begin{aligned}
(1+i)(2+i^2) &= 1 \times 2 + i^2 + 2i + i \times i^2 \\
&= 1 + i^2;
\end{aligned}
$$

$$
\begin{aligned}
(1+2i^2)(1+i+2i^2) &= 1 + i + 2i^2 + 2i^2 + 2i^3 + 2 \times 2i^4 \\
&= 1 + i + 2i^2 + 2i^2 + 2(2+i) + (i \times i^3) \\
&= 1 + i + i^2 + (1+2i) + i \times (2+i) \\
&= 1 + i + i^2 + 1 + 2i + 2i + i^2 \\
&= 2 + 2i + 2i^2.
\end{aligned}
$$

It is straighforward to check that

$$i^{-1} = 1 + 2i^2 \text{ and } (2+i+2i^2)^{-1} = 1 + i^2.$$

□

We conclude this section with the following remarks: For each prime p and each integer $k \geq 2$ there exists a polynomial of degree k with coefficients in Z_p that does not have a nontrivial factorization.

---

[11]This is not a general rule. If a polynomial of degree 2 or 3 is factored nontrivially, one of the factors is linear and the polynomial has a root. But, for instance, a polynomial of degree 4 may be factorable into two polynomials of degree 2, neither of which has a root.

Thus, in the manner illustrated in the preceding two examples, we can construct a field with p^k elements. Conversely, it can be proved that, if there is a field with a finite number m elements—that is, a finite system satisfying the usual rules of arithmetic—then $m = p^k$ for some positive integer k and some prime number p, and it can be obtained from Z_p in the manner previously described (or is Z_p if $k = 1$). Thus, *only for a prime power number of elements do finite fields exist.*

10.2   Block Designs

We begin this section with a simplified motivating example from the design of experiments for statistical analysis.

Example. Suppose there are 7 varieties of a product to be tested for acceptability among consumers. The manufacturer plans to ask some random (or typical) consumers to compare the different varieties. One way to do this is for each of the consumers involved in the testing to do a complete test by comparing all of the 7 varieties. However, the manufacturer, fully aware of the time required for the comparisons and the possible reluctance of individuals to get involved, decides to have each consumer do an incomplete test by comparing only some of the varieties. Thus, the manufacturer asks each person to compare a certain 3 of the varieties. In order to be able to draw meaningful conclusions based on statistical analysis of the results, the test is to have the property that each pair of the 7 varieties is compared by exactly one person. Can such a testing experiment be designed?

We label the different varieties $0, 1, 2, 3, 4, 5$ and 6.[12] There are $\binom{7}{2} = 21$ pairs of the 7 varieties. Each tester gets 3 varieties and thus makes $\binom{3}{2} = 3$ comparisons. Since each pair is to be compared exactly once, the number of testers must be

$$\frac{21}{3} = 7.$$

Thus, in this case, the number of individuals involved in the experiment is the same as the number of varieties being tested. Fortunately, the preceding quotient turned out to be an integer, for otherwise we would have to conclude that it is impossible to design an experiment with the constraints as given. What we now seek is 7 (one for each person

---

[12]Of course, we are free to *label* the varieties in any way we choose. The reason we choose $0, 1, 2, 3, 4, 5, 6$ is that we can think of the varieties as the numbers in Z_7, the integers mod 7.

involved in the test) subsets $B_1, B_2, \ldots, B_7$ of the 7 varieties, which we shall call *blocks*, with the property that each pair of varieties is together in exactly one block. Such a collection of 7 blocks is the following:

$$B_1 = \{0, 1, 3\}, B_2 = \{1, 2, 4\}, B_3 = \{2, 3, 5\}, B_4 = \{3, 4, 6\},$$

$$B_5 = \{0, 4, 5\}, B_6 = \{1, 5, 6\}, B_7 = \{0, 2, 6\}.$$

Another way to present this experimental design is given in the array that follows: In this array, we have one column for each of the 7 varieties and one row for each of the 7 blocks. A 1 in row i and column j ($i = 1, 2, \ldots, 7; j = 0, 1, \ldots, 6$) means that variety j belongs to block B_i, and a 0 means that variety j does not belong to block B_i. The fact that each block contains three varieties is reflected in the table by the fact that each row contains three 1's. The fact that each pair of varieties is together in one block is equivalent to the property of the table that each pair of columns have 1's in exactly one common row. As is evident from the table, each variety occurs in 3 blocks. This array is the incidence array of the experimental design.

	0	1	2	3	4	5	6
B_1	1	1	0	1	0	0	0
B_2	0	1	1	0	1	0	0
B_3	0	0	1	1	0	1	0
B_4	0	0	0	1	1	0	1
B_5	1	0	0	0	1	1	0
B_6	0	1	0	0	0	1	1
B_7	1	0	1	0	0	0	1

□

Before discussing more examples, we define some terms and discuss some elementary properties of designs. Let k, λ, and v be positive integers with

$$2 \leq k \leq v.$$

Let X be any set of v elements, called *varieties*, and let $\mathcal{B}$ be a collection $B_1, B_2, \ldots, B_b$ of k-element subsets of X called *blocks*.[13] Then $\mathcal{B}$ is a *balanced block design* on X, provided that each pair of elements of X occurs together in exactly λ blocks. The number λ is called the *index of the design*. The foregoing assumption that k is at least 2 is to prevent trivial solutions: If $k = 1$, then a block contains no pairs and $\lambda = 0$.

---

[13]We do not rule out the possibility that some of the blocks may be identical, although it is more challenging to find designs all of whose blocks are different. Thus, the collection of blocks is, in general, a multiset of blocks.

Let $\mathcal{B}$ be a balanced block design. If $k = v$, that is, the complete set of varieties occurs in each block, then the design $\mathcal{B}$ is called a *complete* block design. If $k < v$, then $\mathcal{B}$ is a balanced *incomplete* block design, or *BIBD*[14] for short. A complete design corresponds to a testing experiment in which each individual compares each pair of varieties. From a combinatorial point of view, they are trivial, forming a collection of sets all equal to X, and we henceforth deal with incomplete designs—that is, designs for which $k < v$.

Let $\mathcal{B}$ be a BIBD on X. As in the preceding example, we associate with $\mathcal{B}$ an *incidence matrix* or *incidence array* A. The array A has b rows, one corresponding to each of the blocks $B_1, B_2, \ldots, B_b$, and v columns, one corresponding to each of the varieties $x_1, x_2, \ldots, x_v$ in X. The entry a_{ij} at the intersection of row i and column j is 0 or 1:

$$a_{ij} = 1 \text{ if } x_j \text{ is in } B_i,$$

$$a_{ij} = 0 \text{ if } x_j \text{ is not in } B_i.$$

We talk about *the* incidence matrix of $\mathcal{B}$, even though it depends on the order in which we list the blocks and the order in which we list the varieties. The rows of the incidence matrix display the varieties contained in each of the blocks. The columns of the incidence matrix display the blocks containing each of the varieties. Except for the labeling of the varieties and of the blocks, the incidence matrix A contains full information about the BIBD. Since each block contains k varieties, each row of the incidence matrix A contains k 1's. Since there are b blocks the total number of 1's in A equals bk. We now show that each variety is contained in the same number of blocks; that is, each column of A contains the same number of 1's.

Lemma 10.2.1 *In a BIBD, each variety is contained in*

$$r = \frac{\lambda(v - 1)}{k - 1}$$

blocks.

Proof. We use the important technique of counting in two ways and then equating the two counts. Let x_i be any one of the varieties, and suppose that x_i is contained in r blocks

$$B_{i_1}, B_{i_2}, \ldots, B_{i_r}. \tag{10.2}$$

---

[14]BIBDs were introduced by F. Yates: Complex experiments (with discussion), *J. Royal Statistical Society*, Suppl. 2, (1935), 181-247.

Since each block contains k elements, each of these blocks contains $k-1$ varieties other than x_i. We now consider each of the $v-1$ pairs $\{x_i, y\}$, where y is a variety different from x_i, and for each such pair, we count the number of blocks in which both varieties are contained. Each pair $\{x_i, y\}$ is contained in λ blocks (these blocks must be λ of the blocks in (10.2) since they are all the blocks containing x_i). Adding, we get

$$\lambda(v - 1).$$

On the other hand, each of the blocks in (10.2) contains $k - 1$ pairs, one element of which is x_i. Adding, we now get

$$(k - 1)r.$$

Equating these two counts, we obtain

$$\lambda(v - 1) = (k - 1)r.$$

Hence, x_i is contained in $\lambda(v - 1)/(k - 1)$ blocks. This is true for each variety x_i, and thus each variety is contained in $r = \lambda(v - 1)/(k - 1)$ blocks.   □

Corollary 10.2.2 *In a BIBD, we have*

$$bk = vr.$$

Proof. We have already observed that counting by rows, the number of 1's in the incidence matrix A of a BIBD is bk. By Lemma 10.2.1, we know that each column of A contains r 1's. Thus, counting by columns, the number of 1's in A equals vr. Equating the two counts, we obtain $bk = vr$.   □

Corollary 10.2.3 *In a BIBD, we have*

$$\lambda < r.$$

Proof. In a BIBD, we have, by definition, $k < v$; hence, $k - 1 < v - 1$. Using Lemma 10.2.1 we conclude that $\lambda < r$.   □

As a consequence of Lemma 10.2.1, we now have 5 parameters, not all independent, that are associated with a BIBD:

b : the number of blocks;

v : the number of varieties;

k : the number of varieties in each block;

r : the number of blocks containing each variety;

λ : the number of blocks containing each pair of varieties.

We call b, v, k, r, λ the *parameters* of the BIBD. The parameters of the design in our introductory example are: $b = 7, v = 7, k = 3, r = 3$, and $\lambda = 1$.

Example. Is there a BIBD with parameters $b = 12, k = 4, v = 16$, and $r = 3$ (the parameter λ is not specified)?

The equation $bk = vr$ in Corollary 10.2.2 holds, since both sides have the value 48. By Lemma 10.2.1, if there is such a design, its index λ satisfies

$$\lambda = \frac{r(k-1)}{v-1} = \frac{3(3)}{15} = \frac{9}{15}.$$

Since this is not an integer, there can be no such design with four of its parameters as given. $\quad\square$

Example. In this example, we display a design with parameters $b = 12, v = 9, k = 3, r = 4$, and $\lambda = 1$. It is most convenient to define the design by its 12-by-9 incidence matrix:

$$A = \begin{bmatrix}
1 & 1 & 1 & 0 & 0 & 0 & 0 & 0 & 0 \\
0 & 0 & 0 & 1 & 1 & 1 & 0 & 0 & 0 \\
0 & 0 & 0 & 0 & 0 & 0 & 1 & 1 & 1 \\
1 & 0 & 0 & 1 & 0 & 0 & 1 & 0 & 0 \\
0 & 1 & 0 & 0 & 1 & 0 & 0 & 1 & 0 \\
0 & 0 & 1 & 0 & 0 & 1 & 0 & 0 & 1 \\
1 & 0 & 0 & 0 & 1 & 0 & 0 & 0 & 1 \\
0 & 0 & 1 & 1 & 0 & 0 & 0 & 1 & 0 \\
0 & 1 & 0 & 0 & 0 & 1 & 1 & 0 & 0 \\
1 & 0 & 0 & 0 & 0 & 1 & 0 & 1 & 0 \\
0 & 1 & 0 & 1 & 0 & 0 & 0 & 0 & 1 \\
0 & 0 & 1 & 0 & 1 & 0 & 1 & 0 & 0
\end{bmatrix}.$$

It is straightforward to check that this matrix defines a BIBD with parameters as given. $\quad\square$

Example. Consider the squares of a 4-by-4 board:

Let the varieties be the 16 squares of the board. We define blocks as follows: For each given square, we take the 6 other squares that are either in its row or in its column (so not the given square itself).[15] Therefore, each of the 16 squares on the board determines a block in this way. We thus have $b = 16, v = 16$, and $k = 6$. Each square belongs to 6 blocks, since each square lies in a row with 3 other squares and in a column with 3 more squares. Thus, we also have $r = 6$. But we haven't yet shown we have a BIBD. So let's take a pair of squares x and y. There are three possibilities:

1: x and y are in the same row. Then x and y are together in the 2 blocks determined by the other 2 squares in their row.

2: x and y are in the same column. Then x and y are together in the 2 blocks determined by the other 2 squares in their column.

3: x and y are in different rows and in different columns. Then x and y are together in 2 blocks, one determined by the square at the intersection of the row of x and the column of y, the other determined by the intersection of the column of x and the row of y. The following array, where the blocks are those determined by the squares marked with an asterisk ($*$) is illustrative:

	$*$	x	
	y	$*$	

Since each pair of varieties is together in 2 blocks, we have a BIBD with $\lambda = 2$. $\square$

The basic property of designs presented in the next theorem says that, in a BIBD, the number of blocks must be at least as large as the number of varieties and is known as Fisher's inequality.[16]

Theorem 10.2.4 *In a BIBD, $b \geq v$.*

Proof. We outline a linear algebraic proof for those familiar with the ideas it uses. Let A be the b-by-v incidence matrix of a BIBD.

---

[15]We can think of the varieties as a rook on the 4-by-4 board and the blocks as all the squares that a rook on the board can attack.

[16]R.A. Fisher: An examination of the different possible solutions of a problem in incomplete blocks, *Annals of Eugenics*, 10(1940), 52-75.

Since each variety is in r blocks and since each pair of varieties is in λ blocks, the v-by-v matrix $A^T A$, obtained by multiplying[17] the transpose[18] A^T of A by A, has each main diagonal entry equal to r and each off-diagonal element equal to λ:

$$A^T A = \begin{bmatrix} r & \lambda & \cdots & \lambda \\ \lambda & r & \cdots & \lambda \\ \vdots & \vdots & \ddots & \vdots \\ \lambda & \lambda & \cdots & r \end{bmatrix}.$$

Since $\lambda < r$, by Corollary 10.2.3, the matrix $A^T A$ can be shown to have a nonzero determinant[19] and hence is invertible. Thus, $A^T A$ has rank equal to v. Therefore A has rank at least v, and since A is a b-by-v matrix, we have $b \geq v$.[20] $\square$

A BIBD for which equality holds in Theorem 10.2.4, that is, for which the number b of blocks equals the number v of varieties, is called *symmetric*,[21] and this is shortened to SBIBD. Since a BIBD satisfies $bk = vr$, we conclude by cancellation that, for an SBIBD, we also have $k = r$. By Lemma 10.2.1, the index λ for an SBIBD is determined by v and k by

$$\lambda = \frac{k(k-1)}{v-1}. \tag{10.3}$$

Thus, the parameters associated with an SBIBD are as follows:

v : the number of blocks;

v : the number of varieties;

k : the number of varieties in each block;

---

[17]The product of an m-by-n matrix X with typical entry x_{ij} and an n-by-p matrix Y with typical entry y_{jk} is the m-by-p matrix Z whose typical entry is $z_{ik} = \sum_{j=1}^{n} x_{ij} y_{jk}$.

[18]The transpose of an m-by-n matrix X is the n-by-m matrix X^T obtained by letting the rows of X "become" the columns of X^T and the columns of X "become" the rows of X^T. If, as in A in the proof of the theorem, the entries of X are 0's and 1's, then the typical entry of $X^T X$ in row i and column j (by the definition of product, it is determined by column i and column j of X) equals the number of rows in which *both* column i and column j have a 1.

[19]The value of the determinant is $(r - \lambda)^{v-1}(r + (v-1)\lambda)$, which is nonzero by Corollary 10.2.3.

[20]If you didn't understand this proof because you never studied elementary linear algebra, I hope you will now do so. Only then can you appreciate what an elegant and simple proof has just been shown you!

[21]The symmetry has to do with the parameters satisfying $b = v$ and, as shown in the next few lines, $k = r$.

k : the number of blocks containing each variety;

λ : the number of blocks containing each pair of varieties, where λ is given by (10.3).

Some of our examples have been SBIBDs.

We now discuss a method for constructing SBIBDs that uses the arithmetic of the integers mod n. In this method, the varieties are the integers in Z_n, so, to agree with our notation, we use v instead of n.

Thus, let $v \geq 2$ be an integer, and consider the set of integers mod v:

$$Z_v = \{0, 1, 2, \ldots, v - 1\}.$$

Note that addition and multiplication in Z_v are denoted by the usual symbols $+$ and $\times$. Let $B = \{i_1, i_2, \ldots, i_k\}$ be a subset of Z_v consisting of k integers. For each integer j in Z_v, we define

$$B + j = \{i_1 + j, i_2 + j, \ldots, i_k + j\}$$

to be the subset of Z_v obtained by adding mod v the integer j to each of the integers in B. The set $B + j$ also contains k integers. This is because, if

$$i_p + j = i_q + j \quad (\text{in } Z_v),$$

then cancelling j (by adding the additive inverse $-j$ to both sides) we get $i_p = i_q$. The v sets

$$B = B + 0, B + 1, \ldots, B + v - 1$$

so obtained are called the *blocks developed from the block B* and B is called the *starter block*.

Example. Let $v = 7$ and consider

$$Z_7 = \{0, 1, 2, 3, 4, 5, 6\}.$$

Now consider the starter block

$$B = \{0, 1, 3\}.$$

Then we have

$$B + 0 = \{0, 1, 3\}$$
$$B + 1 = \{1, 2, 4\}$$
$$B + 2 = \{2, 3, 5\}$$
$$B + 3 = \{3, 4, 6\}$$
$$B + 4 = \{4, 5, 0\}$$
$$B + 5 = \{5, 6, 1\}$$
$$B + 6 = \{6, 0, 2\}.$$

(Each set in this list, other than the first, is obtained by adding 1 mod 7 to the previous set. In addition the first set B on the list can be gotten from the last by adding 1 mod 7.) This is a BIBD, indeed, the same one in the introductory example of this section. Since $b = v$, we have an SBIBD with $b = v = 7$, $k = r = 3$, and $\lambda = 1$.   □

Example. Let $v = 7$ as in the previous example, but now let the starter block be

$$B = \{0, 1, 4\}.$$

Then we have

$$B + 0 = \{0, 1, 4\}$$
$$B + 1 = \{1, 2, 5\}$$
$$B + 2 = \{2, 3, 6\}$$
$$B + 3 = \{3, 4, 0\}$$
$$B + 4 = \{4, 5, 1\}$$
$$B + 5 = \{5, 6, 2\}$$
$$B + 6 = \{6, 0, 3\}.$$

In this case, we do not obtain a BIBD because, for instance, the varieties 1 and 2 occur together in one block, while the varieties 1 and 5 are together in two blocks.   □

It follows from these two examples that sometimes, but not always, the blocks developed from a starter block are the blocks of an SBIBD. The property that we need in order to obtain an SBIBD in this way is contained in the next definition. Let B be a subset of k integers in Z_v. Then B is called a *difference set mod v*, provided that each nonzero integer in Z_v occurs the same number λ of times among the $k(k-1)$ differences among distinct elements of B (in both orders):

$$x - y \quad (x, y \text{ in } B; x \neq y).$$

Since there are $v - 1$ nonzero integers in Z_v, each nonzero integer in Z_v must occur

$$\lambda = \frac{k(k-1)}{v-1}$$

times as a difference in a difference set.

Example. Let $v = 7$ and $k = 3$ and consider $B = \{0, 1, 3\}$. We compute the subtraction table for the integers in B, ignoring the 0's in the diagonal positions:

$-$	0	1	3
0	0	6	4
1	1	0	5
3	3	2	0

Examining this table, we see that the nonzero integers 1,2,3,4,5,6 in Z_7 each occur exactly once in the off-diagonal positions and hence exactly once as a difference. Hence, B is a difference set mod 7.   □

Example. Again, let $v = 7$ and $k = 3$, but now let $B = \{0, 1, 4\}$. Computing the subtraction table, we now get

$-$	0	1	4
0	0	6	3
1	1	0	4
4	4	3	0

We see that 1 and 6 each occur once as a difference, 3 and 4 each occur twice, and 2 and 5 do not occur at all. Thus, B is not a difference set in this case.   □

Theorem 10.2.5 *Let B be a subset of $k < v$ elements of Z_v that forms a difference set mod v. Then the blocks developed from B as a starter block form an SBIBD with index*

$$\lambda = \frac{k(k-1)}{v-1}.$$

Proof. Since $k < v$, the blocks are not complete. Each block contains k elements. Moreover, the number of blocks is the same as the number v of varieties. Thus, it remains to be shown that each pair of elements of Z_v is together in the same number of blocks. Since B is a difference set, each nonzero integer in Z_v occurs as a difference exactly $\lambda = k(k-1)/(v-1)$ times. We show that each pair of elements of Z_v is in λ blocks and hence λ is the index of the SBIBD.

Let p and q be distinct integers in Z_v. Then $p - q \neq 0$, and since B is a difference set mod v, the equation

$$x - y = p - q$$

has λ solutions with x and y in B. For each such solution x and y, let $j = p - x$. Then

$$p = x + j \text{ and } q = y - x + p = y + j.$$

Thus, p and q are together in the block $B + j$ for each of the λ j's. Hence, p and q are together in λ blocks. Since

$$v(v-1)\lambda = v(v-1)\frac{k(k-1)}{v-1} = vk(k-1),$$

it follows that each pair of distinct integers in Z_v is together in exactly λ blocks.   □

Example. Find a difference set of size 5 in Z_{11}, and use it as a starter block in order to construct an SBIBD.

We show that $B = \{0, 2, 3, 4, 8\}$ is a difference set with $\lambda = 2$. We compute the subtraction table to obtain

$-$	0	2	3	4	8
0	0	9	8	7	3
2	2	0	10	9	5
3	3	1	0	10	6
4	4	2	1	0	7
8	8	6	5	4	0

Examining all the off-diagonal positions, we see that each nonzero integer in Z_{11} occurs twice as a difference and hence B is a difference set. Using B as a starter block, we obtain the following blocks for an SBIBD with parameters $b = v = 11$, $k = r = 5$, and $\lambda = 2$:

$$
\begin{aligned}
B + 0 &= \{0, 2, 3, 4, 8\}\\
B + 1 &= \{1, 3, 4, 5, 9\}\\
B + 2 &= \{2, 4, 5, 6, 10\}\\
B + 3 &= \{0, 3, 5, 6, 7\}\\
B + 4 &= \{1, 4, 6, 7, 8\}\\
B + 5 &= \{2, 5, 7, 8, 9\}\\
B + 6 &= \{3, 6, 8, 9, 10\}\\
B + 7 &= \{0, 4, 7, 9, 10\}\\
B + 8 &= \{0, 1, 5, 8, 10\}\\
B + 9 &= \{0, 1, 2, 6, 9\}\\
B + 10 &= \{1, 2, 3, 7, 10\}.
\end{aligned}
$$

                                                                              □

10.3 Steiner Triple Systems

Let $\mathcal{B}$ be a balanced incomplete block design whose parameters are b, v, k, r, λ. Since $\mathcal{B}$ is incomplete, we know, by definition, that $k < v$; that is, the number of varieties in each block is less than the total number of varieties. Suppose $k = 2$. Then each block in $\mathcal{B}$ contains exactly 2 varieties. In order for each pair of varieties to occur in the same number λ of blocks of $\mathcal{B}$, each subset of 2 varieties must occur as a block exactly λ times. Thus, for BIBDs, with $k = 2$, we have no choice but to take each subset of 2 varieties and write it down λ times.

Example. A BIBD with $v = 6$, $k = 2$, and $\lambda = 1$ is given by

$$\begin{array}{lll} \{0,1\} & \{0,2\} & \{0,3\} \\ \{0,4\} & \{0,5\} & \{1,2\} \\ \{1,3\} & \{1,4\} & \{1,5\} \\ \{2,3\} & \{2,4\} & \{2,5\} \\ \{3,4\} & \{3,5\} & \{4,5\}. \end{array}$$

To get a BIBD with $\lambda = 2$, simply take each of the blocks twice. To get one with $\lambda = 3$, take each of the blocks three times. □

So BIBDs with block size 2 are trivial. The smallest (in terms of block size) interesting case occurs when $k = 3$. Balanced block designs with block size $k = 3$ are called *Steiner triple systems*.[22] The first example given in Section 10.2 is a Steiner triple system. It has 7 varieties and 7 blocks of size 3. Also, each pair of varieties is contained in $\lambda = 1$ block. This is the only instance of a Steiner triple system that forms an SBIBD—that is, one for which the number of blocks equals the number of varieties.

Another example of a Steiner triple system is obtained by taking $v = 3$ varieties 0, 1, and 2 and the one block $\{0, 1, 2\}$. We thus have $b = 1$ and clearly each pair of varieties is contained in $\lambda = 1$ block. This Steiner system is not an incomplete design since $v = k = 3$.[23] Every other Steiner triple system is a BIBD.

Example. The following is an example of a Steiner triple system of index $\lambda = 1$ with 9 varieties:

$$\begin{array}{lll} \{0,1,2\} & \{3,4,5\} & \{6,7,8\} \\ \{0,3,6\} & \{1,4,7\} & \{2,5,8\} \\ \{0,4,8\} & \{2,3,7\} & \{1,5,6\} \\ \{0,5,7\} & \{1,3,8\} & \{2,4,6\}. \end{array}$$

□

In the next theorem, we obtain some relationships that must hold between the parameters of a Steiner triple system.

Theorem 10.3.1 *Let $\mathcal{B}$ be a Steiner triple system with parameters $b, v, k = 3, r, \lambda$. Then*

$$r = \frac{\lambda(v-1)}{2} \tag{10.4}$$

---

[22] After J. Steiner, who was one of the first to consider them: Combinatorische Aufgabe, *Journal für die reine und angewandte Mathematik*, 45 (1853), 181-182.

[23] There is a reason to consider it as a Steiner triple system since we shall use it to construct Steiner triple systems that are incomplete designs.

and

$$b = \frac{\lambda v(v-1)}{6}. \tag{10.5}$$

If the index is $\lambda = 1$, then there is a nonnegative integer n such that $v = 6n + 1$ or $v = 6n + 3$.

Proof. By Theorem 10.2.1, we have

$$r = \frac{\lambda(v-1)}{k-1}$$

for any BIBD. Since a Steiner triple system is a BIBD with $k = 3$, we get (10.4). For a BIBD, we also have, by Corollary 10.2.2,

$$bk = vr.$$

Substituting the value of r, as given by (10.4), and using $k = 3$ again, we get (10.5).

The equations (10.4) and (10.5) tell us that, if there is a Steiner triple system of index λ with v varieties, then $\lambda(v-1)$ is even and $\lambda v(v-1)$ is divisible by 6. Now assume that $\lambda = 1$. Then $v - 1$ is even and hence v is odd, and $v(v-1)$ is divisible by 6. The latter implies that either v or $v-1$ is divisible by 3. First, suppose that v is divisible by 3. Since v is odd, this means that v is 3 times an odd number:

$$v = 3 \times (2n+1) = 6n + 3.$$

Now suppose that $v - 1$ is divisible by 3. Since v is odd, $v - 1$ is even and we find that $v - 1$ is 3 times an even number:

$$v - 1 = 3 \times (2n) = 6n \text{ and so } v = 6n + 1.$$

$\square$

In the remainder of this section we consider only Steiner triple systems of index $\lambda = 1$. By Theorem 10.3.1, the number of varieties in a Steiner triple system of index $\lambda = 1$ is either $v = 6n + 1$ or $v = 6n + 3$, where n is a nonnegative integer. This raises the question as to whether, for all nonnegative integers n, there exist Steiner triple systems with $v = 6n + 1$ and $v = 6n + 3$ varieties. The case $n = 0$ and $v = 6n + 1$ has to be eliminated, since, in that case, $v = 1$ and no triples are possible. For all other cases, it was shown by T.P. Kirkman[24] that Steiner triple systems can be constructed. The proof

---

[24]T.P. Kirkman: On a problem in combinations, *Cambridge and Dublin Mathematics Journal*, 2 (1847), 191-204. This question was also raised later by J. Steiner, who was unaware of Kirkman's work (cf. footnote 20). It was only later that Kirkman's work became known, and this was long after the name *Steiner* (and not *Kirkman*) triple systems had become common.

is beyond the scope of this book. We shall be satisfied to give a method for constructing a Steiner triple system from two known (possibly the same) Steiner systems of smaller order.

Theorem 10.3.2 *If there are Steiner triple systems of index $\lambda = 1$ with v and w varieties, respectively, then there is a Steiner triple system of index $\lambda = 1$ with vw varieties.*

Proof. Let $\mathcal{B}_1$ be a Steiner triple system of index $\lambda = 1$ with the v varieties $a_1, a_2, \ldots, a_v$ and let $\mathcal{B}_2$ be a Steiner triple system of index $\lambda = 1$ with the w varieties $b_1, b_2, \ldots, b_w$. We consider a set X of vw varieties $c_{ij}, (i = 1, \ldots, v; j = 1, \ldots, w)$, which we may think of as the entries (or positions) of a v-by-w array whose rows correspond to $a_1, a_2, \ldots, a_v$ and whose columns correspond to $b_1, b_2, \ldots, b_w$:[25]

$$
\begin{array}{cc}
 & \begin{array}{cccc} b_1 & b_2 & \cdots & b_w \end{array} \\
\begin{array}{c} a_1 \\ a_2 \\ \vdots \\ a_v \end{array} &
\left[\begin{array}{cccc}
c_{11} & c_{12} & \cdots & c_{1w} \\
c_{21} & c_{22} & \cdots & c_{2w} \\
\vdots & \vdots & \ddots & \vdots \\
c_{v1} & c_{v2} & \cdots & c_{vw}
\end{array} \right].
\end{array}
\tag{10.6}
$$

We define a set $\mathcal{B}$ of triples of the elements of X. Let $\{c_{ir}, c_{js}, c_{kt}\}$ be a set of 3 elements of X. Then $\{c_{ir}, c_{js}, c_{kt}\}$ is a triple of $\mathcal{B}$ if and only if one of the following holds:

(i) $r = s = t$, and $\{a_i, a_j, a_k\}$ is a triple in $\mathcal{B}_1$. Put another way, the elements c_{ir}, c_{js}, and c_{kt} are in the same column of the array (10.6) and the rows in which they lie correspond to a triple of $\mathcal{B}_1$.

(ii) $i = j = k$, and $\{b_r, b_s, b_t\}$ is a triple of $\mathcal{B}_2$. Put another way, the elements c_{ir}, c_{js}, and c_{kt} are in the same row of the array (10.6) and the columns in which they lie correspond to a triple of $\mathcal{B}_2$.

(iii) i, j, and k are all different and $\{a_i, a_j, a_k\}$ is a triple of $\mathcal{B}_1$, and r, s, and t are all different and $\{b_r, b_s, b_t\}$ is a triple of $\mathcal{B}_2$. Put another way, the elements c_{ir}, c_{js}, and c_{kt} are in 3 different rows and 3 different columns of the array (10.6), and the rows in which they lie correspond to a triple of $\mathcal{B}_1$ and the columns in which they lie correspond to a triple of $\mathcal{B}_2$.

---

[25]We could think of c_{ij} as the ordered pair (a_i, b_j) but, since we are going to be discussing unordered pairs and triples, it seems less confusing to invent new symbols c_{ij}.

For the rest of the proof we shall implicitly use the fact that no triple of $\mathcal{B}$ lies either in exactly 2 rows or exactly 2 columns of the array (10.6). We now show that this set $\mathcal{B}$ of triples of X defines a Steiner triple system of index $\lambda = 1$. Thus, let c_{ir}, c_{js} be a pair of distinct elements of X. We need to show that there is exactly one triple of $\mathcal{B}$ containing both c_{ir} and c_{js}; that is, we need to show that there is exactly one element c_{kt} of X such that $\{c_{ir}, c_{js}, c_{kt}\}$ is a triple of $\mathcal{B}$. We consider three cases:

Case 1: $r = s$ and thus $i \neq j$. Our pair of elements in this case is c_{ir}, c_{jr} lying in the same column of (10.6). Since $\mathcal{B}_1$ is a Steiner triple system of index $\lambda = 1$, there is a unique triple $\{a_i, a_j, a_k\}$ containing the distinct pair a_i, a_j. Hence, $\{c_{ir}, c_{jr}, c_{kr}\}$ is the unique triple of $\mathcal{B}$ containing the pair c_{ir}, c_{jr}.

Case 2: $i = j$ and thus $r \neq s$. Our pair of elements is now c_{ir}, c_{is} lying in the same row of (10.6). Since $\mathcal{B}_2$ is a Steiner triple system of index $\lambda = 1$, there is a unique triple $\{b_r, b_s, b_t\}$ containing the distinct pair b_r, b_s. Hence, $\{c_{ir}, c_{is}, c_{it}\}$ is the unique triple of $\mathcal{B}$ containing the pair c_{ir}, c_{is}.

Case 3: $i \neq j$ and $r \neq s$. There is a unique triple $\{a_i, a_j, a_k\}$ of $\mathcal{B}_1$ containing the distinct pair a_i, a_j and a unique triple $\{b_r, b_s, b_t\}$ of $\mathcal{B}_2$ containing the distinct pair b_r, b_s. The triple $\{c_{ir}, c_{js}, c_{kt}\}$ is then the unique triple of $\mathcal{B}$ containing the pair c_{ir}, c_{js}.

We have thus shown that $\mathcal{B}$ is a Steiner triple system of index $\lambda = 1$ with vw varieties.   □

Example. The simplest instance in which we may apply Theorem 10.3.2 is that obtained by choosing $\mathcal{B}_1$ and $\mathcal{B}_2$ to be Steiner triple systems with 3 varieties. The result should be a Steiner triple system with $3 \times 3 = 9$ varieties.

Let $\mathcal{B}_1$ be the Steiner triple system with the 3 varieties a_1, a_2, a_3 and unique triple $\{a_1, a_2, a_3\}$, and let $\mathcal{B}_2$ be the Steiner triple system with the 3 varieties b_1, b_2, b_3 and unique triple $\{b_1, b_2, b_3\}$. We consider the set X of 9 varieties comprising the entries of the following array:

$$
\begin{array}{c}
\begin{array}{ccc} b_1 & b_2 & b_3 \end{array} \\
\begin{array}{c} a_1 \\ a_2 \\ a_3 \end{array}
\left[\begin{array}{ccc}
c_{11} & c_{12} & c_{13} \\
c_{21} & c_{22} & c_{23} \\
c_{31} & c_{32} & c_{33}
\end{array}\right].
\end{array}
$$

Following the construction in the proof of Theorem 10.3.2, we obtain the following set of 12 triples, which constitute a Steiner triple system of index 1 with 9 varieties:

(i) The entries in each of the 3 rows:

$$\{c_{11}, c_{12}, c_{13}\}, \{c_{21}, c_{22}, c_{23}\}, \{c_{31}, c_{32}, c_{33}\}.$$

(ii) The entries in each of the 3 columns:

$$\{c_{11}, c_{21}, c_{31}\}, \{c_{12}, c_{22}, c_{32}\}, \{c_{13}, c_{23}, c_{33}\}.$$

(iii) Three entries, no two from the same row or column:[26]

$$\{c_{11}, c_{22}, c_{33}\}, \{c_{12}, c_{23}, c_{31}\}, \{c_{13}, c_{21}, c_{32}\}$$

$$\{c_{13}, c_{22}, c_{31}\}, \{c_{12}, c_{21}, c_{33}\}, \{c_{11}, c_{23}, c_{32}\}.$$

If we replace $c_{11}, c_{21}, c_{31}, c_{12}, c_{22}, c_{32}, c_{13}, c_{23}, c_{33}$ by $0, 1, 2, 3, 4, 5, 6, 7, 8$, respectively, we obtain the Steiner triple system $\mathcal{B}$ with 9 varieties given earlier in this section:

$$
\begin{array}{llll}
\{0,1,2\} & \{0,3,6\} & \{0,4,8\} & \{2,4,6\} \\
\{3,4,5\} & \{1,4,7\} & \{2,3,7\} & \{1,3,8\} \\
\{6,7,8\} & \{2,5,8\} & \{1,5,6\} & \{0,5,7\}
\end{array}
\qquad (10.7)
$$

$\square$

The columns of (10.7) partition the triples of $\mathcal{B}$ into parts so that each variety occurs in exactly one triple in each part. A Steiner triple system of index $\lambda = 1$ with this property is called *resolvable* and each part is called a *resolvability class*. Note that each resolvability class is a partition of the set of varieties into triples. The notion of resolvability of Steiner triple systems arose in the following problem, first posed by Kirkman:[27]

Kirkman's schoolgirl problem: A schoolmistress takes her class of 15 girls on a daily walk. The girls are arranged in 5 rows, with 3 girls in each row, so that each girl has 2 companions. Is it possible to plan a walk for 7 consecutive days so that no girl will walk with any of her classmates in a triplet more than once?

A solution to this problem consists of $7 \times 5 = 35$ triples of the 15 girls, with each pair of girls together in exactly one triple. Moreover, it should be possible to partition the 35 triples into 7 groups of 5 triples each so that, in each group, each girl appears in exactly 1 triple. Now,

---

[26]Considering the array as a 3-by-3 board, these correspond to positions for 3 nonattacking rooks on the board!

[27]T.P. Kirkman: Note on an unanswered prize question, *Cambridge and Dublin Mathematics Journal*, 5 (1850), 255-262, and Query VI, *Lady's and Gentleman's Diary* No. 147, 48.

the number of triples of a Steiner triple system of index $\lambda = 1$ with $v = 15$ varieties is

$$b = \frac{v(v-1)}{6} = 35.$$

Thus, Kirkman's schoolgirl problem asks for a resolvable Steiner triple system of index $\lambda = 1$ with $v = 15$ varieties. The preceding example contains a solution for the Kirkman's schoolgirls problem in the case of 9 girls. In this case, there are 9 girls and arrangements for a daily walk for 4 days with each girl having different companions on all 4 days.

Example. *Solution of Kirkman's schoolgirl problem* What is required is a resolvable Steiner triple system of index $\lambda = 1$ with 15 varieties. Such a Steiner system, along with its resolution into 7 parts (one corresponding to each of the 7 days), is as follows:

$$\begin{array}{llll}
\{0,1,2\} & \{0,3,4\} & \{0,5,6\} & \{0,7,8\} \\
\{3,7,11\} & \{1,7,9\} & \{1,8,10\} & \{1,11,13\} \\
\{4,9,14\} & \{2,12,13\} & \{2,11,14\} & \{2,4,5\} \\
\{5,10,12\} & \{5,8,14\} & \{3,9,13\} & \{3,10,14\} \\
\{6,8,13\} & \{6,10,11\} & \{4,7,12\} & \{6,9,12\}
\end{array}$$

$$\begin{array}{lll}
\{0,9,10\} & \{0,11,12\} & \{0,13,14\} \\
\{1,12,14\} & \{1,3,5\} & \{1,4,6\} \\
\{2,3,6\} & \{2,8,9\} & \{2,7,10\} \\
\{4,8,11\} & \{4,10,13\} & \{3,8,12\} \\
\{5,7,13\} & \{6,7,10\} & \{5,9,11\}
\end{array}$$

$\square$

A resolvable Steiner triple system of index $\lambda = 1$ is also called a *Kirkman triple system*. Suppose $\mathcal{B}$ is a Kirkman triple system with v varieties. Since we have to be able to partition the v varieties into triples, v must be divisible by 3. Hence, by Theorem 10.3.1, in order for a Kirkman system with v varieties to exist, v must be of the form $6n + 3$. The parameters of a Kirkman system are thus of the form

$$\begin{aligned}
v &= 6n + 3, \\
b &= v(v-1)/6 = (2n+1)(3n+1), \\
k &= 3, \\
r &= (v-1)/2 = 3n + 1, \\
\lambda &= 1.
\end{aligned}$$

The number of triples in each resolvability class is

$$\frac{v}{3} = 2n + 1,$$

which fortunately is an integer. (If this number were not an integer for some n, then we would have to conclude that, for such n, a Kirkman triple system with $v = 6n + 3$ could not exist.) It was an unsolved problem for over a hundred years to determine, for each nonnegative integer n, whether there is a Kirkman triple system with $v = 6n + 3$ varieties when, in 1971, Ray-Chaudhuri and Wilson[28] showed how to construct such a system for all n.

10.4 Latin Squares

Latin squares were introduced in Section 1.5 in connection with Euler's problem of the 36 officers, and the reader may wish to review that section before proceeding. A formal definition is the following: Let n be a positive integer and let S be a set of n distinct elements. A *Latin square of order n*, based on the set S, is an n-by-n array, each of whose entries is an element of S such that each of the n elements of S occurs once (and hence exactly once) in each row and once in each column. Thus each of the rows and each of the columns of a Latin square is a permutation of the elements of S. It follows from the pigeonhole principle that we can check whether an n-by-n array based on a set S of n elements is a Latin square in *either* of two ways: (i) check that each element of S occurs at least once in each row and at least once in each column, or (ii) check that no element of S occurs more than once in each row and no more than once in each column.

The actual nature of the elements of S is of no importance and usually we take S to be $Z_n = \{0, 1, \ldots, n-1\}$. In this case, we number the rows and the columns of the Latin square as $0, 1, \ldots, n-1$, rather than the more conventional $1, 2, \ldots, n$. A 1-by-1 array is always a Latin square based on the set consisting of its unique element. Other examples of Latin squares are the following:

$$\begin{bmatrix} 0 & 1 \\ 1 & 0 \end{bmatrix}, \quad \begin{bmatrix} 0 & 1 & 2 \\ 1 & 2 & 0 \\ 2 & 0 & 1 \end{bmatrix}, \quad \begin{bmatrix} 0 & 1 & 2 & 3 \\ 1 & 2 & 3 & 0 \\ 2 & 3 & 0 & 1 \\ 3 & 0 & 1 & 2 \end{bmatrix}. \tag{10.8}$$

To confirm our stated convention, row 0 of the last square is the permutation $0, 1, 2, 3$, and row 2 is the permutation $2, 3, 0, 1$.

---

[28]D.K. Ray-Chaudhuri and R.M. Wilson: Solution of Kirkman's schoolgirl problem, *American Mathematical Society Proceedings, Symposium on Pure Mathematics*, 19 (1971), 187-204.

Consider a Latin square of order n based on Z_n, and let k be any element of Z_n. Then k occurs n times in A, once in each row and once in each column. Thinking of an n-by-n array as n-by-n board, the positions occupied by k are positions for n nonattacking rooks on an n-by-n board. Let $A(k)$ be the set of positions occupied by k's, $(k = 0, 1, \ldots, n-1)$. Then $A(0), A(1), \ldots, A(n-1)$ is a partition of the set of n^2 positions of the board. Thus, a Latin square of order n corresponds to a partition of the positions of an n-by-n array into n sets

$$A(0), A(1), \ldots, A(n-1),$$

each consisting of n positions for nonattacking rooks. This observation is readily verified in the preceding examples. Note that, if, in a Latin square, we replace, say, all the 1's with 2's and all the 2's with 1's, the result is a Latin square. The resulting partition previously described is the same, except that now the set $A(1)$ has become $A(2)$ and $A(2)$ has become $A(1)$. More generally, we can interchange $A(0), A(1), \ldots, A(n-1)$ at will and the result will always be a Latin square. There are $n!$ Latin squares that result in this way. For instance, consider the 4-by-4 Latin square A in (10.8). For this A, we have

$$A(0) = \{(0,0), (1,3), (2,2), (3,1)\} \quad A(1) = \{(0,1), (1,0), (2,3), (3,2)\}$$

$$A(2) = \{(0,2), (1,1), (2,0), (3,3)\} \quad A(3) = \{(0,3), (1,2), (2,1), (3,0)\}.$$

We obtain a new Latin square A' by letting

$$A'(0) = A(2), \quad A'(1) = A(3), \quad A'(2) = A(0), \quad A'(3) = A(1).$$

The result is

$$A' = \begin{bmatrix} 2 & 3 & 0 & 1 \\ 3 & 0 & 1 & 2 \\ 0 & 1 & 2 & 3 \\ 1 & 2 & 3 & 0 \end{bmatrix}.$$

Using this idea of interchanging the positions occupied by the various elements $0, 1, \ldots, n-1$, we can always bring a Latin square to *standard form*, whereby in row 0 the integers $0, 1, \ldots, n-1$ occur in their natural order. The three Latin squares in (10.8) are in standard form.

The three examples of Latin squares in (10.8) are instances of a general construction of a Latin square of order n coming from the addition table of the integers mod n.

Theorem 10.4.1 *Let n be a positive integer. Let A be the n-by-n array whose entry a_{ij} in row i and column j is*

$$a_{ij} = i + j \ (addition \ mod \ n), \ (i, j = 0, 1, \ldots, n - 1).$$

Then A is a Latin square of order n based on Z_n.

Proof. The Latin property of this array is a consequence of the properties of addition in Z_n. Suppose, for some row i of the array, the elements in positions in row i, column j and row i, column k are identical; that is,

$$i + j = i + k.$$

Then, adding the additive inverse $-i$ of i in Z_n to both sides, we get $j = k$, showing that there is no element repeated in row i. In a similar way one shows that there is no element repeated in any column. $\square$

The Latin square of order n constructed in Theorem 10.4.1 is nothing but the addition table of Z_n. There is a more general construction using the integers mod n that produces a wider class of Latin squares. It rests on the existence of multiplicative inverses of some elements of Z_n. (See Theorem 10.1.2.)

Example. We consider Z_5, the integers mod 5. By Theorem 10.1.2, 3 has a multiplicative inverse in Z_5; in fact, $3 \times 2 = 1$ in Z_5. Using the arithmetic of Z_5, we construct a 5-by-5 array whose entry in row i and column j is $a_{ij} = 3 \times i + j$. The result is

$$
\begin{array}{c|ccccc}
 & 0 & 1 & 2 & 3 & 4 \\
\hline
0 & 0 & 1 & 2 & 3 & 4 \\
1 & 3 & 4 & 0 & 1 & 2 \\
2 & 1 & 2 & 3 & 4 & 0 \\
3 & 4 & 0 & 1 & 2 & 3 \\
4 & 2 & 3 & 4 & 0 & 1
\end{array}
\tag{10.9}
$$

Inspection reveals that we have a Latin square of order 5. $\square$

Theorem 10.4.2 *Let n be a positive integer and let r be a nonzero integer in Z_n such that the GCD of r and n is 1. Let A be the n-by-n array whose entry a_{ij} in row i and column j is*

$$a_{ij} = r \times i + j \ (arithmetic \ mod \ n), \ (i, j = 0, 1, \ldots, n - 1).$$

Then A is a Latin square of order n based on Z_n.

Proof. The Latin property of this array follows from the properties of addition and multiplication in Z_n. Suppose, for some row i of the array, the elements in positions (i, j) and (i, k) are identical; that is,

$$r \times i + j = r \times i + k.$$

In a manner similar to that used in the proof of Theorem 10.4.1, by adding the additive inverse of $r \times i$ to both sides we conclude that $j = k$ and there is no repeated element in row i. To show that there is no repeated element in any column, we also have to use the fact that the GCD of r and n is 1. By Theorem 10.1.2, r has a multiplicative inverse r^{-1} in Z_n. Suppose that the elements in positions row i, column j and row k, column j are identical; that is,

$$r \times i + j = r \times k + j.$$

Subtracting j from both sides and rewriting, we get

$$r \times (i - k) = 0.$$

Multiplying by r^{-1}, we get $i = k$, implying that there is no repeated element in column j. Hence, A is a Latin square.   □

Theorem 10.4.1 is the special case of Theorem 10.4.2 obtained by taking $r = 1$.

The Latin square of order n constructed in Theorem 10.4.2, using an integer r with a multiplicative inverse in Z_n, will be denoted by

$$L_n^r.$$

Thus, the Latin square in (10.9) is L_5^3. If r does not have a multiplicative inverse, then the resulting array L_n^r will not be a Latin square. (See Exercise 39.)

There is another way to think of the Latin property of a Latin square. Let

$$R_n = \begin{bmatrix} 0 & 0 & \cdots & 0 \\ 1 & 1 & \cdots & 1 \\ \vdots & \vdots & \cdots & \vdots \\ n-1 & n-1 & \cdots & n-1 \end{bmatrix} \tag{10.10}$$

and

$$S_n = \begin{bmatrix} 0 & 1 & \cdots & n-1 \\ 0 & 1 & \cdots & n-1 \\ \vdots & \vdots & \vdots & \vdots \\ 0 & 1 & \cdots & n-1 \end{bmatrix} \qquad (10.11)$$

be two n-by-n arrays based on Z_n with identical columns and rows, respectively, as shown. Let A be any n-by-n array based on Z_n. Then A is a Latin square if and only if the following conditions are satisfied:

(i) When the arrays R_n and A are juxtaposed[29] to form an array $R_n \times A$, the set of ordered pairs thus obtained equals the set of *all* ordered pairs (i, j) that can be formed using the elements of Z_n;

(ii) When the arrays S_n and A are juxtaposed to form an array $S_n \times A$, the set of ordered pairs thus obtained equals the set of *all* ordered pairs (i, j) that can be formed using the elements of Z_n.

Since the juxtaposed arrays contain n^2 ordered pairs, which is exactly the number of ordered pairs that can be formed using the elements of Z_n, it follows from the pigeonhole principle that the preceding properties can be expressed by saying that the ordered pairs in $R_n \times A$ are all distinct, and the ordered pairs in $S_n \times A$ are all distinct.

Example. We illustrate the foregoing discussion with a Latin square of order 3:

$$\begin{bmatrix} 0 & 0 & 0 \\ 1 & 1 & 1 \\ 2 & 2 & 2 \end{bmatrix}, \begin{bmatrix} 0 & 1 & 2 \\ 1 & 2 & 0 \\ 2 & 0 & 1 \end{bmatrix} \rightarrow \begin{bmatrix} (0,0) & (0,1) & (0,2) \\ (1,1) & (1,2) & (1,0) \\ (2,2) & (2,0) & (2,1) \end{bmatrix},$$

$$\begin{bmatrix} 0 & 1 & 2 \\ 0 & 1 & 2 \\ 0 & 1 & 2 \end{bmatrix}, \begin{bmatrix} 0 & 1 & 2 \\ 1 & 2 & 0 \\ 2 & 0 & 1 \end{bmatrix} \rightarrow \begin{bmatrix} (0,0) & (1,1) & (2,2) \\ (0,1) & (1,2) & (2,0) \\ (0,2) & (1,0) & (2,1) \end{bmatrix}.$$

In each of the two juxtaposed arrays, each ordered pair occurs exactly once. □

We now apply the preceding ideas to two Latin squares. Let A and B be Latin squares based, for instance, on the integers in Z_n.[30]

---

[29]Corresponding entries side by side.

[30]It is not necessary that the two Latin squares be based on the same set of elements. The choice makes for convenience in the exposition.

Then A and B are called *orthogonal*, provided that in the juxtaposed array $A \times B$, each of the ordered pairs (i, j) of integers in Z_n occurs exactly once.[31] This notion of orthogonality was introduced in Section 1.5 in connection with Euler's problem of the 36 officers, where two orthogonal Latin squares of order 3 were given. It is simple to check that there do not exist two orthogonal latin squares of order 2.

Example. The following two Latin squares of order 4 are orthogonal, as is seen by examining their juxtaposed array:

$$
\begin{bmatrix} 0 & 1 & 2 & 3 \\ 1 & 0 & 3 & 2 \\ 2 & 3 & 0 & 1 \\ 3 & 2 & 1 & 0 \end{bmatrix} ,
\begin{bmatrix} 0 & 1 & 2 & 3 \\ 3 & 2 & 1 & 0 \\ 1 & 0 & 3 & 2 \\ 2 & 3 & 0 & 1 \end{bmatrix}
\rightarrow
\begin{bmatrix} (0,0) & (1,1) & (2,2) & (3,3) \\ (1,3) & (0,2) & (3,1) & (2,0) \\ (2,1) & (3,0) & (0,3) & (1,2) \\ (3,2) & (2,3) & (1,0) & (0,1) \end{bmatrix} .
$$

□

Orthogonal Latin squares have application to the design of experiments in which variational differences need to be kept at a minimum in order to be able to draw meaningful conclusions. We illustrate their use with an example from agriculture.

Example. It is desired to test the effects of various quantities of water and various types (or quantities) of fertilizer on the yield of wheat on a certain type of soil. Suppose there are n quantities of water and n types of fertilizer to be tested, so that there are n^2 possible combinations of water and fertilizer. We have at our disposal a rectangular field that is subdivided into n^2 plots, one for each of the n^2 possible water–fertilizer combinations. There is no reason to expect that soil fertility is the same throughout the field. Thus, it may very well be that the first row is of high fertility, and therefore a higher yield of wheat will occur, which is not due solely to the quantity of water and the type of fertilizer used on it. We are likely to minimize the influence of soil fertility on the yield of wheat if we insist that each quantity of water occur no more than once in any row and in any column, and similarly that each type of fertilizer occur no more than once in any row and in any column. Thus, the application of the n quantities of water on the n^2 plots should determine a Latin square A of order n, and also the application of the n types of fertilizer should determine a Latin square B of order n. Since all n^2 possible water–fertilizer combinations are to be treated, when the two Latin squares A and B are juxtaposed all n^2 combinations should occur once. Thus, the Latin squares A and B are to be orthogonal. Two orthogonal Latin squares of order n, one

---

[31] For emphasis, we repeat that, by the pigeonhole principle, we can instead say that each ordered pair occurs *at most* once.

for the application of the n quantities of water and one for the n types of fertilizer, determine a design for an experiment to test the effects of water and fertilizer on the production of wheat. The two orthogonal Latin squares of order 4 in the previous example give us a design for four quantities of water (labeled 0,1,2, and 3) and four types of fertilzer (also labeled 0,1,2, and 3). $\quad\square$.

We now extend our notion of orthogonality from two Latin squares to any number of Latin squares. Let $A_1, A_2, \ldots, A_k$ be Latin squares of order n. Without loss of generality, we assume that each of these Latin squares is based on Z_n. We say that $A_1, A_2, \ldots, A_k$ are *mutually orthogonal*, provided that each pair A_i, A_j $(i \neq j)$ of them is orthogonal. We refer to mutually orthogonal latin squares as *MOLS*. If n is a prime number, we can construct a set of $n - 1$ MOLS of order n.

Theorem 10.4.3 *Let n be a prime number. Then $L_n^1, L_n^2, \ldots, L_n^{n-1}$ are $n - 1$ MOLS of order n.*

Proof. By Corollary 10.1.3, since n is prime, each nonzero integer in Z_n has a multiplicative inverse. By Theorem 10.4.2, the arrays $L_n^1, L_n^2, \ldots, L_n^{n-1}$ are Latin squares of order n. Let r and s be distinct nonzero integers in Z_n. We show that L_n^r and L_n^s are orthogonal. Suppose, that in the juxtaposed array, $L_n^r \times L_n^s$ some ordered pair occurs twice–say, the pair in row i and column j and the pair in row k and column l are the same. Recalling the definition of the Latin squares L_n^r and L_n^s, we see that this means that

$$r \times i + j = r \times k + l \text{ and } s \times i + j = s \times k + l.$$

We rewrite these equations, obtaining

$$r \times (i - k) = (l - j) \text{ and } s \times (i - k) = (l - j);$$

hence

$$r \times (i - k) = s \times (i - k).$$

Suppose that $i \neq k$. Then $(i - k) \neq 0$ and hence has a multiplicative inverse in Z_n. Multiplying the preceding equation by $(i - k)^{-1}$, that is cancelling $(i - k)$, we get $r = s$, a contradiction. Thus, we must have $i = k$, and then, substituting into the first equation, we get $j = l$. It follows that the only way two positions in $L_n^r \times L_n^s$ can contain the same ordered pair is for the two positions to be the same position! This means that L_n^r and L_n^s are orthogonal for all $r \neq s$ and hence $L_n^1, L_n^2, \ldots, L_n^{n-1}$ are MOLS. $\quad\square$

At the end of Section 10.1, we discussed briefy the arithmetical system called a field, which satisfies the usual laws of arithmetic. We remarked that, for each prime number p and each positive integer k, there exists a field with the finite number p^k of elements (and the number of elements in a finite field is always a power of a prime). Theorems 10.4.2 and 10.4.3 generalize to each finite field. We briefly discuss this now.

Let F be a finite field with $n = p^k$ elements for some prime p and positive integer k. Let

$$\alpha_0 = 0, \alpha_1, \ldots, \alpha_{n-1}$$

be the elements of F with α_0, as indicated, the zero element of F. Consider any nonzero element α_r, $(r \neq 0)$ of F, and define an n-by-n array A such that the element a_{ij} in row i and column j of A is

$$a_{ij} = \alpha_r \times \alpha_i + \alpha_j, \quad (i, j = 0, 1, \ldots, n - 1),$$

where the arithmetic is that of the field F. Then a proof like that given for Theorem 10.4.2 (using only the usual laws of arithmetic, which, since F is a field, are satisfied) shows that A is a Latin square of order n based on the elements of F. Denote the Latin square A constructed in this way by $L_n^{\alpha_i}$. Then, following the proof of Theorem 10.4.3,[32] we find that

$$L_n^{\alpha_1}, L_n^{\alpha_2}, \ldots, L_n^{\alpha_{n-1}} \tag{10.12}$$

are $n - 1$ MOLS of order n. We summarize these facts in the next theorem.

Theorem 10.4.4 *Let $n = p^k$ be an integer that is a power of a prime number p. Then there exist $n - 1$ MOLS of order n. In fact, the $n - 1$ Latin squares (10.12) of order n constructed from a finite field with $n = p^k$ elements are $n - 1$ MOLS of order n.*

Example. We illustrate the preceding construction by obtaining three MOLS of order 4. In Section 10.1 we constructed a field with four elements. The elements of this field are

$$\alpha_0 = 0, \alpha_1 = 1, \alpha_2 = i, \alpha_3 = 1 + i.$$

---

[32] Again, only the usual laws of arithmetic were used.

Using the arithmetic of this field (the addition and multiplication tables are given in Section 10.1) we obtain the following Latin squares:

$$L_4^1 = \begin{bmatrix} 0 & 1 & i & 1+i \\ 1 & 0 & 1+i & i \\ i & 1+i & 0 & 1 \\ 1+i & i & 1 & 0 \end{bmatrix}$$

(L_4^1 is just the addition table of F.)

$$L_4^i = \begin{bmatrix} 0 & 1 & i & 1+i \\ i & 1+i & 0 & 1 \\ 1+i & i & 1 & 0 \\ 1 & 0 & 1+i & i \end{bmatrix}$$

$$L_4^{1+i} = \begin{bmatrix} 0 & 1 & i & 1+i \\ 1+i & i & 1 & 0 \\ 1 & 0 & 1+i & i \\ i & 1+i & 0 & 1 \end{bmatrix}.$$

It is straightforward to check that L_4^1, L_4^i, L_4^{1+i} are three MOLS of order 4 based on F. □

By Theorem 10.4.4, there exist $n-1$ MOLS of order n whenever n is a prime power. Is it possible to have a collection of more than $n-1$ MOLS of order n? The negative answer to this question is given in the next theorem.

Theorem 10.4.5 *Let $n \geq 2$ be an integer, and let $A_1, A_2, \ldots, A_k$ be k MOLS of order n. Then $k \leq n-1$; that is, the largest number of MOLS of order n is at most $n-1$.*

Proof. We may assume without loss of generality that each of the given Latin squares is based on the elements of Z_n. We first observe the following: Each of the Latin squares $A_1, A_2, \ldots, A_k$ can be brought to standard form and this does not affect their mutual orthogonality. This latter fact is easy to check for: If after bringing two Latin squares to standard form, their juxtaposed array had a repeated ordered pair, then the juxtaposed array must have had a repeated ordered pair to begin with. Thus, we may assume that each of $A_1, A_2, \ldots, A_k$ is in standard form. Then, for each pair A_i, A_j, the juxtaposed array $A_i \times A_j$ has first row equal to $(0,0), (1,1), \ldots, (n-1, n-1)$. Now consider the entry in the position of row 1, column 0 of each A_i. None of these entries can equal 0, since 0 is already occurring in the position directly above it in column 0. Therefore, in each of $A_1, A_2, \ldots, A_k$, the entry in row 1, column 0 is one of $1, 2, \ldots, n - 1$. Moreover, no two of $A_1, A_2, \ldots, A_k$ can have the same integer in this position. For, if A_i and A_j both had, say, r in this position, then the juxtaposed array $A_i \times A_j$ would contain the pair (r, r) twice, since it is already occurring in row 0. Thus, each of $A_1, A_2, \ldots, A_k$ contains one of the integers $1, 2, \ldots, n - 1$ in the row 1, column 0 position, and no two of them contain the same integer in this position. By the pigeonhole principle, we have $k \leq n - 1$, and the theorem is proved. $\qquad \square$

For n a positive integer, let $N(n)$ denote the largest number of MOLS of order n. We have $N(1) = 2$ because a Latin square of order 1 is orthogonal to itself.[33] Since no two Latin squares of order 2 are orthogonal, we have $N(2) = 1$. It follows from Theorems 10.4.4 and and 10.4.5 that

$$N(n) = n - 1 \text{ if } n \text{ is a prime power.}$$

It is natural to wonder whether $N(n) = n - 1$ for all integers $n \geq 2$. Unfortunately, $N(n)$ may be less than $n - 1$. (By Theorem 10.4.4, n cannot be a prime power if this happens.) The smallest integer that is not a prime power is $n = 6$, and not only do we have $N(6) \neq 5$, but we also have $N(6) = 1$; that is, there do not even exist two orthogonal Latin squares of order 6! This was verified[34] by Tarry[35] around 1900. We can use the integers mod n to show that for each odd integer n there exists a pair of MOLS of order n.

Theorem 10.4.6 $N(n) \geq 2$ *for each odd integer n.*

Proof. Let n be an odd integer. We shall show that the addition table A and the subtraction table B of Z_n are MOLS. The entry a_{ij} in row i and column j of A is $a_{ij} = i + j$ (addition mod n) and we know by Theorem 10.4.1 that A is a Latin square of order n. The entry b_{ij} in row i and column j of B is $b_{ij} = i - j$ (subtraction mod n) and we first show that B is a Latin square. This is straighforward and is like the proof of Theorem 10.4.1. Suppose that the integers in row i of B and columns j and k are the same. This means that

$$i - j = i - k.$$

---

[33] A Latin square of order $n \geq 2$ can never be orthogonal to itself.

[34] Not a trivial verification indeed!

[35] G. Tarry: Le problème de 36 officeurs, *Compte Rendu de l'Association Française pour l'Avancement de Science Naturel*, 1 (1900), 122-123 and 2 (1901), 170-203.

Adding $-i$ to both sides, we obtain $-j = -k$ and hence $j = k$. Hence, there are no repeated elements in a row and, in a similar way, one shows that there are no repeated elements in a column. Thus, B is a Latin square.

We now show that A and B are orthogonal. Suppose that in the juxtaposed array $A \times B$, some ordered pair occurs twice, say,

$$(a_{ij}, b_{ij}) = (a_{kl}, b_{kl}).$$

This means that

$$i + j = k + l \text{ and } i - j = k - l.$$

Adding and subtracting these two equations, we get

$$2i = 2k \text{ and } 2j = 2l.$$

Now, remembering that n is odd, we observe that the GCD of 2 and n is 1, and hence by Theorem 10.1.2, 2 has a multiplicative inverse 2^{-1} in Z_n. Cancelling the 2 in the preceding equations, we get $i = k$ and $j = l$. Hence, the only way $A \times B$ can have the same ordered pair in two positions is for the positions to be the same. We thus conclude that A and B are orthogonal. $\square$

There is a way to combine MOLS in order to get MOLS of larger order. The notation for carrying out and verifying this construction is a little cumbersome, since one has to deal with ordered pairs of ordered pairs. But the idea of the construction is very simple. We illustrate it by obtaining two MOLS of order 12 from two MOLS of order 3 and two MOLS of order 4. Consider the two MOLS of order 3 given by

$$A_1 = \begin{bmatrix} 0 & 1 & 2 \\ 1 & 2 & 0 \\ 2 & 0 & 1 \end{bmatrix} \qquad A_2 = \begin{bmatrix} 0 & 2 & 1 \\ 1 & 0 & 2 \\ 2 & 1 & 0 \end{bmatrix}.$$

These are the addition table and subtraction table of Z_3, respectively. Consider also the two MOLS of order 4 given by

$$B_1 = \begin{bmatrix} 0 & 1 & 2 & 3 \\ 1 & 0 & 3 & 2 \\ 2 & 3 & 0 & 1 \\ 3 & 2 & 1 & 0 \end{bmatrix} \qquad B_2 = \begin{bmatrix} 0 & 1 & 2 & 3 \\ 2 & 3 & 0 & 1 \\ 3 & 2 & 1 & 0 \\ 1 & 0 & 3 & 2 \end{bmatrix}.$$

These are the first two MOLS of order 4 constructed following Theorem 10.4.4 with i replaced by 2 and $1 + i$ replaced by 3. We now form the

12-by-12 arrays $A_1 \otimes B_1$ and $A_2 \otimes B_2$, which are defined as follows: First we replace each entry a^1_{ij} of A_1 by the 4-by-4 array

$$
(a^1_{ij}, B_1) = \begin{bmatrix}
(a^1_{ij}, b^1_{00}) & (a^1_{ij}, b^1_{01}) & (a^1_{ij}, b^1_{02}) & (a^1_{ij}, b^1_{03}) \\
(a^1_{ij}, b^1_{10}) & (a^1_{ij}, b^1_{11}) & (a^1_{ij}, b^1_{12}) & (a^1_{ij}, b^1_{13}) \\
(a^1_{ij}, b^1_{20}) & (a^1_{ij}, b^1_{21}) & (a^1_{ij}, b^1_{22}) & (a^1_{ij}, b^1_{23}) \\
(a^1_{ij}, b^1_{30}) & (a^1_{ij}, b^1_{31}) & (a^1_{ij}, b^1_{32}) & (a^1_{ij}, b^1_{33})
\end{bmatrix}.
$$

The result is the 12-by-12 array $A_1 \otimes B_1$ based on the 12 ordered pairs of integers (p, q) with p in Z_3 and q in Z_4. We obtain the 12-by-12 array $A_2 \otimes B_2$ in a similar way from A_2 and B_2. It is elementary to check that $A_1 \otimes B_1$ and $A_2 \otimes B_2$ are Latin squares, based on the set of 12 ordered pairs and that they are orthogonal. We leave this verification for the exercises. Now, in order to have these 12-by-12 arrays based on Z_{12},[36] we set up a one-to-one correspondence between Z_{12} and the ordered pairs (p, q). Any of the 12! such correspondences will do. One is the following (the one obtained by taking the ordered pairs in lexicographic order):

$$(0,0) \to 0, \quad (0,1) \to 1, \quad (0,2) \to 2, \quad (0,3) \to 3,$$

$$(1,0) \to 4, \quad (1,1) \to 5, \quad (1,2) \to 6, \quad (1,3) \to 7,$$

$$(2,0) \to 8, \quad (2,1) \to 9, \quad (2,2) \to 10, \quad (2,3) \to 11.$$

The two MOLS of order 12 obtained in this way are as follows:

$$
\begin{bmatrix}
0 & 1 & 2 & 3 & 4 & 5 & 6 & 7 & 8 & 9 & 10 & 11 \\
1 & 0 & 3 & 2 & 5 & 4 & 7 & 6 & 9 & 8 & 11 & 10 \\
2 & 3 & 0 & 1 & 6 & 7 & 4 & 5 & 10 & 11 & 8 & 9 \\
3 & 2 & 1 & 0 & 7 & 6 & 5 & 4 & 11 & 10 & 9 & 8 \\
4 & 5 & 6 & 7 & 8 & 9 & 10 & 11 & 0 & 1 & 2 & 3 \\
5 & 4 & 7 & 6 & 9 & 8 & 11 & 10 & 1 & 0 & 3 & 2 \\
6 & 7 & 4 & 5 & 10 & 11 & 8 & 9 & 2 & 3 & 0 & 1 \\
7 & 6 & 5 & 4 & 11 & 10 & 9 & 8 & 3 & 2 & 1 & 0 \\
8 & 9 & 10 & 11 & 0 & 1 & 2 & 3 & 4 & 5 & 6 & 7 \\
9 & 8 & 11 & 10 & 1 & 0 & 3 & 2 & 5 & 4 & 7 & 6 \\
10 & 11 & 8 & 9 & 2 & 3 & 0 & 1 & 6 & 7 & 4 & 5 \\
11 & 10 & 9 & 8 & 3 & 2 & 1 & 0 & 7 & 6 & 5 & 4
\end{bmatrix}.
$$

---

[36]This is, of course, not necessary. We do it only to avoid having Latin squares based on a set of elements that are ordered pairs.

$$\begin{bmatrix}
0 & 1 & 2 & 3 & 8 & 9 & 10 & 11 & 4 & 5 & 6 & 7 \\
2 & 3 & 0 & 1 & 10 & 11 & 8 & 9 & 6 & 7 & 4 & 5 \\
3 & 2 & 1 & 0 & 11 & 10 & 9 & 8 & 7 & 6 & 5 & 4 \\
1 & 0 & 3 & 2 & 9 & 8 & 11 & 10 & 5 & 4 & 7 & 6 \\
4 & 5 & 6 & 7 & 0 & 1 & 2 & 3 & 8 & 9 & 10 & 11 \\
6 & 7 & 4 & 5 & 2 & 3 & 0 & 1 & 10 & 11 & 8 & 9 \\
7 & 6 & 5 & 4 & 3 & 2 & 1 & 0 & 11 & 10 & 9 & 8 \\
5 & 4 & 7 & 6 & 1 & 0 & 3 & 2 & 9 & 8 & 11 & 10 \\
8 & 9 & 10 & 11 & 4 & 5 & 6 & 7 & 0 & 1 & 2 & 3 \\
10 & 11 & 8 & 9 & 6 & 7 & 4 & 5 & 2 & 3 & 0 & 1 \\
11 & 10 & 9 & 8 & 7 & 6 & 5 & 4 & 3 & 2 & 1 & 0 \\
9 & 8 & 11 & 10 & 5 & 4 & 7 & 6 & 1 & 0 & 3 & 2
\end{bmatrix}.$$

The preceding construction works in general, and it yields the following result.

Theorem 10.4.7 *If there is a pair of MOLS of order m and there is a pair of MOLS of order k, then there is a pair of MOLS of order mk. More generally,*

$$N(mk) \geq \min\{N(m), N(k)\}.$$

We can combine Theorem 10.4.7 with Theorem 10.4.4 in order to obtain the next result.

Theorem 10.4.8 *Let $n \geq 2$ be an integer and let*

$$n = p_1^{e_1} \times p_2^{e_2} \times \cdots \times p_k^{e_k}$$

be the factorization of n into distinct prime numbers $p_1, p_2, \ldots, p_k$. Then

$$N(n) \geq \min\{p_i^{e_i} - 1 : i = 1, 2, \ldots, k\}.$$

Proof. Using Theorem 10.4.7 and a simple induction argument on the number k of distinct prime factors of n, we get

$$N(n) \geq \min\{N(p_i^{e_i}) : i = 1, 2, \ldots, k\}.$$

By Theorem 10.4.4, we have

$$N(p_i^{e_i}) = p_i^{e_i} - 1$$

and the theorem follows.   □

Corollary 10.4.9 *Let $n \geq 2$ be an integer that is not twice an odd number. Then there exists a pair of orthogonal Latin squares of order n.*

Proof. If p is a prime number and e is a positive integer, we have $p^e - 1 \geq 2$ unless $p = 2$ and $e = 1$. Hence, by Theorem 10.4.8, we have $N(n) \geq 2$, provided that the prime factorization of n does not contain exactly one 2, that is, provided n is not twice an odd number.   □

The integers n for which Corollary 10.4.9 does *not* guarantee the existence of a pair of MOLS of order n are the integers

$$2, 6, 10, 14, 18, \ldots, 4k + 2, \ldots. \tag{10.13}$$

We have already remarked that there do not exist pairs of MOLS of order 2 and of order 6. Thus, the first undecided n is $n = 10$. It was conjectured by Euler in 1782 that for *no* integer n in the sequence (10.13) does there exist a pair of MOLS of order n. The combined efforts of Bose, Shrikhande, and Parker[37] succeeded in showing that Euler's conjecture holds only for $n = 2$ and $n = 6$; that is, except for 2 and 6 for each integer n in the sequence (10.13), there exists a pair of MOLS of order n. We do not prove this result, but the following is a pair of MOLS of order 10 constructed by Parker[38] in 1959:

$$
\begin{bmatrix}
0 & 6 & 5 & 4 & 7 & 8 & 9 & 1 & 2 & 3 \\
9 & 1 & 0 & 6 & 5 & 7 & 8 & 2 & 3 & 4 \\
8 & 9 & 2 & 1 & 0 & 6 & 7 & 3 & 4 & 5 \\
7 & 8 & 9 & 3 & 2 & 1 & 0 & 4 & 5 & 6 \\
1 & 7 & 8 & 9 & 4 & 3 & 2 & 5 & 6 & 0 \\
3 & 2 & 7 & 8 & 9 & 5 & 4 & 6 & 0 & 1 \\
5 & 4 & 3 & 7 & 8 & 9 & 6 & 0 & 1 & 2 \\
2 & 3 & 4 & 5 & 6 & 0 & 1 & 7 & 8 & 9 \\
4 & 5 & 6 & 0 & 1 & 2 & 3 & 9 & 7 & 8 \\
6 & 0 & 1 & 2 & 3 & 4 & 5 & 8 & 9 & 7
\end{bmatrix}
$$

[37]R.C. Bose, S.S. Shrikhande, and E.T. Parker: Further results on the construction of mutually orthogonal Latin squares and the falsity of Euler's conjecture, *Canadian J. Math.*, 12 (1960), 189-203. See also the account written by Martin Gardner in his Mathematical Games column in the *Scientific American* (November, 1959).

[38]E.T. Parker: Orthogonal Latin squares, *Proc. Nat. Acad. Sciences*, 45 (1959), 859-862.

$$\begin{bmatrix}
0 & 9 & 8 & 7 & 1 & 3 & 5 & 2 & 4 & 6 \\
6 & 1 & 9 & 8 & 7 & 2 & 4 & 3 & 5 & 0 \\
5 & 0 & 2 & 9 & 8 & 7 & 3 & 4 & 6 & 1 \\
4 & 6 & 1 & 3 & 9 & 8 & 7 & 5 & 0 & 2 \\
7 & 5 & 0 & 2 & 4 & 9 & 8 & 6 & 1 & 3 \\
8 & 7 & 6 & 1 & 3 & 5 & 9 & 0 & 2 & 4 \\
9 & 8 & 7 & 0 & 2 & 4 & 6 & 1 & 3 & 5 \\
1 & 2 & 3 & 4 & 5 & 6 & 0 & 7 & 8 & 9 \\
2 & 3 & 4 & 5 & 6 & 0 & 1 & 8 & 9 & 7 \\
3 & 4 & 5 & 6 & 0 & 1 & 2 & 9 & 7 & 8
\end{bmatrix}.$$

For nearly 200 years, 10 was the smallest undecided case of Euler's conjecture.

By Theorem 10.4.5, for each integer $n \geq 2$, we have $N(n) \leq n - 1$, and by Theorem 10.4.4, we have equality if n is a power of a prime. There are no other known values of n for which $N(n) = n - 1$. We establish a connection between $n - 1$ MOLS of order n and the block designs of Section 10.2. Let $A_1, A_2, \ldots, A_{n-1}$ denote $n - 1$ MOLS of order n. We use the $n + 1$ arrays

$$R_n, \ S_n, \ A_1, \ A_2, \ \ldots, \ A_{n-1}, \tag{10.14}$$

where R_n and S_n are defined in (10.10) and (10.11), to construct a block design $\mathcal{B}$ with parameters

$$b = n^2 + n, \ v = n^2, \ k = n, \ r = n + 1, \ \lambda = 1.$$

Recall that $A_i(k)$ denotes the set of positions of A_i occupied by k, $(k = 0, 1, \ldots, n - 1)$. Since A_i is a Latin square, $A_i(k)$ contains one position from each row and each column; in particular, no two positions in $A_i(k)$ belong to the same row or to the same column. We also use this notation for R_n and S_n. For instance, $R_n(0)$ denotes the set of positions of R_n that are occupied by 0's, and this set is the set of positions of row 0, and $S_n(1)$ denotes the set of positions of S_n that are occupied by 1's and this is the set of positions of column 1.

We take the set X of varieties to be the set of $v = n^2$ positions of an n-by-n array; that is,

$$X = \{(i, j) : i = 0, 1, \ldots, n - 1; j = 0, 1, \ldots, n - 1\}.$$

Each of the $n + 1$ arrays in (10.14) determines n blocks:

$$R_n(0) \ \ R_n(1) \ \ \ldots \ \ R_n(n - 1) \tag{10.15}$$

$$S_n(0) \quad S_n(1) \quad \cdots \quad S_n(n-1) \tag{10.16}$$

$$
\begin{array}{cccc}
A_1(0) & A_1(1) & \cdots & A_1(n-1) \\
\vdots & \vdots & \cdots & \vdots \\
A_{n-1}(0) & A_{n-1}(1) & \cdots & A_{n-1}(n-1).
\end{array}
\tag{10.17}
$$

Thus, we have $b = n \times (n+1) = n^2 + n$ blocks, each containing $k = n$ varieties. Let $\mathcal{B}$ denote this collection of blocks. In order to be able to conclude that $\mathcal{B}$ is a BIBD with the specified parameters, we need only check that each pair of varieties occur together in exactly $\lambda = 1$ block. There are three possibilities to consider:

(i) Two varieties in the same row: These are together in precisely one of the blocks in (10.15) and in no other blocks.

(ii) Two varieties in the same column: These are together in precisely one of the blocks in (10.16) and in no other blocks.

(iii) Two varieties (i, j) and (p, q) belonging to different rows and to different columns: These two varieties are not together in any of the blocks in (10.15) and (10.16). Suppose that they are together in blocks $A_r(e)$ and $A_s(f)$. This means that there is an e in positions row i, column j and row p, column q of A_r and an f in the same positions of A_s. If $r \neq s$, then, in the juxtaposed array $A_r \times A_s$, the ordered pair (e, f) appears twice, contradicting the orthogonality of A_r and A_s. Thus, $r = s$, which implies that A_r has both an e and an f in positions row i, column j and row p, column q. We conclude that also $e = f$. Hence, $A_r(e)$ and $A_s(f)$ are the same block, and we now conclude that (i, j) and (p, q) are together in *at most* one block.

At this point, we know that each pair of varieties are together in, at most, one block. This is now enough for us to conclude that each pair of varieties are together in exactly one block. This follows by a counting argument similar to one we have made in Section 10.2: There are n^2 varieties, we can form $n^2(n^2 - 1)/2$ pairs of them, and we know that each pair is in, at most, one of the $n^2 + n$ blocks. Each block has n varieties and thus contains $n(n-1)/2$ pairs. For all blocks, this gives a total of

$$(n^2 + n) \times \frac{n(n-1)}{2} = \frac{n^2(n^2 - 1)}{2}$$

pairs, which is exactly the total number of pairs of varieties. Hence, by the pigeonhole principle, each pair of varieties must be in exactly one block. Thus, $\mathcal{B}$ is a BIBD of index $\lambda = 1$.

We note that the design $\mathcal{B}$ constructed is *resolvable* in the sense used in Section 10.2 for Steiner systems. The collection of $n^2 + n$ blocks is partitioned into $n + 1$ parts (*resolvability classes*) of n blocks each (see (10.15), (10.16), and (10.17)), and each resolvability class is a partition of the n^2 varieties.

Example. We illustrate the preceding construction of a BIBD using the two Latin squares of order 3:

$$A_1 = \begin{bmatrix} 0 & 1 & 2 \\ 1 & 2 & 0 \\ 2 & 0 & 1 \end{bmatrix} \qquad A_2 = \begin{bmatrix} 0 & 2 & 1 \\ 1 & 0 & 2 \\ 2 & 1 & 0 \end{bmatrix}.$$

The varieties are the 9 positions of a 9-by-9 array, and the blocks are pictured geometrically by resolvability classes as follows:

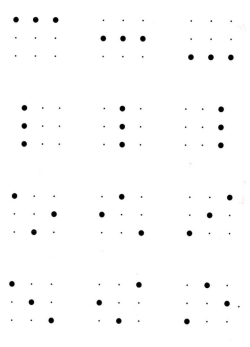

If we think of the varieties as *points* and the blocks as *lines*, and, as usual, call two lines *parallel*, provided that they have no point in common, then each of the preceding displays (the resolvability classes) consists of 3 parallel lines. Each pair of varieties being together in exactly one block translates to two points determining exactly one line. The resolvability of the design also translates to the property that, given a line and a point not on it, there is exactly one line parallel to the first containing the given point. This is the so-called *parallel postulate* of Euclidean geometry.

Theorem 10.4.10 *Let $n \geq 2$ be an integer. If there exist $n-1$ MOLS of order n, then there exists a resolvable BIBD with parameters*

$$b = n^2 + n, v = n^2, \ k = n, \ r = n+1, \ \lambda = 1. \qquad (10.18)$$

Conversely, if there exists a resolvable BIBD with parameters (10.18), then there exist $n-1$ MOLS of order n.

Proof. Previously, we showed how to construct a resolvable BIBD with parameters (10.18) from $n-1$ MOLS of order n. This process can be reversed. We outline how, and leave some of the details to be checked for the exercises. Suppose we have a resolvable BIBD $\mathcal{B}$ with parameters (10.18). Since there are n^2 varieties and each block contains n varieties, each resolvability class contains n blocks. Moreover, since there are $n^2 + n$ blocks, there are $n + 1$ resolvability classes

$$\mathcal{B}_1, \mathcal{B}_2, \ldots, \mathcal{B}_{n+1}.$$

We use two of the resolvability classes $\mathcal{B}_n$ and $\mathcal{B}_{n+1}$ in order to "coordinatize" the varieties. Let the blocks in $\mathcal{B}_n$ be

$$H_0, H_1, \ldots, H_{n-1}$$

and let the blocks in $\mathcal{B}_{n+1}$ be

$$V_0, V_1, \ldots, V_{n-1}.$$

(H is for *horizontal* and V is for *vertical*.) Given any variety x, there is a unique i between 0 and $n-1$ such that x is in H_i and a unique j between 0 and $n-1$ such that x is in V_j. This gives an ordered pair of coordinates (i, j) to each variety x. Moreover, since the index λ equals 1, two different varieties do not get the same coordinates (if x and y both had coordinates (i, j), then x and y would be together in the two blocks H_i and V_j). We may now think of the set X of varieties as the coordinate pairs themselves:[39]

$$X = \{(i, j) : i = 0, 1, \ldots, n-1; j = 0, 1, \ldots, n-1\}.$$

Now consider any other resolvability class $\mathcal{B}_p$, ($p = 0, 1, \ldots, n-1$). Let the blocks in $\mathcal{B}_p$ be labeled

$$A_p(0), \ A_p(1), \ \ldots, \ A_p(n-1).$$

---

[39]We make a similar identification in analytic geometry when we give the points of the plane coordinates and the coordinates "become" the points.

These blocks partition X into n sets of size n. Also, as the notation suggests, let A_p be the n-by-n array that has a k in each position of $A_p(k)$. If, for instance, there were two k's in row i of A_p, this would imply that there are two varieties (i, a) and (i, b) that are in both of the blocks H_i and $A_i(k)$. Thus, A_p is a Latin square. Moreover, for $p \neq q$, A_p and A_q are orthogonal: If the juxtaposed array $A_p \times A_q$ contained the same ordered pair in both positions row i, column j and row u, column v, then the two varieties (i, j) and (u, v) would be in two blocks. Hence, $A_1, A_2, \ldots, A_{n-1}$ are MOLS of order n. $\qquad \square$

We conclude this section with some questions that naturally arise when one attempts to construct a Latin square.

There are three natural ways to construct a Latin square of order n:

1. row-by-row,

2. column-by-column, and

3. element-by-element.

The first two ways are quite similar, and we consider only the first.

To construct a Latin square row-by-row means to put in one complete row at a time. Thus, we can construct a Latin square of order 3 by first choosing a permutation of $\{0, 1, 2\}$ for row 0, say, 2,1,0, then a permutation for row 1 (which doesn't give a repeated integer in any column), say, 0,2,1, and then choosing a permutation for row 2, say, 1,0,2 (actually, if one knows all but the last row of a Latin square, then the last row can be filled in uniquely because we must put in each column the integer that is not yet there). The result is

$$\begin{bmatrix} 2 & 1 & 0 \\ 0 & 2 & 1 \\ 1 & 0 & 2 \end{bmatrix}.$$

Will we ever get stuck if we construct a Latin square in this way, at each step choosing an allowable permutation for the next row?

To construct a Latin square element-by-element means to put in all the occurrences of each of the elements, one element at a time. Thus, we could have constructed the preceding Latin square of order 3 by first choosing 3 positions for the 0's (three positions for nonattacking rooks), then 3 positions for the 1's, and finally three positions for the 2's, (as in the row-by-row construction, the last step is uniquely determined).

Will we ever get stuck if we construct a Latin square in this way, at each step choosing the positions for the next integer?

We show that Theorem 9.2.5 of the last chapter allows us to answer both of these questions.[40] First, we make a definition that is suggested by the first question.

Let m and n be integers with $m \leq n$. An m-by-n *Latin rectangle*, based on the integers in Z_n, is an m-by-n array such that no integer is repeated in any row or in any column. Each of the rows of an m-by-n Latin rectangle is a permutation of $\{0, 1, \ldots, n - 1\}$, and no column contains a repeated integer. If $m = n$, then our definition of a Latin rectangle is equivalent to that of a Latin square.[41] An example of a 3-by-5 Latin rectangle is

$$\begin{bmatrix} 0 & 1 & 2 & 3 & 4 \\ 3 & 4 & 0 & 1 & 2 \\ 4 & 0 & 3 & 2 & 1 \end{bmatrix}.$$

We say that an m-by-n Latin rectangle L can be *completed*, provided it is possible to attach $n - m$ rows to L and obtain a Latin square L^* of order n. Such a Latin square L^* is called a *completion* of L. For example, a completion of the previous Latin rectangle L is

$$\begin{bmatrix} 0 & 1 & 2 & 3 & 4 \\ 3 & 4 & 0 & 1 & 2 \\ 4 & 0 & 3 & 2 & 1 \\ 2 & 3 & 1 & 4 & 0 \\ 1 & 2 & 4 & 0 & 3 \end{bmatrix}$$

The answer to our first question is a consequence of the next theorem.

Theorem 10.4.11 *Let L be an m-by-n Latin rectangle based on Z_n with $m < n$. Then L has a completion.*

Proof. It suffices to show that we can adjoin one new row to L to get an $(m + 1)$-by-n Latin rectangle because then we can proceed inductively. We define a bipartite graph $G = (X, \Delta, Y)$ as follows: The set of left vertices is $X = \{x_0, x_1, \ldots, x_{n-1}\}$, which we think of as corresponding to columns $0, 1, \ldots, n - 1$; the set of right vertices is $Y = \{0, 1, \ldots, n - 1\}$, which of course are the elements on which L is based. There is an edge $\{x_i, j\}$ in Δ joining vertex x_i and vertex j if

---

[40]Letting the "cat out of the bag," we never get stuck.
[41]The pigeonhole principle again!

and only if j does not occur in column i of L. Thus, x_i is joined by an edge to all those integers which are candidates for the element in column i of the new row. Since L is an m-by-n Latin rectangle, column i contains m different integers and hence there are $n-m$ candidates for position i of the new row. Moreover, each row of L is a permutation of $\{0, 1, \ldots, n-1\}$, and hence each integer j occurs m times in L in m different columns and hence is a candidate for $n-m$ columns. What this means is that G is regular of degree $n-m \geq 1$. By Theorem 9.2.5, G has a perfect matching. Suppose the edges of a perfect matching are

$$\{x_0, i_0\}, \{x_1, i_1\}, \ldots, \{x_{n-1}, i_{n-1}\}.$$

Then $i_0, i_1, \ldots, i_{n-1}$ is a permutation of $\{0, 1, \ldots, n-1\}$. The $(m+1)$-by-n array obtained by adjoining

$$i_0 \quad i_1 \quad \cdots \quad i_{n-1}$$

as a new row is a Latin rectangle.[42]   □

The following definition is motivated by our second question. Consider an n-by-n array L in which some positions are unoccupied and other positions are occupied by one of the integers $\{0, 1, \ldots, n-1\}$. Suppose that, if an integer k occurs in L, then it occurs n times and no two k's belong to the same row or column. Then we call L a *semi-Latin square*. If m different integers occur in L, then we say L has *index m*. An example of a semi-Latin square of order 5 and index 3 is

$$\begin{bmatrix} 1 & & 0 & & 2 \\ & 2 & 1 & & 0 \\ 0 & 1 & & 2 & \\ 2 & 0 & & 1 & \\ & & 2 & 0 & 1 \end{bmatrix}.$$

We can think of this example as a 5-by-5 board on which there are 5 red nonattacking rooks (the 0's), 5 white nonattacking rooks (the 1's), and 5 blue nonattacking rooks (the 2's). What we seek are positions for 5 green nonattacking rooks and 5 yellow nonattacking rooks on this board. If we think of 3 as green and 4 as yellow, then a solution is

---

[42]By Exercise 8 of Chapter 9, the edges of G can be partitioned into $n-m$ perfect matchings, and these perfect matchings show how to complete L to a Latin square of order n all at once.

given by

$$\begin{bmatrix} 1 & 4 & 0 & 3 & 2 \\ 3 & 2 & 1 & 4 & 0 \\ 0 & 1 & 4 & 2 & 3 \\ 2 & 0 & 3 & 1 & 4 \\ 4 & 3 & 2 & 0 & 1 \end{bmatrix}.$$

We say that a semi-Latin square L of order n can be *completed* to a Latin square, provided that it is possible to fill in the unoccupied positions in order to obtain a Latin square $L^{\#}$ of order n. Such a Latin square $L^{\#}$ is called a *completion* of L. The answer to our second question is a consequence of the final theorem of this chapter.

Theorem 10.4.12 *Let L be a semi-Latin square of order n and index m where $m < n$. Then L has a completion.*

Proof. Suppose the integers that occur in L are $0, 1, \ldots, m-1$. It suffices to show that we can find n unoccupied positions to put m to get a Latin square of order n of index $m + 1$, because then we can proceed inductively. We define a bipartite graph $G = (X, \Delta, Y)$ again. The set of left vertices is $X = \{x_0, x_1, \ldots, x_{n-1}\}$, which we think of as corresponding to rows $0, 1, \ldots, n-1$; the set of right vertices is $Y = \{y_0, y_1, \ldots, y_{n-1}\}$, which we think of as correponding to the columns. There is an edge $\{x_i, y_j\}$ in Δ joining vertex x_i and vertex y_j if and only if the position at row i, column j is unoccupied in L. Since each of the integers $0, 1, \ldots, m-1$ occurs once in each row and once in each column of L, G is regular of degree $n - m$. By Theorem 9.2.1, again, G has a perfect matching and this perfect matching identifies the desired positions for m. $\square$

The similarity between Theorems 10.4.11 and 10.4.12 is not accidental. There is a one-to-one correspondence between m-by-n Latin rectangles and semi-Latin squares of order n and index m that transforms the proof of Theorem 10.4.11 into that of Theorem 10.4.12 and vice versa. This correspondence is the following: Let L be an m-by-n Latin rectangle (based on Z_n) and let the entry in position row i, column j be denoted by a_{ij}. We define an n-by-n array B by letting the entry b_{ij} in position row i, column j be k, provided that i occurs in column j of row k of L. Thus,

$$b_{ij} = k \text{ if and only if } a_{kj} = i.$$

Some positions in B are unoccupied since, if $m < n$, some integers are missing in the columns of L. We leave it as an exercise to show that the array B constructed from L in this way is a semi-Latin square of index m.

Example. Consider the 3-by-5 Latin rectangle

$$A = \begin{bmatrix} 0 & 1 & 2 & 3 & 4 \\ 3 & 4 & 1 & 0 & 2 \\ 1 & 0 & 4 & 2 & 3 \end{bmatrix}.$$

Then, following the preceding construction, we obtain the semi-Latin square B of order 5 and index 3:

$$B = \begin{bmatrix} 0\ 2 & & 1 & & \\ 2\ 0 & 1 & & & \\ & 0 & 2\ 1 & & \\ 1 & & 0\ 2 & & \\ & 1\ 2 & & 0 & \end{bmatrix}.$$

$\square$

10.5 Exercises

1. Compute the addition table and the multiplication table for the integers mod 4.

2. Compute the subtraction table for the integers mod 4. How does it compare with the addition table computed in Exercise 1?

3. Compute the addition table and the multiplication table for the integers mod 5.

4. Compute the subtraction table of the integers mod 5. How does it compare with the addition table computed in Exercise 3?

5. Prove that no two integers in Z_n, arithmetic mod n, have the same additive inverse. Conclude from the pigeonhole principle that

$$\{-0, -1, -2, \ldots, -(n-1)\} = \{0, 1, 2, \ldots, n-1\}.$$

(Remember that $-a$ is the integer which, when added to a in Z_n, gives 0.)

6. Prove that the columns of the subtraction table of Z_n are a rearrangement of the columns of the addition table of Z_n (Cf. Exercises 2 and 4).

7. Compute the addition table and multiplication table for the integers mod 6.

8. Determine the additive inverses of the integers in Z_8, with arithmetic mod 8.

9. Determine the additive inverses of 3, 7, 8, and 19 in the integers mod 20.

10. Determine which integers in Z_{12} have multiplicative inverses, and find the multiplicative inverses when they exist.

11. For each of the following integers in Z_{24}, determine the multiplicative inverse if a multiplicative inverse exists:

$$4, \quad 9, \quad 11, \quad 15, \quad 17, \quad 23.$$

12. Prove that $n-1$ always has a multiplicative inverse in Z_n, $(n \geq 2)$.

13. Let $n = 2m + 1$ be an odd integer with $m \geq 2$. Prove that the multiplicative inverse of $m + 1$ in Z_n is 2.

14. Use the algorithm in Section 10.1 to find the GCD of the following pairs of integers:

    (i) 12 and 31
    (ii) 24 and 82
    (iii) 26 and 97
    (iv) 186 and 334
    (v) 423 and 618

15. For each of the pairs of integers in Exercise 14, let m denote the first integer and let n denote the second integer of the pair. When it exists, determine the multiplicative inverse of m in Z_n.

16. Apply the algorithm for the GCD in Section 10.1 to 15 and 46, and then use the results to determine the multiplicative inverse of 15 in Z_{46}.

17. Start with the field Z_2 and show that $x^3 + x + 1$ cannot be factored in a nontrivial way (into polynomials with coefficients in Z_2), and then use this polynomial to construct a field with $2^3 = 8$ elements. Let i be the root of this polynomial adjoined to Z_2, and then do the following computations:

(i) $(1+i) + (1+i+i^2)$

(ii) $(1+i^2) + (1+i^2)$

(iii) i^{-1}

(iv) $i^2 \times (1+i+i^2)$

(v) $(1+i)(1+i+i^2)$

(vi) $(1+i)^{-1}$

18. Does there exist a BIBD with parameters $b = 10$, $v = 8$ $r = 5$, and $k = 4$?

19. Does there exist a BIBD whose parameters satisfy $b = 20$, $v = 18$, $k = 9$, and $r = 10$?

20. Let $\mathcal{B}$ be a BIBD with parameters b, v, k, r, λ whose set of varieties is $X = \{x_1, x_2, \ldots, x_v\}$ and whose blocks are $B_1, B_2, \ldots, B_b$. For each block B_i, let $\overline{B_i}$ denote the set of varieties which do *not* belong to B_i. Let $\mathcal{B}^c$ be the collection of subsets $\overline{B_1}, \overline{B_2}, \ldots, \overline{B_b}$ of X. Prove that $\mathcal{B}^c$ is a block design with parameters

$$ b' = b, \ v' = v, \ k' = v - k, \ r' = b - r, \ \lambda' = b - 2r + \lambda, $$

provided that we have $b - 2r + \lambda > 0$. The BIBD $\mathcal{B}^c$ is called the *complementary design* of $\mathcal{B}$.

21. Determine the complementary design of the BIBD with parameters $b = v = 7, k = r = 3, \lambda = 1$ in Section 10.2.

22. Determine the complementary design of the BIBD with parameters $b = v = 16, k = r = 6, \lambda = 2$ given in Section 10.2.

23. How are the incidence matrices of a BIBD and its complement related?

24. Show that a BIBD, with v varieties whose block size k equals $v - 1$, does not have a complementary design.

25. Prove that a BIBD with parameters b, v, k, r, λ has a complementary design if and only if $2 \leq k \leq v - 2$ (Cf. Exercises 20 and 24).

26. Let B be a difference set in Z_n. Show that, for each integer k in Z_n, $B+k$ is also a difference set. (This implies that we can always assume without loss of generality that a difference set contains 0 for, if it did not, we can replace it by $B + k$, where k is the additive inverse of any integer in B.)

27. Prove that Z_v is itself a difference set in Z_v. (These are *trivial difference sets*.)

28. Show that $B = \{0, 1, 3, 9\}$ is a difference set in Z_{13}, and use this difference set as a starter block to construct an SBIBD. Identify the parameters of the block design.

29. Is $B = \{0, 2, 5, 11\}$ a difference set in Z_{12}?

30. Show that $B = \{0, 2, 3, 4, 8\}$ is a difference set in Z_{11}. What are the parameters of the SBIBD developed from B?

31. Prove that $B = \{0, 3, 4, 9, 11\}$ is a difference set in Z_{21}.

32. Use Theorem 10.3.2 to construct a Steiner triple system of index 1 having 21 varieties.

33. Let t be a positive integer. Use Theorem 10.3.2 to prove that there exists a Steiner triple system of index 1 having 3^t varieties.

34. Let t be a positive integer. Prove that, if there exists a Steiner triple system of index 1 having v varieties, then there exists a Steiner triple system having v^t varieties (Cf. Exercise 33).

35. Assume a Steiner triple system exists with parameters b, v, k, r, λ where $k = 3$. Let a be the remainder when λ is divided by 6. Use Theorem 10.3.1 to show the following:

    (i) If $a = 1$ or 5, then v has remainder 1 or 3 when divided by 6.

    (ii) If $a = 2$ or 4, then v has remainder 0 or 1 when divided by 3.

    (iii) If $a = 3$, then v is odd.

36. Verify that the following three steps construct a Steiner triple system of index 1 with 13 varieties (we begin with Z_{13}).

    (i) Each of the integers $1, 3, 4, 9, 10, 12$ occurs exactly once as a difference of two integers in $B_1 = \{0, 1, 4\}$.

    (ii) Each of the integers $2, 5, 6, 7, 8, 11$ occurs exactly once as a difference of two integers in $B_2 = \{0, 2, 7\}$.

    (iii) The 12 blocks developed from B_1 together with the 12 blocks developed from B_2 are the blocks of a Steiner triple system of index 1 with 13 varieties.

37. Prove that, if we interchange the rows of a Latin square in any way and interchange the columns in any way, the result is always a Latin square.

38. Use Theorem 10.4.2 with $n = 6$ and $r = 5$ to construct a Latin square of order 6.

39. Let n be a positive integer and let r be a nonzero integer in Z_n such that the GCD of r and n is not 1. Prove that the array constructed using the prescription in Theorem 10.4.2 is not a Latin square.

40. Let n be a positive integer and let r and r' be distinct nonzero integers in Z_n such that the GCD of r and n is 1 and the GCD of r' and n is 1. Show that the Latin squares constructed by using Theorem 10.4.2 need not be orthogonal.

41. Use Theorem 10.4.2 with $n = 8$ and $r = 3$ to construct a Latin square of order 8.

42. Construct 4 MOLS of order 5.

43. Construct 3 MOLS of order 7.

44. Construct 2 MOLS of order 9.

45. Construct 2 MOLS of order 15.

46. Construct 2 MOLS of order 8.

47. Let A be a Latin square of order n for which there exists a Latin square B of order n such that A and B are orthogonal. B is called an *orthogonal mate* of A. Think of the 0's in A as rooks of color red, the 1's as rooks of color white, the 2's as rooks of color blue, and so on. Prove that there are n nonattacking rooks in A, no two of which have the same color. Indeed, prove that the entire set of n^2 rooks can be partitioned into n sets of n nonattacking rooks each, with no two rooks in the same set having the same color.

48. Prove that the addition table of Z_4 is a Latin square without an orthogonal mate (Cf. Exercise 47).

49. First construct 4 MOLS of order 5, and then construct the resolvable BIBD corresponding to them as given in Theorem 10.4.10.

50. Let A_1 and A_2 be MOLS of order m and let B_1 and B_2 be MOLS of order n. Prove that $A_1 \otimes B_1$ and $A_2 \otimes B_2$ are MOLS of order mn.

51. Fill in the details in the proof of Theorem 10.4.10.

52. Construct a completion of the 3-by-6 Latin rectangle

$$\begin{bmatrix} 0 & 1 & 2 & 3 & 4 & 5 \\ 4 & 3 & 1 & 5 & 2 & 0 \\ 5 & 4 & 3 & 0 & 1 & 2 \end{bmatrix}.$$

53. Construct a completion of the 3-by-7 Latin rectangle

$$\begin{bmatrix} 0 & 1 & 2 & 3 & 4 & 5 & 6 \\ 2 & 3 & 0 & 6 & 5 & 4 & 1 \\ 1 & 4 & 6 & 0 & 2 & 3 & 5 \end{bmatrix}.$$

54. How many 2-by-n Latin rectangles have first row equal to

$$0 \quad 1 \quad 2 \quad \cdots \quad n-1 \ ?$$

55. Construct a completion of the semi-Latin square

$$\begin{bmatrix} & 2 & 0 & & & 1 & \\ 2 & 0 & & & & 1 & \\ 0 & & & 2 & 1 & & \\ & & & 1 & 2 & & 0 \\ & & 1 & & & 0 & 2 \\ 1 & & & & & 0 & 2 \end{bmatrix}.$$

56. Construct a completion of the semi-Latin square

$$\begin{bmatrix} 0 & 2 & 1 & & & & 3 \\ 2 & 0 & & 1 & & 3 & \\ 3 & & 0 & 2 & 1 & & \\ & 3 & 2 & 0 & & 1 & \\ & & 3 & & 0 & 2 & 1 \\ 1 & & & & 3 & 0 & 2 \\ & 1 & & 3 & 2 & & 0 \end{bmatrix}.$$

57. Let $n \geq 2$ be an integer. Prove that an $(n-2)$-by-n Latin rectangle has at least 2 completions, and, for each n, find an example that has exactly 2 completions.

58. A Latin square A of order n is *symmetric*, provided the entry a_{ij} at row i, column j equals the entry a_{ji} at column j, row i for all $i \neq j$. Prove that the addition table of Z_n is a symmetric Latin square.

59. A Latin square of order n (based on Z_n) is *idempotent*, provided that its entries on the diagonal running from upper left to lower right are $0, 1, 2, \ldots, n - 1$.

    (i) Construct an example of an idempotent Latin square of order 5.

    (ii) Construct an example of a symmetric, idempotent Latin square of order 5.

60. Prove that a symmetric, idempotent Latin square has odd order.

61. Let $n = 2m + 1$, where m is a positive integer. Prove that the n-by-n array A whose entry a_{ij} in row i, column j satisfies

$$a_{ij} = (m + 1) \times (i + j) \text{ (arithmetic mod } n)$$

is a symmetric, idempotent Latin square of order n. [Remark: The integer $m + 1$ is the multiplicative inverse of 2 in Z_n. Thus, our prescription for a_{ij} is to "average" i and j.]

62. Let L be an m-by-n Latin rectangle (based on Z_n) and let the entry in row i, column j be denoted by a_{ij}. We define an n-by-n array B whose entry b_{ij} in position row i, column j satisfies

$$b_{ij} = k, \text{ provided } a_{kj} = i$$

and is blank otherwise. Prove that B is a semi-Latin square of order n and index m. In particular, if A is a Latin square of order n so is B.

Chapter 11

Introduction to Graph Theory

Take a street map of your favorite city[1] and put a bold dot • at each place where two or more streets come together or at a dead-end street. What you get is an example of what is called a (combinatorial) *graph.* Most likely, some of the streets in your favorite city are one-way streets, which permit traffic in only one direction. Put an arrow ($\rightarrow$) on each one-way street, which indicates the permitted direction of traffic flow, and a double arrow ($\leftrightarrow$) on two-way streets. You now have an example of what is called a *directed graph*, or *digraph.* Now consider the people in your favorite city. Run a string between each pair of people that like each other. You have another example of a graph. Recognizing the fact that sometimes one's fondness for another person is not always reciprocated, you may have to put arrows on your strings as you did for streets, with the result being a digraph. Now take your favorite chemical molecule,[2] made up of atoms, some of which are chemically bound to others. You've got another graph, with the bonds playing the role of the streets or strings. Finally, consider all the different types of animals, insects, and plants that inhabit your favorite city. Put an arrow from one type to another, provided the first preys on the second. This time you get a digraph. Two species may share a common prey. Putting a string between each pair that does, you get a graph which that displays competition between species.

As the preceding discussion suggests, graphs and digraphs provide mathematical models for a set of objects that are related or bound

---

[1]Mine is Madison, Wisconsin.
[2]Play along, and suppose you do have a favorite chemical molecule!

together in some way or other. The first paper on graph theory was written by the famous Swiss mathematician Leonhard Euler, in 1736, and dealt with the well-known Königsberg bridge problem. Graph theory has its historic roots in puzzles and games, but today it provides a natural and very important language and framework for investigations in many disciplines, such as networks, chemistry, psychology, social science, ecology, and genetics. Graphs are also some of the most useful models in computer science, since many questions that arise there can be most easily expressed, investigated, and solved by graph algorithms. In Chapter 9 we discussed a particular class of graphs, called *bipartite graphs*, as they pertain to matchings. The current chapter does not depend on Chapter 9 in any significant way, and each can be read independently of the other. We treat digraphs in Chapter 12.

11.1 Basic properties

A *graph G* (also called a *simple graph*) is composed of two types of objects. It has a finite set

$$V = \{a, b, c, \ldots\}$$

of elements called *vertices* (sometimes also called *nodes*) and a set E of pairs of distinct vertices called *edges*. We denote the graph whose vertex set is V and whose edge set is E by

$$G = (V, E).$$

The number n of vertices in the set V is called the *order* of the graph G. If

$$\alpha = \{x, y\} = \{y, x\}$$

is an edge of G, then we say that α *joins* x and y, and that x and y are *adjacent*; we also say that x and α are *incident*, and y and α are *incident*. We also call x and y the *vertices of the edge* α. A graph is, by definition, an abstract mathematical entity. But we can also think of a graph as a geometrical entity, by representing it with a diagram in the plane. We take one distinct point, a *vertex-point*, for each vertex x (labeling the vertex-point with the vertex) and connect two vertex-points by a simple curve[3] if and only if the corresponding vertices determine an edge α of G. We call such a curve an *edge-curve* and

---

[3] A non-self intersecting curve.

label it with α. In our diagrams, we must take care that *an edge-curve α passes through a vertex-point x only if x is a vertex of the edge α,* for otherwise our diagram will be ambiguous.

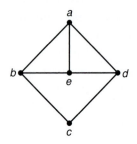

Figure 11.1

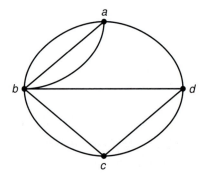

Figure 11.2

Example. Let a graph G of order 5 be defined by

$$V = \{a, b, c, d, e\}$$

and

$$E = \{\{a, b\}, \{b, c\}, \{c, d\}, \{d, a\}, \{e, a\}, \{e, b\}, \{e, d\}\}.$$

A geometric illustration of this graph is shown in Figure 11.1. □

If we alter the definition of a graph to allow a pair of vertices to form more than one edge, then the resulting structure is called a *multigraph*. In a multigraph $G = (V, E)$, E is a multiset. The *multiplicity of an edge* $\alpha = \{x, y\}$ is the number of times, $m\{x, y\}$, it occurs in E. The further generalization by allowing *loops*, edges of the form $\{x, x\}$ making a vertex adjacent to itself,[4] is called a *general graph*.

---

[4]Thus, a loop is a multiset consisting of one vertex with repetition number 2.

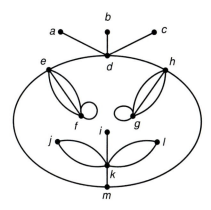

Figure 11.3

Example. In Figure 11.2 we have represented a multigraph of order 4 with 9 edges. In Figure 11.3 we have a general graph of order 13 with 21 edges, called *GraphBuster*.[5]   □

Sometimes, in drawing a geometrical representation of a graph (or multigraph or general graph), we may be forced to draw a curve that intersects another.[6]

A graph of order n is called *complete*, provided that each pair of distinct vertices forms an edge. Thus, in a complete graph, each vertex is adjacent to every other vertex. A complete graph of order n has $n(n-1)/2$ edges and is denoted K_n.

Example. The complete graphs K_1, K_2, K_3, K_4, and K_5 are drawn in Figure 11.4. It is not difficult to convince oneself that, in each drawing of K_5, there are always at least two edge-curves which intersect at a point that is not a vertex-point. Another way to draw K_5 is as a pentagon with an inscribed pentagram.   □

---

[5] "Who you gonna call?" **GraphBuster!** (aka Ghostbuster)

[6] But remember our rule that does not allow an edge-curve α to contain a vertex-point x unless vertex x is incident with edge α.

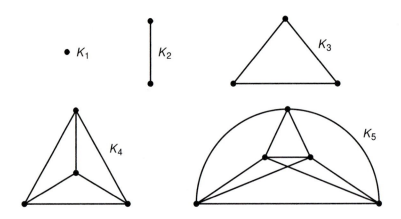

Figure 11.4

A general graph G is called *planar*, provided that it can be represented by a drawing in the plane in the manner just described in such a way that two edge-curves intersect only at vertex-points. Such a drawing is called a *plane-graph* and is a *planar representation* of G. The drawings of K_1, K_2, K_3, and K_4 in Figure 11.4 are plane-graphs, and consequently, those graphs are planar. The drawing of K_5 is not a plane-graph, because two edge-curves intersect at a point that is not a vertex-point, and, indeed, K_5 is not planar. Planar graphs are discussed in more detail in Chapter 13.

The *degree* (*valence*) of a vertex x in a general graph G is the number $\deg(x)$ of edges that are incident with x. If $\alpha = \{x, x\}$ is a loop joining x to itself, then α contributes 2 to the degree of x.[7] To each general graph G, we associate a sequence of numbers that is the list of the degrees of the graph's vertices in nonincreasing order:

$$(d_1, d_2, \ldots, d_n), \quad d_1 \geq d_2 \geq \cdots \geq d_n \geq 0.$$

We call this sequence the *degree sequence* of G.

The degree sequence of the general graph in Figure 11.3 is

$$(6, 5, 5, 5, 5, 5, 3, 2, 2, 1, 1, 1, 1).$$

The degree sequence of a complete graph K_n is

$$(n-1, n-1, \ldots, n-1), \quad (n-1) \text{ repeated } n \text{ times}).$$

The result stated in Theorem 13.2.1 appeared in Euler's first paper on graphs.

---

[7]Because both vertices of $\alpha = \{x, x\}$ equal x, α is incident "twice" with x.

Theorem 11.1.1 *Let G be a general graph. The sum*

$$d_1 + d_2 + \cdots + d_n$$

of the degrees of all the vertices of G is an even number, and consequently, the number of vertices of G with odd degree is even.

Proof. Each edge of G contributes 2 to the sum of the vertex degrees, 1 to each of its two vertices, or 2 to one vertex in the case of a loop. If a sum of integers is even, then the number of odd integer summands must also be even. □

Example. At a party, a lot of handshaking takes place between the guests. Show that, at the end of the party, the number of guests who have shaken hands an odd number of times is even.

The handshaking at the party can be modeled by a multigraph. The vertices are the guests. Each time two guests shake hands we join them by a new edge. The result is a multigraph to which we can apply Theorem 11.1.1. □

Two general graphs $G = (V, E)$ and $G' = (V', E')$ are called *isomorphic*, provided that there is a one-to-one correspondence

$$\theta : V \to V'$$

between their vertex sets such that, for each pair of vertices x and y of V, there are as many edges of G joining x and y as there are edges of G' joining $\theta(x)$ and $\theta(y)$. The one-to-one correspondence θ is called an *isomorphism* of G and G'. The notion of isomorphism is one of "sameness." Two general graphs are isomorphic if and only if, apart from the labeling of their vertices, they are the same.[8] If G and G' are graphs, then we can express the fact that the two graphs G and G' are isomorphic by asserting that there is a one-to-one correspondence between their vertex sets V and V', such that two vertices of V are adjacent in G if and only if the corresponding vertices are adjacent in G'. This relationship holds because, in graphs, two vertices are joined by either 1 or 0 edges.

---

[8]Put another way, two general graphs are isomorphic, provided that one is the other in disguise. The one-to-one correspondence θ is the "unmasking" of G' to reveal that G' is really G: If $\theta(x) = x'$, then under the "mask" sits x.

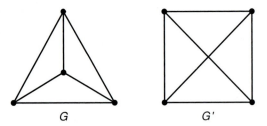

G G'

Figure 11.5

Example. Isomorphic graphs have the same order and the same number of edges, but these properties do not guarantee that two graphs are isomorphic.

First, consider the two graphs G and G' shown in Figure 11.5. These graphs are isomorphic since each is a graph of order 4 with each pair of distinct vertices adjacent, and thus each graph is a complete graph of order 4. This example illustrates the fact that a graph may be drawn in various ways (as in this example one drawing may be a plane-graph and the other not) and the actual way in which it is drawn is of no significance insofar as isomorphism is concerned. What matters is only whether two vertices are adjacent or not (or, in the case of general graphs, how many edges join each pair of vertices). □

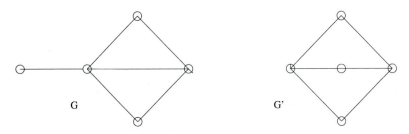

G G'

Figure 11.6

Now consider the two graphs G and G' drawn in Figure 11.6. Are these graphs isomorphic? They have the same order and they have the same number of edges. But the graph G has a vertex whose degree equals 1, while there is no vertex of G' with degree equal to 1. Such a situation cannot occur if two graphs are isomorphic. For, suppose that there is an isomorphism θ between G and G'. Then, for each vertex x of G, the vertex $\theta(x)$ of G' has the same degree as x. In particular, if a number occurs as the degree of a vertex of G, then it must also occur as the degree of a vertex of G'. We conclude that G and G' are not isomorphic. More generally, the same kind of reasoning shows that isomorphic graphs must have the same degree sequence. □

Example. In this example we show that two graphs may not be isomorphic, even if they have the same degree sequence. Consider the two graphs in Figure 11.7. Each of the graphs has degree sequence equal to (3,3,3,3,3,3). Yet these graphs are not isomorphic. This can be seen as follows: In the first graph, G in Figure 11.7, there are 3 vertices x, y, and z, the members of each pair of which are adjacent.[9] In the second graph, G' of that figure, no set of 3 vertices has this property. If θ were an isomorphism between the two graphs, then $\theta(x), \theta(y)$, and $\theta(z)$ would be 3 vertices of G', the members of each pair of which were adjacent. We conclude that G and G' are not isomorphic. □

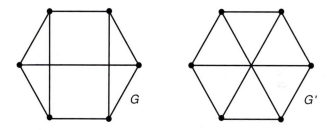

Figure 11.7

We summarize our observations in the next theorem.

Theorem 11.1.2 *Two isomorphic general graphs have the same degree sequence, but two graphs with the same degree sequence need not be isomorphic.*

In the example preceding the theorem we used another necessary condition for two graphs to be isomorphic. Before recording it, we introduce more basic concepts.

Let $G = (V, E)$ be a general graph. A sequence of m edges of the form

$$\{x_0, x_1\}, \{x_1, x_2\}, \ldots, \{x_{m-1}, x_m\} \tag{11.1}$$

is called a *walk of length m*, and this walk *joins the vertices* x_0 and x_m. We also denote the walk (11.1) by

$$x_0 - x_1 - x_2 - \cdots - x_m. \tag{11.2}$$

The walk (11.1) is *closed* or *open* depending on whether $x_0 = x_m$ or $x_0 \neq x_m$. A walk may have repeated edges.[10] If a walk has distinct

---

[9]They form a K_3.

[10]This comment requires further explanation in case we are dealing with a general graph that is not a graph. In a general graph G, each edge has a multiplicity that

edges, then it is called a *trail*.[11] If, in addition, a walk has distinct
vertices (except, possibly, $x_0 = x_m$), then the walk is called a *path*.
A closed path is called a *cycle*. It is easy to show, and is left as an
exercise, that the edges of a trail joining vertices x_0 and x_m can be
partitioned so that one part of the partition determines a path joining
x_0 and x_m, and the other parts determine cycles. In particular, the
edges of a closed trail can be partitioned into cycles. The length of a
cycle of a graph is at least 3. In a general graph, a loop forms a cycle
of length 1, and an edge $\{a, b\}$ of multiplicity $m \geq 2$ determines a cycle
$\{a, b\}, \{b, a\}$ (or $a - b - a$) of length 2.

Example Consider the general graph GraphBuster in Figure 11.3.
Then we have the following statements:

(i) $a - d - b - d - c - d - h - g - h - m - k - i$ is a walk of length
11 joining vertex a and vertex i, but it is not a trail.

(ii) $a - d - e - f - e - m - k - l - k - i$ is a trail of length 9 joining
a and i, but it is not a path.

(iii) $a - d - e - m - k - i$ is a path of length 5 joining a and i.

(iv) $d - e - f - e - m - h - d$ is a closed trail of length 6, but it is
not a cycle.

(v) Each of $f - f$, $e - f - e$,   and $d - e - m - h - d$ is a cycle.□

A general graph G is called *connected*, provided that, for each pair
of vertices x and y, there is a walk joining x and y (equivalently, a
path joining x and y). Otherwise, G is *disconnected*. In a disconnected
general graph there is at least one pair of vertices x and y for which
there is no way to get from x to y (or from y to x) by "walking"
along the edges of G. For most purposes, it suffices to consider only
connected graphs. In a connected graph, $d(x, y)$ denotes the shortest
length of a walk joining the vertices x and y and is called the *distance*
between x and y. We define $d(x, x) = 0$ for each vertex x. It is clear
that a walk joining x and y of length $d(x, y)$ is a path.

---

may be greater than 1. We do not regard an edge as repeated in a walk if the
number of times it occurs in the walk does not exceed its multiplicity. An edge
is repeated only if the number of times it occurs in the walk is greater than the
number of "copies" available in G. This is perfectly reasonable when one considers
a drawing of G, for if an edge $\alpha = \{a, b\}$ has multiplicity 5, say, then in the drawing
there are 5 *different* edge-curves joining the vertex-points a and b.

[11]Thus, in a trail the number of times an edge occurs cannot exceed its multiplicity.

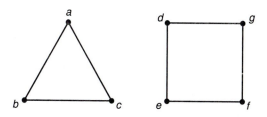

Figure 11.8

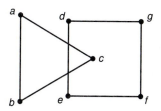

Figure 11.9

Example. The graph drawn in Figure 11.8 is disconnected. There is no walk from vertex a to vertex d. This example illustrates the fact that a disconnected graph can always (and should always!) be drawn so that the resulting geometric entity consists of two disjoint parts. Another way to draw the graph of this example is given in Figure 11.9, but it would be foolish to draw it that way. In general, we try to draw a graph in a way that reveals its structure. $\square$

Let $G = (V, E)$ be a general graph. Let U be a subset of V and let F be a submultiset of E, such that the vertices of each edge in F belong to U. Then $G' = (U, F)$ is also a general graph called a *general subgraph* of G.[12] If F consists of all edges of G that join vertices in U, then G' is called an *induced* general subgraph of G and is denoted by G_U. In case U is the entire set V of vertices of G then G' is called *spanning*. Thus, an induced general subgraph of G is obtained by selecting some of the vertices of G, and *all* of the edges of G that join them. A spanning general subgraph is obtained by taking all the vertices of G and *some* (possibly all) of the edges of G.

---

[12]If G is a graph, then G' is also a graph and is called a *subgraph*. In all definitions like this one, we shall drop the modifier "general" when we are dealing with graphs.

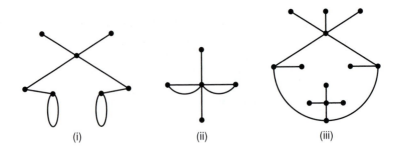

$$\text{(i)} \qquad\qquad\qquad \text{(ii)} \qquad\qquad\qquad \text{(iii)}$$

Figure 11.10

Example. Let G be the general graph GraphBuster in Figure 11.3. In Figure 11.10, there is given

(i) a general subgraph that is neither induced nor spanning.

(ii) a general subgraph that is induced but not spanning.

(iii) a general subgraph (which happens to be a graph) that is spanning, but not induced. □

The next theorem, which states that a general graph consists of one or more connected general graphs, is clear intuitively. We leave the formal verfication for the exercises.

Theorem 11.1.3 *Let $G = (V, E)$ be a general graph. Then the vertex set V can be uniquely partitioned into nonempty parts $V_1, V_2, \ldots, V_k$ so that the following conditions are satisfied:*

(i) *The general subgraphs $G_1 = (V_1, E_1), G_2 = (V_2, E_2), \ldots, G_k = (V_k, E_k)$ induced by $V_1, V_2, \ldots, V_k$, respectively, are connected.*

(ii) *For each $i \neq j$ and each pair of vertices x in V_i and y in V_j, there is no walk that joins x and y.*

The general graphs $G_1, G_2, \ldots, G_k$ in Theorem 11.1.3 are the *connected components* of G. Part (i) of the theorem says that the connected components are indeed connected; part (ii) asserts that the connected components are *maximal* connected induced general subgraphs; that is, for each i and for each set U of vertices, such that V_i is contained in U but $V_i \neq U$, the general subgraph induced by U is disconnected.

In the next theorem we formulate additional necessary conditions in order that general graphs be isomorphic. Its proof should now be obvious and formal verification is left for the exercises.

Theorem 11.1.4 *Let G and G' be two general graphs. Then the following are necessary conditions for G and G' to be isomorphic:*

(i) *If G is a graph, so is G'.*

(ii) *If G is connected, so is G'. More generally, G and G' have the same number of connected components.*

(iii) *If G has a cycle of length equal to some integer k, then so does G'.*

(iv) *If G has an (induced) general subgraph that is a complete graph K_m of order m, so does G'.*

The graphs G and G' in Figure 11.7 are not isomorphic since one has a cycle of length 3 (a subgraph isomorphic to K_3) and the other doesn't.

We conclude this section by showing that a general graph may also be described by a matrix whose entries are nonnegative integers.

Let G be a general graph of order n and let its vertices be, in some order, $a_1, a_2, \ldots, a_n$. Let A be the n-by-n array such that the entry a_{ij} in row i, column j equals the number of edges joining the vertices a_i and a_j, ($1 \leq i, j \leq n$). We always have[13] $a_{ij} = a_{ji}$, and a_{ii} counts the number of loops at vertex a_i. The matrix A is called the *adjacency matrix* of G. In case G is a graph, then A is a matrix of 0's and 1's and the entry a_{ij} equals 1 if and only if a_i and a_j are adjacent in G.

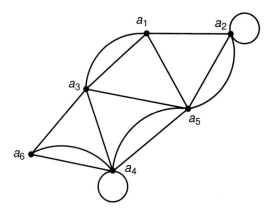

Figure 11.11

---

[13]The matrix is *symmetric*.

Example. Figure 11.11 shows a general graph of order 6 whose 6-by-6 adjacency matrix is

$$\begin{bmatrix} 0 & 1 & 2 & 0 & 1 & 0 \\ 1 & 1 & 0 & 0 & 2 & 0 \\ 2 & 0 & 0 & 1 & 1 & 1 \\ 0 & 0 & 1 & 1 & 2 & 2 \\ 1 & 2 & 1 & 2 & 0 & 0 \\ 0 & 0 & 1 & 2 & 0 & 0 \end{bmatrix}.$$

We can start with either the general graph or the adjacency matrix and then construct the other.   □

The adjacency matrix is uniquely determined by a general graph, apart from the ordering of its rows and columns. This is because, before we can form the adjacency matrix, we must first list the vertices in some order. Conversely, the adjacency matrix of a general graph uniquely determines the general graph up to isomorphism; that is, any two general graphs with the same adjacency matrix are isomorphic.

11.2   Eulerian Trails

In his paper on graph theory published in 1736, Euler solved the now-famous *Königsberg bridge problem*:

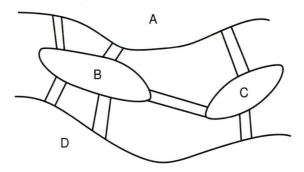

Figure 11.12

The old city of Königsberg in East Prussia was located along the banks and on two islands of the Pregel River, with the four parts of the city connected by seven bridges as shown in Figure 11.12. On Sundays, the citizens of Königsberg would promenade about town, and the problem arose as to whether it was possible to plan a promenade in such a way that each bridge is crossed once and only once.

Euler replaced the map of Königsberg with the general graph drawn in Figure 11.13. In terms of that general graph G and the terminology we have now introduced, the problem is to determine whether there exists a closed trail that contains all the edges of G.

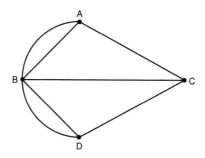

Figure 11.13

Example. Consider the plight of the mail carrier[14] who, starting at the post office, wishes to deliver the mail to the houses on the preassigned streets and then end up back at the post office at the end of the day. What the mail carrier would like is a way to deliver all the mail without having to walk over any street after having already delivered the mail on that street. Can we help the mail carrier out?

Well, maybe we can and maybe we can't. But we surely should recognize her or his problem as a problem in graph theory. Let G be the general graph that can be associated with the street map of a city. (See the introductory remarks for this chapter.) Let G' be the general subgraph consisting of the vertices and edges of G that correspond to the mail carrier's assigned streets. What the mail carrier desires is a closed trail in G' that contains each edge of G' exactly once. Thus, we have the same mathematical problem as the citizens of Königsberg had over 200 years ago, but relative to a different general graph.   □

Motivated by these problems, we make some definitions. A trail in a general graph G is called *Eulerian*, provided that it contains every edge of G. Recall that a trail in a general graph by definition contains each edge at most once, where we interpret this to mean that the number of times that an edge occurs on the trail does not exceed its multiplicity. Both the citizens of Königsberg and the mail carrier seek a closed Eulerian trail. We can easily see that the Königsberg bridge general graph in Figure 11.13 does not have a closed Eulerian trail. We reason as follows: Imagine actually promenading on a closed Eulerian

---

[14]Change mail carrier to street sweeper or snowplow operator to obtain a different formulation of the same mathematical problem.

trail in a general graph. Except for the first time you leave the vertex at which you begin, every time you go into a vertex you leave it (by a new edge, that is, by one that you had not yet gone over). When you finish up, you go into the beginning vertex but don't leave it. What this means is that the edges which are incident with a given vertex can be paired up: One edge of each pair is used to enter the vertex and one is used to leave it.[15] If the edges incident with a vertex can be paired up, that means that there must be an even number of edges at each vertex. We thus conclude that a necessary condition in order that a general graph have a closed Eulerian trail is that the degree of each vertex is even. Since the four vertices of the general graph for the Königsberg bridge problem have odd degree, it does not have a closed Eulerian trail.

Theorem 11.2.2 asserts that the necessary condition for a closed Eulerian trail derived in the preceding discussion is also sufficient for a connected general graph. Before proving it, we establish a lemma, which is also of independent interest.

Lemma 11.2.1 *Let $G = (V, E)$ be a general graph and assume that the degree of each vertex is even. Then each edge of G belongs to a closed trail and hence to a cycle.*

Proof. We can find a closed trail containing any prescribed edge $\alpha_1 = \{x_0, x_1\}$ using the next algorithm. In this algorithm, we construct a set W of vertices and a set F of edges.

Algorithm for a closed trail

(1) Put $i = 1$.

(2) Put $W = \{x_0, x_1\}$.

(3) Put $F = \{\alpha_1\}$.

(4) While $x_i \neq x_0$, do the following:

    (i) Locate an edge $\alpha_{i+1} = \{x_i, x_{i+1}\}$ not in F.

    (ii) Put x_{i+1} in W (x_{i+1} may already be in W).

    (iii) Put α_{i+1} in F.

    (iv) Increase i by 1.

---

[15] If we think of starting our promenade in the "middle" of an edge, then we do not need to distinguish a beginning vertex: each time we enter a vertex we also leave it.

Thus, after the initialization in (1)–(3), at each stage of the algorithm we locate a new edge[16] $\alpha_{i+1} = \{x_i, x_{i+1}\}$ incident with the most recent vertex x_i put in W, add x_{i+1} to W and α_{i+1} to F, and then increase i by 1 and repeat until we finally arrive at x_0 again.

Suppose that an edge α_{i+1} satisfying (4)(i) exists whenever $x_i \neq x_0$. Let the terminal value of i be k, giving the set $W = \{x_0, x_1, \ldots, x_k\}$ of vertices and the multiset $F = \{\alpha_1, \ldots, \alpha_k\}$ of edges. It then follows from the description of the algorithm that

$$\alpha_1, \ldots, \alpha_k \tag{11.3}$$

is a closed trail containing the initial edge α_0. Thus, we have only to show that, if $x_i \neq x_0$, then there is an edge not in F that is incident with x_i. It is here where the hypothesis of even degrees comes in.

It is readily seen that, at the end of each step (4)(iv) of the algorithm, each vertex of the general graph $H = (W, F)$ has even degree, except possibly for the vertex x_0 (which starts out with odd degree 1) and the most recent new vertex x_i (whose degree has just been increased by 1). Moreover, x_0 and x_i have even degree if and only if $x_0 = x_i$. Thus, if $x_i \neq x_0$, x_i has odd degree in the general graph H. Since x_i has even degree in G, there must be an edge $\alpha_{i+1} = \{x_i, x_{i+1}\}$ not yet in F that is incident with x_i. Thus, at the end of the algorithm, $x_k = x_0$ and (11.3) is a closed trail.

The edges of a closed trail can be partitioned into cycles and the proof of the lemma is complete. ☐

Example. We apply the algorithm for a closed trail to the general graph G drawn in Figure 11.14. One way to carry out the algorithm[17] is illustrated in the following table with the initial edge being $\{a, b\}$:

i	x_i	α_i	W	F
1	b	$\{a, b\}$	a, b	$\{a, b\}$
2	c	$\{b, c\}$	a, b, c	$\{a, b\}, \{b, c\}$
3	d	$\{c, d\}$	a, b, c, d	$\{a, b\}, \{b, c\}, \{c, d\}$
4	b	$\{d, b\}$	a, b, c, d	$\{a, b\}, \{b, c\}, \{c, d\}, \{d, b\}$
5	h	$\{b, h\}$	a, b, c, d, h	$\{a, b\}, \{b, c\}, \{c, d\}, \{d, b\}, \{b, h\}$
6	a	$\{h, a\}$	a, b, c, d, h	$\{a, b\}, \{b, c\}, \{c, d\}, \{d, b\}, \{h, b\}, \{h, a\}$

---

[16]More precisely, one whose multiplicity in F is less than that in the edge set E of our graph G.

[17]Since at each stage of the algorithm there may be more than one choice for a new edge, there will, in general, be many ways in which to carry out the algorithm.

We thus obtain the closed trail

$$\{a,b\}, \{b,c\}, \{c,d\}, \{d,b\}, \{b,h\}, \{h,a\}$$

and the cycle

$$\{a,b\}, \{b,h\}, \{h,a\}$$

containing the edge $\{a,b\}$.   □

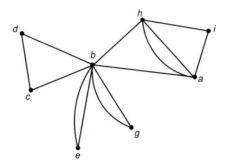

Figure 11.14

Theorem 11.2.2 *Let G be a connected general graph. Then G has a closed Eulerian trail if and only if the degree of each vertex is even.*

Proof. We have already observed that, if G has a closed Eulerian trail, then each vertex has even degree. Now let $G_1 = (V_1, E_1)$ be a connected general graph in which each vertex has even degree. We choose any edge α_1 of G_1 and apply the algorithm for a closed trail given in the proof of Lemma 11.2.1 to obtain a closed trail γ_1 containing the edge α_1. Let $G_2 = (V_2, E_2)$ be the general graph obtained by removing from E_1 the edges that belong to the closed trail γ_1. All vertices have even degree in G_2. If E_2 contains at least one edge, then since we started with G_1 connected, there must be an edge α_2 of G_2 that is incident with a vertex z_1 on the closed trail γ_1. We apply the algorithm for a closed trail to G_2 and the edge α_2 and obtain a closed trail γ_2 containing the edge α_2. We now patch[18] γ_1 and γ_2 together at the vertex z_1 and obtain a closed trail $\gamma_1 \overset{z_1}{*} \gamma_2$ that includes all the edges of both γ_1 and γ_2. Let $G_3 = (V_3, E_3)$ be the general graph obtained by removing the edges of γ_2 from E_2. If E_3 contains at least one edge, then it contains an edge α_3 which is incident with a vertex z_2 on the closed trail $\gamma_1 \overset{z_1}{*} \gamma_2$. We apply the algorithm for a closed trail to G_3 and the edge α_3 and obtain

---

[18]We traverse γ_1 until we first come to the vertex z_1, completely traverse γ_2 ending up at vertex z_1, and then finish our traversal of γ_1.

a closed trail γ_2 containing the edge α_3. We then patch $\gamma_1 \overset{z_1}{*} \gamma_2$ and γ_3 together at vertex z_2 and obtain the closed trail $\gamma_1 \overset{z_1}{*} \gamma_2 \overset{z_2}{*} \gamma_3$, which[19] includes all the edges of γ_1, γ_2 and γ_3. We continue like this until all edges have been included in a closed trail $\gamma_1 \overset{z_1}{*} \gamma_2 \overset{z_2}{*} \cdots \overset{z_{k-1}}{*} \gamma_k$. Thus, repeated calls to our algorithm for a closed trail give an algorithm to construct a closed Eulerian trail in a connected general graph, each of whose vertices has even degree. $\qquad\square$

Example. We continue with the preceding example and obtain a closed Eulerian trail in the general graph G of Figure 11.14, using the algorithm in the proof of Theorem 11.2.2. Since the algorithm requires us to make choices, there are several ways to carry out the algorithm. One possible result is the following:

$$\gamma_1 = a - b - c - d - b - h - a,$$
$$\gamma_2 = b - e - b, (z_1 = b),$$

$$\gamma_1 \overset{b}{*} \gamma_2 = a - b - e - b - c - d - b - h - a,$$

$$\gamma_3 = b - g - b, (z_2 = b),$$

$$\gamma_1 \overset{b}{*} \gamma_2 \overset{b}{*} \gamma_3 = a - b - g - b - e - b - c - d - b - h - a,$$

$$\gamma_4 = h - i - a - h, (z_3 = h),$$

$$\gamma_1 \overset{b}{*} \gamma_2 \overset{b}{*} \gamma_3 \overset{h}{*} \gamma_4 =$$
$$a - b - g - b - e - b - c - d - b - h - i - a - h - a.$$

$\qquad\square$

Theorem 11.2.2 and its proof furnish a characterization of general graphs with a closed Eulerian trail and an algorithm for constructing a closed Eulerian trail if one exists. For an open Eulerian trail we have the next theorem.

Theorem 11.2.3 *Let G be a connected general graph. Then G has an open Eulerian trail if and only if there are exactly two vertices u and v of odd degree. Every open Eulerian trail in G joins u and v.*

Proof. First, we recall from Theorem 11.1.1 that the number of vertices of G of odd degree is even. If there is in G an open Eulerian trail, then it must join two vertices u and v of G of odd degree, and every

---

[19]This notation is a little ambiguous. Do you know why?

other vertex of G must have even degree (since each time the Eulerian trail goes into a vertex x different from u and v it leaves, resulting in a pairing of the edges incident with x). Now assume that G is connected and has exactly two vertices u and v of odd degree. Let G' be the general graph obtained from G by adding a new edge $\{u, v\}$ joining u and v. Then G' is connected and each vertex now has even degree. Hence, by Theorem 11.2.2, G' has an Eulerian trail γ'. We can think of γ' as beginning at the vertex v with first edge being the new edge $\{u, v\}$ joining u and v. Removing this edge from γ' and starting at the vertex u we obtain an open Eulerian trail γ in G joining u and v. We can apply the algorithm for a closed Eulerian trail to G' and thereby obtain an algorithm for an open Eulerian trail in G.   □

The previous theorem is further generalized in the next theorem. We leave the proof for the exercises.

Theorem 11.2.4 *Let G be a connected general graph and suppose that the number of vertices of G with odd degree is $m > 0$. Then the edges of G can be partitioned into $m/2$ open trails. It is impossible to partition the edges of G into fewer than $m/2$ open trails.*

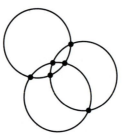

Figure 11.15

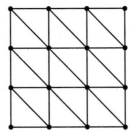

Figure 11.16

Example. Consider the graphs drawn in Figures 11.15, 11.16, and 11.17. Is it possible to trace these plane graphs with a pencil without removing the pencil from the paper?

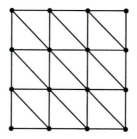

Figure 11.16

In order to be able to trace a plane graph without removing one's pencil from the paper, it is necessary and sufficient that there is an Eulerian trail, either open or closed. The vertices of the graph drawn in Figure 11.15 all have degree equal to 4 and hence, by Theorem 11.2.2, the graph is traceable. The graph drawn in Figure 11.16 has 2 vertices of odd degree and hence, by Theorem 11.2.3, has an open Eulerian trail joining the 2 vertices of odd degree. The graph drawn in Figure 11.17 has 4 vertices of odd degree and hence, by Theorem 11.2.3, is not traceable. However it follows from Theorem 11.2.4 that it can be traced if one is allowed to lift the pencil once from the paper. The proof of Theorem 11.2.2 contains an algorithm to trace a plane graph when a tracing exists. □

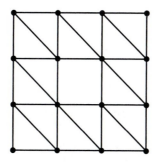

Figure 11.17

By Theorem 11.2.4, if a general graph G has $m > 0$ vertices of odd degree, then the edges can be partitioned into $m/2$ open trails, each trail joining 2 vertices of odd degree. If one wants to trace out G as discussed in the previous example, then it is necessary to lift the pencil only $(m/2) - 1$ times. In tracing out G, lifting one's pencil is no great hardship, but if G represents the route of a mail carrier (as discussed in the example at the beginning of this section) who has to deliver mail on foot on each of the streets corresponding to the edges of G, then what's the mail carrier to do? Fly? If the mail carrier's route does not contain a closed Eulerian trail, then in order for all the mail to be delivered and for the mail carrier to return to the post office, the mail carrier will have to walk over some streets more than once. How can we minimize the number of streets that the mail carrier will have to walk over, after already having delivered the mail at the houses on those streets? This problem is known as the *Chinese postman problem*.[20] A precise formulation is the following:

> *Chinese postman problem*: Let G be a connected general graph. Find a closed walk of shortest length which uses each edge of G at least once.[21]

We close this section with a simple observation concerning the solution of the Chinese postman problem.

Theorem 11.2.5 *Let G be a connected general graph having K edges. Then there is a closed walk in G of length $2K$ in which the number of times an edge is used equals twice its multiplicity.*

Proof. Let G^* be the general graph obtained from G by doubling the multiplicity of each edge of G. Then G^* is a connected graph with $2K$ edges. Moreover, each vertex of G^* has even degree (twice its degree in G). Applying Theorem 11.2.2 to G^*, we see that G^* has a closed Eulerian trail. This closed trail in G^* is a closed walk in G of the required type.   □

Example. Consider a graph G with vertices $1, 2, \ldots, n$ and $n-1$ edges $\{1, 2\}, \{2, 3\}, \ldots, \{n-1, n\}$. Thus, the edges of G form a path joining vertex 1 to vertex n. Any closed walk in G that includes each edge must include each edge at least twice. Thus, if the post office is at vertex k, our Chinese postman can do no better than to walk to vertex 1, retrace his steps back to the post office, walk to vertex n, and retrace his steps back to the post office. The length of such a walk is $2(n-1)$, that is, twice the number of edges. The graph G is a simple instance of a tree. Trees are studied in Sections 11.5 and 11.7. For a tree, the smallest length of a closed walk that includes each edge at least once equals twice the number of edges. (See Exercise 78.)   □

The reader has perhaps already noticed that, while the Chinese postman problem, as we have phrased it, may be interesting from a

---

[20] Not because it has particular relevance to China, but because it was introduced by the Chinese mathematician M.K. Kwan in a paper: Graphic programming using odd or even points, *Chinese Math.*, 1 (1962), 273-277.

[21] A solution to this problem is given in J. Edmonds and E.L. Johnson: Matching, Euler tours and the Chinese postman, *Math. Programming*, 5 (1973), 88-124.

purely mathematical point of view, it is not very practical. This is because we have not taken into account the length of the streets. Some streets may be very long, while others are very short. If the mail carrier has to repeat some streets, obviously the shorter ones are to be preferred. To make the problem practical, we should attach a nonnegative *weight* to each edge and then measure a walk not by its length (the number of its edges) but by its total weight (the sum of the weights of its edges, counting the weight of an edge the number of times that it is used in the walk). The practical Chinese postman problem is to determine a walk of smallest weight which includes each edge at least once. This problem has also been satisfactorily solved from an algorithmic point of view.[22]

11.3 Hamilton Paths and Cycles

In the 19th century, Sir William Rowan Hamilton invented a puzzle whose object was to determine a route on the sides of a dodecahedron[23] that started at some corner and returned there after having visited every other corner exactly once. The corners and sides of a dodecahedron determine a graph with 20 vertices and 30 edges, which is drawn in Figure 11.18. There are many readily discovered solutions to Hamilton's puzzle.[24]

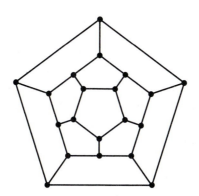

Figure 11.18

Hamilton's puzzle can be formulated for any graph:

---

[22]Ibid.

[23]The dodecahedron is one of the regular solids. It is bounded by 12 regular pentagons which come together at 30 sides, determining 20 corner points.

[24]And this perhaps explains why Hamilton's puzzle was not a great commercial success!

> Given a graph G can one determine a route along the edges
> of G that begins at some vertex and then returns there after
> having visited every other vertex exactly once?

Today, a solution to Hamilton's puzzle for a graph G is called a Hamilton cycle. More precisely, a *Hamilton cycle* of a graph G of order n
is a cycle of G of length n. Recall that a cycle is a closed path all of
whose vertices are distinct, except that the first vertex is the same as
the last. Hence, a Hamilton cycle in the graph G of order n is a cycle

$$x_1 - x_2 - \cdots - x_n - x_1$$

of length n where $x_1, x_2, \ldots, x_n$ are the n vertices of G in some order.
A *Hamilton path* in G joining vertices a and b is a path

$$a = x_1 - x_2 - \cdots - x_n = b$$

of length $n - 1$ of G. Thus, a Hamilton path in G is given by a
permutation of the n vertices of G in which consecutive vertices are
joined by an edge of G. The Hamilton path joins the first vertex of
the permutation to the last. The edges of a Hamilton path and of a
Hamilton cycle are necessarily distinct.

We can also consider Hamilton paths and cycles in general graphs,
but higher multiplicities of edges have no impact on the existence and
nonexistence of Hamilton paths and cycles. Whether or not there is a
Hamilton path or Hamilton cycle is determined solely by which pairs
of vertices are joined by an edge and not on the multiplicity of an edge
joining a pair of vertices. It is for this reason that *we consider only
graphs, and not general graphs, in this section.*

Example. A complete graph K_n of order $n \geq 3$ has a Hamilton
cycle. In fact, since each pair of distinct vertices of K_n form an edge,
each permutation of the n vertices of K_n is a Hamilton path. Since
the first vertex and last vertex are joined by an edge, each Hamilton
path can be extended to a Hamilton cycle. We thus see that K_n has $n!$
Hamilton paths and $(n-1)!$ Hamilton cycles (corresponding to circular
permutations of length n).   □

Example. For each of the two graphs drawn in Figure 11.19, determine whether there is a Hamilton path or cycle.

First, consider the graph on the left. Then $a - b - c - d - f - e - a$
is a Hamilton cycle, and thus $a - b - c - d - f - e$ is a Hamilton path.
Another Hamilton path is $a - b - c - d - e - f$, but this Hamilton path
cannot be extended to a Hamilton cycle since a and f are not joined
by an edge.

Now consider the "dumbbell" graph on the right. A Hamilton path is $a - b - c - d - e - f$, but this graph does not have a Hamilton cycle. The reason is that a Hamilton cycle is closed, and thus would have to cross the "bar" of the dumbbell twice, but this is not allowed in a Hamilton cycle.   □

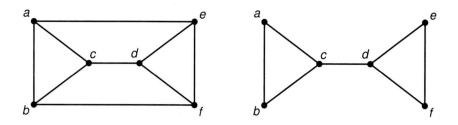

Figure 11.19

At first glance, the question of the existence of a Hamilton cycle in a graph seems similar to the question of the existence of a closed Eulerian trail in a graph. For the latter, we seek a closed trail that includes every edge exactly once. For the former, we seek a closed path that includes every vertex exactly once. Beyond this superficial resemblance, the two questions are very much different. In Theorem 11.1.1 an easily verifiable characterization of (general) graphs with a closed Eulerian trail is given, and we have a satisfactory algorithm for constructing one when those conditions are met. No such characterization exists for graphs with a Hamilton cycle, nor is there a satisfactory algorithm for constructing a Hamilton cycle in a graph, should one exist. The question of the existence and construction of Hamilton cycles (and paths) in graphs is today a major unsolved question in graph theory.

So if we cannot characterize graphs with Hamilton cycles (that is, find conditions which are both necessary and sufficient for their existence in a graph), we have to be content to find conditions that are sufficient for their existence (that is, guarantee a Hamilton cycle) and, separately, conditions that are necessary for their existence (so if they are not met, guarantee that there is no Hamilton cycle). One obvious necessary condition for a Hamilton cycle is that the graph has to be connected. Another less obvious condition was hinted at in our analysis of the dumbbell graph in Figure 11.19. An edge of a connected graph is called a *bridge*, provided its removal from the graph leaves a disconnected graph. In a certain sense, a connected graph with a bridge is just barely connected: remove the bridge and the graph "breaks apart." The bar of the dumbbell graph in Figure 11.19 is a bridge.

Theorem 11.3.1 *A connected graph with a bridge does not have a Hamilton cycle.*[25]

Proof. Suppose that $\alpha = \{x, y\}$ is a bridge of a connected graph G. Let G' be the graph obtained from G by removing the edge α but not any vertices. Since G is connected, G' has two connected components.[26] Suppose G has a Hamilton cycle γ. Then γ would, say, begin in one of the components in G', would eventually cross over to the other, via α, and then would have to cross back to the first, also via α. But then γ is not a Hamilton cycle since it would include a vertex other than the initial and terminal vertex more than once.   □

We now discuss a simple suffecent condition for a Hamilton cycle in a graph which is due to Ore.[27]

Let G be a graph of order n and consider the following property which may or may not be satisfied in G:

> *Ore property*: For all pairs of distinct vertices x and y that are not adjacent,
>
> $$\deg(x) + \deg(y) \geq n.$$

What are the implications for a graph which satisfies the Ore property? First of all, a graph all of whose vertices have "large" degree[28] must have a lot of edges, and these edges are distributed somewhat uniformly throughout the graph. We would hope that such a graph has a Hamilton cycle.[29] Now suppose, for instance, that G is a graph with $n = 50$ vertices which satisfies the Ore property. If G had a vertex x of small degree, say, 4, this implies that there are 45 vertices different from x that are not adjacent to x. By the Ore property, each of these 45 vertices has degree at least 46. Thus, the Ore property implies either that all vertices have large degree or that there are some vertices of small degree and *very* many vertices of *very* large degree. Therefore, the Ore property compensates for the possible presence of vertices of small degree (which might keep a graph from having a Hamilton cycle) by forcing there to be a lot of vertices of high degree (which might help a graph to have a Hamilton cycle).

---

[25] Although it might have a Hamilton path.

[26] If G' had more than two connected components, then putting the edge α back could only combine two of these components and the resulting graph (namely, G) is disconnected, contrary to assumption.

[27] Ö. Ore: A note on Hamilton circuits, *Amer. Math. Monthly*, 67 (1960), 55.

[28] This will be made precise in Corollary 11.3.3.

[29] If having a lot of edges well distributed over the graph did not guarantee a Hamilton cycle, what chance would we ever have of finding a condition that would?

Theorem 11.3.2 *Let G be a graph of order $n \geq 3$ that satisfies the Ore property. Then G has a Hamilton cycle.*

Proof. Suppose G is not connected. Then the vertices of G can be partitioned into two parts, U and W, such that there are no edges joining a vertex in U with a vertex in W. Let r be the number of vertices in U and let s be the number in W. Then $r + s = n$, and each vertex in U has degree at most $r - 1$, and each vertex in W has degree at most $s - 1$. Let x be any vertex in U and let y be any vertex in W. Then x and y are not adjacent, but the sum of their degrees is, at most,

$$(r - 1) + (s - 1) = r + s - 2 = n - 2,$$

and this contradicts the Ore property. Hence G is connected.

To complete the proof of the theorem, we give an algorithm[30] for constructing a Hamilton cycle in a graph. We first describe the algorithm and then show that, if the graph satisfies the Ore property, the result of the algorithm is always a Hamilton cycle.

Algorithm for a Hamilton cycle

(1) Start with any vertex and, by attaching adjacent vertices at either end, construct a longer and longer path until it is not possible to make it any longer. Let the path be

$$\gamma : y_1 - y_2 - \cdots - y_m. \qquad (11.4)$$

(2) Check to see if y_1 and y_m are adjacent.

   (i) If y_1 and y_m are not adjacent, go to (3). Else y_1 and y_m are adjacent, and go to (ii).

   (ii) If $m = n$, then stop and output the Hamilton cycle $y_1 - y_2 - \cdots - y_m - y_1$. Else, y_1 and y_m are adjacent and $m < n$, and go to (iii).

   (iii) Locate a vertex z not on γ and a vertex y_k on γ such that z is adjacent to y_k. Replace γ with the path of length $m + 1$ given by

   $$z - y_k - \cdots - y_m - y_1 \cdots - y_{k-1},$$

   and go back to (2).

---

[30]This algorithm is implicit in Ore's original proof of his theorem and was explicitly formulated by M.O. Albertson.

(3) Locate a vertex y_k with $1 < k < m$ such that y_1 and y_k are adjacent and y_{k-1} and y_m are adjacent. Replace γ with the path

$$y_1 - \cdots - y_{k-1} - y_m - \cdots - y_k.$$

The two ends of this path, namely, y_1 and y_k are adjacent, and go back to (2)(ii).

To prove that the algorithm does construct a Hamilton cycle, we have to show that in (2)(iii) we can locate the specified vertex z, and in (3) we can locate the specified vertex y_k.

First, consider (2)(iii). We have $m < n$. Since we have already shown that the Ore property implies that G is connected, there must be some vertex z not on the cycle γ which is adjacent to one of the vertices $y_1, \ldots, y_m$.

Now consider (3). We know that y_1 and y_m are not adjacent. Let the degree of y_1 be r and let the degree of y_m be s. By the Ore property, we have $r + s \geq n$. Since γ is a longest path from step (1), y_1 is adjacent to only vertices on γ and hence to r of the vertices $y_2, \ldots, y_{m-1}$. Similarly y_m is adjacent to s of the vertices $y_2, \ldots, y_{m-1}$. Each of the r vertices joined to y_1 is preceded in the path γ by some vertex, and one of these must be adjacent to y_m. For, if not, then y_m is adjacent to at most $(m-1) - r$ vertices and hence $s \leq m - 1 - r$. This means that

$$r + s \leq m - 1 \leq n - 1,$$

contrary to the Ore property. Thus, there is a vertex y_k such that y_1 is adjacent to y_k and y_m is adjacent to y_{k-1}. Hence, the algorithm stops after having constructed a Hamilton cycle in G. $\qquad\square$

One way to guarantee the Ore property in a graph is to assume that all vertices have degree equal to or greater than half the order of the graph. This results in a theorem of Dirac[31], which although proved in 1952 before Theorem 11.3.2, is a consequence of it.

Corollary 11.3.3 *A graph of order $n \geq 3$, in which each vertex has degree at least $n/2$, has a Hamilton cycle.*

A proof with algorithm similar to that given for Theorem 11.3.2 can be constructed for the next theorem in which a sufficient condition is given for a Hamilton path in a graph. We leave the proof as an exercise for the reader.

---

[31]G.A. Dirac: Some theorems on abstract graphs, *Proc. London Math. Soc.*, 2 (1952), 69-81.

Theorem 11.3.4 *A graph of order n, in which the sum of the degrees of each pair of nonadjacent vertices is at least $n-1$, has a Hamilton path.*

Example. *The traveling salesperson problem.* Consider a salesperson who is planning a business trip that takes him to certain cities in which he has customers and then brings him back home to the city from whence he started. Between some of the pairs of cities he has to visit there is direct air service; between others there is not. Can he plan the trip so that he flies into each city to be visited exactly once?

Let the number of cities to be visited, including his home city, be n. We let these cities be the vertices of a graph G of order n, in which there is an edge between 2 cities, provided that there is direct air service between them. Then what the salesperson seeks is a Hamilton cycle in G. If the graph G has the Ore property, then we know from Theorem 11.3.2 that there is a Hamilton cycle and, from its proof, a good way to construct one. But, in general, there is no good algorithm known which will construct a Hamilton cycle for the salesperson or which will tell him that no Hamilton cycle exists. The problem as formulated is not the real problem that a traveling salesperson faces. This is because distances between the cities he has to visit will in general vary, and what he would like is a Hamilton cycle in which the total distance travelled is as small as possible.[32] □

11.4 Bipartite Multigraphs

Let $G = (V, E)$ be a multigraph. Then G is called *bipartite*, provided that the vertex set V may be partitioned into two subsets X and Y so that each edge of G has one vertex in X and one vertex in Y. A pair X, Y with this property is called a *bipartition* of G (and of its vertex set V). Two vertices in the same part of the bipartition are not adjacent. As we did in Chapter 9 for bipartite graphs, we usually picture a bipartite multigraph so that the vertices in X are on the left (thus called *left vertices*) and the vertices in Y are on the right (thus called *right vertices*).[33] Note that a bipartite multigraph does not have

---

[32]On the other hand, he may want a Hamilton cycle that minimizes the total cost of his trip. Mathematically, there is no difference since, rather than attaching a weight to each edge that represents the distance between the cities it joins, we attach a weight that represents costs. In both cases we want a Hamilton cycle in which the sum of the weights attached to the edges of the cycle is minimum.

[33]Of course, *left* and *right* are interchangeable.

any loops. A multigraph that is isomorphic to a bipartite multigraph is also bipartite.

Example. A bipartite multigraph with bipartition X, Y, where $X = \{a, b, c, d\}$ and $Y = \{u, v, w\}$, is shown in Figure 11.20.   □

Example. Consider the graph G shown in Figure 11.21. Although it is not apparent from the drawing, G is a bipartite graph. This is because we may also draw G as in Figure 11.22, which reveals that G has a bipartition $X = \{a, c, g, h, j, k\}, Y = \{b, d, e, f, i\}$. □

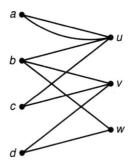

Figure 11.20

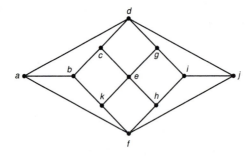

Figure 11.21

The previous example demonstrates that a drawing of a bipartite graph or a listing of its edges may not directly reveal the bipartite property. A description of the edges of a graph may reveal a bipartition of its vertices.

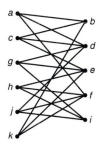

Figure 11.22

Example. Let G be the graph whose vertices are the integers from
1 to 20, with two integers joined by an edge if and only if their difference is an odd integer. The vertices of G are naturally partitioned
into the even integers and the odd integers. Since the difference between two odd integers is even and so is the difference between two
even integers, two integers are adjacent in G if and only if one is
odd and one is even. Thus, G is a bipartite graph with bipartition
$X = \{1, 3, \ldots, 17, 19\}, Y = \{2, 4, \ldots, 18, 20\}$.   □

A bipartite graph[34] G with bipartition X, Y is called *complete*,
provided that each vertex in X is adjacent to each vertex in Y. Accordingly, if X contains m vertices and Y contains n vertices, then
G has $m \times n$ edges. A complete bipartite graph with m left vertices
and n right vertices is denoted by $K_{m,n}$. The graph G in the previous
example is a $K_{10,10}$.

Since the bipartiteness of a multigraph may not be apparent from
the way it is presented, we would like to have some alternative way to
recognize bipartite multigraphs.

Theorem 11.4.1 *A multigraph is bipartite if and only if each of its
cycles has even length.*

Proof. First, assume that G is a bipartite multigraph with bipartition
X, Y. The vertices of a walk of G must alternate between X and Y.
Since a cycle is closed, this implies that a cycle contains as many left
vertices as it does right vertices and hence has even length.

Now suppose that each cycle of G has even length. First, assume
that G is connected. Let x be any vertex of G. Let X be the set
consisting of those vertices whose distance to x is even and let Y be
the set consisting of those vertices whose distance to x is odd. Since
G is assumed to be connected, X, Y is a partition of the vertices of

---
[34]Not bipartite multigraph.

G. We show that X, Y is a bipartition, that is, that no two vertices in X, respectively Y, are adjacent. Suppose, to the contrary, that there exists an edge $\{a, b\}$ where a and b are both in X. Let

$$x - \cdots - a \text{ and } x - \cdots - b \qquad (11.5)$$

be walks of shortest length from x to a and x to b, respectively. Since the first vertex of each of these walks is x, there is a vertex z that is the last common vertex of these two walks. Thus, the walks in (11.5) are of the form

$$x - \cdots - z - \cdots - a \text{ and } x - \cdots - z - \cdots - b \qquad (11.6)$$

where the walks

$$z - \cdots - a \text{ and } z - \cdots - b \qquad (11.7)$$

have no vertex in common other than z. Since the walks from x to a and x to b in (11.6) are shortest walks, the walks from x to z contained in them must have the same length. Therefore, the two walks in (11.7) are both of odd length or both of even length. The edge $\{a, b\}$ now implies the existence of a cycle

$$z - \cdots - a - b - \cdots - z$$

of odd length, contrary to hypothesis. Thus, there cannot be an edge joining two vertices in X, and similarly, one shows that there can be no edge joining two vertices in Y. Hence G is bipartite.

If G is not connected, then we apply the preceding argument to each connected component of G and conclude that each component is bipartite. But this implies that G is bipartite as well.   □

In Section 11.7 we give a simple algorithm for determining the distances from a specified vertex x of a connected graph to every other vertex. Referring to the proof of Theorem 11.4.1, this will determine a bipartition of G if G is bipartite.

Example. Let n be a positive integer. We consider the set of all n-tuples of 0's and 1's as the vertices of a graph Q_n with two vertices joined by an edge if and only if they differ in exactly one coordinate. If $x = (x_1, \ldots, x_n)$ and $y = (y_1, \ldots, y_n)$ are joined by an edge, then the number of 1's in y is either 1 more or 1 less than the number of 1's in x. Let X consist of those n-tuples that have an even number of 1's; let Y consist of those n-tuples that have an odd number of 1's. Then two distinct vertices in X differ in at least two coordinates and hence are not adjacent. Similarly, two distinct vertices in Y are not adjacent. Hence, Q_n is a bipartite graph with bipartition X, Y.

Q_n is the graph of vertices and edges of an n-dimensional cube. The graphs Q_2 and Q_3 are shown in Figures 4.2-4.3, however, in a way that does not automatically reveal their bipartite nature; the drawings given in Figure 11.23 do. The reflected Gray code constructed in Section 4.3 is a Hamilton cycle in the graph Q_n. Thus, the search for a method to generate all the combinations of an n-element set with consecutive combinations differing as little as possible (one new element in or one old element out) is the same as the search for a Hamilton cycle (or path) in the n-cube graph Q_n. □

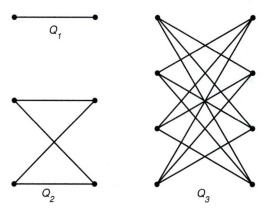

Figure 11.23

Example. Consider an n-by-n chessboard. Define a graph B_n whose vertices are the 64 squares of the board, where two squares are joined by an edge if and only if they have a common side.[35] This graph is the same as the domino bipartite graph that we associated with a general board with forbidden positions in Section 9.1. If we think of the squares of the board as alternately colored black and white, then we see that no two black squares are adjacent and no two white squares are adjacent. Thus, the usual coloring of a chessboard determines a bipartition of B_n, and hence B_n is a bipartite graph. We refer to Exercise 3 of Chapter 1, which asked whether it is possible to walk from one corner of an 8-by-8 board to the opposite corner, passing through each square exactly once. We now recognize this problem as asking whether the graph B_8 has a Hamilton path. Now B_8 is a bipartite graph with 32 white (or left) vertices and 32 black (or right) vertices. The desired Hamilton path starts and ends at a vertex of the same color, say, black. Since B_8

---

[35]That is, two squares are adjacent as vertices if and only if they are adjacent squares on the board.

is bipartite the colors of the vertices in a path must alternate. Thus, it is impossible to include all the vertices in a Hamilton path from one corner to its opposite corner, since such a path must include one more black square than white square.   □

Reasoning similar to that used in the preceding example establishes the following elementary result.

Theorem 11.4.2 *Let G be a bipartite graph with bipartition X, Y. If $|X| \neq |Y|$, then G does not have a Hamilton cycle. If $|X| = |Y|$, then G does not have a Hamilton path that begins at a vertex in X and ends at a vertex in X. If X and Y differ by at least 2, then G does not have a Hamilton path. If $|X| = |Y| + 1$, then G does not have a Hamilton path that begins at X and ends at Y, or vice versa.*

We close this section by discussing another old recreational problem[36] which, in modern language, also asks for a Hamilton cycle in a certain graph.

Example. (*The Knight's Tour Problem*). Consider an 8-by-8 chessboard and the chess piece known as a *knight*. A knight moves by moving 2 squares vertically and 1 square horizontally from its current location or 1 square vertically and 2 squares horizontally. Is it possible to place the knight on the board so that, with legal moves, the knight lands in each square exactly once? Such a tour is called a *knight's tour*, and one can ask for a knight's tour which has the property that the move from the last square to the first square is also a legal knight's move. A knight's tour with this property is called *reentrant*.

A solution of the problem, due to Euler, is

58	43	60	37	52	41	62	35
49	46	57	42	61	36	53	40
44	59	48	51	38	55	34	63
47	50	45	56	33	64	39	54
22	7	32	1	24	13	18	15
31	2	23	6	19	16	27	12
8	21	4	29	10	25	14	17
3	30	9	20	5	28	11	26

where the numbers indicate the order in which the squares are visited by the knight. In particular, square number 1 is the initial position of

---

[36]This problem was apparently first posed and solved by Indian chess players around 200 B.C.

the knight, and square 64 is the last. Since the move from square 1 to square 64 is a legal knight's move, this tour is reentrant. Note that, in this tour, the knight first visits all the squares on the lower half of the board before entering the upper half.

The problem of the knight's tour can be considered on any m-by-n board, and we recognize it as a problem of the existence of a Hamilton path in a graph. Consider the squares of an m-by-n board to be the vertices of a graph $\mathcal{K}_{m,n}$ in which two squares are joined by an edge if and only if the move from one to the other is a legal knight's move. A Hamilton path in $\mathcal{K}_{m,n}$ represents a knight's tour on the m-by-n board, and a Hamilton cycle represents a reentrant tour. Considering the squares of the board to be alternately colored black and white, as usual, we see that a knight always moves from a square of one color to a square of the other color. Thus, the graph $\mathcal{K}_{m,n}$ is a bipartite graph of order $m \times n$. If m and n are both odd, then there is one more square of one color than the other and hence, by Theorem 11.4.2, a reentrant knight's tour cannot exist. If at least one of m and n is even, then there is an equal number of black and white squares, and hence a reentrant tour is possible.

On a 1-by-n board, a knight cannot move at all. On a 2-by-n board, each of the 4 corner squares is accesssible by a knight from only one square. This means that in the graph $\mathcal{K}_{m,n}$, the corner squares each have degree equal to 1, and hence a knight's tour is impossible. What about a 3-by-3 board? On such a board the square in the middle is accessible by a knight from no other square. Hence, in the graph $\mathcal{K}_{m,m}$ the middle square has degree 0, and no tour is possible. Do not despair, for here is a nonreentrant tour, for a knight on a 3-by-4 board:

1	4	7	10
12	9	2	5
3	6	11	8

The labeling of the squares from 1 to n^2, using a knight's tour on an n-by-n board, results in a square array of numbers in which each of the numbers from 1 to n^2 appears exactly once. An unsolved problem is to determine whether there is a knight's tour on an 8-by-8 board in which the resulting array is a magic square.[37]   □

---

[37] See H.E. Dudeney: *Amusements in Mathematics*, Dover Publishing Co., New York, 1958.

11.5   Trees

Suppose we want to build a connected graph of order n, using the smallest number of edges that we can "get away with."[38] One simple method of construction is to select one vertex and join it by an edge to each of the other $n-1$ vertices. The result is a complete bipartite graph $K_{1,n-1}$, called a *star*. The star $K_{1,n-1}$ is connected and has $n-1$ edges. If we remove any edge from it, we obtain a disconnected graph with a vertex meeting no edges. Another simple method of construction is to join the n vertices in a path. The resulting graph also is connected, has $n - 1$ edges, and if we remove any edge, we obtain a disconnected graph. Can we construct a connected graph with n vertices that has fewer than $n - 1$ edges?

Suppose we have a connected graph G of order n. Let's think of putting in the edges of G one by one. Thus, we start with n vertices and no edges and hence with a graph with n connected components. Each time we put in an edge we can decrease the number of connected components by, at most, 1: If the new edge joins 2 vertices that were already in the same component, then the number of components stays the same; if the new edge joins 2 vertices that were in different components, then those two components become one and all others are unaltered. Since we start with n components, and an edge can decrease the number of components by at most 1, we require at least $n - 1$ edges in order to reduce the number of components to 1, that is, in order to get a connected graph. So we have proved the next elementary result.

Theorem 11.5.1 *A connected graph of order n has at least $n-1$ edges. Moreover, for each positive integer n, there exist connected graphs with exactly $n - 1$ edges. Removing any edge from a connected graph of order n with exactly $n-1$ edges leaves a disconnected graph, and hence each edge is a bridge.*

A *tree* is defined to be a connected graph that becomes disconnected upon the removal of any edge. Thus, a tree is a connected graph, each of whose edges is a bridge: Each edge is essential for the connectedness of the graph. We now prove that a connected graph can be shown to be a tree, simply by counting the number of its edges.

Theorem 11.5.2 *A connected graph of order $n \geq 1$ is a tree if and only if it has exactly $n - 1$ edges.*

---

[38]For example, connect n cities by roads, using the fewest number of roads, in such a way that it is possible to get from each city to every other one.

Proof. By Theorem 11.5.1 a connected graph of order n with exactly $n - 1$ edges is a tree (each of its edges is a bridge). Conversely, we prove by induction on n that a tree G of order n has exactly $n - 1$ edges. If $n = 1$, then G has no edges, and the conclusion is vacuously true. Assume that $n \geq 2$. Let α be any edge of G and let G' be the graph obtained from G by removing α. Since α is a bridge, G' has two connected components, G'_1 and G'_2, consisting of k and l vertices, respectively, where k and l are positive integers with $k + l = n$. Each edge of G'_1 is a bridge of G'_1, for, otherwise, its removal from G would clearly leave a connected graph, contrary to our assumption that G is a tree. Similarly, each edge of G'_2 is a bridge of G'_2. Thus, G'_1 and G'_2 are trees, and by the induction hypothesis, G'_1 has $k - 1$ edges, and G'_2 has $l - 1$. Hence, G has $(k - 1) + (l - 1) + 1 = n - 1$ edges, as desired. $\square$

Another characterization of a tree is given in the next theorem. First, we prove a lemma.

Lemma 11.5.3 *Let G be a connected graph and let $\alpha = \{x, y\}$ be an edge of G. Then α is a bridge if and only if no cycle of G contains α.*

Proof. First suppose that α is a bridge. Then G consists of two connected graphs held together by α, and there can be no cycle containing α.[39] Now suppose that α is not a bridge. Then removing α from G leaves a connected graph G'. Hence, there is in G', and hence in G, a path

$$x - \cdots - y$$

that joins x and y and that does not contain the edge α. Then

$$x - \cdots - y - x$$

is a cycle containing the edge α. $\square$

Theorem 11.5.4 *Let G be a connected graph of order n. Then G is a tree if and only if there are no cycles in G.*

Proof. We know that each edge of a tree is a bridge and hence, by Lemma 11.5.3, is not contained in any cycle. Thus, if G is a tree, then G does not have any cycle. Now suppose that G does not have any cycles. Since there are no cycles, it follows from Lemma 11.5.3, again, that each edge of G is a bridge and hence that G is a tree. $\square$

Theorem 11.5.4 implies another characterization of trees.

---

[39]Keep in mind that the edges of a cycle are all different.

Theorem 11.5.5 *A graph G is a tree if and only if every pair of distinct vertices x and y is joined by a unique path. This path is necessarily a shortest path joining x and y, that is, a path of length $d(x, y)$.*

Proof. First, suppose that G is a tree. Since G is connected, each pair of distinct vertices is joined by some path. If some pair of vertices is joined by two different paths, then G contains a cycle, contradicting Theorem 11.5.4.

Now suppose that each pair of distinct vertices of G is joined by a unique path. Then G is connected. Since each pair of vertices of a cycle is joined by 2 different paths, G cannot have any cycles and, once again by Theorem 11.5.4, G is a tree. □

Let G be a graph. A *pendent vertex* of G is a vertex whose degree is equal to 1. Thus, a pendent vertex is incident with exactly one edge, and any edge incident with a pendent vertex is called a *pendent edge*.

Example. The graph G of order $n = 7$, shown in Figure 11.24, has three pendent vertices, namely, a, b, and g, and three pendent edges. This graph is not a tree. This is because the edge $\{c, d\}$ is not a bridge, or because it has $7 > 6$ edges (cf. Theorem 11.5.2), or because it has a cycle (cf. Theorem 11.5.4). □

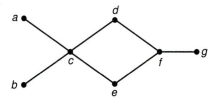

Figure 11.24

Theorem 11.5.6 *Let G be a tree of order $n \geq 2$. Then G has at least 2 pendent vertices.*

Proof. Let the degrees of the vertices of G be $d_1, d_2, \ldots, d_n$. Since G has $n - 1$ edges, it follows from Theorem 11.1.1 that

$$d_1 + d_2 + \cdots + d_n = 2(n - 1).$$

If at most one of the d_i equals 1, we have

$$d_1 + d_2 + \cdots + d_n \geq 1 + 2(n - 1),$$

a contradiction. Hence, at least 2 of the d_i equal 1; that is, there are at least 2 pendent vertices. □

Example. What is the smallest and largest number of pendent vertices a tree G of order $n \geq 2$ can have?

Each of the two vertices of a tree of order 2 is pendent. Now let $n \geq 3$. If all the vertices of a tree were pendent, then the tree would not be connected (in fact, n would have to be even and no two edges would be incident). A star $K_{1,n-1}$ has $n - 1$ pendent vertices, and hence $n - 1$ is the largest number of pendent vertices a tree of order $n \geq 3$ can have. A tree whose edges are arranged in a path has exactly 2 pendent vertices. Thus, by Theorem 11.5.6, 2 is the smallest number of pendent vertices for a tree of order $n \geq 2$. $\square$

Example. (*How to grow trees*). By Theorem 11.5.6, a tree has a pendent vertex and hence a pendent edge. If we remove an edge from a tree, G, then we get a graph with two connected components each of which is also a tree. If the edge removed is pendent, then one of the smaller trees consists of a single vertex, and the other is a tree G' of order $n - 1$. Conversely, if we have a tree G' of order $n - 1$, then selecting a new vertex u and joining it by an edge $\{u, x\}$ to a vertex x of G' we get a tree G in which u is a pendent vertex. This implies that *every* tree can be constructed as follows: Start with a single vertex and iteratively choose a new vertex, and put in a new edge joining the new vertex to any old vertex. A tree of order 5 is constructed in Figure 11.25 in this way. $\square$

Figure 11.25

Using the method of construction of the previous example, it is not difficult to now show that the number t_n of nonisomorphic trees of order n satisfies $t_1 = 1, t_2 = 1, t_3 = 1, t_4 = 2, t_5 = 3$, and $t_6 = 6$. The different trees with 6 vertices are shown Figure 11.26.

We have defined a tree to be a connected graph, each of whose edges is a bridge. Thus, if a connected graph G is not a tree, then it has a nonbridge; that is, an edge whose removal does not disconnect the graph. If we iteratively remove nonbridge edges until every edge is a bridge of the remaining graph, we get a tree with the same set of vertices as G and some of its edges; that is, we get a spanning subgraph that is a tree. A tree that is a spanning subgraph of a graph G is called a *spanning tree* of G.

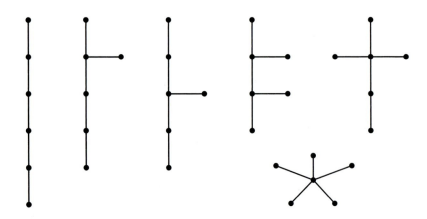

Figure 11.26

Theorem 11.5.7 *Every connected graph has a spanning tree.*

Proof. The algorithmic proof is contained in the preceding paragraph. We give a more precise formulation of the algorithm. Recall from Lemma 11.3.1 that an edge of a connected graph is a bridge if and only if it is not contained in any cycle.

Algorithm for a spanning tree

Let $G = (V, E)$ be a connected graph of order n.

(i) Set F equal to E.

(ii) While there is an edge α of F such that α is not a bridge of the graph $T = (V, F)$, remove α from F.

The terminal graph $T = (V, F)$ is a spanning tree of G.

As we have argued, the terminal graph $T = (V, F)$ is connected and does not have any bridges; hence, it is a tree.   □

We remark that our restriction to graphs in Theorem 11.5.7 is not essential. If G is a general graph, then we can immediately remove all loops, and all but one copy of each edge in G, and then apply Theorem 11.5.7 and the algorithm in its proof. Thus, every connected general graph has a spanning tree as well.

Example. Let G be the connected graph of order 7, shown on the left in Figure 11.27. This graph has exactly one bridge; hence, we can begin the algorithm for a spanning tree by removing any other edge, say the edge $\{1, 2\}$. The edges $\{1, 4\}, \{4, 5\}, \{2, 5\}$, and $\{2, 3\}$ are now bridges and can no longer be removed. Removing the edge $\{6, 7\}$ leaves the spanning tree shown on the right.   □

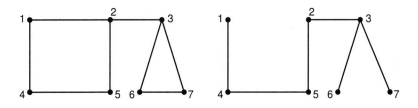

Figure 11.27

We conclude this section with two properties of spanning trees that will be used in subsequent sections of this chapter.

Theorem 11.5.8 *Let T be a spanning tree of a connected graph G. Let $\alpha = \{a, b\}$ be an edge of G that is not an edge of T. Then there is an edge β of T such that the graph T' obtained from T by inserting α and deleting β is also a spanning tree of G.*

Proof. Let the graph G, and hence the graph T, have n vertices. First, consider the graph T' obtained from T by inserting the given edge α. Since T' is not a tree, it has, by Theorem 11.5.4, a cycle γ which necessarily contains at least one edge of T. By Lemma 11.3.1, each edge of γ is not a bridge of T'. Let β be any edge of γ other than α. Removing β from T' results in a graph with n vertices and $n-1$ edges that is connected and hence is a tree. □

Theorem 11.5.9 *Let T_1 and T_2 be spanning trees of a connected graph G. Let β be an edge of T_1. Then there is an edge α of T_2 such that the graph obtained from T_1 by inserting α and deleting β is a spanning tree of G.*

Proof. We first remark on the difference between Theorems 11.5.8 and 11.5.9. In Theorem 11.5.8 we are given a spanning tree and some edge α not in it, and we want to put α in and take out *any* edge β of T as long as the result is a spanning tree. In Theorem 11.5.9 we are given a spanning tree T_1 and we want to take out a *specific* edge β of T_1 and put in any edge of T_2 as long as the result is a spanning tree.

To prove the theorem, first remove the edge β from the spanning tree T_1 of G. The result is a graph with two connected components T_1' and T_2'' (both of which must be trees). Since T_2 is also a spanning tree of G, T_2 is connected with the same set of vertices as T_1, and hence there must be some edge α of T_2 that joins a vertex of T_1' and a vertex of T_2''. The graph obtained from T_1 by inserting the edge α and removing the edge β is a connected graph with $n-1$ edges; hence, it is a tree. (We note that if β is not an edge of T_2, then α is not an edge of T_1, for otherwise we would get a connected graph of order n with fewer than $n-1$ edges.)   □

It is natural for us to ask for the number of spanning trees of a connected graph. The number of spanning trees of any connected graph can be computed by an algebraic formula,[40] but such a formula is beyond the scope of this book.

Example. The number of spanning trees of the graph of order 4 (a cycle of length 4) shown in Figure 11.28 is 4 as shown. Each of these spanning trees is a path of length 3. Consequently, all are isomorphic. □

A famous formula of Cayley asserts that the number of spanning trees of a complete graph K_n is n^{n-2}, a surprisingly simple formula.

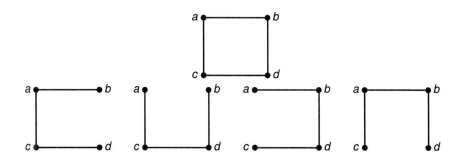

Figure 11.28

As illustrated in the preceding example, many of these trees may be isomorphic to each other. Thus, while each tree of order n occurs as a spanning tree of K_n, it may occur many times (with different labels on its vertices). Thus, n^{n-2} does not represent the number of nonisomorphic trees of order n. The latter number is a more complicated function of n.

11.6   The Shannon Switching Game

We discuss in this section a game that can be played on any multigraph. It was invented by C. Shannon[41] and its elegant solution was found

---

[40]It is the absolute value of the determinant of any submatrix of order $n-1$ of the Laplacian matrix of a graph!

[41]Shannon is generally recognized as the founder of modern communication theory.

by A. Lehman.[42] The remainder of this book is independent of this section.

Shannon's game is played by two people, called here the *positive player* P and the *negative player* N, who alternate turns.[43] Let $G = (V, E)$ be a multigraph in which two of its vertices u and v have been distinguished. Thus, the "gameboard" consists of a multigraph with two distinguished vertices. The goal of the positive player is to construct a path between the distinguished vertices u and v. The goal of the negative player is to deny the positive player his goal, that is, to destroy all paths between u and v. The play of the game proceeds as follows: When it is N's turn, N destroys some edge of G by putting a negative sign $-$ on it.[44] When it is P's turn, P puts a positive sign $+$ on some edge of G, which now cannot be destroyed by N. Play proceeds until one of the players achieves his goal:

(i) There is a path between u and v that has only $+$ signs on its edges. In this case, *the positive player has won.*

(ii) Every path in G between u and v contains a $-$ sign on at least one of its edges; that is, N has destroyed all paths between u and v. In this case *the negative player has won.*

It is evident that, after all edges of the multigraph G have been played, that is, have either a $+$ or a $-$ on them, exactly one of the players will have won. In particular, the game never ends in a draw. If G is not connected and u and v lie in different connected components of G, then we can immediately declare N the winner.[45]

We consider the following questions:

(a) Does there exist a strategy that P can follow which will guarantee him a win, *no matter how well N plays?* If so, determine such a winning strategy for P.

(b) Does there exist a strategy that N can follow which will guarantee him a win, *no matter how well P plays?* If so, determine such a winning strategy for N.

---

[42] A. Lehman: A solution of the Shannon switching game, *J. Society Industrial and Applied Mathematics*, 12 (1964), 687-725. Our description of the game and its solution is based on Section 3 of the author's article *Networks and the Shannon switching game*, Delta, 4 (1974), 1-23.

[43] Or the *constructive* and *destructive* player, respectively.

[44] If the game is played by drawing G on paper with a pencil, then N can destroy an edge by erasing it.

[45] And P should be embarrassed for getting involved in a game in which it was impossible for him to win!

The answers to these questions may sometimes depend on whether the positive or negative player has the first move.

Example. First, consider the multigraph on the left in Figure 11.29, with distinguished vertices u and v as shown. In this game the positive player P wins, whether he plays first or second. This is because a $+$ on either edge determines a path between u and v. Now consider the middle graph in Figure 11.29. In this game the negative player N wins, whether he plays first or second. This is because a $-$ on either of the two edges destroys all paths between u and v. Finally, consider the right graph in Figure 11.29. In this game, whichever player goes first, and thereby claims the only edge of the graph, is the winner.   □

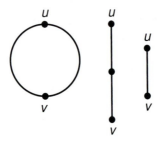

Figure 11.29

Motivated by the preceding example, we make the following definitions: A game is called a *positive game*, provided that the positive player has a winning strategy, whether he plays first or second. A game is called a *negative game*, provided that the negative player has a winning strategy, whether he plays first or second. A game is called a *neutral game*, provided that the player who plays first has a winning strategy. We note that, if the positive player has a winning strategy when he plays second, then he also has a winning strategy when he plays first. This is because the positive player can ignore his first move[46] and play according to the winning strategy as the second player. If the strategy calls for him to put a $+$ on an edge that already has one, he then has a "free move" and can put a $+$ on any available edge. Similarly, if the negative player has a winning strategy when he plays second, then he has a winning strategy when he plays first.

---

[46]But the negative player cannot.

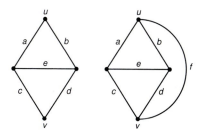

Figure 11.30

Example. Consider the game determined by the left graph in Figure 11.30, with distinguished vertices u and v as shown. Assume that P has first move and puts a $+$ on edge e. We pair up the remaining edges by pairing a with b and c with d. If P counters a move by N on an edge, by a move on the other edge of its pair, then P is guaranteed a win. Thus, P can win this game, provided he has first move. Now assume that N has first move and puts a $-$ on edge e. We now pair up the remaining edges by pairing a with c and b with d. If N counters a move by P on an edge by a move on the other edge of its pair, then N is guaranteed a win. Hence, N can win this game, provided he has first move. We conclude that the game determined by Figure 11.30 is a neutral game.

Now suppose that we add a new edge f, which joins the distinguished vertices u and v, resulting in the graph shown on the right in Figure 11.30. Suppose the negative player makes the first move in this new game. If N does not put a $-$ on the new edge f, then the positive player can put a $+$ on that edge, thereby winning the game. If N does put a $-$ on f, then the rest of the game is the same as the previous game, with P making the first move and hence P can win. Thus, P has a winning strategy as second player, and this game is a positive game.   □

The principle illustrated in the previous example holds in general.

Theorem 11.6.1 *A neutral game is converted into a positive game if a new edge joining the distinguished vertices u and v is added to the multigraph of the game.*

A characterization of positive games is given in the next theorem. Recall that, if $G = (V, E)$ is a multigraph and U is a subset of the vertex set V, then G_U denotes the multisubgraph of G induced by U— that is, the multigraph with vertex set U whose edges are all the edges of G that join two vertices in U. Put another way, G_U is obtained from

G by deleting all vertices in $\overline{U} = V - U$ and all edges that are incident with at least one vertex in $\overline{U}$.

Theorem 11.6.2 *The game determined by a multigraph $G = (V, E)$ with distinguished vertices u and v is a positive game if and only if there is a subset U containing u and v of the vertex set V such that the induced multisubgraph G_U has two spanning trees, T_1 and T_2, with no common edges.*

Otherwise stated, a game is a positive game if and only if there are two trees T_1 and T_2 in G such that T_1 and T_2 have the same set of vertices, both u and v are vertices of T_1 and T_2, and T_1 and T_2 have no edges in common. The game determined by the right graph in Figure 11.30 was shown to be a positive game. For T_1 and T_2 we can take the two trees in Figure 11.31. In this case T_1 and T_2 are spanning trees of G; that is, $U = V$, but this need not always be so. It is possible that the set U contain only some of the vertices of V.

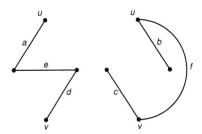

Figure 11.31

We shall not give a complete proof of Theorem 11.6.2. Rather, we shall show only how to use the pair of trees T_1 and T_2 in order to devise a winning strategy for the positive player P when the negative player N makes the first move. After each sequence of play, consisting of a move by the negative player followed by a move by the positive player, we shall construct a new pair of spanning trees of G_U that have one more edge in common than the previous pair. Initially, we have the spanning trees T_1 and T_2 of G_U with no edges in common, and we now label these trees as

$$T_1^{(0)} = T_1 \text{ and } T_2^{(0)} = T_2.$$

The first sequence of play

Player N goes first and puts a $-$ on some edge β. We consider two cases:

Case 1: β is an edge of one of the trees $T_1^{(0)}$ and $T_2^{(0)}$, say, the tree $T_1^{(0)}$.

Since $T_1^{(0)}$ and $T_2^{(0)}$ are spanning trees of G_U, it follows from Theorem 11.5.9 that there is an edge α of $T_2^{(0)}$ such that the graph obtained from $T_1^{(0)}$ by inserting α and deleting β is a spanning tree $T_1^{(1)}$ of G_U. Our instructions to P are to put a $+$ on the edge α. We let $T_2^{(1)} = T_2^{(0)}$. The trees $T_1^{(1)}$ and $T_2^{(1)}$ have exactly one edge in common, namely, the edge α with a $+$ on it.

Case 2: β is neither an edge of $T_1^{(0)}$ nor an edge of $T_2^{(0)}$.

Our instructions to P are now to place a $+$ on any edge α of $T_1^{(0)}$ or of $T_2^{(0)}$, say, an edge α of $T_1^{(0)}$.[47] Since $T_2^{(0)}$ is a spanning tree of G_U and α is an edge of G_U, it follows from Theorem 11.5.9 that there is an edge γ of $T_2^{(0)}$ such that the graph obtained from $T_2^{(0)}$ by inserting α and deleting γ is a spanning tree $T_2^{(1)}$ of G_U. We let $T_1^{(1)} = T_1^{(0)}$. The trees $T_1^{(1)}$ and $T_2^{(1)}$ have only the edge α with a $+$ in common.

We conclude that, at the end of the first sequence of play, there are two spanning trees, $T_1^{(1)}$ and $T_2^{(1)}$, of G_U that have exactly one edge in common, namely, the edge with a $+$ on it that was played by P.

The second sequence of play

Player N puts a $-$ on a second edge δ of G, and we seek a counter-move for P. The determination of an edge ρ on which P should put a $+$ is very much like that in the first sequence of play, and we shall be briefer in our description:

Case 1: δ is an edge of one of the two trees $T_1^{(1)}$ and $T_2^{(1)}$, say, the tree $T_2^{(1)}$.

There is an edge ρ of $T_1^{(1)}$ such that the graph $T_1^{(2)}$ obtained from $T_1^{(1)}$ by inserting the edge δ and deleting the edge ρ is a spanning tree of G_U. Our instructions to P are to place a $+$ on the edge ρ. We let $T_2^{(2)} = T_2^{(1)}$.

Case 2: δ is neither an edge of $T_1^{(1)}$ nor of $T_2^{(1)}$.

---

[47] In this case N has "wasted" his move and P gets a "free" move anywhere on one of the trees $T_1^{(0)}$ and $T_2^{(0)}$.

Our instructions to P are to place a $+$ on any available edge[48] of $T_1^{(1)}$ and $T_2^{(1)}$, say, an edge ρ of $T_1^{(1)}$. There exists an edge ϵ of $T_2^{(1)}$ such that the graph $T_2^{(2)}$ obtained from $T_2^{(1)}$ by inserting the edge ρ and deleting the edge ϵ is a spanning tree of G_U. We let $T_1^{(2)} = T_1^{(1)}$.

We conclude that, at the end of the second sequence of play, there are two spanning trees, $T_1^{(2)}$ and $T_2^{(2)}$, of G_U that have exactly two edges in common, namely, the two edges with a $+$ on them that were played by P.

The description of the remainder of the strategy for P is very similar to that given for the first and second sequence of play. At the end of the kth sequence of play there are two spanning trees, $T_1^{(k)}$ and $T_2^{(k)}$ of G_U, which have exactly k edges in common, namely, the k edges with a $+$ on them that have been played up to this point by P. Let the number of vertices in U be m. Then, at the end of the $(m-1)$st sequence of play, the spanning trees $T_1^{(m-1)}$ and $T_2^{(m-1)}$ of G_U have exactly $m-1$ edges in common. Since a tree with m vertices has only $m-1$ edges this means that $T_1^{(m-1)}$ is the same tree as $T_2^{(m-1)}$, and thus the edges with a $+$ on them are the edges of a spanning tree of G_U. Because u and v belong to U, there is a path of edges with a $+$ on them joining the distinguished vertices u and v. We therefore conclude that, had the positive player P followed our instructions, then, at the end of the $(m-1)$st sequence of play, if not before, he would have put $+$ signs on a set of edges that contains a path joining u and v and thus would have won the game. Our instructions to P are thus a winning strategy for him.

Theorem 11.6.2 can be used to classify neutral and negative games as follows: Let $G = (V, E)$ be a multigraph with distinguished vertices u and v. Let G^* be the multigraph obtained from G by inserting a new edge joining u and v. Then the following conclusions can be drawn:

1. The game played with G, u, and v is a neutral game if and only if it is not a positive game, but the game played with G^*, u, and v is a positive game.

2. The game played with G, u, and v is a negative game if and only if neither the game played with G, u, and v nor the game played with G^*, u, and v are positive games.

---

[48]That is, an edge that has not yet been "signed."

Thus, by Theorem 11.6.2, the game played with G, u, and v is a neutral game if and only if G does not contain two disjoint trees with the same set of vertices including u and v, but by inserting a new edge joining u and v we are able to find two such trees. The game played with G, u, and v is a negative game if and only if, even with the new edge joining u and v, two such trees do not exist. In a neutral game G, the positive player can win when he goes first by pretending that the game is being played with G^* with N going first and that N's first move was to put a $-$ on the new edge joining u and v. In general, there is no easily describable winning strategy for negative games in which N goes second or for neutral games in which N goes first.

11.7 More on Trees

In the proof of Theorem 11.5.7, we have given an algorithm for obtaining a spanning tree of a connected graph. Reviewing this algorithm, we see that it is more "destructive" than it is constructive: Iteratively, we locate an edge that is in a cycle—a nonbridge edge—of the current graph and remove or "destroy" it. Implicit in this algorithm is the assumption that we have some subalgorithm for locating a nonbridge edge. In Section 11.5, we have also described a procedure that will construct any tree with n vertices, equivalently, any spanning tree of the complete graph K_n of order n. This procedure can be refined to apply to any graph[49] in order to grow all of its spanning trees. We formalize the resulting algorithm now. It need not be assumed that the initial graph G is connected. A by-product of the algorithm is an algorithm to determine whether a graph is connected.

Algorithm to grow a spanning tree

Let $G = (V, E)$ be a graph of order n and let u be any vertex.

(1) Put $U = \{u\}$ and $F = \emptyset$.

(2) While there exists a vertex x in U and a vertex y not in U such that $\alpha = \{x, y\}$ is an edge of G,

    (i) Put the vertex y in U.

    (ii) Put the edge α in F.

---

[49]There is no loss in generality in considering only graphs in this section. If we have a general graph, we can immediately remove all loops and all but one copy of each edge and apply the results and algorithms of this section to the resulting graph.

(3) Put $T = (U, F)$.

In step (2) there will, in general, be many choices for the vertices x and y, and thus we have considerable latitude in carrying out the algorithm. Two special and important rules for choosing x and y are described after the next theorem.

Theorem 11.7.1 *Let $G = (V, E)$ be a graph. Then G is connected if and only if the graph $T = (U, F)$ constructed by carrying out the preceding algorithm is a spanning tree of G.*

Proof. If T is a spanning tree of G, then surely G is connected. Now assume that G is connected. Initially, T has one vertex and no edges and is therefore connected. Each application of (2) adds one new vertex to U and one new edge to F, which joins the new vertex to an old vertex. It then follows inductively that, at each stage of the algorithm, the current $T = (U, F)$ is connected with $|F| = |U| - 1$, and hence T is a tree. Suppose that, upon termination of the algorithm, we have $U \neq V$. Since G is connected, there must be an edge from some vertex in U to some vertex not in U, contradicting the assumption that the algorithm has terminated. Thus, upon termination, we have $U = V$, and $T = (U, F)$ is a spanning tree of G. $\qquad \square$

It should be clear that each spanning tree of a connected graph can be constructed by making the right choices for x and y in carrying out the algorithm for growing a spanning tree. We now describe one way to make choices that result in a spanning tree with a special property. The resulting algorithm is described next and it constructs what is called a *breadth-first spanning tree* rooted at a prescribed vertex, the initial vertex u in the set U. A connected graph G has, in general, many breadth-first spanning trees T rooted at a vertex u. Their common feature is that the distance between u and x in G is the same as the distance between u and x in T for each vertex x. For convenience, we call a breadth-first spanning tree a *BFS-tree*. In the algorithm, we attach two numbers to each vertex x. One of these is called its *breadth-first number*, denoted $bf(x)$. The breadth-first numbers represent the order in which vertices are put into the BFS-tree. The other number represents the distance between u and x in the BFS-tree, and is denoted by $D(x)$.[50]

---

[50]The number $D(x)$ depends on the choice of root u, but otherwise depends only on the graph G and not on the particular BFS-tree rooted at u. The number $bf(x)$ does depend on the BFS-tree.

BF-algorithm to grow a BFS-tree rooted at u

Let $G = (V, E)$ be a graph of order n and let u be any vertex.

(1) Put $i = 1$, $U = \{u\}$, $D(u) = 0$, $bf(u) = 1$, $F = \emptyset$, and $T = (U, F)$.

(2) If there is no edge in G that joins a vertex x in U to a vertex y not in U, then stop. Otherwise, determine an edge $\alpha = \{x, y\}$ with x in U and y not in U such that x has the smallest breadth-first number $bf(x)$, and do the following:

   (i) Put $bf(y) = i + 1$.

   (ii) Put $D(y) = D(x) + 1$.

   (iii) Put the vertex y into U.

   (iv) Put the edge $\alpha = \{x, y\}$ into F.

   (v) Put $T = (U, F)$.

   (vi) Increase i by 1 and go back to (2).

Theorem 11.7.2 *Let $G = (V, E)$ be a graph and let u be any vertex of G. Then G is connected if and only if the graph $T = (U, F)$ constructed by carrying out the preceding BF-algorithm is a spanning tree of G. If G is connected, then, for each vertex y of G, the distance in G between u and y equals $D(y)$, and this is the same as the distance between u and y in T.*

Proof. The BF-algorithm is a special way of carrying out the general algorithm for growing a spanning tree. It thus follows from Theorem 11.7.1 that G is connected if and only if the terminal graph $T = (U, F)$ is a spanning tree.

Now assume that G is connected so that at the termination of the algorithm $T = (U, F)$ is a spanning tree of G. It should be clear from the algorithm that $D(y)$ equals the distance between u and y in the tree T. Trivially, $D(u) = 0$ is the distance between u and itself in G. Suppose that there is some vertex y such that $D(y) = l$ is greater than the distance k between u and y in G. We may assume that k is the smallest number with this property. Then there is a path

$$\gamma: \quad u = x_0 - x_1 - \cdots - x_{k-1} - x_k = y$$

in G joining u and y whose length k satisfies

$$k < l = D(y).$$

The distance between u and the vertex x_{k-1} of γ is, at most, $k-1$ and hence, by the minimality of k, $D(x_{k-1}) \le k-1$. Since $y = x_k$ is adjacent to x_{k-1}, it follows from the BF-algorithm that we would put $D(y) = k$ unless $D(y)$ had already been assigned a smaller number. Hence, $D(y) \le k < l$, a contradiction. Therefore, the function D gives the distance in G (and in T) from u to each vertex. □

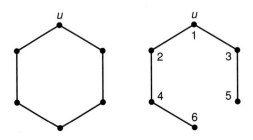

Figure 11.32

Example. Each BFS-tree of a complete graph K_n is a star $K_{1,n}$. A BFS-tree of the cycle of length 6 on the left in Figure 11.32 is the tree on the right in that figure. A BFS-trees of the graph Q_3 of vertices and edges of a three-dimensional cube is shown in Figure 11.33. (Recall from Section 11.4 that the vertices of this graph are the 3-tuples of 0's and 1's and that two vertices are adjacent if and only if they differ in exactly one coordinate.) In each case, the breadth-first numbers are noted next to the vertices of the tree. The distances $D(x)$ are readily determined. □

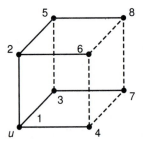

Figure 11.33

A breadth-first spanning tree rooted at u of a connected graph G is a spanning tree that is as "broad" as possible; each vertex is as close to the root as G will allow. The algorithm for a BFS-tree can be regarded as a systematic way to search (or list) all the vertices of G without repetition. According to this algorithm, one visits the vertices closest to the root first (breadth takes precedence over depth). We now describe a way to carry out the algorithm to grow a tree that produces a spanning tree that is as deep as possible. A spanning tree produced by this algorithm is called a *depth-first spanning tree*, abbreviated a *DFS-tree*, rooted at a vertex u. In this case, depth takes precedence over breadth. In the algorithm, we attach a number to each vertex x, called its *depth-first number*, and denoted by $df(x)$. The depth-first algorithm is also known as *backtracking*. In backtracking one proceeds in the forward direction as long as one is able; when it is no longer possible to advance, then one backtracks to the first vertex from which one can go forward.

DF-algorithm to grow a DFS-tree rooted at u

Let $G = (V, E)$ be a graph of order n and let u be any vertex.

(1) Put $i = 1$, $U = \{u\}$, $df(u) = 1$, $F = \emptyset$, and $T = (U, F)$.

(2) If there is no edge in G that joins a vertex x in U to a vertex y not in U, then stop. Otherwise, determine an edge $\alpha = \{x, y\}$ with x in U and y not in U such that x has the largest depth-first number $df(x)$, and do the following:

   (i) Put $df(y) = i + 1$.
   (ii) Put the vertex y into U.
   (iii) Put the edge $\alpha = \{x, y\}$ into F.
   (iv) Put $T = (U, F)$.
   (vi) Increase i by 1 and go back to (2).

Theorem 11.7.3 *Let $G = (V, E)$ be a graph and let u be any vertex of G. Then G is connected if and only if the graph $T = (U, F)$, constructed by carrying out the preceding DF-algorithm, is a spanning tree of G.*

Proof. The DF-algorithm is a special way of carrying out the general algorithm for growing a spanning tree. It thus follows from Theorem 11.7.1 that G is connected if and only if the constructed graph $T = (U, F)$ is a spanning tree. □

Example. Each DFS-tree of a complete graph K_n is a path. A DFS-tree of a cycle of any length is also a path. A DFS-tree of the graph Q_3 of vertices and edges of a three-dimensional cube is shown in Figure 11.34. In each case, the depth-first numbers are noted next to the vertices of the tree. □

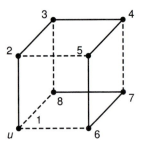

Figure 11.34

Example. If G is a tree, then each BFS-tree and DFS-tree of G is G itself, with its vertices ordered in the order they are visited. In this case, one often speaks of a *breadth-first search* of G and a *depth-first search* of G. The tree G may represent a data structure for a computer file in which information is stored at places corresponding to the vertices of G. In order to find a particular piece of information, one needs to "search" each vertex of the tree until one finds the desired information. Both a breadth-first search and a depth-first search provide an algorithm for searching each vertex, at most, once. If we think of a tree as a system of roads connecting various cities, then a depth-first search of G can be visualized as a walk along the edges, in which each vertex is visited at least once.[51] Starting at the root u we walk in the forward direction as long as possible and go backward only until we locate a vertex from which we can again go forward. Such a walk is illustrated in Figure 11.35, where we have returned to the root u (so our walk is a closed walk in which we traverse each edge exactly twice).

$\square$

According to Theorem 11.7.2, the number $D(x)$ computed by the breadth-first algorithm starting with a vertex u equals the distance from u to x in a connected graph. However, in graphs that model various physical situations, some edges are more "costly" than others. An edge might represent a road connecting two cities, and the physical distance between these cities should be taken into account if the graph is to provide an accurate model. An edge might also represent a potential new road between two cities, and the cost of constructing that road must be considered. These two situations motivate us to consider graphs in which a weight is attached to each edge.[52]

---

[51]But we search each vertex only the first time it is visited.

[52]The physical significance of the weight is irrelevant for the mathematical problems that we solve. However, the fact that weight may have relevant physical significance leads to important applications of the mathematical results obtained.

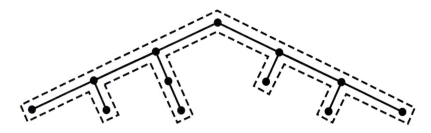

Figure 11.35

Let $G = (V, E)$ be a graph in which to each edge $\alpha = \{x, y\}$ there is associated a nonnegative number $c(\alpha) = c\{x, y\}$, called its *weight*. We call G a *weighted graph* with weight function c. The *weight of a walk*

$$\gamma : \{x_0, x_1\}, \{x_1, x_2\}, \ldots, \{x_{k-1}, x_k\}$$

in G is defined to be

$$c(\gamma) = c\{x_0, x_1\} + c\{x_1, x_2\} + \cdots + c\{x_{k-1}, x_k\},$$

the sum of the weights of the edges of γ. The *weighted-distance* $d_c(x, y)$ between a pair of vertices x and y of G is the smallest weight of all the walks joining x and y. If there is no walk joining x and y, then we define $d_c(x, y) = \infty$. We also define $d_c(x, x) = 0$ for each vertex x. Since all weights are nonnegative, if $d_c(x, y) \neq \infty$, then there is a path of weight $d_c(x, y)$ joining the pair of distinct vertices x and y. Starting with a vertex u in a connected graph G, we show how to compute $d_c(u, x)$ for each vertex x and construct a spanning tree rooted at u such that the weighted-distance between u and each vertex x equals $d_c(u, x)$. We call such a spanning tree a *distance-tree for u*. The algorithm presented next is usually called *Dijkstra's algorithm*[53] and can be regarded as a weighted generalization of the BF-algorithm.

Algorithm for a distance-tree for u

Let $G = (V, E)$ be a weighted graph of order n and let u be any vertex.

(1) Put $U = \{u\}$, $D(u) = 0$, $F = \emptyset$, and $T = (U, F)$.

(2) If there is no edge in G that joins a vertex x in U to a vertex y not in U, then stop. Otherwise, determine an edge $\alpha = \{x, y\}$ with x in U and y not in U such that $D(x) + c\{x, y\}$ is as small as possible, and do the following:

---

[53]E.W. Dijkstra: A note on two problems in connection with graphs, *Numerische Math.*, 1 (1959), 285-292.

(i) Put the vertex y into U.

(ii) Put the edge $\alpha = \{x, y\}$ into F.

(iii) Put $D(y) = D(x) + c\{x, y\}$ and go back to (2).

Theorem 11.7.4 *Let $G = (V, E)$ be a weighted graph and let u be any vertex of G. Then G is connected if and only if the graph $T = (U, F)$ obtained by carrying out the preceding algorithm is a spanning tree of G. If G is connected, then for each vertex y of G, the distance between u and y equals $D(y)$, and this is the same as the distance between u and y in the weighted tree T.*

Proof. The algorithm for a distance tree is a special way of carrying out our general algorithm for growing a spanning tree. It thus follows from Theorem 11.7.1 that G is connected if and only if the constructed graph $T = (U, F)$ is a spanning tree, that is, if and only if the terminal value of U is V.

Now, assume that G is connected, so that at the termination of the algorithm, $U = V$, and $T = (U, F)$ is a spanning tree of G. It is clear from the algorithm that $D(y)$ equals the distance between u and y in the tree T. Trivially, $D(u) = 0$ is the distance between u and itself in G. Suppose, to the contrary, that there is some vertex y such that $D(y)$ is greater than the distance d between u and y in G. We may assume that y is the first vertex put in U with this property. There is a path

$$\gamma : \quad u = x_0 - x_1 - \cdots - x_k = y$$

in G joining u and y whose weight is $d < D(y)$. Let x_j be the last vertex of γ which is put into U before y. (Since u is the first vertex put into U, the vertex x_j exists.) It follows from our choice of y that $D(x_j)$ equals the weighted-distance from u to x_j in G. The subpath

$$\gamma' : \quad u = x_0 - x_1 - \cdots - x_j - x_{j+1}$$

of γ has weight

$$D(x_j) + c\{x_j, x_{j+1}\} \leq d < D(y).$$

Hence, by the algorithm, x_{j+1} is put into U before y, contradicting our choice of x_j. This contradiction implies that $D(y)$ is the weighted-distance between u and y for all vertices y. $\square$

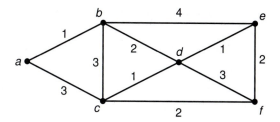

Figure 11.36

Example. Let G be the weighted graph in Figure 11.36, where the numbers next to an edge denote its weight. If we carry out the algorithm for a distance-tree for u, we obtain the tree drawn in Figure 11.37, with the vertices and edges selected in the following order:

$$\text{vertices}: a,\ b,\ d,\ c,\ e,\ f,$$

$$\text{edges}: \{a,b\},\ \{b,d\},\ \{a,c\},\ \{d,e\},\ \{c,f\}.$$

$\square$

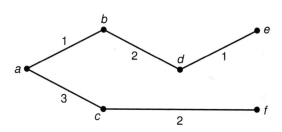

Figure 11.37

We conclude this section by discussing another practical problem, called the *minimum connector problem*. Its practicality is illustrated in the next example.

Example. There are n cities $A_1, A_2, \ldots, A_n$, and it is desired to connect some of them by highways so that each city is accesssible from any other. The cost of constructing a direct highway between city A_i and city A_j is estimated to be $c\{A_i, A_j\}$. Determine which cities should be directly connected by highways in order to minimize the total construction costs.

Since we are to minimize the total construction costs, a solution of the problem corresponds to a tree[54] with vertices $A_1, A_2, \ldots, A_n$, in

---

[54] If we did not have a tree, we could eliminate one or more of the highways without destroying the accessibility feature and thereby reduce costs.

which there is an edge joining cities A_i and A_j if and only if we put a direct highway between A_i and A_j. Indeed, if we consider the complete graph K_n with the n vertices $A_1, A_2, \ldots, A_n$, whose edges are weighted by the construction costs in the problem, then we seek a spanning tree the sum of whose edge weights is as small as possible. In what follows, we give two algorithms to solve the "minimum-weight spanning tree problem" for any weighted connected graph. □

Let $G = (V, E)$ be a weighted connected graph with weight function c. We define the *weight of a subgraph H* of G to be

$$c(H) = \sum_{\{\alpha \text{ an edge of } H\}} c(\alpha),$$

the sum of the weights of the edges of H. A spanning tree of G that has the smallest weight of all spanning trees of G is a *minimum-weight spanning tree*. If all the edges of G have the same weight, then every spanning tree of G is a minimum-weight spanning tree. Given any connected graph, by appropriately assigning weights to its edges, we can make any spanning tree the unique minimum-weight spanning tree. We now describe an algorithm known as Kruskal's algorithm.[55] This algorithm is also known as a *greedy algorithm*, since, at each stage, we choose an edge of smallest weight consistent with the fact that, upon termination, the chosen edges are to be the edges of a spanning tree. Consistency is simply the idea that we should never choose edges which can be used to create a cycle.

Greedy algorithm for a minimum-weight spanning tree

Let $G = (V, E)$ be a weighted connected graph with weight function c.

(1) Put $F = \emptyset$.

(2) While there exists an edge α not in F such that $F \cup \{\alpha\}$ does not contain the edges of a cycle of G, determine such an edge α of minimum weight and put α in F.

(3) Put $T = (V, F)$.

Theorem 11.7.5 *Let $G = (V, E)$ be a weighted connected graph with weight function c. Then the preceding greedy algorithm constructs a minimum-weight spanning tree $T = (V, F)$ of G.*

---

[55] J.B. Kruskal, Jr.: On the shortest spanning subtree of a graph and the traveling salesman problem, *Proc. Amer. Math. Soc.*, 7 (1956), 48-50.

Proof. In the greedy algorithm, we begin with $n = |V|$ vertices and no edges (initially $F = \emptyset$), and hence with a spanning graph (V, F) with n connected components. Choosing an edge α that does not create a cycle means that α joins vertices in different components of (V, F), and hence putting α in F decreases the number of connected components by 1. On termination, we have $n - 1$ edges in F, and hence $T = (V, F)$ is a spanning tree. We now show that T is a minimum-cost spanning tree.

Let the $n - 1$ edges of F be $\alpha_1, \alpha_2, \ldots, \alpha_{n-1}$ in the order that they are put in F. Let $T^* = (V, F^*)$ be a minimum-weight spanning tree, which has the largest number of edges in common with T. Thus, no minimum-weight spanning tree has more edges in common with F than F^* does. If we can show that $F^* = F$, then it follows that T is a minimum-weight spanning tree. Suppose, to the contrary, that $F^* \neq F$. Let α_k be the first edge of F that is not in F^*. Thus, the edges $\alpha_1, \ldots, \alpha_{k-1}$ all belong to F^*. By Theorem 11.5.8, there is an edge β of T^* such that the graph T^{**}, obtained from T^* by inserting α_k and deleting β, is a spanning tree of G. The edge β is an edge of the cycle that is created by inserting the edge α_k into T^*; since T is a tree, at least one of the edges of the cycle does not belong to T, and we choose such an edge β. We have

$$c(T^{**}) = c(T^*) - c(\beta) + c(\alpha_k). \qquad (11.8)$$

Since T is a minimum-weight spanning tree, we conclude that

$$c(\alpha_k) \geq c(\beta). \qquad (11.9)$$

Because $L = \{\alpha_1, \ldots, \alpha_{k-1}, \beta\}$ is a subset of the edges of T^*, no cycle has all its edges contained in L. Hence, in determining the kth edge to be put in F in carrying out the greedy algorithm, β is a possible choice. It thus follows from (11.9) that

$$c(\alpha_k) = c(\beta)$$

and from (11.7.5) that T^{**} is also a minimum-weight spanning tree. Since T^{**} has one more edge[56] in common with T than T^* has, we contradict our choice of T^*; the proof of the theorem is complete. □

Example. Let G be the weighted graph of order 7, shown in Figure 11.38, where the numbers next to the edges are their weights. In applying the greedy algorithm to determine a minimum-weight spanning

---

[56] The edge α_k.

tree of G, we often have more than one good choice for the next edge. One way to carry out the greedy algorithm for the weighted graph in Figure 11.38 is to choose, in order, the edges

$$\{a,b\}, \{c,d\}, \{e,f\}, \{d,g\}, \{e,g\}, \{a,g\}.$$

The weight of the resulting spanning tree T is

$$C(T) = 1 + 1 + 2 + 3 + 4 + 4 = 15.$$

Note that the algorithm does not grow the tree T in the sense that we have previously used that term.   □

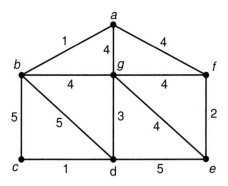

Figure 11.38

The best way to carry out the greedy algorithm is to arrange the edges in a sequence from smallest to largest weight and then iteratively select the first edge[57] that does not create a cycle. A disadvantage of the greedy algorithm is that one has to be able to recognize when a new edge creates a cycle and thus cannot be chosen. Prim[58] modified the greedy algorithm by showing how to grow a minimum-weight spanning tree, thereby making it unnecessary to deal with cycles.

Prim's algorithm for a minimum-weight spanning tree

Let $G = (V, E)$ be a weighted connected graph with weight function c and let u be any vertex of G.

(1) Put $i = 1$, $U_1 = \{u\}$, $F_1 = \emptyset$ and $T_1 = (U_1, F_1)$.

(2) For $i = 1, 2, \ldots, n - 1$, do the following:

---

[57] This is the greedy feature of the algorithm.

[58] R.C. Prim: *Shortest connection networks and some generalizations*, Bell Systems Tech. J., 36 (1957), 1389-1401.

(i) Locate an edge $\alpha_i = \{x, y\}$ of smallest weight such that x is in U_i and y is not in U_i.

(ii) Put $U_{i+1} = U_i \cup \{y\}$, $F_{i+1} = F_i \cup \{\alpha_{i+1}\}$ and $T_{i+1} = (U_{i+1}, F_{i+1})$.

(iii) Increase i to $i + 1$.

(3) Output $T_{n-1} = (U_{n-1}, F_{n-1})$. (Here $U_{n-1} = V$.)

Theorem 11.7.6 *Let $G = (V, E)$ be a weighted graph with weight function c. Then Prim's algorithm constructs a minimum-weight spanning tree $T = (V, F)$ of G.*

Proof. The proof is similar to the proof of Theorem 11.7.5. We use the same notation as in that proof, and we shall also be brief. At the end of each stage of the algorithm, we have grown a tree on a subset of the vertices of G. The theorem asserts that the tree $T = T_{n-1} = (V, F_{n-1})$ at termination of the algorithm, is a minimum-weight spanning tree. Of all the minimum-weight spanning trees of G, let $T^* = (V, F^*)$ be one for which the edges $\alpha_1, \ldots, \alpha_{k-1}$ are in T^* and k is largest. Suppose that $k \neq n$, that is, that $T^* \neq T$. Then α_k is not in F^* where α_k joins a vertex in U_k to a vertex in its complement $\overline{U}_k$. Since T^* is a spanning tree, there is an edge β of T^* that joins a vertex in U_k to a vertex in $\overline{U}_k$ such that inserting α_k in T^* and deleting β gives a spanning tree T^{**}. We have $c(\beta) \leq c(\alpha_k)$. Since α_k has the smallest weight of all edges with one vertex in U_k and the other in $\overline{U}_k$, it follows that $c(\beta) = c(\alpha_k)$ and T^{**} is a minimum-weight spanning tree with one more edge in common with T. $\square$

Example. We apply Prim's algorithm to the weighted graph G in Figure 11.38, with the initial vertex equal to a. One way of carrying out the algorithm results in the edges (in the order they are chosen)

$$\{a, b\}, \{a, f\}, \{f, e\}, \{e, g\}, \{g, d\}, \{d, c\},$$

which gives a spanning tree of weight 15. The advantage of Prim's algorithm over the greedy algorithm is clear in that, at each stage, we have only to determine an edge of smallest weight which joins a vertex that has already been reached to a vertex not yet reached. In the algorithm, cycles are automatically avoided in contrast to the greedy algorithm in which cycles must be explicitly avoided. $\square$

11.8 Exercises

1. How many nonisomorphic graphs of order 1 are there? of order 2? of order 3? Explain why the answer to each of the preceding questions is ∞ for general graphs.

2. Determine each of the 11 nonisomorphic graphs of order 4, and give a planar representation of each.

3. Does there exist a graph of order 5 whose degree sequence equals $(4, 4, 3, 2, 2)$?

4. Does there exist a graph of order 5 whose degree sequence equals $(4, 4, 4, 2, 2)$? a multigraph?

5. Use the pigeonhole principle to prove that a graph of order $n \geq 2$ always has two vertices of the same degree. Does the same conclusion hold for multigraphs?

6. Let $(d_1, d_2, \ldots, d_n)$ be a sequence of n nonnegative even integers. Prove that there exists a general graph with this sequence as its degree sequence.

7. Let $(d_1, d_2, \ldots, d_n)$ be a sequence of n nonnegative integers whose sum $d_1 + d_2 + \cdots + d_n$ is even. Prove that there exists a general graph with this sequence as its degree sequence. Devise an algorithm to construct such a general graph.

8. Let G be a graph with degree sequence $(d_1, d_2, \ldots, d_n)$. Prove that, for each k with $0 < k < n$,

$$\sum_{i=1}^{k} d_i \leq k(k-1) + \sum_{i=k+1}^{n} \min\{k, d_i\}.$$

9. Draw a connected graph whose degree sequence equals

$$(5, 4, 3, 3, 3, 3, 3, 2, 2).$$

10. Prove that any two connected graphs of order n with degree sequence $(2, 2, \ldots, 2)$ are isomorphic.

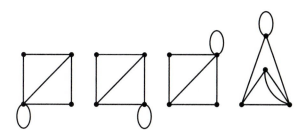

Figure 11.39

11. Determine which pairs of the general graphs in Figure 11.39 are isomorphic and, if isomorphic, find an isomorphism.

12. Determine which pairs of the multigraphs in Figure 11.40 are isomorphic, and for those that are isomorphic, find an isomorphism.

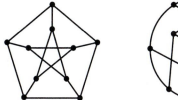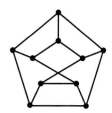

Figure 11.40

13. Prove that, if two vertices of a general graph are joined by a walk, then they are joined by a path.

14. Let x and y be vertices of a general graph and suppose that there is a closed walk containing both x and y. Must there be a closed trail containing both x and y?

15. Let x and y be vertices of a general graph and suppose that there is a closed trail containing both x and y. Must there be a cycle containing both x and y?

16. Let G be a connected graph of order 6 with degree sequence $(2, 2, 2, 2, 2, 2)$.

    (a) Determine all the nonisomorphic induced subgraphs of G.

    (b) Determine all the nonisomorphic spanning subgraphs of G.

    (b) Determine all the nonisomorphic subgraphs of order 6 of G.

17. First, prove that any two multigraphs G of order 3 with degree sequence $(4, 4, 4)$ are isomorphic. Then

    (a) determine all the nonisomorphic induced subgraphs of G.

    (b) determine all the nonisomorphic spanning subgraphs of G.

    (b) determine all the nonisomorphic subgraphs of order 3 of G.

18. Let γ be a trail joining vertices x and y in a general graph. Prove that the edges of γ can be partitioned so that one part of the partition determines a path joining x and y and the other parts determine cycles.

19. Let G be a general graph and let G' be the graph obtained from G by deleting all loops and all but one copy of each edge with multiplicity greater than 1. Prove that G is connected if and only if G' is connected. Also prove that G is planar if and only if G' is planar.

20. Prove that a graph of order n with at least

$$\frac{(n-1)(n-2)}{2} + 1$$

    edges must be connected. Give an example of a disconnected graph of order n with one fewer edge.

21. Let G be a general graph with exactly two vertices x and y of odd degree. Let G^* be the general graph obtained by putting a new edge $\{x, y\}$ joining x and y. Prove that G is connected if and only if G^* is connected.

22. (This and the following two exercises prove Theorem 11.1.3.) Let $G = (V, E)$ be a general graph. If x and y are in V, define $x \sim y$ to mean that either $x = y$ or there is a walk joining x and y. Prove that, for all vertices x, y, and z, we have

    (i) $x \sim x$.

    (ii) $x \sim y$ if and only if $y \sim x$.

    (iii) if $x \sim y$ and $y \sim z$, then $x \sim z$.

23. (Continuation of Exercise 22.) For each vertex x, let

$$C(x) = \{z : x \sim z\}.$$

    Prove the following:

(i) For all vertices x and y, either $C(x) = C(y)$ or else $C(x) \cap C(y) = \emptyset$. In other words two of the sets $C(x)$ and $C(y)$ cannot intersect unless they are equal.

(ii) If $C(x) \cap C(y) = \emptyset$, then there does not exist an edge joining a vertex in $C(x)$ to a vertex in $C(y)$.

24. (Continuation of Exercise 23.) Let $V_1, V_2, \ldots, V_k$ be the different sets that occur among the $C(x)$'s. Prove that

(i) $V_1, V_2, \ldots, V_k$ form a partition of the vertex set V of G.

(ii) the general subgraphs $G_1 = (V_1, E_1), G_2 = (V_2, E_2), \ldots, G_k = (V_k, E_k)$ of G induced by $V_1, V_2, \ldots, V_k$, respectively, are connected.

The induced subgraphs $G_1, G_2, \ldots, G_k$ are the *connected components* of G.

25. Prove Theorem 11.1.4.

26. Determine the adjacency matrices of the first and second general graphs in Figure 11.39.

27. Determine the adjacency matrices of the first and second multi-graphs in Figure 11.40.

28. Let A and B be two n-by-n matrices of numbers whose entries are denoted by a_{ij} and b_{ij}, $(1 \le i, j \le n)$, respectively. Define the product $A \times B$ to be the n-by-n matrix C whose entry c_{ij} in row i and column j is given by

$$c_{ij} = \sum_{p=1}^{n} a_{ip} b_{pj}, \quad (1 \le i, j \le n).$$

If k is a positive integer, define

$$A^k = A \times A \times \cdots \times A \quad (k \ A's).$$

Now let A denote the adjacency matrix of a general graph of order n with vertices $a_1, a_2, \ldots, a_n$. Prove that the entry in row i, column j of A^k equals the number of walks of length k in G joining vertices a_i and a_j.

29. Determine if the multigraphs in Figure 11.41 have Eulerian trails (closed or open). In case there is an Eulerian trail, use our algorithms to construct one.

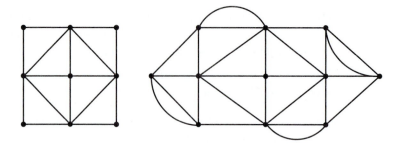

Figure 11.41

30. Which complete graphs K_n have closed Eulerian trails? open Eulerian trails?

31. Prove Theorem 11.2.4.

32. What is the fewest number of open trails into which the edges of GraphBuster can be partitioned?

33. Show how, removing one's pencil from the paper the fewest number of times, to trace the plane graphs in Figures 11.15, 11.16, and 11.17.

34. Determine all nonisomorphic graphs of order at most 6 that have a closed Eulerian trail.

35. Show how, removing one's pencil from the paper the fewest number of times, to trace out the graph of the regular dodecahedron shown in Figure 11.18.

36. Let G be a connected graph. Let γ be a closed walk that contains each edge of G at least once. Let G^* be the multigraph obtained from G by increasing the multiplicity of each edge from 1 to the number of times it occurs in γ. Prove that γ is a closed Eulerian trail in G^*. Conversely, suppose we increase the multiplicity of some of the edges of G and obtain a multigraph with m edges, each of whose vertices has even degree. Prove that there is a closed walk in G of length m which contains each edge of G at least once. This exercise shows that the Chinese postman problem for G is equivalent to determining the smallest number of copies of the edges of G that need to be inserted so as to obtain a multigraph all of whose vertices have even degree.

37. Solve the Chinese postman problem for the complete graph K_6.

38. Solve the Chinese postman problem for the graph obtained from K_6 by removing any edge.

39. Call a graph *cubic* if each vertex has degree equal to 3. The complete graph K_4 is the smallest example of a cubic graph. Find an example of a connected, cubic graph that does not have a Hamilton path.

40. * Let G be a graph of order n having at least

$$\frac{(n-1)(n-2)}{2} + 2$$

edges. Prove that G has a Hamilton cycle. Exhibit a graph of order n with one fewer edge that does not have a Hamilton cycle.

41. Let $n \geq 3$ be an integer. Let G_n be the graph whose vertices are the $n!$ permutations of $\{1, 2, \ldots, n\}$, wherein two permutations are joined by an edge if and only if one can be obtained from the other by the interchange of two numbers (an arbitrary transposition). Deduce from the results of Section 4.1 that G_n has a Hamilton cycle.

42. Prove Theorem 11.3.4.

43. Devise an algorithm analogous to our algorithm for a Hamilton cycle that constructs a Hamilton path in graphs satisfying the condition given in Theorem 11.3.4.

44. Which complete bipartite graphs $K_{m,n}$ have Hamilton cycles? Which have Hamilton paths?

45. Prove that a multigraph is bipartite if and only if each of its connected components is.

46. Prove that $K_{m,n}$ is isomorphic to $K_{n,m}$.

47. Prove that a bipartite multigraph with an odd number of vertices does not have a Hamilton cycle.

48. Is GraphBuster a bipartite graph? If so, find a bipartition of its vertices. What if we delete the loops?

49. Let $V = \{1, 2, \ldots, 20\}$ be the set of the first 20 positive integers. Consider the graphs whose vertex set is V and whose edge sets are defined below. For each graph, investigate whether the graph

(i) is connected (if not connected, determine the connected components), (ii) is bipartite, (iii) has an Eulerian trail, and (iv) has a Hamilton path.

(a) $\{a, b\}$ is an edge if and only if $a + b$ is even.

(b) $\{a, b\}$ is an edge if and only if $a + b$ is odd.

(c) $\{a, b\}$ is an edge if and only if $a \times b$ is even.

(d) $\{a, b\}$ is an edge if and only if $a \times b$ is odd.

(e) $\{a, b\}$ is an edge if and only if $a \times b$ is a perfect square.

(f) $\{a, b\}$ is an edge if and only if $a - b$ is divisible by 3.

50. What is the smallest number of edges that can be removed from K_5 in order to leave a bipartite graph?

51. Find a knight's tour on the boards of the following sizes:

(a) 5-by-5

(b) 6-by-6

(c) 7-by-7

52. $*$ Prove that there does not exist a knight's tour on a 4-by-4 board.

53. Prove that a graph is a tree if and only if it does not contain any cycles, but the insertion of any new edge always creates exactly one cycle.

54. Which trees have an Eulerian path?

55. Which trees have a Hamilton path?

56. Grow all the nonisomorphic trees of order 7.

57. Let $(d_1, d_2, \ldots, d_n)$ be a sequence of integers.

(a) Prove that there is a tree of order n with this degree sequence if and only if $d_1, d_2, \ldots, d_n$ are positive integers with sum $d_1 + d_2 + \cdots + d_n = 2(n - 1)$.

(b) Write an algorithm that, starting with a sequence $(d_1, d_2, \ldots, d_n)$ of positive integers, either constructs a tree with this degree sequence or concludes that none is possible.

58. A *forest* is a graph each of whose connected components is a tree. In particular, a tree is a forest. Prove that a graph is a forest if and only if it does not have any cycles.

59. Prove that the removal of an edge from a tree leaves a forest of two trees.

60. Let G be a forest of k trees. What is the fewest number of edges that can be inserted in G in order to obtain a tree?

61. Determine a spanning tree for GraphBuster.

62. Prove that, if a tree has a vertex of degree p, then it has at least p pendent vertices.

63. Determine a spanning tree for each of the graphs in Figures 11.15, 11.16, and 11.17.

64. For each integer $n \geq 3$ and for each integer k with $2 \leq k \leq n-1$, construct a tree of order n with exactly k pendent vertices.

65. Use the algorithm for a spanning tree in Section 11.5 in order to construct a spanning tree of the graph of the dodecahedron.

66. How many cycles does a connected graph of order n with n edges have?

67. Let G be a graph of order n that is not necessarily connected. A forest is defined in Exercise 58. A *spanning forest* of G is a forest consisting of a spanning tree of each of the connected components of G. Modify the algorithm for a spanning tree given in Section 11.5 so that it constructs a spanning forest of G.

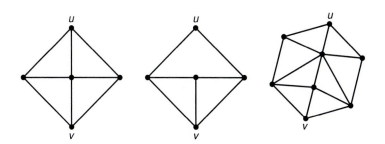

Figure 11.42

68. Determine whether the Shannon switching games played on the graphs in Figure 11.42 are positive, negative or neutral games.

69. Let G be a connected multigraph. An *edge-cut* of G is a set F of edges whose removal disconnects G. An edge-cut F is *minimal*,

provided that no subset of F other than F itself is an edge-cut. Prove that a bridge is always a minimal edge-cut, and conclude that the only minimal edge-cuts of a tree are the sets consisting of a single edge.

70. Let G be a connected multigraph having a vertex of degree k. Prove that G has a minimal edge-cut F with $|F| \leq k$.

71. Let F be a minimal edge-cut of a connected multigraph $G = (V, E)$. Prove that there exists a subset U of V such that F is precisely the set of edges that join a vertex in U to a vertex in the complement $\overline{U}$ of U.

72. (Continuation of Exercise 71.) Prove that a spanning tree of a connected multigraph contains at least one edge of every edge-cut.

73. Use the algorithm for growing a spanning tree in Section 11.7 in order to grow a spanning tree of GraphBuster. (Note: Graph-Buster is a general graph and has loops and edges of multiplicity greater than 1. The loops can be ignored and only one copy of each edge need be considered.)

74. Use the algorithm for growing a spanning tree in order to grow a spanning tree of the graph of the regular dodecahedron.

75. Apply the BF-algorithm of Section 11.7 to determine a BFS-tree for the following:

    (a) The graph of the regular dodecahedron (any root).
    (b) GraphBuster (any root).
    (c) A graph of order n whose edges are arranged in a cycle (any root).
    (d) A complete graph K_n (any root).
    (e) A complete bipartite graph $K_{m,n}$ (a left-vertex root and a right-vertex root).

    In each case, determine the breadth-first numbers and the distance of each vertex from the root chosen.

76. Apply the DF-algorithm of Section 11.7 to determine a DFS-tree for (a), (b), (c), (d), and (e) as in Exercise 75. In each case, determine the depth-first numbers.

77. Let G be a graph that has a Hamilton path which joins two vertices u and v. Is the Hamilton path a DFS-tree rooted at u for G? Could there be other DFS-trees?

78. (Solution of the Chinese postman problem for trees.) Let G be a tree of order n. Prove that the length of a shortest closed walk that includes each edge of G at least once is $2(n-1)$. Show how the depth-first algorithm finds a walk of length $2(n-1)$ that includes each edge exactly twice.

79. Use Dijkstra's algorithm in order to construct a distance tree for u for the weighted graph in Figure 11.43, with specified vertex u as shown.

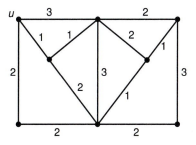

Figure 11.43

80. Consider the complete graph K_n with labeled vertices $1, 2, \ldots, n$, in which the edge joining vertices i and j is weighted by $c\{i, j\} = i+j$ for all $i \neq j$. Use Dijkstra's algorithm to construct a distance tree rooted at vertex $u = 1$ for

   (a) K_4.
   (b) K_6.
   (c) K_8.

81. Consider the complete graph K_n with labeled vertices $1, 2, \ldots, n$, with the weight function $c\{i, j\} = |i - j|$ for all $i \neq j$. Use Dijkstra's algorithm to construct a distance tree rooted at vertex $u = 1$ for

   (a) K_4.
   (b) K_6.
   (c) K_8.

82. Consider the complete graph K_n whose edges are weighted as in Exercise 80. Apply the greedy algorithm to determine a minimum-weight spanning tree for

    (a) K_4.

    (b) K_6.

    (c) K_8.

83. Consider the complete graph K_n whose edges are weighted as in Exercise 81. Apply the greedy algorithm to determine a minimum-weight spanning tree for

    (a) K_4.

    (b) K_6.

    (c) K_8.

84. Same as Exercise 82, using Prim's algorithm in place of the greedy algorithm.

85. Same as Exercise 83, using Prim's algorithm in place of the greedy algorithm.

86. Let G be a weighted connected graph in which all edge weights are different. Prove that there is exactly one spanning tree of minimum weight.

87. Define a *caterpillar* to be a tree T that has a path γ such that every edge of T is either an edge of γ or has one of its vertices on γ.

    (a) Verify that all trees with 6 or fewer vertices are caterpillars.

    (b) Let T_7 be the tree on 7 vertices consisting of three paths of length 2 meeting at a central vertex c. Prove that T_7 is the only tree on 7 vertices that is not a caterpillar.

    (c) Prove that a tree is a caterpillar if and only if it does not contain T_7 as a spanning subgraph.

88. Let $d_1, d_2, \ldots, d_n$ be positive integers. Prove that there is a caterpillar with degree sequence $(d_1, d_2, \ldots, d_n)$ if and only if $d_1 + d_2 + \cdots + d_n = 2(n-1)$. Compare with Exercise 57.

89. A *graceful labeling* of a graph G with vertex set V and with m edges is an injective function $g : V \to \{0, 1, 2, \ldots, m\}$ such that the labels $|g(x) - g(y)|$ corresponding to the m edges $\{x, y\}$ of G

are $1, 2, \ldots, m$ in some order. It has been *conjectured* by Kotzig and Ringel (1964) that every tree has a graceful labeling. Find a graceful labeling of the tree T_7 in the previous exercise, any path, and the graph $K_{1,n}$.

90. Verify that cycles of lengths 5 and 6 cannot be gracefully labeled. Then find graceful labelings of cycles of lengths 7 and 8.

Chapter 12

Digraphs and Networks

In this chapter we briefly discuss directed graphs (abbreviated, digraphs). As already pointed out in the opening paragraphs of Chapter 11, digraphs are similar to graphs, the difference being that in digraphs, the edges have directions and are called arcs. Thus, digraphs model nonsymmetric relations, in the same sense that graphs model symmetric relations. Many of the results we prove are directed analogues of results already proved for graphs.

A network is a digraph with two distinguished vertices s and t, in which each arc has a nonnegative weight, called its *capacity*. Thinking of each arc as a conduit over which flows some substance, and the capacity of an arc as the amount that can flow through the conduit per unit time (say), one important problem is that of finding the maximum possible flow from the "source" s to the "target" t, subject to the given capacities. The answer to this problem, along with an efficient algorithm for constructing a maximum flow, is given by the so-called *max-flow min-cut theorem*. We then use the max-flow min-cut theorem to give another proof of the basic result, Corollary 9.2.4, about matchings in bipartite graphs.

12.1   Digraphs

A *digraph* $D = (V, A)$ has a set V of elements called *vertices* and a set A of *ordered* pairs of not necessarily distinct vertices called *arcs*. Each arc is of the form

$$\alpha = (a, b), \tag{12.1}$$

where a and b are vertices. We think of the arc α as *leaving* a and *entering* b, that is, pointed (or directed) from a to b.

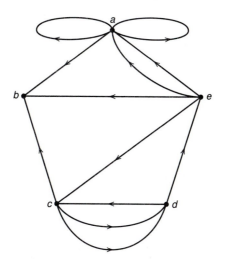

Figure 12.1

In contrast to graphs, (a, b) is not the same as (b, a). We shall use terminology which is similar to that used for graphs, but there are distinctions that apply to digraphs which don't apply to graphs. Thus, the arc α in (12.1) has *initial vertex* $\iota(\alpha) = a$ and *terminal vertex* $\tau(\alpha) = b$. A digraph may contain both of the arcs (a, b) and (b, a) as well as loops of the form (a, a). A loop (a, a) enters and exits the same vertex a. We may generalize a digraph to a *general digraph* in which multiple arcs are allowed.[1] We draw general digraphs as we draw graphs, but for digraphs we put an arrow on each edge-curve in order to indicate its direction.

Example. A general digraph is shown in Figure 12.1. It is not a digraph, since some of the arcs have multiplicities greater than 1. $\square$

A vertex x of a general digraph $D = (V, A)$ has two degrees. The *outdegree* of x is the number of arcs of which x is the initial vertex:

$$|\{\alpha | \iota(\alpha) = x\}|.$$

The *indegree* of x is the number of arcs of which x is the terminal vertex:

$$|\{\alpha | \tau(\alpha) = x\}|.$$

A loop (x, x) contributes 1 to both the indegree and outdegree of the vertex x. A proof similar to the one given for Theorem 11.1.1 establishes the next elementary result.

---

[1] The number of arcs, counting multiplicities, however, should always be finite.

Theorem 12.1.1 *In a general digraph the sum of the indegrees of the vertices equals the sum of the outdegrees, and each is equal to the number of arcs.*

Example. In the general digraph of Figure 12.1 the indegrees of the vertices a, b, c, d, e are

$$4, 3, 2, 2, 1;$$

the outdegrees are

$$3, 0, 3, 2, 4.$$

In each case the sum is 12, the number of arcs.   □

With any general graph $G = (V, E)$, we can obtain a general digraph $D = (V, A)$ by giving each edge $\{a, b\}$ of E an orientation, that is, by replacing $\{a, b\}$ with either (a, b) or (b, a).[2] Such a digraph D is called an *orientation* of G. A general graph has many different orientations. Conversely, given a general digraph $D = (V, A)$ we can remove the directions of its arcs, thereby obtaining a general graph $G = (V, E)$. Such a graph is called the *underlying general graph of G*. A general digraph has exactly one underlying general graph.

Example. The underlying general graph of the general digraph in Figure 12.1 is shown in Figure 12.2.   □

An orientation of a complete graph K_n with n vertices is called a *tournament*. It is a digraph such that each distinct pair of vertices is joined by exactly one arc. This arc may have either of the two possible directions. A tournament can be regarded as the record of who beat whom in a round-robin tournament in which each player plays every other player exactly once, and there are no ties. The nicest kinds of tournaments[3] are those in which it is possible to order the players in a list

$$p_1, p_2, \ldots, p_n$$

so that each player beats all those further down on the list. Such tournaments are called *transitive tournaments*. In a transitive tournament there is a consistent ranking of the players.

---

[2]If the multiplicity of $\{a, b\}$ is greater than 1, then some copies of $\{a, b\}$ can be replaced by (a, b), and others can be replaced by (b, a).

[3]From the point of view of ranking the players at the end.

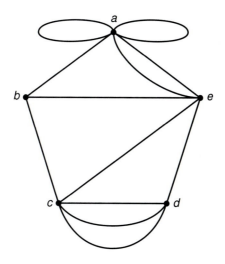

Figure 12.2

We modify our definitions of walk, path, and cycle in a general graph in order to obtain analogous concepts for general digraphs. Let $D = (V, A)$ be a general digraph. A sequence of m arcs of the form

$$(x_0, x_1), (x_1, x_2), \ldots, (x_{m-1}, x_m) \qquad (12.2)$$

is called a *directed walk of length m from vertex x_0 to vertex x_m*. The *initial vertex* of the walk (12.2) is x_0 and the *terminal vertex* is x_m. The directed walk is *closed* if $x_0 = x_m$ and *open* otherwise. We also denote the walk (12.2) by

$$x_0 \to x_1 \to x_2 \to \cdots \to x_m.$$

A directed walk with distinct arcs is a *directed trail*; a directed trail with distinct vertices (except possibly the initial and terminal vertices) is a *path*;[4] a closed path is a *directed cycle*.

Example. Consider the general digraph of order 5 in Figure 12.1. Then

$$d \to e \to c \to d \to e$$

is a directed walk,

$$c \to d \to e \to c \to b$$

is a directed trail,

$$c \to d \to e \to a \to b$$

---

[4]In contrast to walks and cycles, we use 'path' instead of 'directed path.'

is a path, and each of

$$c \to d \to e \to c, \quad c \to d \to c, \quad a \to a$$

is a directed cycle. □

A general digraph is *connected*, provided that its underlying general graph is connected. A general digraph is *strongly connected*, provided that, for each pair of distinct vertices a and b, there is a directed walk[5] from a to b and a directed walk from b to a. Thinking of a general digraph as a network of one-way streets connecting the various parts of a city, we see that strong connectivity means that one can get from any part of the city to any other part, traveling along streets only in their given direction.

Example. The general digraph in Figure 12.1 is connected, but it is not strongly connected. The easiest way to see that it is not strongly connected is to observe that vertex b has outdegree equal to 0. Thus, it is not possible to leave b. □

A directed trail in a general digraph D is called *Eulerian*, provided that it contains every arc of D. A *Hamilton path* is a path that contains every vertex. A *directed Hamilton cycle* is a directed cycle that contains every vertex.

The next two theorems are the directed analogues of Theorems 11.2.2 and 11.2.3. Since their proofs are similar we omit them.

Theorem 12.1.2 *Let D be a connected digraph. Then D has a closed Eulerian directed trail if and only if the indegree of each vertex equals the outdegree.*

Theorem 12.1.3 *Let D be a connected digraph and let x and y be distinct vertices of D. Then there is a directed Eulerian trail from x to y if and only if*

(*i*) *the outdegree of x exceeds its indegree by 1;*

(*ii*) *the indegree of y exceeds its outdegree by 1;*

(*iii*) *for each vertex $z \neq x, y$, the indegree of z equals its outdegree.*

---

[5]And thus a path.

There is also a directed analogue of Theorem 11.3.2 due to Ghouila-Houri[6] giving a sufficient condition for the existence of a directed Hamilton cycle, but it is much more difficult to prove. We shall be content simply to state the theorem. In the theorem, D is a digraph (and not a general digraph) without loops.[7]

Theorem 12.1.4 *Let D be a strongly connected digraph without any loops. If, for each vertex x, we have*

$$(outdegree\ of\ x) + (indegree\ of\ x) \geq n,$$

then D has a directed Hamilton cycle.

We now show that a tournament always has a Hamilton path. This implies that it is always possible to rank the players in the order

$$p_1, p_2, \ldots, p_n, \tag{12.3}$$

so that p_1 beats p_2, p_2 beats p_3, ... , p_{n-1} beats p_n. This does not imply that we have a consistent ranking of the players, since we are not asserting that each player beats all players further down on the list. Indeed, a tournament may even have a directed Hamilton cycle, thereby implying that for each player there is a ranking (12.3) in which he is ranked first!

Theorem 12.1.5 *Every tournament has a Hamilton path.*

Proof. Let D be a tournament of order n. Let

$$\gamma : x_1 \to x_2 \to \cdots \to x_p \tag{12.4}$$

be a longest path in D. We show that a longest path (12.4) is a Hamilton path by showing that, if $p < n$, then we can find a longer path. Suppose that $p < n$ so that the set U of vertices not on the path (12.4) is nonempty. Let u be any vertex in U. If there is an arc from u to x_1 or an arc from x_p to u, then we can find a longer path. Thus, we assume that the arc between x_1 and u has u as its terminal vertex. Similarly, we assume that the arc between x_p and u has u as its initial vertex. It follows that there must be consecutive vertices x_k and x_{k+1} on the path γ such that the arc between x_k and u has u as its terminal

---

[6]A. Ghouila-Houri: Une condition suffisante d'existence d'un circuit hamiltonien, *C.R. Acad. Sci.*, 251 (1960), 494.

[7]More than one arc from one vertex to another is of no help in locating a Hamilton directed cycle, nor is a loop of any help.

vertex, and the arc between x_{k+1} and u has u as its initial vertex. But then

$$x_1 \to \cdots \to x_k \to u \to x_{k+1} \to \cdots \to x_p$$

is a longer path than γ. We leave it as an exercise to use this proof to determine an algorithm for a Hamilton path in a tournament.   □

We conclude this brief introduction to digraphs by proving two theorems of some practical importance. The first of these is a theorem of Robbins[8] which characterizes those general graphs that have a strongly connected orientation. Thus, this theorem will tell the traffic engineer of a city with no one-way streets whether it is possible (and how) to make all streets into one-way streets in such a way that one can get from any part of the city to any other part.[9]

Theorem 12.1.6 *Let $G = (V, E)$ be a connected graph. Then G has a strongly connected orientation if and only if G does not have any bridges.*

Proof. First, assume that G has a bridge α. The removal of α from G results in a disconnected graph with two connected components $G_1 = (V_1, E_1)$ and $G_2 = (V_2, E_2)$. If we orient α from G_1 to G_2, then there is no directed walk from a vertex of G_2 to a vertex of G_1. If we orient α from G_2 to G_1, there is no directed walk from a vertex in G_1 to a vertex in G_2. Hence, G does not have a strongly connected orientation.

Now assume that G does not have any bridges. By Lemma 11.5.3, each edge of G is contained in some cycle. The next algorithm determines a strong orientation of G.

Algorithm for a strongly connected orientation of a bridgeless connected graph

Let $G = (V, E)$ be a connected graph without bridges.

(1) Put $U = \emptyset$.

(2) Locate a cycle γ of G.

   (i) Orient the edges of γ so that it becomes a directed cycle.

   (ii) Add the vertices of γ to U.

---

[8]H.E. Robbins: A theorem on graphs, with an application to a problem in traffic control, *Amer. Math. Monthly*, 46 (1939), 281-283.

[9]The consequences to the traffic engineer if he fails to achieve this property are obvious!

(3) While $U \neq V$, do the following:

    (i) Locate an edge $\alpha = \{x, y\}$ joining a vertex x in U to a vertex y not in U.

    (ii) Locate a cycle γ containing the edge α.

    (iii) Orient the edge α from x to y and continue to orient the edges of γ as if to form a directed cycle until arriving at a vertex z in U.

    (iv) Add to U all the vertices of γ from x to z.

(4) Orient in either direction every edge that has not yet received an orientation.

We note that a cycle containing the edge $\alpha = \{x, y\}$ in (3)(ii) can be located by finding a path (for instance, a shortest path) joining x and y in the graph obtained by deleting the edge α. It should be clear that the digraph obtained by applying the preceding algorithm is a strongly connected orientation of G, provided that step (3) terminates—that is, provided that the set U does achieve the value V. But if $U \neq V$, then since G is connected, there must be an edge α joining a vertex in U to a vertex not in U. Since no edge of G is a bridge, the edge α is contained in some cycle γ by Lemma 11.5.3. From this, it follows that the terminal value of U is V. $\qquad\qquad\qquad\qquad\qquad\qquad\qquad\square$

Example. (*A trading problem.*)[10] There are n traders $t_1, t_2, \ldots, t_n$ who enter a market, each with an indivisible item[11] to offer in trade. We assume for simplicity that a trader never has any use for more than one of the items, but except for this assumption, the items are freely transferable from one trader to another. Each trader ranks the n items brought to the market (including his own) according to his preference for them. There are no ties, and thus each trader ranks the items from 1 to n. The effect of the market activity is to redistribute (or permute) the ownership of the items among the n traders. Such a permutation is called an *allocation*. We regard an allocation as a one-to-one function

$$\rho : \{t_1, t_2, \ldots, t_n\} \to \{t_1, t_2, \ldots, t_n\},$$

where $\rho(t_i) = t_j$ means that trader t_i receives the item of trader t_j in the allocation. An allocation ρ is called a *core allocation*, provided

---

[10]This example and its subsequent analysis is partly based on the article "On cores and indivisibility" by L. Shapely and H. Scarf, in *Studies in Optimization*, (MAA Studies in Mathematics, vol. 10), 1974, Mathematical Association of America, Washington, D.C., 104-123.

[11]For instance, a car or a house.

that it has the following property: There does not exist a subset S of fewer than n traders such that, by trading amongst themselves, each receives an item that he ranks more highly than in the allocation ρ.[12] For example, suppose that $n = 5$ and the preferences of the traders are as given by the table

$$
\begin{array}{c|ccccc}
 & t_1 & t_2 & t_3 & t_4 & t_5 \\
\hline
t_1 & 4 & 3 & 1 & 2 & 5 \\
t_2 & 4 & 3 & 1 & 2 & 5 \\
t_3 & 4 & 3 & 5 & 1 & 2 \\
t_4 & 1 & 4 & 3 & 5 & 2 \\
t_5 & 4 & 5 & 2 & 1 & 3 \\
\end{array}
\qquad (12.5)
$$

The first row of this table gives t_1's ranking of the items. Thus, t_1 values the item of t_3 most highly, then the items of t_4, t_2, t_1, t_5 in this order. The interpretation of the other rows of the table is similar. One possible allocation ρ is

$$\rho(t_1) = t_2, \ \rho(t_2) = t_3, \ \rho(t_3) = t_1, \ \rho(t_4) = t_5, \ \rho(t_5) = t_4.$$

This allocation is not a core allocation since

$$\rho'(t_1) = t_4, \ \rho'(t_4) = t_1$$

defines an allocation for the two traders t_1, t_4 in which each gets an item he values more highly than he gets in ρ. A core allocation in this case is ρ^*:

$$\rho^*(t_1) = t_3, \ \rho^*(t_2) = t_2, \ \rho^*(t_3) = t_4, \ \rho^*(t_4) = t_1, \ \rho^*(t_5) = t_5.$$

Does every trading problem have a core allocation? In the remainder of this section we answer this question.[13]   □

   A digraph furnishes a convenient mathematical model for a trading problem. We consider a digraph $D = (V, A)$ in which the vertices are the n traders. We put an arc from each vertex to every other, including the vertex itself.[14] Each vertex has indegree equal to n and outdegree equal to n. The digraph D is a *complete digraph* of order n. For each vertex t_i, we label (or weight) the arcs leaving t_i with $1, 2, \ldots, n$ in

---

[12]Put another way there does not exist a subset S of fewer than n traders and an allocation ρ' for them such that, for each trader t_i in S, t_i ranks the item of $\rho'(t_i)$ higher than that of $\rho(t_i)$.

[13]In the affirmative.

[14]Thereby creating a loop at each vertex.

accordance with the preferences of t_i. An allocation corresponds to a partition of the vertices into directed cycles. This is a consequence of the next lemma, which implies that a one-to-one function from a set to itself can be thought of as a digraph that consists of one or more directed cycles with no vertices in common.

Lemma 12.1.7 *Let D be a digraph in which each vertex has outdegree at least 1. Then there is a directed cycle in D.*

Proof. An algorithm that constructs a directed cycle is now given:

Algorithm for a directed cycle

Let u be any vertex.

(1) Put $i = 1$ and $x_1 = u$.

(2) If x_i is the same as one of the previously chosen vertices x_j, $(j < i)$, then go to (4). Else, go to (3).

(3) Do the following:

    (i) Choose an arc (x_i, x_{i+1}) leaving vertex x_i.

    (ii) Increase i by 1.

    (iii) Go to (2).

(4) Output the directed cycle

$$x_j \to x_{j+1} \to \cdots \to x_{i-1} \to x_i = x_j.$$

Since each vertex is the initial vertex of at least one arc and since we stop as soon as we obtain a repeated vertex, the algorithm does output a directed cycle as shown. $\square$

Corollary 12.1.8 *Let X be a set of n elements and let $f : X \to X$ be a one-to-one function. Let $D_f = (X, A_f)$ be the digraph whose set of arcs is*

$$A_f = \{(x, f(x)) : x \text{ in } X\}.$$

Then the arcs of D_f can be partitioned into directed cycles with each vertex belonging to exactly one directed cycle.

Proof. Since the function f is one-to-one, it is a consequence of the pigeonhole principle that f is also onto. It now follows from the definition of the set A_f of arcs that each vertex of D_f has its indegree and outdegree equal to 1. By Lemma 12.1.7, D_f has a directed cycle γ. Either each vertex is a vertex of γ, in which case our partition contains only γ, or removing γ (its vertices and arcs), we are left with a digraph, each of whose vertices also have indegree and outdegree equal to 1. We continue to remove directed cycles until we exhaust all of the vertices, and this gives us the desired partition. $\qquad\square$

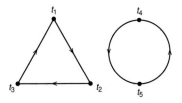

Figure 12.3

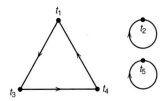

Figure 12.4

Example. The digraphs D_ρ and $D_{\rho'}$ corresponding to the allocations ρ and ρ^* defined in the preceding example give the directed cycle partitions shown in Figures 12.3 and 12.4, respectively. $\qquad\square$

The problem of the existence of core allocations can be regarded as a directed version of the stable marriage problem described in Section 9.4. We now use the digraph model to answer our question about the existence of core allocations.

Theorem 12.1.9 *Every trading problem has a core allocation.*

Proof. The proof shows how successive use of the algorithm for directed cycles, given in the proof of Lemma 12.1.7, results in a core allocation.

Let the set of traders be $V = \{t_1, t_2, \ldots, t_n\}$. Consider the preference digraph $D^1 = (V, A^1)$, where there is an arc (t_i, t_j) from t_i to t_j if and only if t_i prefers the item of t_j over all other items. Each vertex has outdegree 1; hence, by Lemma 12.1.7, there is a directed cycle γ_1 in D^1. Let V^1 be the set of vertices of γ_1. Let $D^2 = (V - V^1, A^2)$ be the preference digraph[15] with vertex set $V - V^1$ in which there is an arc from t_i to t_j if and only if t_i prefers the item of t_j over all the other items of the traders in $V - V^1$. Each vertex of the digraph D^2 has outdegree 1 and again, by Lemma 12.1.7, we can find a directed cycle γ_2. We let V^2 be the set of vertices of γ_2, and we consider the preference digraph $D^3 = (V - (V^1 \cup V^2), A^3)$. Continuing in this way, we obtain $k \geq 1$ directed cycles $\Gamma = \{\gamma_1, \gamma_2, \ldots, \gamma_k\}$ with vertex sets $V^1, V^2, \ldots, V^k$, respectively, where $V^1, V^2, \ldots, V^k$ is a partition of V, the set of traders. The set Γ of cycles determines an allocation ρ: Each trader t_p is a vertex of exactly one of the directed cycles in Γ, and this directed cycle has an arc from t_p to some t_q. Defining $\rho(t_p) = t_q$, we obtain an allocation.

We now show that the allocation ρ is a core allocation. Let U be any subset of fewer than n traders. Let j be the smallest integer such that $U \cap V^j \neq \emptyset$. Then

$$U \subseteq V^j \cup \cdots \cup V^k = V - (V^1 \cup \cdots \cup V^{j-1}),$$

and U is a subset of the vertices of the digraph D^j. Let t_s be any trader in $U \cap V^j$. Then, in the allocation ρ, t_s gets the item he ranks the highest among all the items of traders in $V - (V^1 \cup \cdots \cup V^{j-1})$ and hence among all the traders in S. Thus, by trading among the members of U, t_s cannot obtain an item he ranks higher than the item he was assigned in ρ. Therefore, ρ is a core allocation. $\qquad\square$

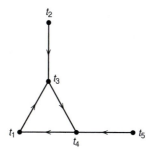

Figure 12.5

Example. Consider the trader problem determined by the table in (12.5). The preference digraph D^1 is pictured in Figure 12.5. This digraph has exactly one directed cycle, namely,

$$t_1 \to t_3 \to t_4 \to t_1.$$

---

[15]Note well that the vertex set of D^2 is only a subset of the traders.

The preference digraph D^2 is pictured in Figure 12.6, and it consists of the two disjoint directed cycles

$$t_2 \rightarrow t_2 \text{ and } t_5 \rightarrow t_5.$$

We can pick either of these directed cycles, and then the other is the preference digraph D^3.[16] A core allocation for our problem is given by

$$\rho(t_1) = t_3, \rho(t_3) = t_4, \rho(t_4) = t_1, \rho(t_2) = t_2, \rho(t_5) = t_5.$$

$\square$

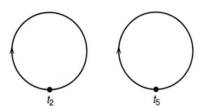

t_2 t_5

Figure 12.6

12.2   Networks

A *network* is a digraph (V, A) in which two vertices—the *source s* and the *target t*—are distinguished, where $s \neq t$, and in which each arc α has a nonnegative weight $c(\alpha)$, called its *capacity*. We denote a network by $N = (V, A, s, t, c)$.

The basic problem to be treated for networks is that of moving a substance from the source to the target, within the constraints provided by the arcs of the digraph and their capacities. Formally, a *flow* in the network N is defined to be a function f that assigns a real number $f(\alpha)$ to each arc α, subject to the following constraints:

(i) $0 \leq f(\alpha) \leq c(\alpha)$. (The flow through an arc is nonnegative and does not exceed its capacity.)

(ii) $\sum_{\iota(\alpha)=x} f(\alpha) = \sum_{\tau(\alpha)=x} f(\alpha)$ for each vertex $x \neq s, t$. (For each vertex x other than the source and the target, the flow into x equals the flow out of x.)

---

[16]In general, when one of the preference digraphs consists of pairwise disjoint, directed cycles, then the core allocation ρ constructed in the proof of Theorem 12.1.9 is determined.

In order to demonstrate that the net flow out of the source,

$$\sum_{\iota(\alpha)=s} f(\alpha) - \sum_{\tau(\alpha)=s} f(\alpha) \qquad (12.6)$$

equals the net flow into the target,

$$\sum_{\tau(\alpha)=t} f(\alpha) - \sum_{\iota(\alpha)=t} f(\alpha) \qquad (12.7)$$

(where the common value is the amount moved from the source to the target), we prove the next result. For each set of vertices U, we let

$$\overrightarrow{U} = \{\alpha : \iota(\alpha) \text{ is in } U, \tau(\alpha) \text{ is not in } U\}$$

and

$$\overleftarrow{U} = \{\alpha : \iota(\alpha) \text{ is not in } U, \tau(\alpha) \text{ is in } U\}.$$

Lemma 12.2.1 *Let f be a flow in the network $N = (V, A, s, t, c)$ and let U be a set of vertices containing the source s but not the target t. Then*

$$\sum_{\alpha \in \overrightarrow{U}} f(\alpha) - \sum_{\alpha \in \overleftarrow{U}} f(\alpha) = \sum_{\iota(\alpha)=s} f(\alpha) - \sum_{\tau(\alpha)=s} f(\alpha).$$

Proof. We evaluate the sum

$$\sum_{x \in U} \left(\sum_{\iota(\alpha)=x} f(\alpha) - \sum_{\tau(\alpha)=x} f(\alpha) \right) \qquad (12.8)$$

in two different ways. On the one hand, it follows from the definition of a flow that all terms in the outer sum are zero except for that one corresponding to the vertex s. Hence, the value is

$$\sum_{\iota(\alpha)=s} f(\alpha) - \sum_{\tau(\alpha)=s} f(\alpha).$$

On the other hand, we can rewrite the expression (12.6) as

$$\sum_{x \in U} \sum_{\iota(\alpha)=x} f(\alpha) - \sum_{x \in U} \sum_{\tau(\alpha)=x} f(\alpha),$$

or equivalently,

$$\sum_{\iota(\alpha) \in U} f(\alpha) - \sum_{\tau(\alpha) \in U} f(\alpha). \qquad (12.9)$$

Each arc α with both its initial and terminal vertx in U makes a net contribution of $f(\alpha) - f(\alpha) = 0$ to the sum (12.9); hence, the sum (12.9) equals

$$\sum_{\alpha \text{ in } \vec{U}} f(\alpha) - \sum_{\alpha \text{ in } \overleftarrow{U}} f(\alpha).$$

Thus, the equation in the statement of the lemma holds. $\qquad\square$

In Lemma 12.2.1, take $U = V - \{t\}$. Then $\vec{U}$ is the set of all arcs whose terminal vertex is t, and $\overleftarrow{U}$ is the set of all arcs whose initial vertex is t. Hence,

$$\sum_{\iota(\alpha)=s} f(\alpha) - \sum_{\tau(\alpha)=s} f(\alpha) = \sum_{\tau(\alpha)=t} f(\alpha) - \sum_{\iota(\alpha)=t} f(\alpha). \qquad (12.10)$$

The common value of the two expressions in (12.10) is called the *value of the flow f* and is denoted by $\operatorname{val}(f)$.

Given a network $N = (V, A, s, t, c)$, a flow in N is a *maximum flow*, provided that it has the largest value among all flows in N. The value of a maximum flow (the maximum value of a flow) equals the minimum value of another quantity associated with a network. We shall prove this important fact only in the case that the capacity function is integer-valued,[17] and in doing so, we obtain an algorithm for constructing a maximum flow.

A *cut* in a network $N = (V, A, s, t, c)$ is a set C of arcs such that each path from the source s to the target t contains at least one arc in C. The *capacity* $\operatorname{cap}(C)$ of a cut C is the sum of the capacities of the arcs in C. A cut is a *minimum cut*, provided that it has the smallest capacity among all cuts in N.

A cut is a *minimal cut*, provided that each set obtained from C by the deletion of one of its arcs is not a cut.[18] (This means that, for each arc α in C, there is a path from s to t that contains α, but no other arc of C.)

We first show that any minimal cut is a cut of the form $\vec{U}$ for some set of vertices U containing s but not containing t. This implies that the smallest capacity of a cut is achieved by a cut of this form $\vec{U}$.

---

[17]It then follows that it is also true for capacity functions, all of whose values are rational numbers, by choosing a common denominator for all the rational values. In case the values of the capacity function are not all rational, one must resort to a limiting process.

[18]So a minimum cut is defined arithmetically, while a minimal cut is defined set-theoretically. If all the arc capacities are positive, then a minimum cut is also a minimal cut.

Lemma 12.2.2 *Let $N = (V, A, s, t, c)$ be a network with C a minimal cut. Let U be the set of all vertices x for which there exists a path from the source s to x that contains no arc in C. Then $\vec{U}$ is a cut and $C = \vec{U}$.*

Proof. Note that s is in U, since the trivial path consisting only of the vertex s contains no arc in C. Since C is a cut, the target t is not contained in U. Hence, $\vec{U}$ is a cut. Each arc (x, y) in $\vec{U}$ is in C, for otherwise there exists a path from s to y containing no arc in U, and y would be in U. Thus, $\vec{U} \subseteq C$.

Now, let $\alpha = (a, b)$ be any arc in C. Since C is a minimal cut, there is a path γ from s to t that contains α, but no other arc of C. This implies that the initial vertex a of α is in U. If there were a path γ' from s to b that contained no arc in C, then γ' followed by the part of γ from b to t would give a path from s to t containing no arc in C. It follows that the terminal vertex b of α is not in U. Therefore, α is in $\vec{U}$ and $C \subseteq \vec{U}$. □

We now prove the very important *max-flow min-cut theorem*.

Theorem 12.2.3 *Let $N = (V, A, s, t, c)$ be a network. Then the maximum value of a flow in N equals the minimum capacity of a cut in N. In other words, the value of a maximum flow equals the capacity of a minimum cut. If the capacities of all the arcs are integers, then there is a maximum flow all of whose values are integers as well.*

Proof. We prove the theorem only under the assumption that the capacity values are all integers. The full theorem can then be established by means of a limiting argument.

The first part of the proof does not use the integrality of the capacity function. We first show that, for each flow f and each cut C,

$$\text{val}(f) \leq \text{cap}(C).$$

By Lemma 12.2.2 it suffices to prove this inequality for cuts of the form $\vec{U}$, where U is a set of vertices with s in U and t not in U. By Lemma 12.2.1 and the fact that flow values are nonnegative, we have

$$
\begin{aligned}
\text{val}(f) &= \sum_{\alpha \in \vec{U}} f(\alpha) - \sum_{\alpha \in \overleftarrow{U}} f(\alpha) \\
&\leq \sum_{\alpha \in \vec{U}} c(\alpha) \\
&= \text{cap } \vec{U} .
\end{aligned}
$$

The remainder of the proof is devoted to showing that there is a flow $\hat{f}$ with only integer values and a cut $\hat{C}$ such that $\text{val}(\hat{f}) = \text{cap}(\hat{C})$. Such a flow $\hat{f}$ is a maximum flow, and the cut $\hat{C}$ is a minimum cut.

We start with an arbitrary integer-valued flow f on N. The *zero flow*, in which all flow values equal zero, will suffice, although in general, it is possible to find an integer-valued flow by trial and error which has a reasonable – for the problems at hand – value. We then describe an algorithm that results in one of the following two possibilities:

> **Breakthrough:** An integer-valued flow f' has been found with $\text{val}(f') = \text{val}(f) + 1$. In this case, we repeat the algorithm with $f = f'$.

> **Nonbreakthrough:** Breakthrough has not occurred. In this case, we exhibit a cut whose capacity equals the value of the flow f. The cut is our desired minimum cut $\hat{C}$ and the flow f is our desired maximum flow $\hat{f}$.

Basic Flow Algorithm

Begin with any integer-valued flow f on the network $N = (V, A, s, t, c)$.

(0) Set $U = \{s\}$.

(1) While there exists an arc $\alpha = (x, y)$ with either

   (a) x in U, y not in U, and $f(\alpha) < c(\alpha)$, or
   (b) x not in U, y in U, and $f(\alpha) > 0$,

   put y in U (in case of (a)) or put x in U (in case of (b)).

(2) Output U.

Thus, in the algorithm, we seek either (a) an arc in $\overrightarrow{U}$ (*flowing away from s and towards t*) whose flow value is less than capacity (and update U by putting its terminal vertex in U) or (b) an arc in $\overleftarrow{U}$ (*flowing towards s and away from t*) with a positive flow value (and update U by putting its initial vertex in U). The algorithm terminates when no such arcs remain, and we then output the current set U.

We consider two cases according to whether or not the target t is in U. As we shall see, these cases correspond to Breakthrough and Nonbreakthrough.

Case 1: The target t is in U.

It follows from the algorithm that, for some integer m, there is a sequence of distinct vertices

$$x_0 = s, x_1, x_2, \ldots, x_{m-1}, x_m = t$$

such that, for each $j = 0, 1, 2, \ldots, m-1$, either

(a) $\alpha_j = (x_j, x_{j+1})$ is an arc of the network with $f(\alpha_j) < c(\alpha_j)$

or

(b) $\alpha_j = (x_{j+1}, x_j)$ is an arc of the network with $f(\alpha_j) > 0$.

We now define an integer-valued function f' on the set A of arcs by

$$f'(\alpha) = \begin{cases} f(\alpha) + 1 & \text{if } \alpha \text{ is one of the arcs } \alpha_j \text{ in (a)}; \\ f(\alpha) - 1 & \text{if } \alpha \text{ is one of the arcs } \alpha_j \text{ in (b)}; \\ f(\alpha) & \text{otherwise.} \end{cases}$$

It follows from the definition of f' and the assumption that all capacities and flow values of f are integers that $0 \le f'(\alpha) \le c(\alpha)$. The fact that f' is a flow can be checked by showing that, for each vertex x_j with $j = 1, 2, \ldots, m-1$, the total flow into x_j equals the total flow out of x_j (e.g., if (x_{j-1}, x_j) and (x_{j+1}, x_j) are both arcs, then the flow into x_i has a net change of $+1 - 1 = 0$). The value val(f') of the flow f' is val(f) $+ 1$, since either $(s, x_1) = (x_0, x_1)$ is an arc, in which case the flow out of s is increased by 1, or $(x_1, s) = (x_1, x_0)$ is an arc, in which case the flow into s is decreased by 1; in either case, there is a net increase of 1 in the flow out of s.

Case 2: The target t is not in U.

In this case, $\overrightarrow{U}$ is a cut, and it follows from the algorithm that

1. $f(\alpha) = c(\alpha)$ for each arc α in $\overrightarrow{U}$ and
2. $f(\alpha) = 0$ for each arc α in $\overleftarrow{U}$.

Hence,

$$\begin{aligned} \text{val}(f) &= \sum_{\alpha \in \overrightarrow{U}} f(\alpha) - \sum_{\alpha \in \overleftarrow{U}} f(\alpha) \\ &= \sum_{\alpha \in \overrightarrow{U}} c(\alpha) \\ &= \text{cap } \overrightarrow{U}. \end{aligned}$$

Hence,, $\hat{f} = f$ is a maximum flow and $\hat{C} = \overrightarrow{U}$ is a minimum cut. $\square$

We conclude this section by deducing from the max-flow min-cut theorem two important combinatorial results, including the theorem of König from Chapter 9.

Example. Let $D = (V, A)$ be a digraph that models a communication network. The vertices represent junctions (relay points) in the network, and the arcs represent direct (one-way) lines of communication. Consider two junctions corresponding to vertices s and t in V. By putting together direct lines we can hope to establish a communication path from s to t. Because communication lines may fail, in order that communication from s to t be possible even in the presence of some failure, it is important to have redundancy in the digraph—that is, arcs whose failure does not prevent communication from s to t. Define an *st-separating set* to be a set S of arcs of D such that every path from s to t uses at least one arc in S. If the arcs of an *st*-separating set all fail, communication from s to t is impossible. Menger's theorem, stated next, characterizes the minimum number of arcs in an *st*-separating set. □

Theorem 12.2.4 *Let s and t be distinct vertices of a digraph $D = (V, A)$. Then the maximum number of pairwise arc-disjoint paths from s to t equals the minimum number of arcs in an st-separating set.*

Proof. Let $N = (V, A, s, t, c)$ be the network in which the capacity of each arc is 1. A cut in N is an *st*-separating set in D (and vice versa), and the capacity of a cut equals the number of its arcs.

Consider an integer-valued flow f in N, and let $\mathrm{val}(f) = p$. Since all the capacity values equal 1, f takes on only the values 0 and 1: For each arc α, f either "chooses" α (if $f(\alpha) = 1$) or not (if $f(\alpha) = 0$). We prove by induction on p that there exist p pairwise arc-disjoint paths from s to t made up of arcs chosen by f. If $p = 0$, this is trivial. Assume $p \geq 1$. There exists a path γ from s to t; otherwise, if U is the set of vertices that can be reached from s by a path, then $\overrightarrow{U} = \emptyset$ is a cut in N with capacity equal to zero, contradicting $p \geq 1$. Let f' be the integer flow of value $p - 1$ obtained from f by reducing by 1 the value of the flow on the arcs of γ. By induction, there exist $p - 1$ pairwise arc-disjoint arcs from s to t made up of arcs chosen by f'. These $p - 1$ paths, together with γ, are p pairwise arc-disjoint paths made up of arcs chosen by f.

Conversely, if there are p pairwise arc-disjoint paths from s to t, then clearly there is an integer flow in N with value p. The theorem now follows from Theorem 12.2.3. □

We recall, from Chapters 9 and 11, that a bipartite graph G is a graph whose vertices can be partitioned into two sets X and Y so that each edge joins a vertex in X and a vertex in Y. The pair X, Y is called a *bipartition* of G. From Chapter 9, a matching in G is a set of pairwise vertex-disjoint edges; a cover of G is a set C of vertices such that each edge of G has at least one of its vertices in C. The maximum number of edges in a matching in G is denoted by $\rho(G)$, and the minimum number of vertices in a cover is denoted by $c(G)$. We show how to deduce Corollary 9.2.4 of Chapter 9 from the Theorem 12.2.4 of Menger.

Theorem 12.2.5 *Let G be a bipartite graph. Then $\rho(G) = c(G)$.*

Proof. Let X, Y be a bipartition of G. We first construct a digraph $D = (X \cup Y \cup \{s, t\}, A)$.

Let s and t be distinct elements not in $X \cup Y$. The arcs of D are those obtained as follows:

1. (s, x) for each x in X;

2. (x, y) for each edge $\{x, y\}$ of G (thus, all arcs of N are directed from X to Y);

3. (y, t) for each y in Y.

Let $\gamma_1, \ldots, \gamma_p$ be a set of pairwise arc-disjoint paths of D from s to t. Each path γ_i is of the form s, x_i, y_i, t for some x_i in X and y_i in Y, and the edges $\{x_1, y_1\}, \ldots, \{x_p, y_p\}$ form a matching in G of size p. Conversely, from a matching in G of size p, we can construct p pairwise arc-disjoint paths in D. Hence, $\rho(G)$ equals the maximum number of pairwise arc-disjoint paths from s to t in D.

Now let $C = X' \cup Y'$ be a cover of G, where $X' \subseteq X$ and $Y' \subseteq Y$. Since each path of D from s to t uses an arc of the form (x, y), where $\{x, y\}$ is an edge of G, it follows that

$$S = \{(s, x') | x' \text{ in } X'\} \cup \{(y', t) | y' \text{ in } Y'\} \qquad (12.11)$$

is an st-separating set in D with $|C| = |S|$. Conversely, if S is an st-separating set in D of the form (12.11), then the set C defined by $C = X' \cup Y'$ is a cover of G. Now let T be *any* st-separating set in D. Then the set $\hat{T}$ obtained from T by replacing each arc in T of the form (x, y) (x in X and y in Y) with the arc (s, x) is also an st-separating set. Moreover, $\hat{T}$ has the form (12.11) for some $X' \subseteq X$ and $Y' \subseteq Y$, $|\hat{T}| \leq |T|$ (because, for instance, there may be several arcs in T of the form $(x, \cdot)$), and $X' \cup Y'$ is a cover of G. It now follows that $c(G)$ equals the the smallest number of arcs in an st-separating set in D. Therefore, $\rho(G) = c(G)$ follows from Theorem 12.2.4.   □

12.3   Exercises

1. Prove Theorem 12.1.2.

2. Prove Theorem 12.1.3.

3. Prove that an orientation of K_n is a transitive tournament if and only if it does not have any directed cycles of length 3.

4. Give an example of a digraph that does not have a closed Eulerian directed trail but whose underlying general graph has a closed Eulerian trail.

5. Prove that a digraph has no directed cycles if and only if its vertices can be labeled from 1 up to n so that the terminal vertex of each arc has a larger label than the initial vertex.

6. Prove that a digraph is strongly connected if and only if there is a closed, directed walk that contains each vertex at least once.

7. Let T be any tournament. Prove that it is possible to change the direction of at most one arc in order to obtain a tournament with a directed Hamilton cycle.

8. Use the proof of Theorem 12.1.5 in order to write an algorithm for determining a Hamilton path in a tournament.

9. Prove that a tournament is strongly connected if and only if it has a directed Hamilton cycle.

10. Prove that every tournament contains a vertex u such that, for every other vertex x, there is a path from u to x of length at most 2.

11. Prove that every graph has the property that it is possible to orient each of its edges so that, for each vertex x, the indegree and outdegree of x differ by at most 1.

12. * Devise an algorithm for constructing a directed Hamilton cycle in a strongly connected tournament.

13. Apply the algorithm in Section 12.1 and determine a strongly connected orientation of the graphs in Figures 11.15 to 11.18.

14. Prove the following generalization of Theorem 12.1.6: Let G be a connected graph. Then, after replacing each bridge $\{a, b\}$ by the two arcs (a, b) and (b, a), one in each direction, it is possible to give the remaining edges of G an orientation so that the resulting digraph is strongly connected.

15. Modify the algorithm for constructing a strongly connected orientation of a bridgeless connected graph in order to accommodate the situation described in Exercise 12.

16. Consider a trader problem in which trader t_1 ranks his item number 1. Prove that, in every core allocation, t_1 gets to keep his own item.

17. Construct an example of a trading problem, with n traders, with the property that, in each core allocation, exactly one trader gets the item he ranks first.

18. Show that, for the trading problem in which the preferences are given by the table

	t_1	t_2	t_3
t_1	2	1	3
t_2	3	2	1
t_3	1	3	2

there are exactly two core allocations. Which of these results from applying the constructive proof of Theorem 12.1.9?

19. Suppose that, in a trading problem, some trader ranks his own item number k. Prove that, in each core allocation, that player obtains an item he ranks no lower than k. (Thus, a player never leaves with an item that he values less than the item he brought to trade.)

20. Prove that, in the core allocation obtained by applying the constructive proof of Theorem 12.1.9, at least one player gets an item he ranks number 1. Show by example that there may be core allocations in which no player gets his first choice.

21. Prove that, in a trading problem, there is a core allocation in which every trader gets the item he ranks number 1 if and only if the digraph D^1 constructed in the proof of Theorem 12.1.9 consists of directed cycles, no two of which have a vertex in common.

22. Construct a core allocation for the trading problem in which the preferences are given by the following table:

	t_1	t_2	t_3	t_4	t_5	t_6	t_7
t_1	2	3	1	4	7	5	6
t_2	1	6	4	3	2	7	5
t_3	2	7	3	5	1	4	6
t_4	3	4	2	7	1	6	5
t_5	1	3	4	2	5	7	6
t_6	2	4	1	5	3	7	6
t_7	7	3	4	2	1	6	5

23. Explicitly write the algorithm for a core allocation that is implicit in the proof of Theorem 12.1.9.

24. Determine a maximum flow and a minimum cut in each of the networks $N = (V, A, s, t, c)$ in Figure 12.7. (The numbers near arcs are their capacities.)

25. Determine the maximum number of pairwise arc-disjoint paths from s to t in the digraphs of the networks in Exercise 22. Verify that the number is maximum by exhibiting an st-separating set with the same number of arcs. (Cf. Theorem 12.2.4.)

26. Consider the network in Figure 12.8, where there are *three* sources s_1, s_2, and s_3 for a certain commodity and *three* targets t_1, t_2, and t_3. Each source has a certain supply of the commodity, and each target has a certain demand for the commodity. These supplies and demands are the numbers in brackets next to the sources and sinks. The supplies are to flow from the sources to the targets, subject to the flow capacities on each arc. Determine whether all the demands can be met simultaneously with the available supplies. (One possible way to approach this problem is to introduce an auxiliary source s and an auxiliarly target t, arcs from s to each s_i with capacity equal to s_i's supply, and arcs from each t_j to t with capacity equal to t_j's demand, and then find a maximum flow from s to t in the augmented network and check whether all demands are met.)

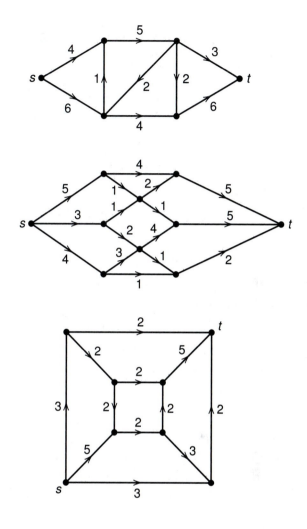

Figure 12.7

27. In Exercise 26, change the supplies at s_1, s_2, and s_3 to a, b, and c, respectively, and determine again whether all the demands can be met simultaneously with the available supplies.

28. * Formulate and prove a theorem that gives necessary and sufficient conditions for a network with multiple sources and sinks, with prescribed supplies and demands, respectively, to have a flow that simultaneously meets all demands with the available supplies.

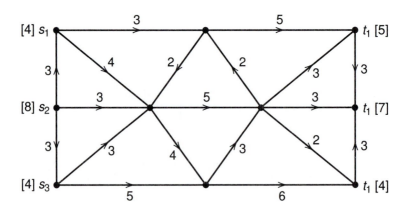

Figure 12.8

29. Consider the set A of the 2^n binary sequences of length n. This exercise concerns the existence of a circular arrangement γ_n of 2^n 0's and 1's, so that the 2^n sequences of n consecutive bits of γ give all of A, that is, are all distinct. Such a circular arrangement is called a *de Bruijn cycle*. For example, if $n = 2$, the circular arrangement $0, 0, 1, 1$ (regarding the first 0 as following the last 1) gives $0, 0$; $0, 1$; $1, 1$; and $1, 0$. For $n = 3$, $0, 0, 0, 1, 0, 1, 1, 1$ (regarded cyclically) is a de Bruijn cycle. Define a digraph Γ_n whose vertices are the 2^{n-1} binary sequences of length $n - 1$. Given two such binary sequences x and y, we put an arc e from x to y, provided that the last $n - 2$ bits of x agree with the first $n - 2$ bits of y, and then we label the arc e with the first bit of x.

    (a) Prove that every vertex of Γ_n has indegree and outdegree equal to 2. Thus, Γ_n has a total of $2 \cdot 2^{n-1} = 2^n$ arcs.

    (b) Prove that Γ_n is strongly connected, and hence Γ_n has a closed Eulerian directed trail (of length 2^n).

    (c) Let $b_1, b_2, \ldots, b_{2^n}$ be the labels of the arcs (considered as a circular arrangement) as one traverses an Eulerian directed trail of Γ_n. Prove that $b_1, b_2, \ldots, b_{2^n}$ is a de Bruijn cycle.

    (d) Prove that, given any two vertices x and y of the digraph Γ_n, there is a path from x to y of length at most $n - 1$.

Chapter 13

More on Graph Theory

In this second chapter on graph theory we study some of the fundamental numbers that are associated with a graph. The most famous
of all these numbers is the *chromatic number*, because of its association with the 4-*color problem*. This problem, which, for over 100 years
was an unsolved problem,[1] asks the following: Consider a map that is
drawn on the plane or on the surface of a sphere in which the countries
are connected regions. We want to color each region with one color so
that neighboring regions are colored differently. Will 4 colors always
suffice to color any map in this way? The short answer is *yes*. The long
answer is that the proof requires[2] an elaborate argument and depends
substantially on calculations by computer. The 4-color problem can
be restated in terms of graphs. Choose a vertex-point in the interior
of each country, and join two vertex-points by an edge-curve whenever
the two countries share a border.[3] In this way, we obtain a plane-graph
(and hence a planar graph) which is called the *dual graph* of the map.
Coloring the regions of a map so that neighboring regions are colored
differently is equivalent to coloring the vertices[4] of its dual graph in
such a way that two vertices which are adjacent are colored differently.
Thus, the 4-color problem can be restated as follows: Every planar

---

[1] A problem being unsolved for over 100 years is not automatically famous. What
made the 4-color problem so famous is that it is easily stated and understood by
almost anyone. And it is also very appealing!

[2] At least the currently known proof does. But a proof that 4 colors do suffice is
beyond an attack by amateur means. The elementary approaches have been tried
and they have failed! For a very brief history of the 4-color problem, see Section
1.4.

[3] Two countries which have only one, or more generally only finitely many points
in common, are not considered to have a common border.

[4] More precisely, we think of assigning colors to the vertices.

graph is 4-colorable. In this chapter, we shall prove that every planar graph is 5-colorable, and more generally, we shall investigate colorings of graphs and other graphical parameters of interest.

13.1   Chromatic Number

In this section we consider only graphs, since the presence of either more than one edge joining a pair of distinct vertices or loops has no essential effect on the types of questions treated here.

Let $G = (V, E)$ be a graph. A *vertex-coloring* of G is an assignment of a color to each of the vertices of G in such a way that adjacent vertices are assigned different colors. If the colors are chosen from a set of k colors, then the vertex-coloring is called a *k-vertex-coloring*, abbreviated *k-coloring*, whether or not all k colors are used. If G has a k-coloring, then G is said to be *k-colorable*. The smallest k, such that G is k-colorable, is called the *chromatic number* of G, denoted by $\chi(G)$. The actual nature[5] of the colors used is of no consequence. Thus, sometimes we describe the colors as red, blue, green, ... , while at other times we simply use the integers 1, 2, 3, ... to designate the colors. Isomorphic graphs have the same chromatic number.

A *null graph* is defined to be a graph without any edges.[6] A null graph of order n is denoted by N_n.

Theorem 13.1.1 *Let G be a graph of order $n \geq 1$. Then*

$$1 \leq \chi(G) \leq n.$$

Moreover, $\chi(G) = n$ if and only if G is a complete graph, and $\chi(G) = 1$ if and only if G is a null graph.

Proof. The inequalities in the theorem are obvious, since any graph with at least one vertex requires at least one color, and any assignment of n distinct colors to the vertices of G is a vertex-coloring. In any vertex-coloring of K_n, no two vertices can be assigned the same color;

---

[5]Should we say color?

[6]A null graph is not necessarily an empty graph, since it may have vertices. The *empty graph* is a graph without any vertices. Thus, a graph $G = (V, E)$ is a null graph if and only if $E = \emptyset$, while G is the empty graph if and only if $V = \emptyset$ (and hence $E = \emptyset$). The empty graph is a very special null graph, namely, the null graph of order 0. Confusing? Not to worry. Just remember that a null graph has no edges.

hence, $\chi(K_n) = n$. Suppose that G is not a complete graph. Then there are two vertices x and y that are not adjacent. Assigning x and y the same color and the remaining $n - 2$ vertices different colors, we obtain an $(n - 1)$-coloring of G, and hence $\chi(G) \leq n - 1$. Assigning all vertices of N_n the same color is a vertex-coloring, and hence $\chi(N_n) = 1$. Suppose that G is not a null graph. Then there are vertices x and y that are adjacent and that thus cannot be assigned the same color in any vertex-coloring of G. Hence, in this case $\chi(G) \geq 2$. $\square$

Corollary 13.1.2 *Let G be a graph and let H be a subgraph of G. Then $\chi(G) \geq \chi(H)$. If G has a subgraph[7] equal to a complete graph K_p of order p, then*

$$\chi(G) \geq p.$$

Proof. It follows from the definition of chromatic number that, if H is any subgraph of G, then $\chi(G) \geq \chi(H)$. Hence, by Theorem 13.1.1, $\chi(G) \geq \chi(K_p) = p$. $\square$

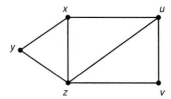

Figure 13.1

Example. Let G be the graph shown in Figure 13.1. Since G has a subgraph equal to K_3, the chromatic number of G is at least 3. Coloring the vertices x and v red, the vertices u and y blue, and the vertex z green, we obtain a 3-coloring of G. Hence, $\chi(G) = 3$. $\square$

Let $G = (V, E)$ be a graph that is k-colored, using the colors $1, 2, \ldots, k$. Let V_i denote the subset of vertices that are assigned the color i, $(i = 1, 2, \ldots, k)$. Then $V_1, V_2, \ldots, V_k$ is a partition of V, called a *color partition* for G. Moreover, the induced subgraphs $G_{V_1}, G_{V_2}, \ldots, G_{V_k}$ are null graphs. Conversely, if we can partition the vertices into k parts, with each part inducing a null graph, then the chromatic number is, at most, k. Hence, another way to describe the chromatic number of G is that $\chi(G)$ *is the smallest integer k such that the vertices of G can be partitioned into k sets with each set inducing a null graph.* In the coloring of the graph in Figure 13.1 described in the preceding example, the partition is $\{x, v\}$ (the red vertices), $\{u, y\}$

---

[7]This subgraph will necessarily be an induced subgraph.

(the blue vertices), and $\{z\}$ (the green vertices). Using these ideas, we can now obtain another lower bound on the chromatic number of a graph.

Corollary 13.1.3 *Let $G = (V, E)$ be a graph of order n and let q be the largest order of an induced subgraph of G equal to a null graph N_q. Then[8]*

$$\chi(G) \geq \left\lceil \frac{n}{q} \right\rceil.$$

Proof. Let $\chi(G) = k$ and let $V_1, V_2, \ldots, V_k$ be a color partition for G. Then $|V_i| \leq q$ for each i, and we obtain

$$n = |V| = \sum_{i=1}^{k} |V_i| \leq \sum_{i=1}^{k} q = k \times q.$$

Hence,

$$\chi(G) = k \geq \frac{n}{q}.$$

Since $\chi(G)$ is an integer, the corollary follows. $\qquad\square$

Example. Continuing with the graph in Figure 13.1, an examination of the graph reveals that the largest order of an induced null subgraph is $q = 2$ (that is, of every 3 vertices at least 2 are adjacent). Thus, by Corollary 13.1.3, we again obtain

$$\chi(G) \geq \left\lceil \frac{5}{2} \right\rceil = 3.$$

$\qquad\square$

According to Theorem 13.1.1, the graphs with chromatic number 1 are the null graphs. It is then natural to ask for a characterization of graphs with chromatic number 2. Graphs with chromatic number 2 have a color partition with 2 sets. This should bring to mind bipartite graphs.

Theorem 13.1.4 *Let G be a graph with at least one edge. Then $\chi(G) = 2$ if and only if G is bipartite.*

Proof. The chromatic number of a graph with at least one edge is at least 2. If G is a bipartite graph, then, coloring the left vertices red and the right vertices blue,[9] we obtain a 2-coloring of G. Conversely,

---

[8] Recall that the ceiling of x, $\lceil x \rceil$, is the smallest integer a such that $a \geq x$.

[9] Of course we could have said "coloring the left vertices left and the right vertices right," using left and right as our two colors.

the color partition arising from a 2-coloring is a bipartition for G, establishing the bipartiteness of G. □

It follows from Theorems 11.4.1 and 13.1.4 that the chromatic number of a graph that is not a null graph equals 2 if and only if each cycle has even length. Graphs with chromatic number 3 can have a very complicated structure and do not admit a simple characterization.

Example. (*A scheduling problem*). Many scheduling problems can be formulated as problems that ask for the chromatic number (but often will settle for a number not much larger than the chromatic number) of a graph. The basic idea is that we associate a graph with a scheduling problem whose vertices are the "tasks" to be scheduled, putting an edge between two tasks whenever they conflict, and hence cannot be scheduled at the same time. A color partition for G furnishes a schedule without any conflicts. The chromatic number of the graph thus equals the smallest number of time slots in a schedule with no conflicts.

For instance, suppose we want to schedule 9 tasks $a, b, c, d, e, f, g,$ h, i, where each task conflicts with the task that follows it in the list, and also i conflicts with a. The "conflict" graph G in this case is a graph of order 9 whose edges are arranged in a cycle of length 9. Of any 5 vertices of this graph, at least 2 are adjacent. Hence, the q in Corollary 13.1.3 is at most 4, and it follows that $\chi(G) \geq 3$. It is easy to find a 3-coloring so that $\chi(G) = 3$. Thus, this scheduling problem requires 3 time slots. □

The determination of the chromatic number of a graph is a difficult problem, and there is no known good algorithm[10] for it. Therefore, it is of importance to have estimates for the chromatic number of a graph and some means for finding a vertex-coloring in which the number of colors used is "not too large." In Corollaries 13.1.2 and 13.1.3, we have given two lower bounds for the chromatic number. Theorem 13.1.1 contains an upper bound, namely, $n - 1$ for a graph of order n, which is not a complete graph, but this bound is rather poor. One would hope to be able to do better. Indeed, we show that a better bound can be obtained from the degrees of the vertices, and there is a simple algorithm for obtaining a vertex-coloring that does not exceed this bound. This algorithm is another example of a greedy algorithm,[11] which pro-

---

[10]One for which the number of steps required grows like a polynomial function of the order of the graph. Most experts believe that no good algorithm is possible.

[11]A greedy algorithm for a minimum-weight spanning tree is given in Section 11.7. Unlike that greedy algorithm, which actually constructed a minimum-weight spanning tree, the current algorithm gives only an upper bound for the chromatic number.

ceeds sequentially by "choosing the first available color," ignoring the consequences this may have for later choices. We use the positive integers to color the vertices, and thus we can speak about one color being smaller than another.

Greedy algorithm for vertex-coloring

Let G be a graph in which the vertices have been listed in some order $x_1, x_2, \ldots, x_n$.

(1) Assign the color 1 to vertex x_1.

(2) For each $i = 2, 3, \ldots, n$, let p be the smallest color such that none of the vertices $x_1, \ldots, x_{i-1}$ which are adjacent to x_i is colored p, and assign the color p to x_i.

Theorem 13.1.5 *Let G be a graph for which the maximum degree of a vertex is Δ. Then the greedy algorithm produces a $(\Delta+1)$-coloring[12] of the vertices of G, and hence*

$$\chi(G) \leq \Delta + 1.$$

Proof. In words, the greedy algorithm considers each vertex in turn, and assigns to it the smallest color which has not already been assigned to a vertex to which it is adjacent. In particular, two adjacent vertices are never assigned the same color, and hence the greedy algorithm does produce a vertex-coloring. There are at most Δ vertices adjacent to vertex x_i, and hence, at most, Δ of the vertices $x_1, \ldots, x_{i-1}$ are adjacent to x_i. Therefore, when we consider vertex x_i in step (2) of the algorithm, at least one of the colors $1, 2, \ldots, \Delta + 1$ has not already been assigned to a vertex adjacent to x_i, and the algorithm assigns the smallest of these to x_i. It follows that the greedy algorithm produces a $(\Delta + 1)$-coloring of the vertices of G.   □

The greedy algorithm just might color the vertices of G in the fewest possible number, namely, $\chi(G)$, of colors. How well or how badly it does depends on the order in which the vertices are listed before the algorithm is applied. Let $V_1, V_2, \ldots, V_{\chi(G)}$ be a color partition arising from a vertex coloring using $\chi(G)$ colors. Suppose we list the vertices of V_1 first, followed by the vertices of V_2, . . . , followed by the vertices

---

[12]Remember that a $(\Delta + 1)$-coloring does not mean that all $\Delta + 1$ colors are actually used.

of $V_{\chi(G)}$.[13] It is easy to see that the greedy algorithm colors the vertices in V_1 with the color 1, the vertices in V_2 with one of the colors 1 or 2, ... , the vertices in $V_{\chi(G)}$ with one of the colors $1, 2, \ldots, \chi(G)$. Thus, with this listing of the vertices, the greedy algorithm colors the vertices with the fewest possible number of colors.

Example. Consider a complete bipartite graph $K_{1,n}$. The largest degree of a vertex is $\Delta = n$. Thus, by Theorem 13.1.5, the greedy algorithm produces an $(n + 1)$-coloring. In fact, it does a lot better. No matter how the vertices are listed, the greedy algorithm colors the vertices with only 2 colors, the minimum possible number of colors. Thus, the greedy algorithm sometimes can give a much better coloring than is suggested by Theorem 13.1.5.

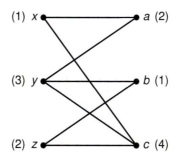

Figure 13.2

Now consider the bipartite graph drawn in Figure 13.2, and list the vertices as x, a, b, y, z, c. Then the colors assigned to these vertices by the greedy algorithm are, respectively, $1, 2, 1, 3, 2, 4$. Hence, the greedy algorithm produces a 4-coloring, yet the chromatic number is 2. □

The upper bound for the chromatic number given in Theorem 13.1.5 can be improved, except for two classes of graphs. These are the complete graphs K_n, for which $\Delta = n - 1$ and $\chi(G) = n$, and the graphs C_n of odd order n whose edges are arranged in a cycle (of odd length), for which $\Delta = 2$ and $\chi(G) = 3$. The proof of the next theorem of Brooks[14] is omitted.

Theorem 13.1.6 *Let G be a connected graph for which the maximum degree of a vertex is Δ. If G is neither a complete graph K_n nor an odd cycle graph C_n, then $\chi(G) \leq \Delta$.*

---

[13] All we want to do is to keep the vertices of the same color together.
[14] R.L. Brooks: On coloring the nodes of a network, *Proc. Cambridge Philos. Soc.*, 37 (1941), 194-197.

A conclusion from our discussion of chromatic number is that coloring the vertices of a graph (so that adjacent vertices are colored differently) is hard if one wants to use the fewest number of colors. We now remove the restriction that the number of colors is minimum, but consider a more difficult question: Given a graph G and a set $\{1, 2, \ldots, k\}$ of k colors, how many k-colorings of G are there? If we know that $\chi(G) > k$, then the question is easy and the answer is 0.[15]

For each nonnegative integer k, the number of k-colorings of the vertices of a graph G is denoted by

$$p_G(k).$$

If $\chi(G) > k$, then $p_G(k) = 0$. For example, for a complete graph, we have

$$p_{K_n}(k) = k(k-1)\ldots(k-(n-1)) = [k]_n,$$

since each vertex must be a different color.[16] For a null graph, we have

$$p_{N_n}(k) = k^n,$$

since we can arbitrarily assign colors to each of the vertices.[17]

Example. We determine $p_G(k)$ for the graph G in Figure 13.1. First we color the vertices x, y, z. These vertices can be colored in

$$k(k-1)(k-2)$$

ways, since each has to receive a different color. Next, we color u and observe that it must receive a color different from that of x and z. There are $k-2$ ways to color u. Finally, v can receive any of the colors other than the (distinct) colors of u and z, and hence there are $k-2$ ways to color v. Thus,

$$p_G(k) = k(k-1)(k-2) \times (k-2) \times (k-2) = k(k-1)(k-2)^3. \qquad \square$$

---

[15]If $\chi(G) > k$, but we do not have that information, then the question is much more difficult. This is because, in answering it, we are implicitly determining whether or not $\chi(G) \leq k$: $\chi(G) \leq k$ if and only if the the number of ways to color G with k colors is not 0.

[16]$[k]_n$ is the function that was introduced in Section 8.2 and counts the number of n-permutations of a set of k distinct objects. In the situation here, the k objects are the k colors and the n-permutations are the assignments of a color to each of the n vertices of K_n. Since each pair of vertices is adjacent in K_n, all vertices have to be colored differently.

[17]We recall from Chapter 3 that k^n counts the number of n-permutations of a set of k objects (the k colors here) in which unlimited repetition is allowed. Since no vertices of N_n are adjacent, we can freely repeat colors.

It is not hard to count the number of ways to color the vertices of a tree. What is surprising is that, for each k, the number of k-colorings of a tree depends only on the number of vertices of the tree, and not on which tree is being considered!

Theorem 13.1.7 *Let T be a tree of order n. Then*

$$p_T(k) = k(k-1)^{n-1}.$$

Proof. We grow T as described in Section 11.5, and color the vertices as we do. The starting vertex can be colored with any one of the k colors. Each new vertex y we add is adjacent to only one of the previous vertices x. Hence, y can be colored with any one of the $k-1$ colors different from the color of x. Thus, each of the $n-1$ vertices, other than the first, can be colored in $k-1$ ways, and the formula follows. □

The observant reader will have noticed that, thus far, each of the formulas obtained for the number of ways to color the vertices of a graph has turned out to be a polynomial function of the number k of colors. Indeed, this is no accident and is a general phenomenon: $p_G(k)$ is always a polynomial function of k. We now turn to proving this fact. As a result of this property, $p_G(k)$ is called the *chromatic polynomial* of the graph G. The chromatic polynomial of G evaluated at k gives the number of k-colorings of G. The chromatic number of G is the smallest nonnegative integer that is not a root of the chromatic polynomial.

The fact that $p_G(k)$ is a polynomial rests on a simple observation. Let x and y be two vertices of G that are adjacent. Let G_1 be the graph obtained from G by removing the edge $\{x, y\}$ joining x and y. The k-colorings of G_1 can be partitioned into two parts, $C(k)$ and $D(k)$. In the first part, $C(k)$, we put those k-colorings of G_1 in which x and y are assigned the same color. In the second part, $D(k)$, we put those k-colorings in which x and y are assigned different colors. Thus,

$$p_{G_1}(k) = |C(k)| + |D(k)|.$$

There is a one-to-one correspondence between the k-colorings of G_1, in which x and y are assigned different colors and the k-colorings of G. Hence,

$$p_G(k) = |D(k)|.$$

Let G_2 be the graph obtained from G by *identifying* the vertices x and y. This means that we delete the edge $\{x, y\}$, replace x and y by one new vertex, denoted $\overline{xy}$, and join $\overline{xy}$ to any vertex that is joined either to x or y in G.[18] There is a one-to-one correspondence between the k-colorings of G_1, in which x and y are assigned the same color, and the k-colorings of G_2. Therefore,

$$p_{G_2}(k) = |C(k)|.$$

Combining the previous three equations, we get

$$p_{G_1}(k) = p_G(k) + p_{G_2}(k),$$

from which it follows that

$$p_G(k) = p_{G_1}(k) - p_{G_2}(k). \tag{13.1}$$

In words, the number of k-colorings of G can be obtained by finding the number of k-colorings of G_1 (in which the edge $\{x, y\}$ has been removed, making it possible for x and y to be assigned the same color) and subtracting the number of k-colorings of G_2 (in which the vertices x and y have been identified so that they must be assigned the same color). Why is this a useful observation?

The order of G_1 is the same as the order of G, and G_1 has one fewer edge than G. The order of G_2 is one less than the order of G, and G_2 has at least one fewer edge than G. Put another way, G_1 and G_2 are closer (in terms of the number of edges) to a null graph than G is. Thus, our observation suggests an algorithm to determine the number of k-colorings of G: Continue to remove edges and identify vertices until all graphs so obtained are null graphs. By (13.1), the number of k-colorings of G can be expressed in terms of the number of k-colorings of each of these null graphs. But we know what the number of k-colorings of a null graph is: The number of k-colorings of a null graph of order p is k^p. Hence, we can obtain the number of k-colorings of G by subtracting and adding the number of k-colorings of null graphs.[19] What's more, since k^p is a polynomial in k, the number of k-colorings of G is a polynomial in k; that is, the chromatic polynomial of G is indeed a polynomial! Before formalizing the previous discussions, we consider an example.

Example. Let G be a cycle graph C_5 of order 5 whose edges are arranged in a cycle. Choosing any edge of G and applying (13.1), we see that

$$p_G(k) = p_{G_1}(k) - p_{G_2}(k),$$

---

[18]We can think of moving x and y together until they coincide. This may create a multiple edge, in which case we delete one copy.

[19]Null graphs may be very uninteresting, but as we have just seen they have an important role to play in graph colorings.

where G_1 is a tree of order 5 whose edges are arranged in a path and G_2 is a cycle graph C_4 of order 4. By Theorem 13.1.7, $p_{G_1}(k) = k(k-1)^4$.[20] We do to G_2 what we did to G and obtain

$$p_{G_2}(k) = k(k-1)^3 - p_{G_3}(k),$$

where G_3 is a cycle graph C_3 of order 3. Since G_3 is a complete graph K_3, and thus $p_{G_3}(k) = k(k-1)(k-2)$, we obtain

$$p_G(k) = k(k-1)^4 - (k(k-1)^3 - k(k-1)(k-2)).$$

This simplifies to

$$p_G(k) = k(k-1)(k-2)(k^2 - 2k + 2).$$

Note that $p_G(0) = 0$, $p_G(1) = 0$, $p_G(2) = 0$ and $p_G(3) > 0$. Hence, $\chi(G) = 3$, a fact that is easy to establish directly.   □

Let G be a graph and let $\alpha = \{x, y\}$ be an edge of G. We now denote the graph obtained from G by deleting the edge α by $G_{\ominus\alpha}$. We also denote the graph obtained from G by identifying x and y (as previously defined) by $G_{\otimes\alpha}$. We say that $G_{\otimes\alpha}$ is obtained from G by *contracting* the edge α. Thus, (13.1) can be rewritten as

$$p_G(k) = p_{G_{\ominus\alpha}}(k) - p_{G_{\otimes\alpha}}(k). \qquad (13.2)$$

As already implied, repeated use of deletion and contraction gives an algorithm for determining $p_G(k)$. In the next algorithm, we consider objects $(\pm, H)$, where H is a graph. For the purposes of the algorithm, we call such an object a *signed graph*, a graph with either a plus sign $+$ or minus sign $-$ associated with it.

Algorithm for computing the chromatic polynomial of a graph

Let $G = (V, E)$ be a graph.

(1) Put $\mathcal{G} = \{(+, G)\}$.

(2) While there exists a signed graph in $\mathcal{G}$ that is not a null graph, do the following:

   (i) Choose a nonnull signed graph (ϵ, H) in $\mathcal{G}$ and an edge α of H.

---

[20]This illustrates an important point in this process, namely, if one obtains a graph whose chromatic polynomial is known, then make use of that information.

(ii) Remove (ϵ, H) from $\mathcal{G}$ and put in the two signed graphs $(\epsilon, H_{\ominus \alpha})$ and $(-\epsilon, H_{\otimes \alpha})$.

(3) Put $p_G(k) = \sum \epsilon k^p$, where the summation extends over all signed graphs (ϵ, H) in $\mathcal{G}$ and p is the order of H.

In words, we start with G with a $+$ attached to it. Using the deletion/contraction process, we reduce G and all resulting graphs to null graphs, keeping track of the associated sign as determined by multiple applications of (13.2). When there are no remaining graphs with an edge, we compute the order p of each null graph so obtained and then form the monomial $\pm k^p$, which is its chromatic polynomial, adjusted for sign. By repeated use of (13.2), adding all these polynomials we obtain the chromatic polynomial of G. In particular, since the sum of monomials is a polynomial, we obtain a polynomial. In the deletion/contraction process, exactly one graph is a null graph of the same order as G. This graph results by successive deletion of all edges of G, without any contraction, and occurs with a $+$ sign. We have thus proved the next theorem.

Theorem 13.1.8 *Let G be a graph of order $n \geq 1$. Then, the number of k-colorings of G is a polynomial in k of degree equal to n (with leading coefficient equal to 1) and this polynomial–the chromatic polynomial of G–is computed correctly by the preceding algorithm.*

It is straightforward to see that, if a graph G is disconnected, then its chromatic polynomial is the product of the chromatic polynomials of its connected components. In particular, the chromatic number is the largest of the chromatic numbers of its connected components. In the next theorem, we generalize this observation. The resulting formula can sometimes be used to shorten the computation of the chromatic polynomial of a graph.

Let $G = (V, E)$ be a connected graph and let U be a subset of the vertices of G. Then U is called an *articulation set* of G, provided that the subgraph G_{V-U} induced[21] by the vertices not in U is disconnected. If G is not complete, then G contains two nonadjacent vertices a and b, and hence $V - \{a, b\}$ is an articulation set. A complete graph does not have an articulation set. Therefore, a connected graph has an articulation set if and only if it is not complete.

---

[21] Recall that the vertices of this subgraph are those in $V - U$, and two vertices are adjacent in G_{V-U} if and only if they are adjacent in G.

Lemma 13.1.9 *Let G be a graph and assume that G contains a subgraph H equal to a complete graph K_r. Then the chromatic polynomial of G is divisible by the chromatic polynomial $[k]_r$ of K_r.*

Proof. In any k-coloring of G, the vertices of H are all colored differently. Moreover, each choice of colors for the vertices of H can be extended to the same number $q(k)$ of colorings for the remaining vertices of G. Hence, $p_G(k) = [k]_r q(k)$. $\qquad\square$

Theorem 13.1.10 *Let U be an articulation set of G and suppose that the induced subgraph G_U is a complete graph K_r. Let the connected components of G_{V-U} be the induced subgraphs $G_{U_1}, \ldots, G_{U_t}$. For $i = 1, \ldots, t$, let $H_i = G_{U \cup U_i}$ be the subgraph of G induced by $U \cup U_i$. Then*

$$p_G(k) = \frac{p_{H_1}(k) \times \cdots \times p_{H_t}(k)}{([k]_r)^{t-1}}.$$

In particular, the chromatic number of G is the largest of the chromatic numbers of $H_1, \ldots, H_t$.

Proof. The graphs $H_1, \ldots, H_t$ all have the vertices of U in common, but are otherwise pairwise disjoint. Each k-coloring of G can be obtained by first choosing a k-coloring of H_1 (there are $p_{H_1}(k)$ such colorings and now all the vertices of U are colored), and then completing the colorings of each H_i, $(i = 2, \ldots, t)$ (each in $p_{H_i}(k)/[k]_r$ ways, by Lemma 13.1.9). $\qquad\square$

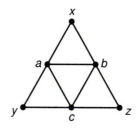

Figure 13.3

Example. Let G be the graph drawn in Figure 13.3. Let $U = \{a, b, c\}$. Applying Theorem 13.1.10, we see that

$$p_G(k) = \frac{(q(k))^3}{(k(k-1)(k-2))^2},$$

where $q(k)$ is the chromatic polynomial of a complete graph G' of order 4 with 1 missing edge. It is simple to calculate (in fact, use Theorem 13.1.10 again) that $q(k) = k(k-1)(k-2)^2$. Hence,

$$p_G(k) = k(k-1)(k-2)^4.$$

□

13.2   Plane and Planar Graphs

Let $G = (V, E)$ be a planar general graph and let G' be a planar representation of G. Thus, G' is a plane-graph and G' consists of a collection of points in the plane, called vertex-points because they correspond to the vertices of G, and a collection of curves, called edge-curves because they correspond to the edges of G. Also, an edge-curve α is a simple curve that passes through a vertex-point x if and only if the vertex x of G is incident with the edge α of G.[22] Only endpoints can be common points of edge-curves.

The plane graph G' divides the plane into a number of regions that are bounded by one or more of the edge-curves.[23] Exactly one of these regions extends infinitely far.

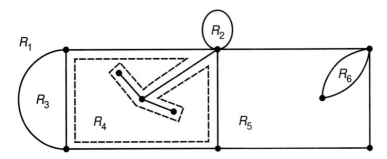

Figure 13.4

Example. The plane-graph shown in Figure 13.4 has 10 vertex-points, 14 edge-curves, and 6 regions. Each of the regions is bordered by some of the edge-curves, but we have to be very careful how we count the edge-curves. The regions R_2, R_3, R_5, and R_6 are bordered by 1, 2, 6, and 2 edge-curves, respectively. The region R_4 is bordered by 10 edge-curves (and not 4 or 7). This is because, as we traverse R_4 by walking around its border, three of the edge-curves are traversed twice

---

[22] Recall that we give the same label to a vertex and its corresponding vertex-point and the same label to an edge and its corresponding edge-curve.

[23] Thus, a plane-graph has points, curves, and now regions.

(see the dashed line in Figure 13.4). The region R_1 is bordered[24] by 7 edge-curves. In sum, we count the number of edge-curves bordering regions in such a way that each edge-curve is counted twice, either because it borders two different regions or because it borders the same region twice. □

Let G' be a plane-graph with n vertex-points, e edge-curves, and r regions. Let the number of edge-curves bordering the regions be, respectively,

$$f_1, f_2, \ldots, f_r.$$

Then, using the convention established in the preceding example, we have

$$f_1 + f_2 + \cdots + f_r = 2e. \tag{13.3}$$

We now derive a relationship between n, e, and r which implies in particular that any two of them determine the third. This relationship is known as *Euler's formula*.

Theorem 13.2.1 *Let G be a plane-graph of order n with e edge-curves and assume that G is connected. Then the number r of regions into which G divides the plane satisfies*

$$r = e - n + 2. \tag{13.4}$$

Proof. First, assume that G is a tree. Then $e = n - 1$ and $r = 1$ (the only region is the infinite region that is bordered twice by each edge-curve). Hence, (13.4) holds in this case. Now assume that G is not a tree. Since G is connected, it has a spanning tree T with $n' = n$ vertices, $e' = n - 1$ edges, and $r' = 1$ regions, where $r' = e' - n' + 2$. We can think of starting with the edge-curves of T and then inserting one new edge-curve at a time until we have G. Each time we insert an edge-curve, we divide an existing region into two regions. Hence, each time we insert another edge-curve, e' increases by 1, r' increases by 1, and n' stays the same (n' is always n). Therefore, starting with $r' = e' - n' + 2$ for a spanning tree, this relationship persists as we include the remaining edge-curves, and the theorem is proved. □

Euler's formula has an important consequence for planar graphs (with no loops and multiple edges).

---

[24]R_1 might appear to be bordered by none of the edge-curves, since it extends infinitely far in all directions. However, a geometrical figure drawn in the plane can also be thought of as drawn on a sphere. Loosely speaking, we put a large sphere on top of the figure and then "wrap" the sphere with the plane. The infinite region is now some finite region on the sphere. Note also that a region may have "interior" border curves as, e.g., R_1 and R_4 do.

Theorem 13.2.2 *Let G be a connected planar graph. Then there is a vertex of G whose degree is at most 5.*

Proof. Let G' be a planar representation of G. Since a graph has no loops, no region of G' is bordered by only one edge-curve. Similarly, since a graph has no multiple edges, no region is bordered by only 2 edge-curves (unless G has exactly one edge). Thus, in (13.3), each f_i satisfies $f_i \geq 3$, and hence we have

$$3r \leq 2e, \text{ or equivalently, } \frac{2e}{3} \geq r.$$

Using this inequality in Euler's formula, we get

$$\frac{2e}{3} \geq r = e - n + 2, \text{ or equivalently, } e \leq 3n - 6. \tag{13.5}$$

Let $d_1, d_2, \ldots, d_n$ be the degrees of the vertices of G. By Theorem 11.1.1, we have

$$d_1 + d_2 + \cdots + d_n = 2e.$$

Hence, the average of the degrees of G satisfies

$$\frac{d_1 + d_2 + \cdots + d_n}{n} \leq \frac{6n - 12}{n} < 6.$$

Since the average of the degrees is less than 6, some vertex must have degree 5 or less. $\qquad\square$

If a graph G has a subgraph that is not planar, then G is not planar. Thus, in attempting to describe planar graphs, it is of interest to find nonplanar graphs G, each of whose subgraphs, other than G itself, is planar.

Example. A complete graph K_n is planar if and only if $n \leq 4$.

If $n \leq 4$, then K_n is planar. Now consider K_5. As shown in the proof of Theorem 13.2.2 (see (13.5)), the number n of vertices and the number e of edges of a planar graph satisfies $e \leq 3n - 6$. Since K_5 has $n = 5$ vertices and $e = 10$ edges, K_5 is not planar. Since K_5 is not planar, K_n is not planar for all $n \geq 5$. $\qquad\square$

Example. A complete bipartite graph $K_{p,q}$ is planar if and only if $p \leq 2$ or $q \leq 2$.

It is easy to draw a planar representations of $K_{p,q}$ if $p \leq 2$ or $q \leq 2$. Now consider $K_{3,3}$. A bipartite graph does not have any cycles of length 3; hence, in a planar representation of a planar bipartite graph, each region is bordered by at least 4 edge-curves. Arguing as in the proof of Theorem 13.2.2, we obtain $r \leq e/2$. Applying Euler's formula, we get

$$\frac{e}{2} \geq e - n + 2; \text{ equivalently, } 2n - 4 \geq e.$$

Since $K_{3,3}$ has $n = 6$ vertices and $e = 9$ edges, $K_{3,3}$ is not planar. Since $K_{3,3}$ is not planar, $K_{p,q}$ is not planar if both $p \geq 3$ and $q \geq 3$. □

Let $G = (V, E)$ be a nonplanar graph and let $\{x, y\}$ be any edge of G. Let G' be obtained from G by choosing a new vertex z not in V and replacing the edge $\{x, y\}$ with the two edges $\{x, z\}$ and $\{z, y\}$. We say that G' is obtained from G by *subdividing the edge* $\{x, y\}$. If G is not planar, then clearly G' is also not planar.[25] A graph H is called a *subdivison* of a graph G, provided that H can be obtained from G by successively subdividing edges. If H is a subdivision of G, then we can think of H as obtained from G by inserting several new vertices (possibly none) on each of its edges. For example, the graphs in Figure 13.5 are subdivisions of $K_{3,3}$ and K_5, respectively. It follows that each of these graphs is not planar.

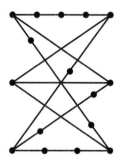 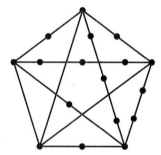

Figure 13.5

A nonplanar graph cannot contain a subdivision of a K_5 or a $K_{3,3}$. It is a remarkable theorem of Kuratowski[26] that the converse holds as well. We state this theorem without proof.

Theorem 13.2.3 *A graph G is planar if and only if it does not have a subgraph that is a subdivision of a K_5 or of a $K_{3,3}$.*

Loosely speaking, Theorem 13.2.3 says that a graph that is not planar has to contain a subgraph that either looks like a K_5 or looks like

---

[25]If there were a planar representation of G', then by "erasing" the vertex-point z we obtain a planar representation of G.

[26]K. Kuratowski: Sur le problème des courbes gauches en topologie, *Fund. Math.*, 15 (1930), 271-283.

a $K_{3,3}$. Thus, the two graphs K_5 and $K_{3,3}$ are the only two "obstructions" to planarity. As noted by Wagner[27] and Harary and Tutte,[28] planar graphs can also be characterized by using the notion of contraction of an edge in place of subdivision of an edge. A graph H is a *contraction* of a graph G, provided that H can be obtained from G by successively contracting edges.

Theorem 13.2.4 *A graph G is planar if and only if it does not contain a subgraph that contracts to a K_5 or a $K_{3,3}$.*

13.3   A 5-color Theorem

In this section we show that the chromatic mumber of a planar graph is, at most, 5. This was first proved by P.J. Heawood in 1890 after he discovered an error in a "proof" published in 1879 by A. Kempe, in which Kempe claimed that the chromatic number of a planar graph is, at most, 4. Although Kempe's proof was wrong, it contained good ideas, which Heawood used to prove his 5-color theorem. As described in the introduction to this section, and also in Section 1.4, a proof that the chromatic number of every planar graph is, at most, 4 has now been obtained, and it relies heavily on computer checking.

There is an easy proof, which uses Theorem 13.2.2, of the fact that the chromatic number of a planar graph G is, at most, 6. Indeed, suppose there is a planar graph whose chromatic number is 7 or more, and let G be such a graph with the smallest number of vertices. By Theorem 13.2.2, G has a vertex x of degree at most 5. Removing x (and all incident edges) from G leaves a planar graph G' with one fewer vertex. The minimal assumption on G implies that G' has a 6-coloring. Since x is adjacent in G to at most 5 vertices, we can take a 6-coloring of G' and assign a color to x in such a way as to produce a 6-coloring of G, a contradiction. It follows that the chromatic number of every planar graph is 6 or less. It is harder, but not terribly so, to prove that a planar graph has a 5-coloring, but the jump from 5 colors to 4 colors is a giant one.

Before proving that 5 colors suffice to color the vertices of any planar graph, we make one observation. In the previous section, we showed that a complete graph K_5 of order 5 is not planar, and hence

---

[27]K. Wagner: Über eine Eigenschaft der ebenen Komplexe, *Math. Ann.*, 114 (1937), 570-590.

[28]F. Harary and W.T. Tutte: A dual form of Kuratowski's theorem, *Canadian Math. Bull.*, 8 (1965), 17-20.

a planar graph cannot contain 5 vertices, the members of every pair of which are adjacent. It is erroneous to conclude from this that every planar graph has a 5-coloring. For instance, with 3 replacing 5, a cycle graph C_5 of order 5 does not have a K_3 as a subgraph, yet its chromatic number is 3 and it does not have a 2-coloring. So it does *not* simply suffice to say that there do not exist 5 vertices such that each must be assigned different colors and hence a 4-coloring is possible.

The next theorem is an important step in the proof of the 5-color theorem. It applies to nonplanar graphs as well as planar graphs.

Theorem 13.3.1 *Let there be given a k-coloring of the vertices of a graph $H = (U, F)$. Let two of the colors be red and blue, and let W be the subset of vertices in U that are assigned either the color red or the color blue. Let $H_{r,b}$ be the subgraph of H induced by the vertices in W and let $C_{r,b}$ be a connected component of $H_{r,b}$. Interchanging the colors red and blue assigned to the vertices of $C_{r,b}$, we obtain another k-coloring of H.*

Proof. Suppose that there are two adjacent vertices which are now colored the same. This color must be either red or blue, say, red. If x and y are both vertices of $C_{r,b}$, then before we switched colors, x and y were colored blue which is impossible. If neither x nor y is a vertex in $C_{r,b}$, then their colors weren't switched and so they both started out with color red, again impossible. Thus, one of x and y is a vertex in $C_{r,b}$ and the other isn't, say, x is in $C_{r,b}$ and y is not. Therefore, x started out with the color blue and y started out with the color red. Since x and y are adjacent and each is assigned one of the colors red and blue, they must be in the same connected component of $H_{r,b}$, contradicting the fact that x is in $C_{r,b}$ and y isn't. $\square$

Theorem 13.3.2 *The chromatic number of a planar graph is, at most, 5.*

Proof. Let G be a planar graph of order n. If $n \leq 5$, then $\chi(G) \leq 5$. We now let $n > 5$ and prove the theorem by induction on n. We assume that G is drawn in the plane as a plane-graph. By Theorem 13.2.2, there is a vertex x whose degree is at most 5. Let H be the subgraph of order $n - 1$ of G induced by the vertices different from x. By the induction hypothesis, there is a 5-coloring of H. If the degree of x is 4 or less, then we can assign one of the 5 colors to x and obtain

a 5-coloring of G.[29] Now suppose that the degree of x is 5. There
are 5 vertices adjacent to x. If two of these vertices are assigned the
same color, then, as before, there is a color we can assign x in order to
obtain a 5-coloring of G. So we now further suppose that each of the
vertices y_1, y_2, y_3, y_4, y_5 adjacent to x is assigned a different color. As
in Figure 13.6, the vertices $y_1, \ldots, y_5$ are labeled consecutively around
vertex x; the colors are the numbers 1, 2, 3, 4, and 5 with y_j colored j,
$(j = 1, 2, 3, 4, 5)$.

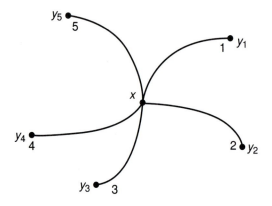

Figure 13.6

We consider the subgraph $H_{1,3}$ of H induced by the vertices of
colors 1 and 3. If y_1 and y_3 are in different connected components of
$H_{1,3}$, then we apply Lemma 13.1.1 to H and obtain a 5-coloring in
which y_1 and y_3 are colored the same. This frees up a color for x, and
we obtain a 5-coloring of G. Now assume that y_1 and y_3 are in the
same connected component of $H_{1,3}$. Then there is a path in H joining
y_1 and y_3 such that the colors of the vertices on the path alternate
between 1 and 3. This path along with the edge-curves joining x and
y_1 and x and y_3 determine a closed curve γ. Of the remaining three
vertices y_2, y_4, and y_5 adjacent to x, one of them is inside γ and two
are outside γ, or the other way around. See Figure 13.7, in which y_2
is inside γ and y_4 and y_5 are outside. We now consider the subgraph
$H_{2,4}$ of H induced by the vertices of colors 2 and 4. But (see Figure
13.7) vertices y_2 and y_4 cannot be in the same connected component
of $H_{2,4}$ since y_2 is in the interior of a simple closed curve and y_4 is in
the exterior of that curve. Switching the colors 2 and 4 of the vertices
in the connected component of $H_{2,4}$ that contains x_2, we obtain by

---

[29]This is just like our proof that 6 colors suffice to color the vertices of a planar
graph. But for a 5-coloring, we are not yet done, since we now have to deal with
the case that x has degree 5.

Lemma 13.1.1 a 5-coloring of H in which none of the vertices adjacent to x is assigned color 2. We now assign the color 2 to x and obtain a 5-coloring of G. □

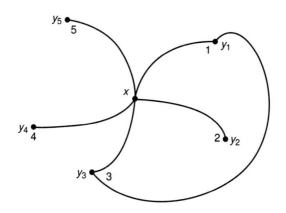

Figure 13.7

In 1943, Hadwiger[30] made a conjecture about the chromatic number of graphs, which, except in a few cases, is still unsolved. This is perhaps not too surprising since the truth of one instance of this conjecture is equivalent to the existence of a 4-coloring of any planar graph. This conjecture asserts: *A connected graph G whose chromatic number satisfies $\chi(G) \geq p$ can be contracted to a K_p. Equivalently, if G cannot be contracted to a K_p then $\chi(G) < p$.* The converse of the conjecture is false; that is, it is possible for a graph to be contractable to a K_p and yet have chromatic number less than p. For instance, a graph of order 4 whose edges are arranged in a cycle has chromatic number 2, yet the graph itself can be contracted to a K_3.

Theorem 13.3.3 *Hadwiger's conjecture holds for $p = 5$ if and only if every planar graph has a 4-coloring.*

Partial Proof. We prove only that if Hadwiger's conjecture holds for $p = 5$, then every planar graph G has a 4-coloring. Let G be a planar graph and suppose that G is contractable to a K_5. A contraction of a planar graph is also planar, and this implies that K_5 is planar, a statement we know to be false. Hence, G is not contractable to a K_5, and hence the truth of Hadwiger's conjecture for $p = 5$ implies that $\chi(G) \leq 4$. □

[30]H. Hadwiger: Über eine Klassifikation der Streckenkomplexe, *Vierteljschr. Naturforsch. Ges., Zurich*, 88 (1943), 133-142.

Hadwiger's conjecture is also known to be true for $p \le 4$ and for $p = 6$. We verify Hadwiger's conjecture for $p = 2$ and 3 in the next theorem and leave its validity for $p = 4$ as a challenging exercise.

Theorem 13.3.4 *Let $p \le 3$. If G is a connected graph with chromatic number $\chi(G) \ge p$, then G can be contracted to a K_p.*

Proof. If $p = 1$, then by contracting each edge, we arrive at a K_1. If $p = 2$, then G has at least one edge α, and by contracting all edges except for α, we arrive at a K_2. Now, suppose $p = 3$ and $\chi(G) \ge 3$. Then G is not bipartite, and by Theorem 11.4.1, G has a cycle of odd length. Let γ be an odd cycle of smallest length in G. Then the only edges joining vertices of γ are the edges of γ, for otherwise we could find an odd cycle of length shorter than γ. By contracting all the edges of G except for the edges of γ we arrive at γ. We may further contract edges to obtain a K_3.   □

13.4 Independence Number and Clique Number

Let $G = (V, E)$ be a graph of order n. A set of vertices U of G is called *independent*,[31] provided that no two of its vertices are adjacent, equivalently, provided the subgraph G_U of G induced by the vertices in U is a null graph. Thus, the chromatic number $\chi(G)$ equals the smallest integer k such that the vertices of G can be partitioned into k independent sets. Each subset of an independent set is also an independent set. Consequently, we seek large independent sets. The largest number of vertices in an independent set is called the *independence number* of the graph G and is denoted by $\alpha(G)$. The independence number is the largest number of vertices that can be colored the same in a vertex-coloring of G. Corollary 13.1.3 can be rephrased as

$$\chi(G) \ge \left\lceil \frac{n}{\alpha(G)} \right\rceil.$$

For a null graph N_n, a complete graph K_n, and a complete bipartite graph $K_{m,n}$, we have

$$\alpha(N_n) = n, \quad \alpha(K_n) = 1, \quad \text{and} \quad \alpha(K_{m,n}) = \max\{m, n\}.$$

The determination of the independence number of a graph is, in general, a difficult computational problem.

---

[31]Sometimes also called *stable*.

Example. Let G be the graph in Figure 13.8. Then $\{a, e\}$ is an independent set that is not a subset of any larger independent set. Also $\{b, c, d\}$ is an independent set with the same property. Of any 4 vertices, two are adjacent, and hence we have $\alpha(G) = 3$.   □

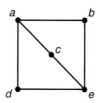

Figure 13.8

Example. A zoo wishes to place various species of animals in the same enclosure. Obviously, if one species preys on another, then both should not be put in the same enclosure. What is the largest number of species that can be placed in one enclosure?

We form the *zoo graph*, G, whose vertices are the different animal species in the zoo, and put an edge between two species if and only if one of them preys on the other. The largest number of species that can be placed in the same enclosure equals the independence number $\alpha(G)$ of G. How many enclosures are required in order to accommodate all the species? The answer is the chromatic number $\chi(G)$ of G.   □

Example. (*The problem of the 8 queens*). Consider an 8-by-8 chessboard and the chess piece known as a *queen*. In chess, a queen can attack any piece that lies in its row or column or in one of the two diagonals containing it. If 9 queens are placed on the board, then necessarily, two lie in the same row and thus can attack one another. Is it possible to place 8 queens on the board so that no queen can attack another?

Figure 13.9

Let G be the *queens graph* of the chessboard. The vertices of G are the squares of the board, with two squares adjacent if and only if a queen placed on one can attack a queen placed on the other. Our question thus asks whether the independence number of the queens graph equals 8. In fact, $\alpha(G) = 8$ and there are 92 different ways to place 8 nonattacking queens on the board. One of these is shown in Figure 13.9.   □

Let $G = (V, E)$ be a graph and let U be an independent set of vertices that is not a subset of any larger independent set. Thus, no two vertices in U are adjacent, and each vertex not in U is adjacent to at least one vertex in U.[32] A set of vertices with the latter property is called a dominating set. More precisely, a set W of vertices of G is a *dominating set*, provided that each vertex not in W is adjacent to at least one vertex in W. Vertices in W may or may not be adjacent. Clearly, if W is a dominating set, then any set of vertices containing W is also a dominating set. The problem is to find the smallest number of vertices in a dominating set. The smallest number of vertices in a dominating set is called the *domination number* of G and is denoted by $\text{dom}(G)$.

Example. Consider a building, perhaps housing an art gallery, consisting of a complicated array of corridors. It is desired to place guards throughout the building so that each part of the building is visible, and therefore protected, by at least one guard. How many guards need to be employed to safeguard our building?

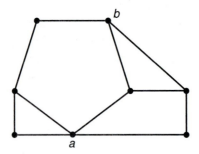

Figure 13.10

We construct a graph G whose vertices are the places where two or more corridors come together or where one corridor ends, and whose edges correspond to the corridors. For example, we might have the corridor graph shown in Figure 13.10. The least number of guards

---

[32]If not, then U could be enlarged.

that can protect the building equals the domination number $\text{dom}(G)$ of G. For the graph G in Figure 13.10, it is not difficult to check that $\text{dom}(G) = 2$ and that $\{a, b\}$ is a dominating set of 2 vertices. □

For a null graph, complete graph, and complete bipartite graph, we have

$$\text{dom}(N_n) = n, \quad \text{dom}(K_n) = 1, \quad \text{and} \quad \text{dom}(K_{m,n}) = 2 \text{ if } m, n \geq 2.$$

In general, it is very difficult to compute the domination number of a graph. The domination number of a disconnected graph is clearly the sum of the domination numbers of its connected components. For a connected graph, we have a simple inequality.

Theorem 13.4.1 *Let G be a connected graph of order $n \geq 2$. Then*

$$\text{dom}(G) \leq \left\lfloor \frac{n}{2} \right\rfloor.$$

Proof. Let T be a spanning tree of G. Then

$$\text{dom}(G) \leq \text{dom}(T),$$

and hence it suffices to prove the inequality for trees of order $n \geq 2$. We use induction on n. If $n = 2$, then either vertex of T is a dominating set and hence $\text{dom}(T) = 1 = \lfloor 2/2 \rfloor$. Now suppose that $n \geq 3$. Let y be a vertex that is adjacent to a pendent vertex x. Let T^* be the graph obtained from T by removing the vertex y and all edges incident with y. The connected components of T^* are trees, at least one of which is a tree of order 1. Let $T_1, \ldots, T_k$ be the trees of order at least 2. Let their orders be $n_1 \geq 2, \ldots, n_k \geq 2$, respectively. Then $n_1 + \cdots + n_k \leq n - 2$. By the induction hypothesis, each T_i has a dominating set of size at most $\lfloor n_i/2 \rfloor$. The union of these dominating sets along with y gives a dominating set of T of size at most

$$1 + \left\lfloor \frac{n_1}{2} \right\rfloor + \cdots + \left\lfloor \frac{n_k}{2} \right\rfloor \leq 1 + \left\lfloor \frac{n_1 + \cdots + n_k}{2} \right\rfloor$$
$$\leq 1 + \left\lfloor \frac{n-2}{2} \right\rfloor = \left\lfloor \frac{n}{2} \right\rfloor.$$

□

A *clique* in a graph G is a subset U of vertices, each pair of which is adjacent, equivalently, the subgraph induced by U is a complete graph. The largest number of vertices in a clique is called the *clique number* of G and is denoted by $\omega(G)$. For a null graph, complete graph, and complete bipartite graph, we have

$$\omega(N_n) = 1, \quad \omega(K_n) = n \quad \text{and } \omega(K_{m,n}) = 2.$$

The notion of a clique of a graph is "complementary" to that of independence in the following sense. Let $\overline{G} = (V, \overline{E})$ be the *complementary graph* of G. Recall that the complementary graph of G has the same set of vertices as G, and two vertices are adjacent in $\overline{G}$ if and only if they are not adjacent in G. It follows from definitions that, for a subset U of V, U is an independent set of G if and only if U is a clique of $\overline{G}$, and U is a clique of G if and only if U is an independent set of $\overline{G}$. In particular, we have

$$\alpha(G) = \omega(\overline{G}) \quad \text{and } \omega(G) = \alpha(\overline{G}).$$

The chromatic number and clique number are related by the inequality (cf. Theorem 13.1.2)

$$\chi(G) \geq \omega(G). \tag{13.6}$$

Every bipartite graph G with at least one edge satisfies $\chi(G) = \omega(G) = 2$. A cycle graph C_n of odd order $n > 3$ with n edges arranged in a cycle satisfies $\chi(C_n) = 3 > 2 = \omega(C_n)$.

Since independence and clique are complementary notions, and since a vertex-coloring is a partition of the vertices of a graph into independent sets, it is natural to consider the notion complementary to that of vertex-coloring. Replacing independent set with clique in the definition of vertex-coloring, we obtain the following definitions. A *clique-partition* of a graph G is a partition of its vertices into cliques. The smallest number of cliques in a clique-partition of G is the *clique-partition number* of G, denoted by $\theta(G)$. We have

$$\chi(G) = \theta(\overline{G}) \quad \text{and } \theta(G) = \chi(\overline{G}).$$

The inequality "complementary" to that in (13.6) is

$$\theta(G) \geq \alpha(G). \tag{13.7}$$

This holds because two nonadjacent vertices cannot be in the same clique.

It is natural to investigate graphs for which equality holds in (13.6) (graphs whose chromatic number equals their clique number), and graphs for which equality holds in (13.7) (graphs whose clique-partition number equals its independence number). Graphs for which equality holds in either of these inequalities need not be too special. For instance, let H be any graph with chromatic number equal to p (thus $\omega(H) \leq p$). Let G be a graph with two connected components, one of which is H and the other of which is a K_p. Then we have $\chi(G) = p$ and $\omega(G) = p$, and hence equality holds in (13.6), no matter what the structure of H. Some structure can be imposed by requiring that (13.6) hold not only for G but for *every* induced subgraph of G.

A graph G is called χ-*perfect*, provided that $\chi(H) = \omega(H)$ for every induced subgraph H of G. The graph G is θ-*perfect*, provided that $\theta(H) = \alpha(H)$ for every induced subgraph H of G. It was conjectured by Berge[33] in 1961 and proved by Lovász[34] in 1972 that there is only one kind of perfection. We state this theorem without proof.

Theorem 13.4.2 *A graph G is χ-perfect if and only if it is θ-perfect. Equivalently, G is χ-perfect if and only if $\overline{G}$ is χ-perfect.*

As a result of this theorem we now refer to *perfect graphs*, and we show the existence of a large class of perfect graphs.

Let $G = (V, E)$ be a graph. A *chord* of a cycle of G is an edge that joins two nonconsecutive vertices of the cycle. A chord is thus an edge that joins two vertices of the cycle, but which is not itself an edge of the cycle. A cycle of length 3 cannot have any chords. A graph is *chordal*, provided that each cycle of length greater than 3 has a chord. A chordal graph has no chordless cycles. An induced subgraph of a chordal graph is also a chordal graph.

Example. Complete graphs and all bipartite graphs are perfect. A complete graph K_n is a chordal graph as is every tree.[35] A complete bipartite graph $K_{m,n}$ with $m \geq 2$ and $n \geq 2$ is not a chordal graph, since such a graph has a chordless cycle of length 4. The graph obtained from a complete graph K_n by removing one edge is a chordal graph, since every cycle of K_n of length greater than 3 has at least two chords. $\square$

A special class of chordal graphs arises by considering intervals on a line. A closed interval on the real line is denoted by

$$[a, b] = \{x : a \leq x \leq b\}.$$

---

[33]C. Berge: Färbung von Graphen, deren sämtliche bzw. deren ungerade Kreise starr sind, *Wiss. Z. Martin-Luther-Univ., Halle-Wittenberg Math.-Natur, Reihe,* (1961), 114-115.

[34]L. Lovász: Normal hypergraphs and the perfect graph conjecture, *Discrete Math.*, 2 (1972), 253-267.

[35]If a graph doesn't have any cycles it surely cannot have a chordless cycle.

Let

$$I_1 = [a_1, b_1], \ I_2 = [a_2, b_2], \ \ldots, \ I_n = [a_n, b_n] \qquad (13.8)$$

be a family of closed intervals. Let G be the graph whose set of vertices is $\{I_1, I_2, \ldots, I_n\}$, where two intervals I_i and I_j are adjacent if and only if $I_i \cap I_j \neq \emptyset$. Such a graph G is called a *graph of intervals,* and any graph isomorphic to a graph of intervals is called an *interval graph.* Thus, the vertices of an interval graph can be thought of as intervals with two vertices adjacent if and only if the intervals have at least one point in common.

Example. A complete graph K_n is an interval graph. We choose the intervals (13.8) with

$$a_1 < a_2 < \cdots < a_n < b_n < \cdots < b_2 < b_1.$$

If $i \neq j$ and $i < j$, then $I_j \subset I_i$, and thus $I_i \cap I_j \neq \emptyset$. Hence, the graph of intervals is a complete graph.

Now let G be the graph of order 4 obtained from K_4 by removing one edge. We choose the intervals (13.8) $(n = 4)$ with

$$a_4 < a_1 < a_3 < b_4 < a_2 < b_1 < b_2 < b_3.$$

Except for the intervals I_2 and I_4, every pair of intervals has a nonempty intersection. $\qquad\square$

Theorem 13.4.3 *Every interval graph is a chordal graph.*

Proof. Let G be an interval graph with intervals $I_1, I_2, \ldots, I_n$ as given in (13.8). Suppose that $k > 3$ and that

$$I_{j_1} - I_{j_2} - \cdots - I_{j_k} - I_{j_1}$$

is a cycle of length k. We show that at least one of the intervals of the cycle has a nonempty intersection with the interval two away from it on the cycle. We assume the contrary and obtain a contradiction. Suppose that I_m, I_p, I_q, I_r are four consecutive intervals on the cycle for which $I_m \cap I_q = \emptyset$ and $I_p \cap I_r = \emptyset$, so that there is no chord joining I_m and I_q and no chord joining I_p and I_r. Then

$$I_m \cap I_p \neq \emptyset, \ I_p \cap I_q \neq \emptyset, \ I_q \cap I_r \neq \emptyset, \ I_m \cap I_q = \emptyset, \ \text{and} \ I_p \cap I_r = \emptyset.$$

If $a_q < a_p$ and $b_p < b_q$, then $I_p \subset I_q$, and hence $\emptyset \neq I_m \cap I_p \subset I_m \cap I_q$, a contradiction. Therefore, either $a_p \leq a_q$ or $b_q \leq b_p$. If $a_p \leq a_q$, then

$a_q \leq a_r$. If $b_q \leq b_p$, then $b_r \leq b_q$. Thus, for three consecutive intervals I_p, I_q, I_r of the cycle, we have one of

$$a_p \leq a_q \leq a_r \text{ or } b_r \leq b_q \leq b_p. \tag{13.9}$$

Now, let $p = j_1$ and first suppose that $a_{j_1} \leq a_{j_2}$. Then, iteratively using (13.9), we obtain

$$a_{j_1} \leq a_{j_2} \leq \cdots \leq a_{j_k} \leq a_{j_1},$$

and we conclude that all of the intervals have the same left endpoint. If $b_{j_2} \leq b_{j_1}$, then, in a similar way, we conclude that all of the intervals have the same right endpoint. In either case, all of the intervals of the cycle have a point in common, contradicting our assumption that intervals two apart on the cycle have no point in common. This contradiction establishes the validity of the theorem. □

To conclude this section we show that chordal graphs, and hence interval graphs, are perfect. We require another lemma for the proof. Recall that a subset U of the vertices of a graph $G = (V, E)$ is an articulation set, provided that the subgraph G_{V-U} induced by the vertices not in U is disconnected. The lemma demonstrates that the chromatic number of a graph equals its clique number if certain smaller induced graphs have this property.

Lemma 13.4.4 *Let $G = (V, E)$ be a connected graph and let U be an articulation set of G such that the subgraph G_U induced by U is a complete graph. Let the connected components of the induced subgraph G_{V-U} be $G_1 = (U_1, E_1), \ldots, G_t = (U_t, E_t)$. Assume that the induced graphs $G_{U_i \cup U}$ satisfy*

$$\chi(G_{U_i \cup U}) = \omega(G_{U_i \cup U}) \quad (i = 1, 2, \ldots, t).$$

Then

$$\chi(G) = \omega(G).$$

Proof. Let $k = \omega(G)$. Because each clique of $G_{U_i \cup U}$ is a clique of G we have

$$\omega(G_{U_i \cup U}) \leq k \quad (i = 1, 2, \ldots, t).$$

Since vertices in different U_i's are not adjacent, each clique of G is a clique of $G_{U_j \cup U}$ for some j. Hence, for at least one j,

$$\omega(G_{U_j} \cup U) = k.$$

We now use the hypotheses and Theorem 13.1.10 to obtain

$$
\begin{aligned}
\chi(G) &= \max\{\chi(G_{U_1 \cup U}), \ldots, \chi(G_{U_t \cup U})\} \\
&= \max\{\omega(G_{U_1 \cup U}), \ldots, \omega(G_{U_t \cup U})\} \\
&= k = \omega(G).
\end{aligned}
$$

$\square$

An articulation set U is a *minimal articulation set*, provided that, for all subsets $W \subseteq U$ with $W \neq U$, W is not an articulation set. In the next theorem we show that minimal articulation sets in chordal graphs induce a complete subgraph.

Theorem 13.4.5 *Let $G = (V, E)$ be a connected chordal graph and let U be a minimal articulation set of G. Then the subgraph G_U induced by U is a complete graph.*

Proof. We assume to the contrary that G_U is not a complete graph and obtain a contradiction. Let a and b be vertices in U that are not adjacent. Since U is an articulation set, the graph G_{V-U} has at least two connected components, $G_1 = (U_1, E_1)$ and $G_2 = (U_2, E_2)$. If a was not adjacent to any vertex of G_1, then it would follow that $U - \{a\}$ is also an articulation set. Since U is a minimal articulation set, we conclude that a is adjacent to at least one vertex in U_1. In a similar way one concludes that a is adjacent to a vertex in U_2, and that b is adjacent to at least one vertex in U_1 and at least one vertex in U_2. Since G_1 and G_2 are connected, there is a path γ_1 joining a to b, all of whose vertices different from a and b belong to U_1, and there is a path γ_2 joining b to a, all of whose vertices different from a and b belong to U_2 . We may choose γ_1 and γ_2 so that they have the shortest possible length. It follows that γ_1 followed by γ_2,

$$
\gamma = \gamma_1, \gamma_2,
$$

is a cycle in G of length at least 4. Moreover, since we have chosen γ_1 and γ_2 to have the shortest length, the only possible chord of γ is an edge joining a and b. Since a and b were chosen to be nonadjacent, we conclude that γ does not have a chord, contradicting the hypothesis that G is a chordal graph. $\square$

We now prove that chordal graphs are perfect.

Theorem 13.4.6 *Every chordal graph is perfect.*

Proof. Since an induced subgraph of a chordal graph is also a chordal graph, it suffices to prove that for a chordal graph G we have $\chi(G) = \omega(G)$.

Let G be a chordal graph of order n. We prove by induction on n that

$$\chi(G) = \omega(G).$$

Since complete graphs are known to be perfect, we assume that G is not complete. Then G has an articulation set, and hence a minimal articulation set U. By Theorem 13.4.5, G_U is a complete graph. Let $G_1 = (U_1, E_1), \ldots, G_t = (V_t, E_t)$ be the connected components of G_{V-U}. By the induction hypothesis, each of the graphs $G_{U_i \cup U}$ satisfies

$$\chi(G_{U_i \cup U}) = \omega(G_{U_i \cup U}) \quad (i = 1, 2, \ldots, t).$$

Now, applying Lemma 13.4.4, we conclude that $\chi(G) = \omega(G)$. □

From Theorems 13.4.3 and 13.4.6 we immediately obtain the next corollary.

Corollary 13.4.7 *Every interval graph is a perfect graph.*

A considerable amount of effort has been expended in attempts to characterize perfect graphs. These efforts have been largely directed toward resolving the following *conjecture* of Berge:[36]

> *A graph G is perfect if and only if neither G nor its complementary graph $\overline{G}$ has an induced subgraph equal to a cycle of odd length greater than three without any chords.*

This conjecture was resolved recently in the affirmative. We leave to the exercises the verification that, if either G or its complementary graph $\overline{G}$ has an induced subgraph equal to a chordless cycle of odd length greater than 3, then G is not perfect.

13.5   Connectivity

Graphs are either connected or disconnected. But it is evident that some connected graphs are "more connected" than others.

---

[36]C. Berge: Färbung von Graphen, deren sämtliche bzw. deren ungerade Kreise starr sind, *Wiss. Z. Martin-Luther-Univ., Halle-Wittenberg Math.-Natur, Reihe,* (1961), 114-115.

Example. We could measure how connected a graph is by measuring how difficult it is to disconnect the graph. But how shall we measure the difficulty required to disconnect a graph? There are two natural ways for doing this. Consider, for instance, a tree of order $n \geq 3$ that forms a path. If we take a vertex other than one of the two end vertices of the path and remove it (and, of course, the two incident edges), the result is a disconnected graph. Indeed, a path is not special among trees in this regard. If we take any tree and remove a vertex other than a pendent vertex, the result is a disconnected graph. Thus, a tree is not very connected. It is necessary to remove only one vertex in order to disconnect it. If, instead of removing vertices (and their incident edges), we remove only edges (and none of the vertices) a tree still comes out as "almost disconnected": removing any edge leaves a disconnected graph. In contrast, a complete graph K_n of order n can never be disconnected by removing vertices because removing vertices always leaves one with a smaller complete graph. If, instead of removing vertices, we remove edges, we can disconnect K_n: if we remove all of the $n-1$ edges incident with a particular vertex, then we are left with a disconnected graph.[37] A simple calculation reveals that K_n cannot be disconnected by removing fewer than $n-1$ edges. Thus, by either manner of reckoning,[38] a complete graph K_n is very connected. The main purpose of this section is to formally define these two notions of connectivity and to discuss some of their implications.   □

In order to simplify our exposition we assume throughout this section that all graphs have order $n \geq 2$.

Let $G = (V, E)$ be a graph of order n. If G is a complete graph K_n, then we define its vertex-connectivity to be

$$\kappa(K_n) = n - 1.$$

Otherwise, we define the *vertex-connectivity* of G to be

$$\kappa(G) = \min\{|U| : G_{V-U} \text{ is disconnected }\},$$

the smallest number of vertices whose removal leaves a disconnected graph. Equivalently, the connectivity of a noncomplete graph equals the smallest size of an articulation set. A noncomplete graph has a pair of nonadjacent vertices a and b. Removing all vertices different from a and b leaves a disconnected graph, and hence $\kappa(G) \leq n-2$ if G

---

[37] Indeed, a K_{n-1} and a vertex separate from it.

[38] And, as one would expect, for any reasonable way to measure how connected a graph is.

is a noncomplete graph of order n. The connectivity of a disconnected graph is clearly 0. Thus, we have the next elementary result.

Theorem 13.5.1 *Let G be a graph of order n. Then*

$$0 \le \kappa(G) \le n - 1,$$

with equality on the left if and only if G is disconnected and with equality on the right if and only if G is a complete graph.

The *edge-connectivity* of a graph G is defined to be the minimum number of edges whose removal disconnects G and is denoted by $\lambda(G)$. The edge-connectivity of a disconnected graph G satisfies $\lambda(G) = 0$. A connected graph G has edge-connectivity equal to 1 if and only if it has a bridge. The edge-connectivity of a complete graph K_n satisfies $\lambda(K_n) = n - 1$. If we remove all the edges of a graph that are incident with a specified vertex x, then we obviously obtain a disconnected graph. Thus, the edge-connectivity of a graph G satisfies $\lambda(G) \le \delta(G)$, where $\delta(G)$ denotes the smallest degree of a vertex of G. The basic relation between vertex- and edge-connectivity is contained in the next theorem.[39]

Theorem 13.5.2 *For each graph G, we have*

$$\kappa(G) \le \lambda(G) \le \delta(G).$$

Proof. We have verified the second inequality in the preceding paragraph. We now verify the first inequality. Let G have order n. If G is a complete graph K_n, then $\kappa(G) = \lambda(G) = n - 1$. We henceforth assume that G is not complete. If G is disconnected the inequality holds since $\kappa(G) = \lambda(G) = 0$. So we assume that G is connected. Let F be a set of $\lambda(G)$ edges whose removal leaves a disconnected graph H. Then H has two connected components,[40] with vertex sets V_1 and V_2, respectively, where $|V_1| + |V_2| = n$. If F consists of all edges joining vertices in V_1 to vertices in V_2, then $|F| \ge n - 1$, and hence $\lambda(G) = n - 1$ and G is complete contrary to assumption. Thus, there exist vertices a in V_1 and b in V_2 such that a and b are not adjacent in G. For each edge α in F, we choose one vertex as follows: If a is a vertex of α, we choose

---

[39] This theorem was first proved by H. Whitney: Congruent graphs and the connectivity of graphs, *American J. Math.*, 54 (1932), 150-168. The proof given here is from R.A. Brualdi and J. Csima, A note on vertex- and edge-connectivity, *Bulletin of the Institute of Combinatorics and its Applications*, 2 (1991), 67-70.

[40] If there were more than two components, we could disconnect G by removing fewer edges.

the other vertex of α (the one in V_2); otherwise, we choose the vertex of α that is in V_1. The resulting set U of vertices satisfies $|U| \leq |F|$. Moreover, removing the vertices U from G results in a disconnected graph, since there can be no path from a to b. Thus,

$$\kappa(G) \leq |U| \leq |F| = \lambda(G),$$

completing the proof of the theorem.   □

Example. Suppose that, in a communication system, there are n stations,[41] some of which are linked by a direct communication line. We assume that the system is connected in the sense that each station can communicate with every other station through intermediary communication links. Thus, we have a natural connected graph G of order n in which the vertices correspond to the stations and the edges to the direct links. Now, links may fail and stations may get shut down, and this affects communication. The vertex-connectivity and edge-connectivity of G are intimately related to the reliability of the system. Indeed, as many as $\kappa(G) - 1$ of the stations may be shut down and the others will still be able to communicate amongst themselves. As many as $\lambda(G) - 1$ of the links may fail and all of the stations will still be able to communicate with each other.   □

Let G be a graph. Then G is connected if and only if its vertex-connectivity satisfies $\kappa(G) \geq 1$. If k is an integer and $\kappa(G) \geq k$, then G is called k-*connected*. Thus, the 1-connected graphs are the connected graphs. Notice that, if a graph is k-connected, then it is also $(k-1)$-connected. The vertex-connectivity of a graph equals the largest integer k such that the graph is k-connected. In the remainder of this section we investigate the structure of 2-connected graphs and show, in particular, that the edges (but not the vertices in general) of a graph are naturally partitioned into its "2-connected parts."[42] We define an *articulation vertex* of a graph G to be a vertex a whose removal disconnects G, that is, a vertex such that $\{a\}$ is an articulation set.

Theorem 13.5.3 *Let G be a graph of order $n \geq 3$. Then the following three assertions are equivalent:*

(*i*) *G is 2-connected.*

---

[41] Or, we might have n chips in a computer.

[42] Since 1-connected means connected, we know that the vertices of a graph, and hence the edges, are naturally partitioned into its 1-connected parts, that is, its connected components. When we consider the 2-connected parts, we get only a natural partition of the edges.

(ii) G is connected and does not have an articulation vertex.

(iii) For each triple of vertices a, b, c, there is a path joining a and b that does not contain c.

Proof. If $\kappa(G) \geq 2$, then G is connected and does not have an articulation vertex. Conversely, since $n \geq 3$, if G is connected and without articulation vertices, then $\kappa(G) \geq 2$. Thus, assertions (i) and (ii) are equivalent.

Now assume that (ii) holds. Let a, b, c be a triple of vertices. Since G has no articulation vertices, removing c does not disconnect G. Hence, there is a path joining a and b that does not contain c, and assertion (iii) holds. Conversely, assume that (iii) holds. Then G is surely connected. Suppose that c is an articulation vertex of G. Removing c disconnects G; choosing a and b in different connected components of the resulting graph, we contradict (iii). Hence, G has no articulation vertex and (ii) holds. Therefore, (ii) and (iii) are also equivalent.   □

The reason for the assumption $n \geq 3$ in Theorem 13.5.3 is that a complete graph K_2 is connected and does not have an articulation vertex, that is, satisfies (ii), but it does not satisfy (i), since we have $\kappa(K_2) = 1$.

Let $G = (V, E)$ be a connected graph of order $n \geq 2$. A *block* of G is a maximal induced subgraph of G that is connected and has no articulation vertex. More precisely, let U be a subset of the vertices of G. Then the induced subgraph G_U is a block of G, provided that G_U is connected and has no articulation vertex, and for all subsets W of the vertices of G with $U \subseteq W$ and $U \neq W$, either the induced subgraph G_W is not connected or it has an articulation vertex. It follows from Theorem 13.5.3 that the blocks of G are either the complete graph K_2 or are 2-connected.

Example. Let G be the graph in Figure 13.11. Then the blocks are the induced subgraphs G_U with U equal to

$$\{a, b\}, \{b, c, d, e\}, \{c, f, g, h\}, \{h, i\}, \{i, j\}, \{i, k\}.$$

Four of the blocks are K_2's, and two of the blocks are 2-connected. Notice that, while some of the blocks may have a vertex in common, each edge of G belongs to exactly one block.   □

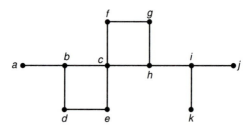

Figure 13.11

Theorem 13.5.4 *Let $G = (V, E)$ be a connected graph of order $n \geq 2$, and let*

$$G_{U_1} = (U_1, E_1), G_{U_2} = (U_2, E_2), \ldots, G_{U_r} = (U_r, E_r)$$

be the blocks of G. Then $E_1, E_2, \ldots, E_r$ is a partition of the set E of edges of G,[43] and each pair of blocks has, at most, one vertex in common.

Proof. Each edge of G belongs to some block, since a block can be a K_2. A block that is a K_2 cannot have an edge in common with any other block, and hence has at most one vertex in common with any other block. Thus, we need consider only blocks G_{U_i} and G_{U_j} $(i \neq j)$ of order at least 3 and hence blocks that are 2-connected. If we show that these blocks can have at most one vertex in common, then it will follow that an edge cannot be in two different blocks.

Suppose that $U_i \cap U_j$ contains at least 2 vertices. Then, since U_i and U_j have a nonempty intersection, the induced graph $G_{U_i \cup U_j}$ is connected. Let x be any vertex in $U_i \cup U_j$. Since G_{U_i} and G_{U_j} are 2-connected, $G_{U_i - \{x\}}$ and $G_{U_j - \{x\}}$ are connected. Moreover, since U_i and U_j have 2 vertices in common, $G_{U_i \cup U_j - \{x\}}$ is connected. It follows that the induced graph $G_{U_i \cup U_j}$ is 2-connected. This gives us a larger 2-connected induced subgraph and contradicts the assumption that G_{U_i} and G_{U_j} are blocks (and hence maximal 2-connected induced subgraphs). Therefore, two distinct blocks can have at most one common vertex. □

We conclude this section with another characterization of graphs that are 2-connected.

Theorem 13.5.5 *Let $G = (V, E)$ be a graph of order $n \geq 3$. Then G is 2-connected if and only if, for each pair a, b of distinct vertices, there is a cycle containing both a and b.*

---

[43]Thus, each edge of G belongs to exactly one block.

Proof. If each pair of distinct vertices of G is on a cycle, then surely G is connected and has no articulation vertex. Hence, by Theorem 13.5.3, G is 2-connected.

Now assume that G is 2-connected. Let a and b be distinct vertices of G. Let U be the set of all vertices x different from a for which there exists a cycle containing both a and x. We first show that $U \neq \emptyset$; that is, there is at least one cycle containing a. Let $\{a, y\}$ be any edge containing a. By Theorem 13.5.1, $\lambda(G) \geq \kappa(G) \geq 2$, and hence the deletion of the edge $\{a, y\}$ does not disconnect G. Consequently, there is a path joining a and y that does not use the edge $\{a, y\}$, and thus a cycle containing both a and y. Therefore, $U \neq \emptyset$.

Suppose, contrary to what we wish to prove, that b is not in U. Let z be a vertex in U whose distance p to b is as small as possible, and let γ be a path from z to b of length p. Since z is in U, there is a cycle γ_1 containing both a and z. The cycle γ_1 contains two paths, γ_1' and γ_1'', joining a to z. Since G is 2-connected, it follows from Theorem 13.5.2 that there is a path γ_2 joining a and b that does not contain the vertex z. Let u be the first vertex of γ that is also a vertex of γ_2.[44] Let v be the last vertex of γ_2 which is also a vertex of γ_1.[45] The vertex v belongs either to γ_1' or to γ_1'', let us say to γ_1'. Then, following a to v along γ_1', v to u along γ_2, u to z along γ, and z back to a along γ_1'', we construct a cycle containing both a and u. Thus, u is in U. But since u is closer to b than z we contradict our choice of z. We conclude that b is in U, and hence there is a cycle containing both a and b. $\quad\square$

An alternative formulation of the characterization of 2-connected graphs in Theorem 13.5.5 is given in the next corollary.

Corollary 13.5.6 *Let G be a graph with at least 3 vertices. Then G is 2-connected if and only if, for each pair a, b of distinct vertices, there are two paths joining a and b whose only common vertices are a and b.*

The corollary is a special case of a theorem of Menger[46] that characterizes k-connected graphs for any k. We state this theorem, the "undirected version" of Menger's theorem for digraphs proved in Section 12.2, without proof.

Theorem 13.5.7 *Let k be a positive integer and let G be a graph of order $n \geq k + 1$. Then G is k-connected if and only if, for each pair*

---

[44]Such a vertex exists, since b is a vertex of γ, that is also a vertex of γ_2.

[45]Such a vertex exists, since a is a vertex of γ_2 that is also a vertex of γ_1.

[46]K. Menger: Zur allgemeinen Kurventheorie, *Fund. Math.*, 10 (1927), 95–115.

a, b of distinct vertices, there are k paths joining a and b such that each pair of paths has only the vertices a and b in common.

If $k = 1$, then the theorem asserts that a graph is 1-connected (i.e., is connected) if and only if each pair of vertices is joined by a path.

13.6   Exercises

1. Prove that isomorphic graphs have the same chromatic number and the same chromatic polynomial.

2. Prove that the chromatic number of a disconnected graph is the largest of the chromatic numbers of its connected components.

3. Prove that the chromatic polynomial of a disconnected graph equals the product of the chromatic polynomials of its connected components.

4. Prove that the chromatic number of a cycle graph C_n of odd length equals 3.

5. Determine the chromatic numbers of the following graphs:

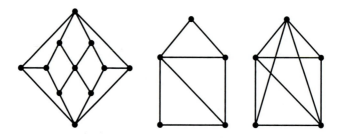

6. Prove that a graph with chromatic number equal to k has at least $\binom{k}{2}$ edges.

7. Prove that the greedy algorithm always produces a coloring of the vertices of $K_{m,n}$ in 2 colors ($m, n \geq 1$).

8. Let G be a graph of order $n \geq 1$ with chromatic polynomial $p_G(k)$.

   (a) Prove that the constant term of $p_G(k)$ equals 0.

   (b) Prove that the coefficient of k in $p_G(k)$ is nonzero if and only if G is connected.

(c) Prove that the coefficient of k^{n-1} in $p_G(k)$ equals $-m$, where m is the number of edges of G.

9. Let G be a graph of order n whose chromatic polynomial is $p_G(k) = k(k-1)^{n-1}$ (i.e., the chromatic polynomial of G is the same as that of a tree of order n). Prove that G is a tree.

10. What is the chromatic number of the graph obtained from K_n by removing one edge?

11. Prove that the chromatic polynomial of the graph obtained from K_n by removing an edge equals

$$[k]_n + [k]_{n-1}.$$

12. What is the chromatic number of the graph obtained from K_n by removing two edges with a common vertex?

13. What is the chromatic number of the graph obtained from K_n by removing two edges without a common vertex?

14. Prove that the chromatic polynomial of a cycle graph C_n equals

$$(k-1)^n + (-1)^n(k-1).$$

15. Prove that the chromatic number of a graph that has exactly one cycle of odd length is 3.

16. Prove that the polynomial $k^4 - 4k^3 + 3k^2$ is not the chromatic polynomial of any graph.

17. Use Theorem 13.1.10 to determine the chromatic number of the following graph:

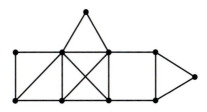

18. Use the algorithm for computing the chromatic polynomial of a graph to determine the chromatic polynomial of the graph Q_3 of vertices and edges of a three-dimensional cube.

19. Find a planar graph that has two different planar representations such that, for some integer f, one has a region bounded by f edge-curves and the other has no such region.

20. Give an example of a planar graph with chromatic number 4 that does not contain a K_4 as an induced subgraph.

21. A plane is divided into regions by a finite number of straight lines. Prove that the regions can be colored with two colors in such a way that regions which share a boundary are colored differently.

22. Repeat Exercise 21, with circles replacing straight lines.

23. Let G be a connected planar graph of order n having $e = 3n - 6$ edges. Prove that, in any planar representation of G, each region is bounded by exactly 3 edge-curves.

24. Prove that a connected graph can always be contracted to a single vertex.

25. Verify that a contraction of a planar graph is planar.

26. Let G be a planar graph of order n in which every vertex has the same degree k. Prove that $k \leq 5$.

27. Let G be a planar graph of order $n \geq 2$. Prove that G has at least two vertices whose degrees are at most 5.

28. A graph is called *color-critical* provided each subgraph obtained by removing a vertex has a smaller chromatic number. Let $G = (V, E)$ be a color-critical graph. Prove the following:

    (a) $\chi(G_{V-\{x\}}) = \chi(G) - 1$ for every vertex x.
    (b) G is connected.
    (c) Each vertex of G has degree at least equal to $\chi(G) - 1$.
    (d) G does not have an articulation set U such that G_U is a complete graph.
    (e) Every graph H has an induced subgraph G such that $\chi(G) = \chi(H)$ and G is color-critical.

29. Let $p \geq 3$ be an integer. Prove that a graph, each of whose vertices has degree at least $p-1$, contains a cycle of length greater than or equal to p. Then use Exercise 28 to show that a graph with chromatic number equal to p contains a cycle of length at least p.

30. $*$ Let G be a graph without any articulation vertices such that each vertex has degree at least 3. Prove that G contains a subgraph that can be contracted to a K_4. (Hint: Begin with a cycle of largest length p. By Exercise 28, we have $p \geq 4$.) Now use Exercise 28 to obtain a proof of Hadwiger's conjecture for $p = 4$.

31. Let G be a connected graph. Let T be a spanning tree of G. Prove that T contains a spanning subgraph T' such that, for each vertex v, the degree of v in G and the degree of v in T' are equal modulo 2.

32. Find a solution to the problem of the 8 queens that is different from that given in Figure 13.9.

33. Prove that the independence number of a tree of order n is at least $\lceil n/2 \rceil$.

34. Prove that the complement of a disconnected graph is connected.

35. Let H be a spanning subgraph of a graph G. Prove that $\mathrm{dom}(G) \leq \mathrm{dom}(H)$.

36. For each integer $n \geq 2$, determine a tree of order n whose domination number equals $\lfloor n/2 \rfloor$.

37. Determine the domination number of the graph Q_3 of vertices and edges of a three-dimensional cube.

38. Determine the domination number of a cycle graph C_n.

39. For $n = 5$ and 6, show that the domination number of the queens graph of an n-by-n chessboard is, at most, 3 by finding 3 squares on which to place queens so that every other square is attacked by at least one of the queens.

40. Show that the domination number of the queens graph of a 7-by-7 chessboard is, at most, 4.

41. $*$ Show that the domination number of the queens graph of an 8-by-8 chessboard is, at most, 5.

42. Prove that an induced subgraph of an interval graph is an interval graph.

43. Prove that an induced subgraph of a chordal graph is chordal.

44. Prove that the only connected bipartite graphs that are chordal are trees.

45. Prove that all bipartite graphs are perfect.

46. Let G be a graph such that either G or its complement $\overline{G}$ has an induced subgraph equal to a chordless cycle of odd length greater than 3. Prove that G is not perfect.

47. Prove that the edge-connectivity of K_n equals $n - 1$.

48. Give an example of a graph G different from a complete graph for which $\kappa(G) = \lambda(G)$.

49. Give an example of a graph G for which $\kappa(G) < \lambda(G)$.

50. Give an example of a graph G for which $\kappa(G) < \lambda(G) < \delta(G)$.

51. Determine the edge-connectivity of the complete bipartite graphs $K_{m,n}$.

52. Let G be a graph of order n with vertex degrees $d_1, d_2, \ldots, d_n$. Assume that the degrees have been arranged so that $d_1 \leq d_2 \leq \cdots \leq d_n$. Prove that, if $d_k \geq k$ for all $k \leq n - d_n - 1$, then G is a connected graph.

53. Let G be a graph of order n in which every vertex has degree equal to d.

    (a) How large must d be in order to *guarantee* that G is connected?

    (b) How large must d be in order to *guarantee* that G is 2-connected?

54. Determine the blocks of the graph given in Figure 13.12 below.

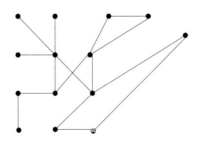

Figure 13.12

55. Prove that the blocks of a tree are all K_2's.

56. Let G be a connected graph. Prove that an edge of G is a bridge if and only if it is the edge of a block equal to a K_2.

57. Let G be a graph. Prove that G is 2-connected if and only if, for each vertex x and each edge α, there is a cycle that contains both the vertex x and the edge α.

58. Let G be a graph each of whose vertices has positive degree. Prove that G is 2-connected if and only if, for each pair of edges α_1, α_2, there is a cycle containing both α_1 and α_2.

59. Prove that a connected graph of order $n \geq 2$ has at least two vertices that are not articulation vertices. (Hint: Take the two end vertices of a longest path.)

Chapter 14

Pólya Counting

Suppose you wish to color the four corners of a regular tetrahedron, and you have just two colors, red and blue. How many different colorings are there? One answer to this question is $2^4 = 16$, since a tetrahedron has 4 corners, and each corner can be colored with either of the two colors. But should we regard all of the 16 colorings to be different? If the tetrahedron is fixed in space, then each corner is distinguished from the others by its position, and it matters which color each corner gets. Thus, in this case, all 16 colorings are different. Now suppose that we are allowed to "move the tetrahedron around." Then, because it is so symmetrical, it matters not which corners are colored red and which are colored blue. The only way two colorings can be distinguished from one another is by the number of corners of each color. Hence, there is one coloring with all red corners, one with three red corners, one with two red corners, one with one red corner, and one with no red corners, giving a total of five different colorings.

Now suppose we color the four corners of a square with the colors red and blue. Again, we have 16 different colorings, provided the square is regarded as fixed in position. How many different colorings are there if we allow the square to move around? The square is also a highly symmetrical figure, although it does not possess the "complete symmetry" of the tetrahedron. As shown in Figure 14.1, there is one coloring with all red corners, one with three red corners, two with two red corners (the red corners can either be consecutive or separated by a blue corner), one with one red corner, and one with no red corners, giving a total of six different colorings.

For both the tetrahedron and square, if allowed to freely move around, the $2^4 = 16$ ways to color its corners are partitioned into parts in such a way that two colorings in the same part are regarded as the same (the colorings are *equivalent*), and two colorings in different parts are regarded as different (the colorings are *nonequivalent*). The number of nonequivalent colorings is thus the number of different parts. The purpose of this chapter is to develop and illustrate a technique for counting nonequivalent colorings in the presence of symmetries.

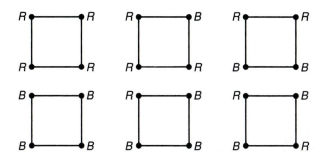

Figure 14.1

14.1   Permutation and Symmetry Groups

Let X be a finite set. Without loss of generality, we take X to be the set

$$\{1, 2, \ldots, n\},$$

consisting of the first n positive integers. Each permutation

$$i_1, i_2, \ldots, i_n$$

of X can be viewed as a one-to-one function from X to itself defined by

$$f : X \to X$$

where

$$f(1) = i_1, f(2) = i_2, \ldots, f(n) = i_n.$$

By the pigeonhole principle, each one-to-one function $f : X \to X$ is onto.[1] To emphasize the view that a permutation can also be viewed as a function, we also denote this permutation by the 2-by-n array

$$\begin{pmatrix} 1 & 2 & \cdots & n \\ i_1 & i_2 & \cdots & i_n \end{pmatrix}. \tag{14.1}$$

---

[1]Thus, one-to-one functions from X to X are one-to-one correspondences.

In (14.1), the value i_k of the function at the integer k is written below k.

Example. The $3! = 6$ permutations of $\{1, 2, 3\}$, regarded as functions, are

$$\begin{pmatrix} 1 & 2 & 3 \\ 1 & 2 & 3 \end{pmatrix}, \quad \begin{pmatrix} 1 & 2 & 3 \\ 1 & 3 & 2 \end{pmatrix}, \quad \begin{pmatrix} 1 & 2 & 3 \\ 2 & 1 & 3 \end{pmatrix},$$

$$\begin{pmatrix} 1 & 2 & 3 \\ 2 & 3 & 1 \end{pmatrix}, \quad \begin{pmatrix} 1 & 2 & 3 \\ 3 & 1 & 2 \end{pmatrix}, \quad \begin{pmatrix} 1 & 2 & 3 \\ 3 & 2 & 1 \end{pmatrix}.$$

□

We denote the set of all $n!$ permutations of $\{1, 2, \ldots, n\}$ by S_n. Thus, S_3 consists of the 6 permutations listed in the previous example. Since permutations are now functions, they can be combined, using composition, that is, following one by another. If

$$f = \begin{pmatrix} 1 & 2 & \cdots & n \\ i_1 & i_2 & \cdots & i_n \end{pmatrix}$$

and

$$g = \begin{pmatrix} 1 & 2 & \cdots & n \\ j_1 & j_2 & \cdots & j_n \end{pmatrix}$$

are two permutations of $\{1, 2, \ldots, n\}$, then their *composition*, in the order f followed by g, is the permutation

$$g \circ f = \begin{pmatrix} 1 & 2 & \cdots & n \\ j_1 & j_2 & \cdots & j_n \end{pmatrix} \circ \begin{pmatrix} 1 & 2 & \cdots & n \\ i_1 & i_2 & \cdots & i_n \end{pmatrix},$$

where

$$(g \circ f)(k) = g(f(k)) = j_{i_k}.$$

Composition of functions defines a *binary operation* on S_n: If f and g are in S_n, then $g \circ f$ is also in S_n.

Example. Let f and g be the permutations in S_4 defined by

$$f = \begin{pmatrix} 1 & 2 & 3 & 4 \\ 3 & 2 & 4 & 1 \end{pmatrix} \qquad g = \begin{pmatrix} 1 & 2 & 3 & 4 \\ 2 & 4 & 3 & 1 \end{pmatrix}.$$

Then

$$(g \circ f)(1) = 3, \quad (g \circ f)(2) = 4, \quad (g \circ f)(3) = 1, \quad (g \circ f)(4) = 2.$$

Thus,

$$g \circ f = \begin{pmatrix} 1 & 2 & 3 & 4 \\ 3 & 4 & 1 & 2 \end{pmatrix}.$$

We also have

$$f \circ g = \begin{pmatrix} 1 & 2 & 3 & 4 \\ 2 & 1 & 4 & 3 \end{pmatrix}.$$

□

The binary operation $\circ$ of composition of permutations in S_n satisfies the *associative law* [2]

$$(f \circ g) \circ h = f \circ (g \circ h),$$

but as the previous example shows, it does not satisfy the commutative law. In general,

$$f \circ g \neq g \circ f,$$

although equality may hold in some instances. We use the usual power notation to denote compositions of a permutation with itself:

$$f^1 = f, \; f^2 = f \circ f, \; f^3 = f \circ f \circ f, \ldots, f^k = f \circ f \circ \cdots \circ f \; (k \; f\text{'s}).$$

The *identity permutation* is the permutation ι of $\{1, 2, \ldots, n\}$ that takes each integer to itself:

$$\iota(k) = k \text{ for all } k = 1, 2, \ldots, n;$$

equivalently,

$$\iota = \begin{pmatrix} 1 & 2 & \cdots & n \\ 1 & 2 & \cdots & n \end{pmatrix}.$$

Obviously,

$$\iota \circ f = f \circ \iota = f$$

for all permutations f in S_n. Each permutation in S_n, since it is a one-to-one function, has an inverse f^{-1} that is also a permutation in S_n:

$$f^{-1}(k) = s, \text{ provided that } f(s) = k.$$

The 2-by-n array for f^{-1} can be gotten from the 2-by-n array for f by interchanging rows 1 and 2 and then rearranging columns so that the integers $1, 2, \ldots, n$ occur in the natural order in the first row. For each permutation f we define $f^0 = \iota$. The inverse of the identity permutation is itself: $\iota^{-1} = \iota$.

---

[2] Composition of functions is always associative.

Example. Consider the permutation in S_6 given by

$$f = \begin{pmatrix} 1 & 2 & 3 & 4 & 5 & 6 \\ 5 & 6 & 3 & 1 & 2 & 4 \end{pmatrix}.$$

Then, interchanging rows 1 and 2, we get

$$\begin{pmatrix} 5 & 6 & 3 & 1 & 2 & 4 \\ 1 & 2 & 3 & 4 & 5 & 6 \end{pmatrix}.$$

Rearranging columns, we get

$$f^{-1} = \begin{pmatrix} 1 & 2 & 3 & 4 & 5 & 6 \\ 4 & 5 & 3 & 6 & 1 & 2 \end{pmatrix}.$$

□

The definition of inverse implies that, for all f in S_n, we have

$$f \circ f^{-1} = f^{-1} \circ f = \iota.$$

A *group of permutations of X*, for short a *permutation group*, is defined to be a nonempty subset G of permutations in S_n satisfying the following three properties:

(i) (*closure under composition*) For all permutations f and g in G, $f \circ g$ is also in G.

(ii) (*identity*) The identity permutation ι of S_n belongs to G.

(iii) (*closure under inverses*) For each permutation f in G the inverse f^{-1} is also in G.

The set S_n of all permutations of $X = \{1, 2, \ldots, n\}$ is a permutation group, called the *symmetric group of order n*. At the other extreme, the set $G = \{\iota\}$ consisting only of the identity permutation is a permutation group.

Every permutation group satisfies the *cancellation law*

$$f \circ g = f \circ h \text{ implies that } g = h.$$

This is because we may apply f^{-1} to both sides of this equation and, using the associative law, obtain:

$$
\begin{aligned}
f^{-1} \circ (f \circ g) &= f^{-1} \circ (f \circ h) \\
(f^{-1} \circ f) \circ g &= (f^{-1} \circ f) \circ h \\
\iota \circ g &= \iota \circ h \\
g &= h.
\end{aligned}
$$

Example. Let n be a positive integer, and let ρ_n denote the permutation of $\{1, 2, \ldots, n\}$ defined by

$$\rho_n = \begin{pmatrix} 1 & 2 & 3 & \cdots & n-1 & n \\ 2 & 3 & 4 & \cdots & n & 1 \end{pmatrix}.$$

Thus, $\rho_n(i) = i+1$ for $i = 1, 2, \ldots, n-1$, and $\rho_n(n) = 1$. Think of the integers from 1 to n as evenly spaced around a circle or on the corners of a regular n-gon, as shown, for $n = 8$, in Figure 14.2.

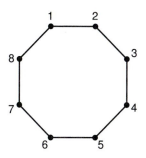

Figure 14.2

Then ρ_n sends each integer to the integer that follows it in the clockwise direction. Indeed, we may consider ρ_n as the rotation of the circle by an angle of $360/n$ degrees. The permutation ρ_n^2 is then the rotation by $2 \times (360/n)$ degrees, and more generally, for each nonnegative integer k, ρ_n^k is the rotation by $k \times (360/n)$ degrees. This implies that

$$\rho_n^k = \begin{pmatrix} 1 & 2 & \cdots & n-k & n-k+1 & \cdots & n \\ k+1 & k+2 & \cdots & n & 1 & \cdots & k-1 \end{pmatrix}.$$

In particular, if r equals $k \bmod n$, then $\rho_n^r = \rho_n^k$. Thus, there are only n distinct powers of ρ_n, namely,

$$\rho_n^0 = \iota, \ \rho_n, \ \rho_n^2, \ \ldots, \rho_n^{n-1}.$$

Also,

$$\rho_n^{-1} = \rho_n^{n-1},$$

and, more generally,

$$(\rho_n^k)^{-1} = \rho_n^{n-k} \text{ for } k = 0, 1, \ldots, n-1.$$

We thus conclude that

$$C_n = \{\rho_n^0 = \iota, \rho_n, \rho_n^2, \ldots, \rho_n^{n-1}\}$$

is a permutation group.[3] It is an example of a *cyclic group* of order n. As the reader may realize, this is the group that was implicitly used for calculating the number of ways to arrange n distinct objects in a circle. More about this later.   □

Let Ω be a geometrical figure. A *symmetry* of Ω is a (geometric) motion or congruence that brings the figure Ω onto itself. The geometric figures that we consider, like a square, a tetrahedron, and a cube, are composed of corners (or vertices) and edges, and in the case of three-dimensional figures, of faces (or sides). As a result, each symmetry acts as a permutation on the corners, on the edges, and, in the case of three-dimensional figures, on the faces. A symmetry of Ω followed by another, that is, the composition of two symmetries, is again a symmetry. Similarly, the inverse of a symmetry is also a symmetry. Finally, the motion that leaves everything fixed[4] is a symmetry, the identity symmetry. Hence, we conclude that the symmetries of Ω act as a permutation group G_C on its corners, a permutation group G_E on its edges, and, in the case where Ω is three-dimensional, a permutation group G_F on its faces.[5] As a result, a set of permutations that results by considering all the symmetries of a figure is automatically a permutation group. Thus, we have a *corner-symmetry group*, an *edge-symmetry group*, a *face-symmetry group*, and so on.

Example. Consider a square Ω with its corners labeled 1, 2, 3, and 4 and its edges labeled a, b, c, and d, as in Figure 14.3. There are 8 symmetries of Ω and they are of two types. There are the 4 rotations about the center of the square through the angles of 0, 90, 180, and 270 degrees. These 4 symmetries constitute the *planar symmetries* of Ω, the symmetries where the motion takes place in the plane containing Ω. The planar symmetries by themselves form a group. The other symmetries are the four reflections about the lines joining opposite corners and the lines joining the midpoints of opposite sides. For these symmetries the motion takes place in space since to "flip" the square one needs to go outside of the plane containing it.

---

[3]In more formal language, the permutation group C_n is isomorphic to the additive group of the integers mod n as discussed in Section 10.1.

[4]So nothing actually moves in this motion!

[5]There is an abstract concept of a group, which is defined to be a nonempty set with a binary operation, which satisfies the associative law and also (i) closure under composition, (ii) identity, and (iii) closure under inverses. Permutation groups are groups, since the associative law is automatic for composition of functions. The symmetries of a figure Ω form a group under this definition, but, as indicated, these symmetries can act as a permutation group of its corners, a permutation group of its edges, and so on.

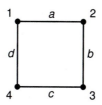

Figure 14.3

The rotations acting on the corners give the four permutations

$$\rho_4^0 = \iota = \begin{pmatrix} 1 & 2 & 3 & 4 \\ 1 & 2 & 3 & 4 \end{pmatrix} \qquad \rho_4 = \begin{pmatrix} 1 & 2 & 3 & 4 \\ 2 & 3 & 4 & 1 \end{pmatrix}$$

$$\rho_4^2 = \begin{pmatrix} 1 & 2 & 3 & 4 \\ 3 & 4 & 1 & 2 \end{pmatrix} \qquad \rho_4^3 = \begin{pmatrix} 1 & 2 & 3 & 4 \\ 4 & 1 & 2 & 3 \end{pmatrix}.$$

The reflections acting on the corners give the four permutations[6]

$$\tau_1 = \begin{pmatrix} 1 & 2 & 3 & 4 \\ 1 & 4 & 3 & 2 \end{pmatrix} \qquad \tau_2 = \begin{pmatrix} 1 & 2 & 3 & 4 \\ 3 & 2 & 1 & 4 \end{pmatrix}$$

$$\tau_3 = \begin{pmatrix} 1 & 2 & 3 & 4 \\ 2 & 1 & 4 & 3 \end{pmatrix} \qquad \tau_4 = \begin{pmatrix} 1 & 2 & 3 & 4 \\ 4 & 3 & 2 & 1 \end{pmatrix}.$$

Thus, the corner-symmetry group of a square is

$$G_C = \{\rho_4^0 = \iota, \rho_4, \rho_4^2, \rho_4^3, \tau_1, \tau_2, \tau_3, \tau_4\}.$$

We check that

$$\tau_3 = \rho_4 \circ \tau_1, \quad \tau_2 = \rho_4^2 \circ \tau_1, \quad \text{and } \tau_4 = \rho_4^3 \circ \tau_1.$$

Hence, we can also write

$$G_C = \{\rho_4^0 = \iota, \rho_4, \rho_4^2, \rho_4^3, \tau_1, \rho_4 \circ \tau_1, \rho_4^2 \circ \tau_1, \rho_4^3 \circ \tau_1\}.$$

Consider the edges of Ω to be labeled a, b, c and d, as in Figure 14.3. The edge-symmetry group G_E is obtained by letting the symmetries of

---

[6]τ_1 comes from the reflection about the line joining vertices 1 and 3, τ_2 comes from the reflection about the line joining vertices 2 and 4, τ_3 comes from the reflection about the line joining the midpoints of the lines a and c, and τ_4 comes from the reflection about the line joining the midpoints of the lines b and d.

Ω act on the edges. For example, the reflection about the line joining
the corners 2 and 4 gives the following permutation of the edges:

$$\begin{pmatrix} a & b & c & d \\ b & a & d & c \end{pmatrix}.$$

The other permutation of the edges in G_C can be obtained in a similar
way.   □

In a similar way we can obtain the symmetry group of a regular n-
gon for any $n \geq 3$. Besides the n rotations $\rho_n^0 = \iota, \rho, \ldots, \rho_n^{n-1}$, we have
n reflections $\tau_1, \tau_2, \ldots, \tau_n$. If n is even, then there are $n/2$ reflections
about opposite corners and $n/2$ reflections about the lines joining the
midpoints of opposite sides. If n is odd, then the reflections are the n
reflections about the lines joining a corner to the side opposite it. The
resulting group

$$D_n = \{\rho_n^0 = \iota, \rho, \ldots, \rho_n^{n-1}, \tau_1, \tau_2, \ldots, \tau_n\}$$

of $2n$ permutations of $\{1, 2, \ldots, n\}$ is an instance of a *dihedral group* of
order $2n$. In the next example we compute D_5.

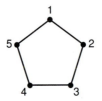

Figure 14.4

Example. (*The dihedral group of order* 10). Consider the regular
pentagon with its vertices labeled $1, 2, 3, 4$, and 5, as in Figure 14.4.
Its (corner) symmetry group D_5 contains 5 rotations and 5 reflections.
The 5 rotations are

$$\rho_5^0 = \iota = \begin{pmatrix} 1 & 2 & 3 & 4 & 5 \\ 1 & 2 & 3 & 4 & 5 \end{pmatrix} \qquad \rho_5^1 = \begin{pmatrix} 1 & 2 & 3 & 4 & 5 \\ 2 & 3 & 4 & 5 & 1 \end{pmatrix}$$

$$\rho_5^2 = \begin{pmatrix} 1 & 2 & 3 & 4 & 5 \\ 3 & 4 & 5 & 1 & 2 \end{pmatrix} \qquad \rho_5^3 = \begin{pmatrix} 1 & 2 & 3 & 4 & 5 \\ 4 & 5 & 1 & 2 & 3 \end{pmatrix}$$

$$\rho_5^4 = \begin{pmatrix} 1 & 2 & 3 & 4 & 5 \\ 5 & 1 & 2 & 3 & 4 \end{pmatrix}.$$

Let τ_i denote the reflection about the line joining corner i to the side opposite it $(i = 1, 2, 3, 4, 5)$. Then we have

$$\tau_1 = \begin{pmatrix} 1 & 2 & 3 & 4 & 5 \\ 1 & 5 & 4 & 3 & 2 \end{pmatrix} \qquad \tau_2 = \begin{pmatrix} 1 & 2 & 3 & 4 & 5 \\ 3 & 2 & 1 & 5 & 4 \end{pmatrix}$$

$$\tau_3 = \begin{pmatrix} 1 & 2 & 3 & 4 & 5 \\ 5 & 4 & 3 & 2 & 1 \end{pmatrix} \qquad \tau_4 = \begin{pmatrix} 1 & 2 & 3 & 4 & 5 \\ 2 & 1 & 5 & 4 & 3 \end{pmatrix}$$

$$\tau_5 = \begin{pmatrix} 1 & 2 & 3 & 4 & 5 \\ 4 & 3 & 2 & 1 & 5 \end{pmatrix}.$$

□

Suppose we have a group G of permutations of a set X, where X is again taken to be the set $\{1, 2, \ldots, n\}$ of the first n positive integers. A *coloring* of X is an assignment of a color to each element of X. Let C be a collection of colorings of X. Usually we have a number of colors, say red and blue, and C consists of all colorings of X with these colors. But this need not be the case. The set C can be *any* collection of colorings of X as long as G takes a coloring in C to another coloring in C in the manner to be described now.

Let **c** be a coloring of X in which the colors of $1, 2, \ldots, n$ are denoted by $c(1), c(2), \ldots, c(n)$, respectively. Let

$$f = \begin{pmatrix} 1 & 2 & \cdots & n \\ i_1 & i_2 & \cdots & i_n \end{pmatrix}$$

be a permutation in G. Then $f * \mathbf{c}$ is defined to be the coloring in which i_k has the color $c(k)$; that is,

$$(f * \mathbf{c})(i_k) = c(k), \quad \text{or using the inverse of } f,$$

$$(f * \mathbf{c})(k) = c(f^{-1}(k)).$$

In words, since f moves k to i_k, the color of k, namely $\mathbf{c}(k)$, moves to $f(k) = i_k$ and becomes the color of i_k. The set C of colorings is required to have the property that, for all f in G and all **c** in C, $f * \mathbf{c}$ is also in C. This implies that f permutes the colorings in C, and thus G *acts* as a permutation group on the set C of colorings. Hence, $f * \mathbf{c}$ denotes the coloring in C into which **c** is sent by f. Note that, if C is the set of *all* colorings of X for a given set of colors, then C automatically has the required property.

The basic relationship that holds between the two operations ∘ (composition of permutations in G) and $*$ (action of permutations in G on colorings in $\mathcal{C}$) is

$$(g \circ f) * \mathbf{c} = g * (f * \mathbf{c}). \tag{14.2}$$

The left side of equation (14.2) is the coloring in which the color of k moves to $(g \circ f)(k)$. The right side is the coloring in which the color of k moves to $f(k)$ and then moves to $g(f(k))$. Since $(g \circ f)(k) = g(f(k))$ by the definition of composition, we have verified (14.2).

Example. We continue with the earlier example in which Ω is the square in Figure 14.3, and G_C is the corner-symmetry group of Ω. Let $\mathcal{C}$ be the set of all colorings of the corners $1, 2, 3, 4$ of Ω in which the colors are either red or blue. The permutation group G_C contains 8 permutations, and there are 16 colorings in $\mathcal{C}$. Let us denote a coloring by writing the colors of the corners in the order $1, 2, 3, 4$, using R to denote red and B to denote blue. For instance,

$$(R, B, B, R) \tag{14.3}$$

is the coloring in which corner 1 is red, corner 2 is blue, corner 3 is blue, and corner 4 is red. The permutation ρ_4 sends this coloring into the coloring

$$(R, R, B, B),$$

in which corners 1 and 2 are red and corners 3 and 4 are blue. In the following table, we list the effect of each permutation in G_C on the coloring (14.3).

Notice that the permutation τ_4 doesn't change the coloring (14.3); that is, τ_4 fixes the coloring (14.3). Of course, the identity ι also doesn't change it. In fact, each coloring on the list appears exactly twice. Let us say that two colorings are equivalent, provided that there is a permutation in G_C which sends one to the other. Thus, the coloring (R, B, B, R) is equivalent to each of

$$(R, B, B, R), \ (R, R, B, B), \ (B, R, R, B), \ \text{and} \ (B, B, R, R).$$

Permutation in G_C	Effect on the coloring (R, B, B, R)
$\rho_4^0 = \iota$	(R, B, B, R)
ρ_4	(R, R, B, B)
ρ_4^2	(B, R, R, B)
ρ_4^3	(B, B, R, R)
τ_1	(R, R, B, B)
τ_2	(B, B, R, R)
τ_3	(B, R, R, B)
τ_4	(R, B, B, R)

Since a permutation cannot change the number of corners of each of the colors, a necessary condition for two colorings to be equivalent is that they contain the same number of R's and the same number of B's.[7] The coloring (R, B, R, B) also has two R's and two B's, but is not equivalent to (R, B, B, R). Indeed, as can now be checked, (R, B, R, B) is equivalent only to (R, B, R, B) and (B, R, B, R), and each of these colorings arise four times as we examine the effect of all the permutations in G_C on it. In particular, we can now conclude that there are 2 nonequivalent colorings among all the colorings with 2 red and 2 blue corners. The coloring (R, R, R, R) is clearly equivalent only to itself, as is the coloring (B, B, B, B). Consider the coloring (R, B, B, B) with 1 red and 3 blue corners. This coloring is equivalent, by a rotation, to each of the colorings $(R, B, B, B), (B, R, B, B), (B, B, R, B)$, and (B, B, B, R), and hence all colorings with 1 red are equivalent. Similarly, all colorings with 3 red (and therefore 1 blue) are equivalent by a rotation. Consequently, there are $2 + 1 + 1 + 1 + 1 = 6$ nonequivalent ways to color the corners of a square with two colors, under the action of the corner-symmetry group G_C of the square. If we don't allow the full symmetry group of the square, but only the group of symmetries consisting of the 4 rotations $\rho_0 = \iota$, ρ_4, ρ_4^2, and ρ_4^3, then the number of nonequivalent colorings is still 6. This is because if two colorings

---

[7] Of course, if two colorings have the same number of R's they must have the same number of B's.

are equivalent by a symmetry of the square, then they are equivalent
by a rotation.   □

We now give the general definition of equivalent colorings. Let G
be a group of permutations acting on a set X, as usual taken to be the
set $\{1, 2, \ldots, n\}$ of the first n positive integers. Let C be a collection of
colorings of X, such that for all f in G and all $\mathbf{c}$ in C, the coloring $f * \mathbf{c}$
of X is also in C. Thus G acts on C in the sense that it takes colorings
in C to colorings in C. Let $\mathbf{c}_1$ and $\mathbf{c}_2$ be two colorings in C. We define
a relation called *equivalence*, denoted by $\overset{G}{\sim}$ (or, more briefly, by $\sim$) on
C as follows: $\mathbf{c}_1$ is *equivalent* (*under the action of G*) to $\mathbf{c}_2$, provided
that there is a permutation f in G such that

$$f * \mathbf{c}_1 = \mathbf{c}_2.$$

Two colorings are *nonequivalent*, provided that they are not equivalent.
We have

(i) (*reflexive property*) $\mathbf{c} \sim \mathbf{c}$ for each coloring $\mathbf{c}$; (because $\iota * \mathbf{c} = \mathbf{c}$).

(ii) (*symmetry property*) If $\mathbf{c}_1 \sim \mathbf{c}_2$, then $\mathbf{c}_2 \sim \mathbf{c}_1$;
(if $f * \mathbf{c}_1 = \mathbf{c}_2$ for some f in G, then $f^{-1} * \mathbf{c}_2 = \mathbf{c}_1$).

(iii) (*transitive property*) If $\mathbf{c}_1 \sim \mathbf{c}_2$ and $\mathbf{c}_2 \sim \mathbf{c}_3$, then $\mathbf{c}_1 \sim \mathbf{c}_3$);
(if $f * \mathbf{c}_1 = \mathbf{c}_2$ and $g * \mathbf{c}_2 = \mathbf{c}_3$, then $(g \circ f) * \mathbf{c}_1 = \mathbf{c}_3$).

It thus follows that $\sim$ is an equivalence relation on C in the sense
defined in Section 4.5, justifying our use of the term "equivalence."
Notice how the three basic properties of a permutation group,
namely, identity, closure under inverses, and closure under composition
are used in the verification of (i)-(iii). By Theorem 4.5.3 of Chapter 4,
equivalence partitions the colorings of C into parts, with two colorings
being in the same part if and only if they are equivalent colorings. In
the next section we derive a general formula for the number of parts—
that is, for the number of nonequivalent colorings—in C under the
action of the permutation group G.

14.2   Burnside's Theorem

In this section we derive and apply a formula of Burnside[8] for counting
the number of nonequivalent colorings of a set X under the action of
a group of permutations of X.

---

[8]W. Burnside: *Theory of Groups of Finite Order*, 2nd edition, Cambridge University Press, London, 1911 (reprinted by Dover, New York, 1955), p. 191.

Let G be a group of permutations of X and let C be a set of colorings of X such that G acts on C. Recall that this means that

$$f * \mathbf{c}$$

is in C for all f in G and all $\mathbf{c}$ in C, and each f in G permutes the colorings in C. It is possible, that for an appropriate choice of f and of $\mathbf{c}$, we have

$$f * \mathbf{c} = \mathbf{c}. \tag{14.4}$$

For example, if in Figure 14.3, we color corners 1 and 3 of the square red and the corners 2 and 4 blue, then reflecting about the line through 1 and 3 or the line through 2 and 4, or rotating by 180 degrees does not alter the coloring; each of these motions fixes the color of each corner and hence fixes the coloring. If, in (14.4), we allow either f to vary over all permutations in G or $\mathbf{c}$ to vary over all colorings in C, then we get both

$$G(\mathbf{c}) = \{f : f \text{ in } G, f * \mathbf{c} = \mathbf{c}\}, \tag{14.5}$$

the set of all permutations in G that fix the coloring $\mathbf{c}$, and

$$C(f) = \{\mathbf{c} : \mathbf{c} \text{ in } C, f * \mathbf{c} = \mathbf{c}\}, \tag{14.6}$$

the set of all colorings in C that are fixed by f. The set $G(\mathbf{c})$ of all permutations that fix the coloring $\mathbf{c}$ is called the *stabilizer*[9] of $\mathbf{c}$. The stabilizer of any coloring also forms a group of permutations.

Theorem 14.2.1 *For each coloring* $\mathbf{c}$*, the stabilizer* $G(\mathbf{c})$ *of* $\mathbf{c}$ *is a permutation group. Moreover, for any permutations* f *and* g *in* G*,* $g * \mathbf{c} = f * \mathbf{c}$ *if and only if* $f^{-1} \circ g$ *is in* $G(\mathbf{c})$*.*

Proof. If f and g both fix $\mathbf{c}$, then f followed by g fixes $\mathbf{c}$; that is, $(g \circ f)(\mathbf{c}) = \mathbf{c}$. Thus, $G(\mathbf{c})$ is closed under composition. Clearly, the identity ι fixes $\mathbf{c}$ since it fixes every coloring. Also, if f fixes $\mathbf{c}$, then so does f^{-1}, and hence $G(\mathbf{c})$ is closed under inverses. All of the defining properties of a permutation group are satisfied; therefore, $G(\mathbf{c})$ is a permutation group.

Suppose that $f * \mathbf{c} = g * \mathbf{c}$. By the basic relationship (14.2), we get

$$(f^{-1} \circ g) * \mathbf{c} = f^{-1} * (g * \mathbf{c}) = f^{-1} * (f * \mathbf{c}) = (f^{-1} \circ f) * \mathbf{c} = \iota * \mathbf{c} = \mathbf{c}.$$

---

[9]A synonym for *fixed* is *stable*.

It follows that $f^{-1} \circ g$ fixes $\mathbf{c}$, and hence $f^{-1} \circ g$ is in $G(\mathbf{c})$. Conversely, suppose that $f^{-1} \circ g$ is in $G(\mathbf{c})$. Then a similar calculation shows that $f * \mathbf{c} = g * \mathbf{c}$.   □

As a corollary of Theorem 14.2.1, starting from a given coloring $\mathbf{c}$, we can determine the number of different colorings we can get under the action of G.

Corollary 14.2.2 *Let $\mathbf{c}$ be a coloring in C. The number*

$$|\{f * \mathbf{c} : f \text{ in } G\}|$$

of different colorings that are equivalent to $\mathbf{c}$ equals the number

$$\frac{|G|}{|G(\mathbf{c})|}$$

obtained by dividing the number of permutations in G by the number of permutations in the stabilizer of $\mathbf{c}$.

Proof. Let f be a permutation in G. By Theorem 14.2.1, the permutations g that satisfy

$$g * \mathbf{c} = f * \mathbf{c}$$

are precisely the permutations in

$$\{f \circ h : h \text{ in } G(\mathbf{c})\}. \tag{14.7}$$

By the cancellation law, $f \circ h = f \circ h'$ implies $h = h'$. Hence, the number of permutations in the set (14.7) equals the number $|G(\mathbf{c})|$ of permutations h in $G(\mathbf{c})$. Thus, for each permutation f, there are exactly $|G(\mathbf{c})|$ permutations that have the same effect on $\mathbf{c}$ as f. Since there are $|G|$ permutations altogether, the number

$$|\{f * \mathbf{c} : f \text{ in } G\}|$$

of colorings equivalent to $\mathbf{c}$ equals

$$\frac{|G|}{|G(\mathbf{c})|},$$

proving the corollary.   □

The next theorem of Burnside gives a formula for counting the number of nonequivalent colorings.

Theorem 14.2.3 *Let G be a group of permutations of X and let $\mathcal{C}$ be a set of colorings of X such that $f * \mathbf{c}$ is in $\mathcal{C}$ for all f in G and all $\mathbf{c}$ in $\mathcal{C}$. Then the number $N(G, \mathcal{C})$ of nonequivalent colorings in $\mathcal{C}$ is given by*

$$N(G, \mathcal{C}) = \frac{1}{|G|} \sum_{f \in G} |\mathcal{C}(f)|. \tag{14.8}$$

In words, the number of nonequivalent colorings in $\mathcal{C}$ equals the average of the number of colorings fixed by the permutations in G.

Proof. With the information we now have, the proof is a simple application of a technique we have experienced many times, namely, counting in two different ways and then equating counts. What do we count? We count the number of pairs $(f, \mathbf{c})$ such that f fixes $\mathbf{c}$, that is, such that $f * \mathbf{c} = \mathbf{c}$. One way to count is to consider each f in G and compute the number of colorings that f fixes, and then add up all quantities. Counting in this way, we get

$$\sum_{f \in G} |\mathcal{C}(f)|, \tag{14.9}$$

since $\mathcal{C}(f)$ is the set of colorings that are fixed by f.

Another way to count is to consider each $\mathbf{c}$ in $\mathcal{C}$ and compute the number of permutations f such that $f * \mathbf{c} = \mathbf{c}$, and then add up all the quantities. For each coloring $\mathbf{c}$, the set of all f such that $f * \mathbf{c} = \mathbf{c}$ is what we have called the stabilizer $G(\mathbf{c})$ of $\mathbf{c}$. Thus, each $\mathbf{c}$ contributes

$$|G(\mathbf{c})|$$

to the sum. By Corollary 14.2.2,

$$|G(\mathbf{c})| = \frac{|G|}{\text{(the number of colorings equivalent to } \mathbf{c})}. \tag{14.10}$$

Hence, counting in this way, we get

$$\sum_{\mathbf{c} \in \mathcal{C}} \frac{|G|}{\text{(the number of colorings equivalent to } \mathbf{c})}. \tag{14.11}$$

But the sum (14.11) can be simplified if we group the colorings by equivalence class. Two colorings in the same equivalence class contribute the same amount (14.10) to this sum, so the total contribution of every equivalence class is $|G|$. Consequently, (14.11) equals

$$N(G, \mathcal{C}) \times |G|, \tag{14.12}$$

since the number of equivalence classes is the number $N(G, \mathcal{C})$ of nonequivalent colorings. Equating (14.9) and (14.12), we get

$$\sum_{f \in G} |\mathcal{C}(f)| = N(G, \mathcal{C}) \times |G|;$$

solving for $N(G, \mathcal{C})$, we obtain (14.8).   □

In the remainder of this section we illustrate Burnside's theorem with several examples.

Example. (*Counting circular permutations*). How many ways are there to arrange n distinct objects in a circle?

As already hinted at in Section 13.1, the answer is the number of ways to color the corners of a regular n-gon Ω with n different colors that are nonequivalent with respect to the group of rotations of Ω. Let $\mathcal{C}$ consist of all $n!$ ways to color the n corners of Ω in which each of the n colors occurs once. Then the cyclic group

$$C_n = \{\rho_n^0 = \iota, \rho_n, \ldots, \rho_n^{n-1}\}$$

acts[10] on $\mathcal{C}$, and the number of circular permutations equals the number of nonequivalent colorings in $\mathcal{C}$. The identity permutation ι in C_n fixes all $n!$ of the colorings in $\mathcal{C}$. Every other permutation in $\mathcal{C}$ does not fix any coloring in $\mathcal{C}$, since, in the colorings of $\mathcal{C}$, every corner has a different color.[11] Hence, using (14.8) of Theorem 14.2.3, we see that the number of nonequivalent colorings is

$$N(C_n, \mathcal{C}) = \frac{1}{n}(n! + 0 + \cdots + 0) = (n-1)!.$$

□

Example. (*Counting necklaces*). How many ways are there to arrange $n \geq 3$ differently colored beads in a necklace?

We have almost the same situation as described in the previous example, except since necklaces can be flipped over, the group G of permutations now has to be taken to be the entire vertex-symmetry group of a regular n-gon. Thus, in this case, G is the dihedral group D_n of order $2n$. The only permutation that can fix a coloring is the identity and it fixes all $n!$ colorings. Hence, the number of nonequivalent colorings–that is, the number of different necklaces–is by (14.8),

$$N(D_n, \mathcal{C}) = \frac{1}{2n}(n! + 0 + \cdots + 0) = \frac{(n-1)!}{2}.$$

□

---

[10]Recall that ρ_n is the rotation by $360/n$ degrees.

[11]In fact, no permutation different from the identity can fix any coloring if all colors are different. This is because, for a permutation different from the identity, at least one color has to move, and hence the coloring is changed.

Example. How many nonequivalent ways are there to color the corners of a regular 5-gon with the colors red and blue?

The group of symmetries of a regular 5-gon is the dihedral group

$$D_5 = \{\rho_5^0 = \iota, \rho_5, \rho_5^2, \rho_5^3, \rho_5^4, \tau_1, \tau_2, \tau_3, \tau_4, \tau_5\},$$

where, as in Section 13.1, τ_j is the reflection about the line joining corner j with the midpoint of the opposite side ($j = 1, 2, 3, 4, 5$). Let $\mathcal{C}$ be the set of all $2^5 = 32$ colorings of the corners of a regular 5-gon. We compute the number of colorings left fixed by each permutation in D_5 and then apply Theorem 14.2.3. The identity ι fixes all colorings. Each of the other 4 rotations fixes only two colorings, namely, the coloring in which all corners are red, and the coloring in which all corners are blue. Thus,

$$|\mathcal{C}(\rho_5^i)| = \begin{cases} 32 & \text{if } i = 0, \\ 2 & \text{if } i = 1, 2, 3, 4. \end{cases}$$

Now consider any of the reflections τ_j, say, τ_1. In order that a coloring be fixed by τ_1, corners 2 and 5 must have the same color and corners 3 and 4 must have the same color. Hence, the colorings fixed by τ_1 are obtained by picking a color for corner 1 (two choices), picking a color for corners 2 and 5 (two choices) and picking a color for corners 3 and 4 (again two choices). Therefore, the number of colorings fixed by τ_1 equals $2 \times 2 \times 2 = 8$. A similar calculation holds for each reflection, and we have

$$|\mathcal{C}(\tau_j)| = 8 \quad \text{for each } j = 1, 2, 3, 4, 5.$$

Therefore, by (14.8), the number of nonequivalent colorings is

$$N(D_5, \mathcal{C}) = \frac{1}{10}(32 + 2 + 2 + 2 + 2 + 8 + 8 + 8 + 8 + 8) = 8.$$

□

Example. How many nonequivalent ways are there to color the corners of a regular 5-gon with the colors red, blue, and green?

We refer to the previous example, but now the set $\mathcal{C}$ of all colorings of the corners of a regular 5-gon numbers $3^5 = 243$. The identity fixes all 243 colorings. Every other rotation fixes only 3 colorings. Each reflection fixes $3 \times 3 \times 3 = 27$ colorings. Hence, the number of nonequivalent colorings is

$$N(D_5, \mathcal{C}) = \frac{1}{10}(243 + 3 + 3 + 3 + 3 + 27 + 27 + 27 + 27 + 27) = 39.$$

How many nonequivalent ways are there to color the corners of a regular 5-gon with p colors? Generalizing the preceding calculations, we find that this number is

$$\frac{1}{10}(p^5 + 4 \times p + 5 \times p^3) = \frac{p(p^2 + 4)(p^2 + 1)}{10}.$$

$\square$

Example. Let $S = \{\infty \cdot r, \infty \cdot b, \infty \cdot g, \infty \cdot y\}$ be a multiset of four distinct objects r, b, g, y, each with an infinite repetition number. How many n-permutations of S are there if we do not distinguish between a permutation read from left to right and the permutation read from right to left? Thus, for instance, r, g, g, g, b, y, y is regarded as equivalent to y, y, b, g, g, g, r.

The answer is the number of nonequivalent ways to color the integers from 1 to n with the four colors red, blue, green, and yellow under the action of the group of permutations

$$G = \{\iota, \tau\},$$

where

$$\iota = \begin{pmatrix} 1 & 2 & \cdots & n \\ 1 & 2 & \cdots & n \end{pmatrix} \text{ and } \tau = \begin{pmatrix} 1 & 2 & \cdots & n-1 & n \\ n & n-1 & \cdots & 2 & 1 \end{pmatrix}.$$

Here, ι is, as usual, the identity permutation. The permutation τ is obtained by listing the integers from 1 to n in reverse order. Note that G does form a group, since $\tau \circ \tau = \iota$ and hence $\tau^{-1} = \tau$.[12] Let $\mathcal{C}$ be the set of all 4^n ways to color the integers from 1 to n with the given 4 colors. Then ι fixes all colorings in $\mathcal{C}$. The number of colorings fixed by τ depends on whether n is even or odd. First, suppose that n is even. Then a coloring is fixed by τ if and only if 1 and n have the same color, 2 and $n-1$ have the same color, $\ldots$, $n/2$ and $(n/2)+1$ have the same color. Hence, τ fixes $4^{n/2}$ colorings in $\mathcal{C}$. Now suppose that n is odd. Then a coloring is fixed by τ if and only if 1 and n have the same color, 2 and $n-1$ have the same color, $\ldots$, $(n-1)/2$ and $(n+3)/2$ have the same color, there being no restriction on the color of $(n+1)/2$. Thus, the number of colorings fixed by τ is $4^{(n-1)/2} \times 4 = 4^{(n+1)/2}$. Using the floor function, we can combine both cases and obtain

$$|\mathcal{C}(\tau)| = 4^{\lfloor \frac{n+1}{2} \rfloor}.$$

---

[12]Think of a line segment consisting of n equally spaced points that are labeled $1, 2, \ldots, n$. Then τ is a rotation of this line segment by 180 degrees. Equivalently, τ is a reflection of this line segment about its perpendicular bisector.

Applying Burnside's formula (14.8), we find that the number of nonequivalent colorings is

$$N(G, \mathcal{C}) = \frac{4^n + 4^{\lfloor \frac{(n+1)}{2} \rfloor}}{2}.$$

If instead of four colors, we have p colors, the number of nonequivalent colorings is

$$N(G, \mathcal{C}) = \frac{p^n + p^{\lfloor \frac{(n+1)}{2} \rfloor}}{2}.$$

$\square$

In the next section, we develop a little more theory that will enable us to more easily solve more difficult counting problems, using Theorem 14.2.3.

14.3   Pólya's Counting Formula

The counting formula to be discussed in this section was developed (and extensively applied) by Pólya in an important, long, and very influential paper.[13] It was only around 1960 that it was recognized that ten years before Pólya's famous paper was published, Redfield published a paper[14] in which he anticipated the basic technique of Pólya.

As we have seen in the previous section, success in using Burnside's theorem for counting the number of nonequivalent colorings in the presence of a permutation group G acting on a set $\mathcal{C}$ of colorings is dependent on being able to compute the number $|\mathcal{C}(f)|$ of colorings in $\mathcal{C}$ fixed by a permutation f in G. This computation can be facilitated by consideration of the cyclic structure of a permutation.

Let f be a permutation of $X = \{1, 2, \ldots, n\}$. Let $D_f = (X, A_f)$ be the digraph whose set of vertices is X and whose set of arcs is

$$A_f = \{(i, f(i)) : i \text{ in } X\}.$$

The digraph has n vertices and n arcs. Moreover, the indegree and outdegree of each vertex equal 1. As shown in Corollary 11.8.8, the set A_f of arcs can be partitioned into directed cycles, with each vertex belonging to exactly one directed cycle. The reason is simply that, starting at any vertex j, we proceed along the unique arc leaving j,

---

[13]G. Pólya: Kombinatorische Anzahlbestimmungen für Gruppen, Graphen und chemische Verbindungen, *Acta Mathematica*, 68 (1937), 145-254.

[14]J.H. Redfield: The theory of group-reduced distributions, *American Journal of Mathematics*, 49 (1927), 433-455.

and arrive at another vertex k; we now repeat with k and continue until we arrive back at vertex i, thereby creating a directed cycle. We must eventually arrive at our starting vertex i since each vertex has indegree and outdegree equal to 1. We remove the vertices and arcs of the directed cycle so obtained, and continue until we exhaust all the vertices and arcs of D_f, thereby partitioning both the vertices and arcs of D_f into directed cycles.

Example. Let

$$f = \begin{pmatrix} 1 & 2 & 3 & 4 & 5 & 6 & 7 & 8 \\ 6 & 8 & 5 & 4 & 1 & 3 & 2 & 7 \end{pmatrix}$$

be a permutation of $\{1, 2, \ldots, 8\}$. Then applying the foregoing procedure, we obtain the following partition of D_f into directed cycles:

$$1 \rightarrow 6 \rightarrow 3 \rightarrow 5 \rightarrow 1, \quad 2 \rightarrow 8 \rightarrow 7 \rightarrow 2, \quad 4 \rightarrow 4.$$

Let us write

$$[1\ 6\ 3\ 5]$$

for the permutation of $\{1, 2, 3, 4, 5, 6, 7, 8\}$ that sends 1 to 6, 6 to 3, 3 to 5, and 5 to 1, and that fixes the remaining integers.[15] Thus,

$$[1\ 6\ 3\ 5] = \begin{pmatrix} 1 & 2 & 3 & 4 & 5 & 6 & 7 & 8 \\ 6 & 2 & 5 & 4 & 1 & 3 & 7 & 8 \end{pmatrix}.$$

The digraph corresponding to the permutation $[1\ 6\ 3\ 5]$ is the digraph consisting of the directed cycles

$$1 \rightarrow 6 \rightarrow 3 \rightarrow 5 \rightarrow 1, \quad 2 \rightarrow 2, \quad 4 \rightarrow 4, \quad 7 \rightarrow 7, \quad 8 \rightarrow 8.$$

We call such a permutation, in which certain of the elements are permuted in a cycle and the remaining elements, if any, are fixed, a *cycle permutation* or, more briefly, a *cycle*. If the number of elements in the cycle is k, then we call it a *k-cycle*. Thus, $[1\ 6\ 3\ 5]$ is a 4-cycle. The other directed cycles in the partition of D_f give the following cycles:

$$[2\ 8\ 7] \text{ and } [4].$$

---

[15]The notation is a little ambiguous because we cannot determine from it the set of elements being permuted. All we can conclude is that it is at least $\{1, 3, 5, 6\}$. But there should be no confusion, since the the set will be implicit in the particular problem treated.

We now observe that the partition of D_f into directed cycles corresponds to a factorization (with respect to the composition ∘) of f into permutation cycles:

$$f = \begin{pmatrix} 1 & 2 & 3 & 4 & 5 & 6 & 7 & 8 \\ 6 & 8 & 5 & 4 & 1 & 3 & 2 & 7 \end{pmatrix} = [1\ 6\ 3\ 5] \circ [2\ 8\ 7] \circ [4]. \quad (14.13)$$

The reason is that each integer in the permutation f moves in, at most, one of the cycles in the factorization.

We make two observations about this factorization. The first is that it doesn't matter in which order we write the cycles.[16] This is because each element occurs in exactly one cycle. The second is that the 1-cycle [4] is just the identity permutation,[17] and thus could be omitted in (14.13) without affecting its validity. But we choose to leave it there since, for our counting problems, it is useful to include all 1-cycles.   □

Let f be any permutation of the set X. Then, generalizing from the previous example, we see that, with respect to the operation of composition, f has a factorization

$$f = [i_1\ i_2\ \cdots\ i_p] \circ [j_1\ j_2\ \cdots\ j_q] \circ \cdots \circ [l_1\ l_2\ \cdots\ l_r] \quad (14.14)$$

into cycles, where each integer in X occurs in exactly one of the cycles. We call (14.14) the *cycle factorization* of f. The cycle factorization of f is unique, apart from the order in which the cycles appear, and this order is arbitrary. In the cycle factorization of a permutation of X, every element of X occurs exactly once.

Example. Determine the cycle factorization of each permutation in the dihedral group D_4 of order 8 (the corner-symmetry group of a square).

The permutations in D_4 were computed in Section 13.1. The cycle factorization of each is given in the next table:

---

[16]That is, "disjoint cycles" satisfy the commutative law.

[17]Recall what [4] means here: 4 goes to 4, and every other integer is fixed. This means that every integer including 4, is fixed, and hence we have the identity permutation. If the permutation f in this example were the identity permutation, then we would write $f = [1] \circ [2] \circ \cdots \circ [8]$.

D_4	Cycle factorization
$\rho_4^0 = \iota$	$[1] \circ [2] \circ [3] \circ [4]$
ρ_4	$[1\ 2\ 3\ 4]$
ρ_4^2	$[1\ 3] \circ [2\ 4]$
ρ_4^3	$[1\ 4\ 3\ 2]$
τ_1	$[1] \circ [2\ 4] \circ [3]$
τ_2	$[1\ 3] \circ [2] \circ [4]$
τ_3	$[1\ 2] \circ [3\ 4]$
τ_4	$[1\ 4] \circ [2\ 3]$

Notice that, in the cycle factorization of the identity permutation ι, all cycles are 1-cycles. This is in agreement with the fact that the identity permutation fixes all elements. In the cycle factorizations of the reflections τ_1 and τ_2, two 1-cycles occur, since each of these reflections is about a line joining two opposite corners of the square, and these corners are thus fixed. For τ_3 and τ_4 we get two 2-cycles, since these are reflections about the line joining the midpoints of opposite sides. The reflections in the corner-symmetry group of a regular n-gon with n even behave similarly. Half of them have two 1-cycles and $((n/2) - 1)$ 2-cycles, and half have $(n/2)$ 2-cycles.   □

Example. Determine the cycle factorization of each permutation in the dihedral group D_5 of order 10 (the corner-symmetry group of a regular 5-gon).

The permutations in D_5 were computed in Section 13.1. The cycle factorization of each is given in the following table:

D_5	Cycle factorization
$\rho_5^0 = \iota$	$[1] \circ [2] \circ [3] \circ [4] \circ [5]$
ρ_5	$[1\ 2\ 3\ 4\ 5]$
ρ_5^2	$[1\ 3\ 5\ 2\ 4]$
ρ_5^3	$[1\ 4\ 2\ 5\ 3]$
ρ_5^4	$[1\ 5\ 4\ 3\ 2]$
τ_1	$[1] \circ [2\ 5] \circ [3\ 4]$
τ_2	$[1\ 3] \circ [2] \circ [4\ 5]$
τ_3	$[1\ 5] \circ [3] \circ [2\ 4]$
τ_4	$[1\ 2] \circ [3\ 5] \circ [4]$
τ_5	$[1\ 4] \circ [2\ 3] \circ [5]$

Notice that, in the cycle factorizations of the reflections τ_i, exactly one 1-cycle occurs since each such reflection is about a line joining a corner to the midpoint of the opposite side, and hence only the one corner is fixed. The reflections in the corner-symmetry group of a regular n-gon with n odd behave similarly. Each has one 1-cycle and $(n-1)/2$ 2-cycles. □

The importance of the cycle decomposition in counting nonequivalent colorings is illustrated by the next example.

Example. Let f be the permutation of $X = \{1, 2, 3, 4, 5, 6, 7, 8, 9\}$ defined by

$$\begin{pmatrix} 1 & 2 & 3 & 4 & 5 & 6 & 7 & 8 & 9 \\ 4 & 9 & 1 & 7 & 6 & 5 & 3 & 8 & 2 \end{pmatrix}.$$

The cycle factorization of f is

$$f = [1\ 4\ 7\ 3] \circ [2\ 9] \circ [5\ 6] \circ [8].$$

Suppose that we color the elements of X with the colors red, white, and blue, and let $\mathcal{C}$ be the set of all such colorings. How many

$$|\mathcal{C}(f)|$$

colorings in $\mathcal{C}$ are left fixed by f?

Let **c** be a coloring such that $f * \mathbf{c} = \mathbf{c}$. First, consider the 4-cycle [1 4 7 3]. This 4-cycle moves the color of 1 to 4, the color of 4 to 7, the color of 7 to 3, and the color of 3 to 1. Since the coloring **c** is fixed by f, following through on this cycle, we see that

$$\text{color of } 1 = \text{color of } 4 = \text{color of } 7 =$$

$$\text{color of } 3 = \text{color of } 1.$$

This means that 1, 4, 7, and 3 have the same color. In a similar way, we see that the elements 2 and 9 of the 2-cycle [2 9] have the same color, and the elements 5 and 6 of the 2-cycle [5 6] have the same color. There is no restriction placed on 8, since it belongs to a 1-cycle. So how many colorings **c** are there which are fixed by f—that is, which satisfy $f * \mathbf{c} = \mathbf{c}$? The answer is clear: We pick any one of the three colors red, white, and blue for $\{1, 4, 7, 3\}$ (3 choices), any of the three colors for $\{2, 9\}$ (3 choices), any of the three colors for $\{5, 6\}$ (3 choices), and any of the three colors for $\{8\}$ (3 choices), for a total of

$$3^4 = 81$$

colorings. Note that the exponent 4 in the answer is the *number* of cycles of f in its cycle factorization, and the answer is independent of the sizes of the cycles.   □

The analysis in the preceding example is quite general. It can be used to find the number of colorings fixed by any permutation no matter what the number of colors available is. We record the result in the next theorem. We denote by

$$\#(f)$$

the *number of cycles in the cycle factorization* of a permutation f.

Theorem 14.3.1 *Let f be a permutation of a set X. Suppose we have k colors available with which to color the elements of X. Let $\mathcal{C}$ be the set of all colorings of X. Then the number*

$$|\mathcal{C}(f)|$$

of colorings of $\mathcal{C}$ that are fixed by f equals

$$k^{\#(f)}.$$

Example. How many nonequivalent ways are there to color the corners of a square with the colors red, white, and blue?

Let $\mathcal{C}$ be the set of all $3^4 = 81$ colorings of the corners of a square with the colors red, white, and blue. The corner-symmetry group of a square is the dihedral group D_4, the cycle factorization of whose elements was already computed. We repeat the results in the following table, with additional columns indicating $\#(f)$ and the number $|\mathcal{C}(f)|$ of colorings left fixed by f for each of the permutations f in D_4.

f in D_4	Cycle factorization	$\#(f)$	$|\mathcal{C}(f)|$
$\rho_4^0 = \iota$	$[1] \circ [2] \circ [3] \circ [4]$	4	$3^4 = 81$
ρ_4	$[1\ 2\ 3\ 4]$	1	$3^1 = 3$
ρ_4^2	$[1\ 3] \circ [2\ 4]$	2	$3^2 = 9$
ρ_4^3	$[1\ 4\ 3\ 2]$	1	$3^1 = 3$
τ_1	$[1] \circ [2\ 4] \circ [3]$	3	$3^3 = 27$
τ_2	$[1\ 3] \circ [2] \circ [4]$	3	$3^3 = 27$
τ_3	$[1\ 2] \circ [3\ 4]$	2	$3^2 = 9$
τ_4	$[1\ 4] \circ [2\ 3]$	2	$3^2 = 9$

Hence, by Theorem 14.2.3, the number of nonequivalent colorings is

$$N(D_4, \mathcal{C}) = \frac{81 + 3 + 9 + 3 + 27 + 27 + 9 + 9}{8} = 21.$$

$\square$

Theorems 14.2.3 and 14.3.1 give us a method to compute, in the presence of a group G of permutations of a set X, the number of nonequivalent colorings in the set $\mathcal{C}$ of all colorings of X with a given set of colors. This method requires that we be able to compute the cycle factorization (or at least the number of cycles in the cycle factorization) of each permutation in G. In order to be able to compute the number of nonequivalent colorings for more general sets $\mathcal{C}$ of colorings, we introduce a generating function for the number of permutations in G whose cycle factorizations have the same number of cycles of each size.

Let f be a permutation of X where X has n elements. Suppose that the cycle factorization of f has e_1 1-cycles, e_2 2-cycles, ... , and e_n n-cycles. Since each element of X occurs in exactly one cycle in the cycle factorization of f, the numbers $e_1, e_2, \cdots, e_n$ are nonnegative integers satisfying

$$1e_1 + 2e_2 + \cdots + ne_n = n. \qquad (14.15)$$

We call the n-tuple $(e_1, e_2, \ldots, e_n)$ the *type* of the permutation f and write

$$\text{type}(f) = (e_1, e_2, \ldots, e_n).$$

Note that the number of cycles in the cycle factorization of f is

$$\#(f) = e_1 + e_2 + \cdots + e_n.$$

Different permutations may have the same type, since the type of a permutation depends only on the size of the cycles in its cycle factorization and not on which elements are in which cycles. Since we now want to distinguish permutations only by type, we introduce n indeterminates

$$z_1, z_2, \ldots, z_n,$$

where z_k is to correspond to a k-cycle $(k = 1, 2, \ldots, n)$. To each permutation f with $\text{type}(f) = (e_1, e_2, \ldots, e_n)$, we associate the monomial of f:

$$\text{mon}(f) = z_1^{e_1} z_2^{e_2} \cdots z_n^{e_n}.$$

Notice that the total degree of the monomial of f is the number $\#(f)$ of cycles in the cycle factorization of f.

Let G be a group of permutations of X. Summing these monomials for each f in G, we get the generating function

$$\sum_{f \in G} \text{mon}(f) = \sum_{f \in G} z_1^{e_1} z_2^{e_2} \cdots z_n^{e_n} \qquad (14.16)$$

for the permutations in G according to type. If we combine like terms in (14.16), the coefficient of $z_1^{e_1} z_2^{e_2} \cdots z_n^{e_n}$ equals the number of permutations in G of type $(e_1, e_2, \ldots, e_n)$. The *cycle index*

$$P_G(z_1, z_2, \ldots, z_n) = \frac{1}{|G|} \sum_{f \in G} z_1^{e_1} z_2^{e_2} \cdots z_n^{e_n}$$

of G is this generating function divided by the number $|G|$ of permutations in G.

Example. Determine the cycle index of the dihedral group D_4.

In the example appearing just after Theorem 14.3.1, we gave a table that included the cycle factorization of each permutation in D_4. Using those factorizations, we give the type of each permutation and its associated monomial in the following table:

D_4	Cycle factorization	Type	Monomial
$\rho_4^0 = \iota$	$[1] \circ [2] \circ [3] \circ [4]$	$(4,0,0,0)$	$z_1^4 z_2^0 z_3^0 z_4^0 = z_1^4$
ρ_4	$[1\ 2\ 3\ 4]$	$(0,0,0,1)$	$z_1^0 z_2^0 z_3^0 z_4^1 = z_4$
ρ_4^2	$[1\ 3] \circ [2\ 4]$	$(0,2,0,0)$	$z_1^0 z_2^2 z_3^0 z_4^0 = z_2^2$
ρ_4^3	$[1\ 4\ 3\ 2]$	$(0,0,0,1)$	$z_1^0 z_2^0 z_3^0 z_4^1 = z_4$
τ_1	$[1] \circ [2\ 4] \circ [3]$	$(2,1,0,0)$	$z_1^2 z_2^1 z_3^0 z_4^0 = z_1^2 z_2$
τ_2	$[1\ 3] \circ [2] \circ [4]$	$(2,1,0,0)$	$z_1^2 z_2^1 z_3^0 z_4^0 = z_1^2 z_2$
τ_3	$[1\ 2] \circ [3\ 4]$	$(0,2,0,0)$	$z_1^0 z_2^2 z_3^0 z_4^0 = z_2^2$
τ_4	$[1\ 4] \circ [2\ 3]$	$(0,2,0,0)$	$z_1^0 z_2^2 z_3^0 z_4^0 = z_2^2$

The cycle index of D_4 is

$$P_{D_4}(z_1, z_2, z_3, z_4) = \frac{1}{8}(z_1^4 + 2z_4 + 3z_2^2 + 2z_1^2 z_2).$$

$\square$

We can now determine the number of nonequivalent colorings among all the colorings of a set X, using a specified set of colors, provided that we know the cycle index of the group G of permutations of X.

Theorem 14.3.2 *Let X be a set of n elements, and suppose we have a set of k colors available with which to color the elements of X. Let $\mathcal{C}$ be the set of all k^n colorings of X. Let G be a group of permutations of X. Then the number of nonequivalent colorings is the number*

$$N(G, \mathcal{C}) = P_G(k, k, \ldots, k)$$

obtained by substituting $z_i = k$, $(i = 1, 2, \ldots, n)$ into the cycle index of G.

Proof. This theorem is a consequence of Theorems 14.2.3 and 14.3.1. The cycle index of G is the average

$$P_G(z_1, z_2, \ldots, z_n) = \frac{1}{|G|} \sum_{f \in G} z_1^{e_1} z_2^{e_2} \cdots z_n^{e_n}$$

of the sum of the monomials associated with the permutations f in G. By Theorem 14.3.1, the number of colorings in $\mathcal{C}$ that are fixed by f equals

$$k^{\#(f)} = k^{e_1 + e_2 + \cdots + e_n} = k^{e_1} k^{e_2} \cdots k^{e_n},$$

where $(e_1, e_2, \ldots, e_n)$ is the type of f. By Theorem 14.2.3, the number of nonequivalent colorings is

$$N(G, \mathcal{C}) = \frac{1}{|G|} \sum_{f \in G} k^{e_1} k^{e_2} \cdots k^{e_n} = P_G(k, k, \ldots, k).$$

$\square$

Example. We are given a set of k colors. What is the number of nonequivalent ways to color the corners of a square?

The cycle index of the dihedral group D_4 has already been determined to be

$$P_{D_4}(z_1, z_2, z_3, z_4) = \frac{1}{8}(z_1^4 + 2z_4 + 3z_2^2 + 2z_1^2 z_2).$$

Hence, by Theorem 14.3.2, the number of nonequivalent colorings is

$$P_{D_4}(k, k, k, k) = \frac{k^4 + 2k + 3k^2 + 2k^2 k}{8} = \frac{k^4 + 2k^3 + 3k^2 + 2k}{8}.$$

If the number of colors is $k = 6$, then the number of nonequivalent colorings is

$$P_{D_4}(6, 6, 6, 6) = \frac{6^4 + 26^3 + 36^2 + 2 \times 6}{8} = 231.$$

$\square$

Theorem 14.3.2 gives a satisfactory way to count the number of nonequivalent colorings in $\mathcal{C}$, provided that $\mathcal{C}$ is the set of *all* colorings possible with k given colors. The formula in the theorem requires one to know the number of permutations of each type in the group G of permutations, and so can be difficult to apply. But it is as simple as one could expect, given the fact that G can be any permutation group on the set X of objects being colored. Our final concern is with more general sets $\mathcal{C}$ of colorings. Recall that, in Theorem 14.2.3, the only restriction on $\mathcal{C}$ is that G acts as a permutation group on $\mathcal{C}$; that is, each permutation f in G takes a coloring $\mathbf{c}$ of $\mathcal{C}$ to another coloring $f * \mathbf{c}$ of $\mathcal{C}$. Under these more general circumstances, the most one might expect is to have some formal way to determine the nonequivalent colorings. We show how the cycle index of G can be used to determine the number of nonequivalent colorings where the number of times each color is used is specified.

Let $\mathcal{C}$ be the set of all colorings of X in which the number of elements in X of each color have been specified. For each permutation f of X and each coloring $\mathbf{c}$ in $\mathcal{C}$, the number of times a particular color appears in $\mathbf{c}$ is the same as the number of times that color appears in $f * \mathbf{c}$. Put another way, permuting the objects in X along with their colors does not change the number of colors of each kind. This means that any group G of permutations of X acts as a permutation group on such a set of colorings $\mathcal{C}$.

Example. How many nonequivalent colorings are there of the corners of a regular 5-gon in which three corners are colored red and two are colored blue?

Let $\mathcal{C}$ be the set of all colorings of the corners of a 5-gon with three corners colored red and two colored blue. The number of colorings in $\mathcal{C}$ is 10, since we can select three corners to be colored red in 10 ways and then color the other two corners blue. The corner-symmetry group D_5 acts as a permutation group on $\mathcal{C}$. We have previously computed the cycle factorization of each permutation in G. In the following table, we again list those factorizations, along with the number of colorings in $\mathcal{C}$ fixed by the permutations in D_5.

The reason that none of the rotations different from the identity fixes any coloring is that, for such a rotation to fix a coloring, all colors in the coloring must be the same (and so we do not have three red and two blue colors as specified). Each reflection fixes two colorings in $\mathcal{C}$. This is because, for the 5-gon, each of the reflections has type $(1, 2, 0, 0, 0)$. In order to have two blue corners in a fixed coloring, we must color blue the corners in one of the two 2-cycles in the factorization. Applying Theorem 14.2.3, we find that the number of nonequivalent colorings of the type being counted is

$$\frac{10 + 0 + 0 + 0 + 0 + 2 + 2 + 2 + 2 + 2}{10} = 2.$$

This answer can easily be arrived at directly, the two nonequivalent colorings being the one with two blue corners that are consecutive and the other with two blue corners that are not consecutive. □

D_5	Cycle factorization	Number of fixed colorings
$\rho_5^0 = \iota$	$[1] \circ [2] \circ [3] \circ [4] \circ [5]$	10
ρ_5	$[1\ 2\ 3\ 4\ 5]$	0
ρ_5^2	$[1\ 3\ 5\ 2\ 4]$	0
ρ_5^3	$[1\ 4\ 2\ 5\ 3]$	0
ρ_5^4	$[1\ 5\ 4\ 3\ 2]$	0
τ_1	$[1] \circ [2\ 5] \circ [3\ 4]$	2
τ_2	$[1\ 3] \circ [2] \circ [4\ 5]$	2
τ_3	$[1\ 5] \circ [3] \circ [2\ 4]$	2
τ_4	$[2] \circ [3\ 5] \circ [4]$	2
τ_5	$[1\ 4] \circ [2\ 3] \circ [5]$	2

In order to apply Burnside's theorem to determine the number of nonequivalent colorings when the number of occurrences of each color is specified, we must be able to determine the number of such colorings fixed by a permutation. Let f be a permutation of the set X, and suppose that

$$\text{type}(f) = (e_1, e_2, \ldots, e_n)$$

and

$$\text{mon}(f) = z_1^{e_1} z_2^{e_2} \cdots z_n^{e_n}.$$

Thus, f has e_1 1-cycles, e_2 2-cycles, $\ldots$, e_n n-cycles in its cycle factorization. To initially keep our discussion simple, suppose we have only two colors: red and blue. Let

$$\mathcal{C}_{p,q}$$

denote the set of all colorings of X with p elements colored red and $q = n - p$ elements colored blue. A coloring in $\mathcal{C}_{p,q}$ is fixed by f if and only if, for each cycle in the cycle factorization of f, all of the elements have the same color. Thus, to determine the number of colorings in $C_{p,q}$ fixed by f, we can think of assigning colors to cycles in such a way that the number of *elements* that get assigned the color red is p (and hence the number assigned the color blue is $n - p = q$). Suppose that t_1 of the 1-cycles get assigned red, t_2 of the 2-cycles get red, ... , t_n of the n-cycles get red. In order that the number of elements assigned red be p we must have

$$p = t_1 1 + t_2 2 + \cdots + t_n n. \tag{14.17}$$

Hence, the number $|C_{p,q}(f)|$ of colorings in $C_{p,q}$ that are fixed by f is obtained as follows: Choose a solution of (14.17) in integers $t_1, t_2, \ldots, t_n$ satisfying

$$0 \le t_1 \le e_1, \ 0 \le t_2 \le e_2, \ \cdots \ , 0 \le t_n \le e_n \tag{14.18}$$

(to determine how many cycles of each length are assigned the color red), and then multiply such a solution by

$$\binom{e_1}{t_1}\binom{e_2}{t_2}\cdots\binom{e_n}{t_n}$$

(to determine which cycles of each of the lengths $1, 2, \ldots, n$ are assigned the color red). Now, consider the color red as a variable r and the color blue as a variable b that we can manipulate algebraically in the usual way. Then the number of solutions of (14.17) satisfying (14.18) is the coefficient of $r^p b^q$ in the expression

$$(r + b)^{e_1}(r^2 + b^2)^{e_2}\cdots(r^n + b^n)^{e_n},$$

obtained by making the substitutions

$$z_1 = r + b, \ z_2 = r^2 + b^2, \ \cdots \ , z_n = r^n + b^n \tag{14.19}$$

in the monomial of f. The cycle index of a permutation group G is the average of the monomials of the permutations f in G. Hence, by Theorem 14.2.3, the number of nonequivalent colorings in $C(p, q)$ equals the coefficient of $r^p b^q$ in the expression

$$P_G(r + b, r^2 + b^2, \cdots, r^n + b^n), \tag{14.20}$$

obtained by making the substitutions (14.19) in the cycle index of G. This means that (14.20) is a two-variable generating function for the number of nonequivalent colorings in $\mathcal{C}(p,q)$ with a specified number of elements colored with each color.[18]

The preceding discussion applies for any number of colors, and it enables us to give a generating function for the number of nonequivalent colorings in which the number of colors of each kind is specified. This provides us with the *FINAL THEOREM* in this book.[19] This theorem is commonly called *Pólya's theorem* and its motivation, derivation, and application has been the primary purpose of this chapter.

As with the case of two colors, we need to think of the colors as variables $u_1, u_2, \ldots, u_k$ to be manipulated algebraically. The only change in the preceding argument is the change from two colors to any number k of colors.

Theorem 14.3.3 *Let X be a set of elements and let G be a group of permutations of X. Let $\{u_1, u_2, \ldots, u_k\}$ be a set of k colors and let $\mathcal{C}$ be any set of colorings of X with the property that G acts as a permutation group on $\mathcal{C}$. Then the generating function for the number of nonequivalent colorings of $\mathcal{C}$ according to the number of colors of each kind is the expression*

$$P_G(u_1 + \cdots + u_k, u_1^2 + \cdots + u_k^2, \ldots, u_1^n + \cdots + u_k^n), \qquad (14.21)$$

obtained from the cycle index $P_G(z_1, z_2, \ldots, z_n)$ by making the substitutions

$$z_j = u_1^j + \cdots + u_k^j \quad (j = 1, 2, \ldots, n).$$

In other words, the coefficient of

$$u_1^{p_1} u_2^{p_2} \cdots u_k^{p_k}$$

in (14.21) equals the number of nonequivalent colorings in $\mathcal{C}$ with p_1 elements of X colored u_1, p_2 elements colored u_2, ... , p_k elements colored u_k.

---

[18]The two variables in the generating function are r and b. We could get a one-variable generating function by setting $b = 1$. Nothing is lost by doing so, since as we have already remarked, once the number of reds is specified, the number of blues is whatever is left. However, since we are about to write down the generating function for any number of colors where we cannot reduce the generating function to one variable, it is better here to use two variables.

[19]If you started on page 1 and worked your way here doing most of the exercises then *CONGRATULATIONS!* You know a lot about combinatorics and graph theory. But there is a lot more to know, and the amount increases every day. The number of research articles on combinatorics and graph theory in journals seems to increase every year. But that is not too surprising. As I hope that you have discovered, the subject is exciting and fascinating, and we have given some hints as to its applicability in the physical world. Following the exercises for this chapter, we include a list of books for further study.

Substituting $u_i = 1$ for $i = 1, 2, \ldots, k$ in (14.21), we get the sum of its coefficients and hence the total number of nonequivalent colorings of X with k available colors. Since this substitution yields

$$P_G(k, k, \ldots, k),$$

it follows that Theorem 14.3.3 is a refinement of Theorem 14.3.2. Theorem 14.3.3 contains more detailed information than Theorem 14.3.2, which is subsequently lost upon replacing each u_i with 1.

Example. Determine the generating function for the number of nonequivalent colorings of the corners of a square with 2 colors and also those with 3 colors.

The cycle index of D_4, the corner-symmetry group of the square, has been previously computed to be

$$P_{D_4}(z_1, z_2, z_3, z_4) = \frac{1}{8}(z_1^4 + 2z_4 + 3z_2^2 + 2z_1^2 z_2).$$

Let the two colors be r and b. Then the generating function is

$$P_{D_4}(r + b, r^2 + b^2, r^3 + b^3, r^4 + b^4) =$$

$$\frac{1}{8}((r+b)^4 + 2(r^4 + b^4) + 3(r^2 + b^2)^2 + 2(r + b)^2(r^2 + b^2))$$

$$= \frac{1}{8}(8r^4 + 8r^3 b + 16r^2 b^2 + 8rb^3 + 8b^4).$$

Hence, we have

$$P_{D_4}(r+b, r^2+b^2, r^3+b^3, r^4+b^4) = r^4 + r^3 b + 2r^2 b^2 + rb^3 + b^4. \quad (14.22)$$

Thus, there is one nonequivalent coloring with all corners red and one with all corners blue. There is also one with three corners red and one blue, and one with one corner red and three blue. Finally, there are two with two corners of each color. The total number of nonequivalent colorings, the sum of the coefficients in (14.22), is 6.

Now, suppose that we have three colors r, b, and g. The generating function for the number of nonequivalent colorings is

$$P_{D_4}(r + b + g, r^2 + b^2 + g^2, r^3 + b^3 + g^3, r^4 + b^4 + g^4)$$

$$= \frac{1}{8}((r+b+g)^4 + 2(r^4+b^4+g^4) + 3(r^2+b^2+g^2)^2 + 2(r+b+g)^2(r^2+b^2+g^2)).$$

This expression can be calculated, using the multinomial theorem in Chapter 5. For instance, the coefficient of $r^1 b^2 g^1$ equals

$$\frac{1}{8}(12 + 0 + 0 + 4) = 2.$$

Thus, there are 2 nonequivalent colorings that have one red, two blue, and one green corners. The total number of nonequivalent colorings equals

$$P_{D_4}(3,3,3) = 21.$$

□

Example. Determine the generating function for the number of nonequivalent colorings of the corners of a regular 5-gon with 2 colors and also those with 3 colors.

From our previous calculations, the cycle index of D_5 is

$$P_{D_5}(z_1, z_2, z_3, z_4, z_5) = \frac{1}{10}(z_1^5 + 4z_5 + 5z_1z_2^2).$$

Notice that neither z_3 nor z_4 appear in any nonzero term in the cycle index. This is because no permutation in D_5 has either a 3-cycle or 4-cycle in its cycle factorization. Suppose that we have two colors r and b. Then the generating function for the number of nonequivalent colorings is

$$P_{D_5}(r+b, r^2+b^2, r^3+b^3, r^4+b^4, r^5+b^5)$$

$$= \frac{1}{10}((r+b)^5 + 4(r^5+b^5) + 5(r+b)(r^2+b^2)^2)$$

$$= r^5 + r^4b + 2r^3b^2 + 2r^2b^3 + rb^4 + b^5.$$

The total number of nonequivalent colorings equals

$$1 + 1 + 2 + 2 + 1 + 1 = 8.$$

The generating function for the number of nonequivalent colorings for three colors is

$$\frac{1}{10}((r+b+g)^5 + 4(r^5+b^5+g^5) + 5(r+b+g)(r^2+b^2+g^2)^2).$$

The total number of nonequivalent colorings equals

$$\frac{1}{10}(3^5 + 4(3) + 5(3)(3^2)) = 39.$$

□

Example. (*Coloring the corners and faces of a cube*). Determine the symmetry group of a cube and the number of nonequivalent ways to color the corners and faces of a cube with a specified number of colors.

There are 24 symmetries of a cube, and they are rotations of four different kinds:

(i) The identity rotation ι (number is 1).

(ii) The rotations about the centers of the three pairs of opposite faces by

    (a) 90 degrees (number is 3).

    (b) 180 degrees (number is 3).

    (c) 270 degrees (number is 3).

(iii) The rotations by 180 degrees about midpoints of opposite edges (number is 6).

(iv) The rotations about opposite corners by

    (a) 120 degrees (number is 4).

    (b) 240 degrees (number is 4).

The total number of symmetries of a cube is 24.

In the next table, we give the type of each symmetry as both a permutation of its 8 corners (as a member of the corner-symmetry group of the cube) and as a permutation of its 6 faces (as a member of the face-symmetry group of the cube). In this table, we refer to the classification of the symmetries previously given.

Kind of symmetry	Number of	Corner type	Face type
(i)	1	$(8,0,0,0,0,0,0,0)$	$(6,0,0,0,0,0)$
(ii)(a)	3	$(0,0,0,2,0,0,0,0)$	$(2,0,0,1,0,0)$
(ii)(b)	3	$(0,4,0,0,0,0,0,0)$	$(2,2,0,0,0,0)$
(ii)(c)	3	$(0,0,0,2,0,0,0,0)$	$(2,0,0,1,0,0)$
(iii)	6	$(0,4,0,0,0,0,0,0)$	$(0,3,0,0,0,0)$
(iv)(a)	4	$(2,0,2,0,0,0,0,0)$	$(0,0,2,0,0,0)$
(iv)(b)	4	$(2,0,2,0,0,0,0,0)$	$(0,0,2,0,0,0)$

From the table, we see that the cycle index of the corner-symmetry group G_C of the cube is

$$P_{G_C}(z_1, z_2, \ldots, z_8) = \frac{1}{24}(z_1^8 + 6z_4^2 + 9z_2^4 + 8z_1^2 z_3^2),$$

and that of the face-symmetry group G_F is

$$P_{G_F}(z_1, z_2, \ldots, z_6) = \frac{1}{24}(z_1^6 + 6z_1^2 z_4 + 3z_1^2 z_2^2 + 6z_2^3 + 8z_3^2).$$

The generating function for the number of nonequivalent colorings of the corners of a cube with the colors red and blue is

$$P_{G_C}(r+b, r^2+b^2, \ldots, r^8+b^8)$$

$$= \frac{1}{24}((r+b)^8 + 6(r^4+b^4)^2 + 9(r^2+b^2)^4 + 8(r+b)^2(r^3+b^3)^2).$$

For the faces of the cube, the generating function is

$$P_{G_F}(r+b, r^2+b^2, \ldots, r^6+b^6) =$$

$$\frac{1}{24}((r+b)^6 + 6(r+b)^2(r^4+b^4) + 3(r+b)^2(r^2+b^2)^2 + 6((r^2+b^2)^3 + 8(r^3+b^3)^2).$$

Some algebraic calculation now shows that the generating function for the number of nonequivalent colorings of the corners is

$$r^8 + r^7 b + 3r^6 b^2 + 3r^5 b^3 + 7r^4 b^4 + 3r^3 b^5 + 3r^2 b^6 + rb^7 + b^8$$

and, for the faces, is

$$r^6 + r^5 b + 2r^4 b^2 + 2r^3 b^3 + 2r^2 b^4 + rb^5 + b^6.$$

The total number of nonequivalent colorings for the corners is 23 and for the faces is 10.

If we have k colors, the number of nonequivalent corner colorings is

$$\frac{1}{24}(k^8 + 6k^2 + 9k^4 + 8k^2 k^2) = \frac{1}{24}(k^8 + 17k^4 + 6k^2),$$

and the number of nonequivalent face colorings is

$$\frac{1}{24}(k^6 + 6k^2 k + 3k^2 k^2 + 6k^3 + 8k^2) = \frac{1}{24}(k^6 + 3k^4 + 12k^3 + 8k^2).$$

$\square$

In our final example we illustrate how Theorem 14.3.3 can be applied to determine the number of nonisomorphic graphs of order n with a specified number of edges.

Example. Determine the number of nonisomorphic graphs of order 4 with each possible number of edges.

The number 4 is small enough for us to solve this problem without recourse to Theorem 14.3.3. But, our purpose in this example is to illustrate how to apply Theorem 14.3.3 to count graphs.

Let $\mathcal{G}_4$ be the set of all graphs of order 4 with vertex set $V = \{1, 2, 3, 4\}$. We seek the generating function for the number of nonisomorphic graphs in $\mathcal{G}_4$ with a specified number of edges. The set E of edges of a graph $H_1 = (V, E_1)$ in $\mathcal{G}_4$ is a subset of the set

$$X = \{\{1,2\}, \{1,3\}, \{1,4\}, \{2,3\}, \{2,4\}, \{3,4\}\}.$$

We can think of H_1 as a coloring of the edges in the set X, with two colors "yes" (or y) and "no" (or n), where the edges in E_1 get the color yes and the edges not in E_1 get the color no. Let $\mathcal{C}$ be the set of all colorings of X with the two colors y and n. Thus, the graphs in $\mathcal{G}_4$ are exactly the colorings in $\mathcal{C}$! This is the first important observation for obtaining our solution.

Let $H_2 = (V, E_2)$ be another graph in $\mathcal{G}_4$. Then H_1 and H_2 are isomorphic if and only if there is a permutation f of $V = \{1, 2, 3, 4\}$ (so, a permutation in S_4), such that $\{i, j\}$ is an edge in E_1 if and only if $\{f(i), f(j)\}$ is an edge in E_2. Each of the 24 permutations f in S_4 also permutes the edges in X, using the rule:

$$\{i, j\} \rightarrow \{f(i), f(j)\} \quad (\{i, j\} \text{ in } X).$$

For example, let

$$f = \begin{pmatrix} 1 & 2 & 3 & 4 \\ 3 & 2 & 4 & 1 \end{pmatrix}.$$

Then f permutes the edges as follows:

$$\begin{pmatrix} \{1,2\} & \{1,3\} & \{1,4\} & \{2,3\} & \{2,4\} & \{3,4\} \\ \{2,3\} & \{3,4\} & \{1,3\} & \{2,4\} & \{1,2\} & \{1,4\} \end{pmatrix}.$$

Let $S_4^{(2)}$ be the group of permutations of X obtained in this way from S_4.[20] Our second important observation is that two graphs in $\mathcal{G}_4$ are isomorphic if and only if, as colorings of X, they are equivalent. This observation is an immediate consequence of the definitions of isomorphic graphs and equivalent colorings.

We have thus reduced our problem to counting the number of colorings in $\mathcal{C}$ that are nonequivalent with respect to the permutation group $S_4^{(2)}$, according to the number of y's and n's. This is exactly the setup of Theorem 14.3.3. It only remains to compute the cycle index of $S_4^{(2)}$.

---

[20] Since S_4 is a group of permutations, it follows readily that $S_4^{(2)}$ is also a group of permutations. S_4 and $S_4^{(2)}$ are isomorphic as abstract groups but not as permutation groups.

To do this we have to compute the type of each of the 24 permutations in $S_4^{(2)}$. The results are summarized in the following table.

Type	Monomial	Number of permutations in $S_4^{(2)}$
$(6,0,0,0,0,0)$	z_1^6	1
$(2,2,0,0,0,0)$	$z_1^2 z_2^2$	9
$(0,0,2,0,0,0)$	z_3^2	8
$(0,1,0,1,0,0)$	$z_2 z_4$	6

The cycle index of $S_4^{(2)}$ is

$$P_{S_4^{(2)}}(z_1, z_2, z_3, z_4, z_5, z_6) = \frac{1}{24}(z_1^6 + 9z_1^2 z_2^2 + 8z_3^2 + 6z_2 z_4). \quad (14.23)$$

By Theorem 14.3.3, the generating function for the number of nonequivalent colorings in $\mathcal{C}$ is obtained by making the substitutions

$$z_j = y^j + n^j \quad (j = 1, 2, 3, 4, 5, 6)$$

in (14.23). A little calculation shows that the result is

$$y^6 + y^5 n + 2y^4 n^2 + 3y^3 n^3 + 2y^2 n^4 + yn^5 + n^6.$$

Remembering that the number of y's equals the number of edges, we see that the number of nonisomorphic graphs of order 4, according to the number of edges, is given as follows:

Number of edges	Number of nonisomorphic graphs
6	1
5	1
4	2
3	3
2	2
1	1
0	1

In particular, the total number of nonisomorphic graphs of order 4 equals 11.

14.4 Exercises

1. Let

$$f = \begin{pmatrix} 1 & 2 & 3 & 4 & 5 & 6 \\ 6 & 4 & 2 & 1 & 5 & 3 \end{pmatrix} \text{ and } g = \begin{pmatrix} 1 & 2 & 3 & 4 & 5 & 6 \\ 3 & 5 & 6 & 2 & 4 & 1 \end{pmatrix}.$$

   Determine

   (a) $f \circ g$ and $g \circ f$
   (b) f^{-1} and g^{-1}
   (c) f^2, f^5
   (d) $f \circ g \circ f$
   (e) g^3 and $f \circ g^3 \circ f^{-1}$.

2. Prove that permutation composition is associative:

$$(f \circ g) \circ h = f \circ (g \circ h).$$

3. Determine the symmetry group and corner-symmetry group of an equilateral triangle.

4. Determine the symmetry group and corner-symmetry group of a triangle that is isoceles but not equilateral.

5. Determine the symmetry group and corner-symmetry group of a triangle that is neither equilateral nor isoceles.

6. Determine the symmetry group of a regular tetrahedron. (Hint: There are 12 symmetries.)

7. Determine the corner-symmetry group of a regular tetrahedron.

8. Determine the edge-symmetry group of a regular tetrahedron.

9. Determine the face-symmetry group of a regular tetrehedron.

10. Determine the symmetry group and the corner-symmetry group of a rectangle that is not a square.

11. Compute the corner-symmetry group of a regular hexagon (the dihedral group D_6 of order 12).

12. Determine all the permutations in the edge-symmetry group of a square.

13. Let f and g be the permutations in Exercise 1. Consider the coloring $\mathbf{c} = (R, B, B, R, R, R)$ of $1, 2, 3, 4, 5, 6$ with the colors R and B. Determine the following actions on $\mathbf{c}$:

   (a) $f * \mathbf{c}$

   (b) $f^{-1} * \mathbf{c}$

   (c) $g * \mathbf{c}$

   (d) $(g \circ f) * \mathbf{c}$ and $(f \circ g) * \mathbf{c}$

   (e) $(g^2 \circ f) * \mathbf{c}$

14. By examining all possibilities, determine the number of nonequivalent colorings of the the corners of an equilateral triangle with the colors red and blue. (With the colors red, white, and blue.)

15. By examining all possibilities, determine the number of nonequivalent colorings of the corners of a regular tetrahedron with the colors red and blue. (With the colors red, white, and blue.)

16. Characterize the cycle factorizations of those permutations f in S_n for which $f^{-1} = f$, that is, for which $f^2 = \iota$.

17. In Section 14.2 it is established that there are 8 nonequivalent colorings of the corners of a regular pentagon with the colors red and blue. Explicitly determine 8 nonequivalent colorings.

18. Use Theorem 14.2.3 to determine the number of nonequivalent colorings of the corners of a square with p colors.

19. Use Theorem 14.2.3 to determine the number of nonequivalent colorings of the corners of an equilateral triangle with the colors red and blue. With p colors (Cf. Exercise 3).

20. Use Theorem 14.2.3 to determine the number of nonequivalent colorings of the corners of a triangle that is isoceles, but not equilateral, with the colors red and blue. With p colors (Cf. Exercise 4).

21. Use Theorem 14.2.3 to determine the number of nonequivalent colorings of the corners of a triangle that is neither equilateral nor isoceles, with the colors red and blue. With p colors (Cf. Exercise 5).

22. Use Theorem 14.2.3 to determine the number of nonequivalent colorings of the corners of a rectangle that is not a square with the colors red and blue. With p colors (Cf. Exercise 10).

23. A (one-sided) *marked-domino* is a piece consisting of two squares joined along an edge, where each square on one side of the piece is marked with $0, 1, 2, 3, 4, 5,$ or 6 dots. The two squares of a marked-domino may receive the same number of dots.

    (a) Use Theorem 14.2.3 to determine the number of different marked-dominoes.

    (b) How many different marked-dominoes are there if we are allowed to mark the squares with $0, 1, \ldots, p-1,$ or p dots?

24. A *two-sided marked-domino* is a piece consisting of two squares joined along an edge, where each square on both sides of the piece is marked with $0, 1, 2, 3, 4, 5,$ or 6 dots.

    (a) Use Theorem 14.2.3 to determine the number of different two-sided marked-dominoes.

    (b) How many different two-sided marked-dominoes are there if we are allowed to mark the squares with $0, 1, \ldots, p-1,$ or p dots?

25. How many different necklaces are there that contain 3 red and 2 blue beads?

26. How many different necklaces are there that contain 4 red and 3 blue beads?

27. Determine the cycle factorization of the permutations f and g in Exercise 1.

28. Let f be a permutation of a set X. Give a simple algorithm for finding the cycle factorization of f^{-1} from the cycle factorization of f.

29. Determine the cycle factorization of each permutation in the dihedral group D_6 (Cf. Exercise 11).

30. Determine permutations f and g of the same set X such that f and g each have 2 cycles in their cycle factorizations but $f \circ g$ has only one.

31. Determine the number of nonequivalent colorings of the corners of a regular 5-gon with k colors.

32. Determine the number of nonequivalent colorings of the corners of a regular hexagon with the colors red, white and blue (Cf. Exercise 29).

33. Prove that a permutation and its inverse have the same type (Cf. Exercise 28).

34. Let $e_1, e_2, \ldots, e_n$ be nonnegative integers such that $1e_1 + 2e_2 + \cdots + ne_n = n$. Show how to construct a permutation f of the set $\{1, 2, \ldots, n\}$ such that $\text{type}(f) = (e_1, e_2, \ldots, e_n)$.

35. Determine the number of nonequivalent colorings of the corners of a regular 6-gon with k colors (Cf. Exercise 29).

36. Determine the number of nonequivalent colorings of the corners of a regular 5-gon with the colors red, white, and blue in which two corners are colored red, two are colored white, and one is colored blue.

37. Determine the cycle index of the dihedral group D_6 (Cf. Exercise 29).

38. Determine the generating function for nonequivalent colorings of the corners of a regular hexagon with 2 colors and also with 3 colors (Cf. Exercise 37).

39. Determine the cycle index of the edge-symmetry group of a square.

40. Determine the generating function for nonequivalent colorings of the edges of a square with the colors red and blue. How many nonequivalent colorings are there with k colors? (Cf. Exercise 39.)

41. Let n be an odd prime number. Prove that each of the permutations, $\rho_n, \rho_n^2, \ldots, \rho_n^{n-1}$, of $\{1, 2, \ldots, n\}$ is an n-cycle. (Recall that ρ_n is the permutation that sends 1 to 2, 2 to 3, $\ldots$, $n-1$ to n, and n to 1.)

42. Let n be a prime number. Determine the number of different necklaces that can be made from n beads of k different colors.

43. The nine squares of a 3-by-3 chessboard are to be colored red and blue. The chessboard is free to rotate but cannot be flipped over. Determine the generating function for the number of nonequivalent colorings and the total number of nonequivalent colorings.

44. A stained glass window in the form of a 3-by-3 chessboard has 9 squares, each of which is colored red or blue (the colors are transparent and the window can be looked at from either side). Determine the generating function for the number of different

stained glass windows and the total number of stained glass windows.

45. Repeat Exercise 44 for stained glass windows in the form of a 4-by-4 chessboard with 16 squares.

46. Find the generating function for the different necklaces that can be made with p beads each of color red or blue if p is a prime number (cf. Exercise 42).

47. Determine the cycle index of the dihedral group D_{2p}, where p is a prime number.

48. Find the generating function for the different necklaces that can be made with $2p$ beads each of color red or blue if p is a prime number.

49. Ten balls are stacked in a triangular array with 1 atop 2 atop 3 atop 4. (Think of billiards.) The triangular array is free to rotate. Find the generating function for the number of nonequivalent colorings with the colors red and blue. Find the generating function if we are also allowed to turn over the array.

50. Use Theorem 14.3.3 to determine the generating function for non-isomorphic graphs of order 5. (Hint: This exercise will require some work and is a fitting last exercise. One needs to obtain the cycle index of the group $S_5^{(2)}$ of permutations of the set X of 10 unordered pairs of distinct integers from $\{1, 2, 3, 4, 5\}$ (the possible edges of a graph of order 5). First, compute the number of permutations f of S_5 of each type. Then use the fact that the type of f as a permutation of X depends only on the type of f as a permutation of $\{1, 2, 3, 4, 5\}$.)

Answers and Hints to Exercises

We give partial solutions, solutions, or hints to selected exercises.

Chapter 1 Exercises

3. No.

4. $f(n) = f(n-1) + f(n-2)$; $f(12) = 233$.

5. 11.

10. Use a 5-by-6 board with 2-by-3 pieces.

16. No.

21. Since each pair of the three countries 1, 2, and 10 have a common border, 3 colors are necessary. There are 12 different colorings using the colors red, white, and blue.

22. No. The common line sum would have to be $(1 + 2 + \cdots + 7)/3$, but this number is not a whole number.

27. Simple experimentation is usually successful.

30. Balanced. Player II should remove 14 coins from the heap of size 22.

32. Hint: Consider the units digit.

35. Second player. Think of 5's.

36. First player.

37. 105.

39. Hint: Consider a pairing in which the total length of the n line segments is as small as possible.

40. Hint: n must be even. Color the squares black and white with all squares in columns $1, 3, \ldots, n-1$ black and all squares in columns $2, 4, \ldots, n$ white, giving an equal number of black and white squares. The l-tetrominoes on the board are of two types: either they cover 3 black squares and 1 white square, or they cover 3 white squares and 1 black square.

Chapter 2 Exercises

2. See D.O. Shklarsky, N.N. Chentzov, and I.M. Yaglom: *The USSR Olympiad Problem Book*, Freeman, San Francisco, 1962, 169-171.

4. Partition the integers $\{1, 2, \ldots, 2n\}$ into the pairs $\{1, 2\}$, $\{3, 4\}$, $\ldots$, $\{2n - 1, 2n\}$.

7. See D.O. Shklarsky, N.N. Chentzov, and I.M. Yaglom: *The USSR Olympiad Problem Book*, Freeman, San Francisco, 1962, 169-171.

8. What are the possible remainders when an integer is divided by n?

9. The number of sums that can be formed with 10 numbers is $2^{10} - 1$. No sum can exceed 600.

14. 45 minutes.

15. Hint: Consider remainders when an integer is divided by n.

18. Partition the square into four squares of side length 1.

19. (a) Partition the triangle into four equilateral triangles of side length 1/2.

20. Consider one point and the line segments to the other sixteen points. At least six of these line segments have the same color.

24. q_3.

27. For each set A, consider the set B of elements not in A.

28. Hint: First show that there is a way to choose the dance lists that works with $a_1 + a_2 + \cdots + a_{100} = 1620 (= 20 + 80 \cdot 20)$. Then show, by using an averaging argument (for $i = 1, 2, \ldots, 20$, let b_i be the number of lists that contain the ith woman and average these numbers), that there is no arrangement with a sum of 1619 that works.

Chapter 3 Exercises

1. $(\{a, b\})$ 48.

2. $4!(13!)^4$.

3. $52 \times 51 \times 50 \times 49 \times 48$; $\binom{52}{5}$.

4. (a) $5 \times 3 \times 7 \times 2$; (c) 121.

5. (a) 12.

6. Partition the integers according to the number of digits they contain.

8. $6!5!$.

10. $\binom{12}{2} \times \binom{10}{3} + \binom{12}{3} \times \binom{10}{2} + \binom{12}{4} \times \binom{10}{1} + \binom{12}{5}$.

11. $\binom{20}{3} - 2 \times 17 - 17 \times 16 - 18$.

13. (a) $\binom{100}{25}\binom{75}{35}$.

15. (a) $20!/5!$; (b) $\binom{15}{10}\binom{20}{10}10!$.

17. $6!$; $6!\binom{6}{2}$.

27. $\binom{7}{4}^2 4! + 7^2 \binom{6}{3}^2 3!$.

30. $2(5!)^2$.

32. $11!\left(\frac{1}{2!4!5!} + \frac{1}{3!3!5!} + \frac{1}{3!4!4!}\right)$.

36. $(n_1 + 1)(n_2 + 1) \cdots (n_k + 1)$.

39. If 6 nonconsecutive sticks are removed we are left with a solution in integers of the equation $x_1 + x_2 + \cdots + x_7 = 14$, where $x_1, x_7 \geq 0$, and $x_i > 0$ for $i = 2, \ldots 6$).

41. $3 \times \binom{12}{2}$.

43. $\binom{r+k-2}{k-2} + \binom{r+k-3}{k-2}$.

47. Hint: Use the subtraction principle. First, count the total number of ways to put the books on the shelves. Then count the number of ways in which one shelf has more books than the other two (so that shelf has at least $n + 1$ books).

54. 3^n.

Chapter 4 Exercises

1. 35124.

2. $\{3, 7, 8\}$.

4. Hint: 1 is never mobile.

6. (a) 2,4,0,4,0,0,1,0.

7. (a) 48165723.

11. (i) 00111000; (ii) 0010101; (iii) 01000000.

15. (a) $\{x_4, x_2\}$; (b) $\{x_7, x_5, x_3, x_0\}$.

16. (a) $\{x_4, x_1\}$; (b) $\{x_7, x_5, x_2, x_1, x_0\}$.

17. 150th is $\{x_7, x_4, x_2, x_1\}$.

23. (a) 010100111.

24. (a) 010100010.

28. 2,3,4,7,8,9 immediately follows 2,3,4,6,9,10; 2,3,4,6,8,10 immediately precedes 2,3,4,6,9,10.

34. (a) $12 \cdots r$, $12 \cdots (r-1)(r+1)$, $\ldots$, $12 \cdots (r-1)n$.

36. The number of relations on X is 2^{n^2}; the number of reflexive relations is $2^{n(n-1)}$.

41. Hint: Consider transitivity.

48. Hint: Something very familiar.

50. 48.

Chapter 5 Exercises

6. $-3^5 2^{13} \binom{18}{5}$; 0.

7. $\sum_{k=0}^{n} \binom{n}{k} r^k = (1+r)^n$.

8. Hint: $2 = 3 - 1$.

9. $(-1)^n 9^n$.

13. $\binom{n+3}{k}$.

15. Differentiate the binomial formula and then replace x by -1.

17. Integrate the binomial formula, but watch out for the constant of integration.

20. To find a, b, and c, multiply out and compare coefficients.

23. (a) $\frac{24!}{10!14!}$; (b) $\frac{15!}{4!5!6!}$; (c) $\frac{(9!)^2}{4!5!(3!)^3}$.

24. $\frac{45!}{10!15!20!}$.

29. Hint: Consider a set of n boys and n girls, and form committees of size n in which a boy is the leader.

30. $\binom{m_1+m_2+m_3}{n}$.

31. First show that a clutter of size 6 cannot contain a 3-combination.

35. Hint: Number of paths with only one combination is $\binom{n}{\lfloor n/2 \rfloor} - \binom{n}{\lceil (n+1)/2 \rceil}$.

38. Replace all the x_i's with 1.

40. $\frac{10!}{3!4!2!}$.

Chapter 6 Exercises

1. 5,334.

3. $10,000 - (100 + 21) + 4 = 9,883$.

4. 34.

7. 456.

9. (a) Use the change of variable $y_1 = x_1 - 1$, $y_2 = x_2$, $y_3 = x_3 - 4$, and $y_4 = x_4 - 2$.

11. $8! - 4 \times 7! + 6 \times 6! - 4 \times 5! + 4!$.

12. $\binom{8}{4} D_4$.

15. (a) D_7; (b) $7! - D_7$; (c) $7! - D_7 - 7 \times D_6$.

16. Hint: Partition the permutations according to the number of integers in their natural position.

17. $\frac{9!}{3!4!2!} - \left(\frac{7!}{4!2!} + \frac{6!}{3!2!} + \frac{8!}{3!4!}\right) + \left(\frac{4!}{2!} + \frac{6!}{4!} + \frac{5!}{3!}\right) - 3!$.

21. $D_1 = 0$ and $D_2 = 1$. Now use induction and one of the recurrences for D_n.

24. (b) $6! - 12 \times 5! + 54 \times 4! - 112 \times 3! + 108 \times 2! - 48 + 8$.

28. $8! - 32 \times 6! + 288 \times 4! - 768 \times 2! + 384$. (The number 32 arises as follows: The original seating pairs up the boys. The number of seating arrangements in which the boys in exactly one of the pairs are opposite each other is obtained as follows: We can choose one pair in 4 ways, choose the two seats that they occupy in 4 ways, and then seat them in 2 ways. We have $4 \times 4 \times 2 = 32$.)

29. $\frac{9!}{3!4!2!} - \left(\frac{7!}{4!2!} + \frac{6!}{3!2!} + \frac{8!}{3!4!}\right) + \left(\frac{4!}{2!} + \frac{6!}{4!} + \frac{5!}{3!}\right) - 3!$.

31. Hint: Let A_i be the set of integers between 1 and n that are divisible by p_i.

35. The answer is 6, but this is the hard way to do this problem. It's easier just to list all the solutions.

Chapter 7 Exercises

1. (a) f_{2n}; (b) $f_{2n+1} - 1$.

2. Hint: Show that the absolute value of $\frac{1}{\sqrt{5}}\left(\frac{1-\sqrt{5}}{2}^n\right)$ is less than $\frac{1}{2}$.

3. (a) $f_n = f_{n-1} + f_{n-2} = 2f_{n-2} + f_{n-3}$. Now use induction.
   (b) $f_n = 3f_{n-3} + 2f_{n-4}$. Now use induction.

6. First prove by induction on m that $f_{a+b} = f_{a-1}f_b + f_a f_{b+1}$. Now let $m = nk$ and prove that f_m is divisible by f_n by induction on k.

7. Let $m = qn + r$. Then, by the partial solution given for Exercise 5, $f_m = f_{qn-1}f_r + f_{qn}f_{r+1}$. Since, by Exercise 4, f_{qn} is divisible by f_n, the GCD of f_m and f_n equals the GCD of $f_{qn-1}f_r$ and f_n. Now use the standard algorithm for computing GCD (Cf. Section 10.1).

8. $h_n = h_{n-1} + h_{n-2}$.

9. $h_n = 2h_{n-1} + 2h_{n-2}$.

12. $2^{n-2} - (-2)^{n-2}$.

13. $(n+2)!$.

16. $\frac{8}{9} - \frac{2}{3}n + \frac{1}{9}(-2)^n$.

19. (a) 3^n; (c) $\frac{(-1)^{n+1}+1}{2}$.

20. $h_n = h_{n-1} + h_{n-3}$.

22. See Exercise 1 of Chapter 8.

23. $4^{n+1} - 3 \times 2^n$.

25. $3 \times 2^n - n - 2$.

28. (a) $\frac{1}{1-cx}$; (d) e^x.

29. (a) $\frac{x^4}{(1-x^2)^4}$; (c) $\frac{1+x}{(1-x)^2}$.

30. (a) $h_n = 0$ if n is even and $= 4^{(n-1)/2}$ if n is odd; (c) $h_n = \frac{1}{12}(-3 + 4 \times 3^n - (-3)^n)$; (e) $h_n = \frac{8}{9} - \frac{2}{3}n + \frac{1}{9}(-2)^n$.

32. Start with the series $1/(1-x) = 1 + x + x^2 + \cdots$ and differentiate, multiply by x and differentiate, multiply by x and differentiate again.

33. Hint: Use $n = (n-1) + 1$.

36. $\frac{1}{(1-x^2)(1-x^5)(1-x)(1-x^7)}$.

37. Hint: $h_n = \frac{1}{2}(n^2 - n)$.

38. Write h_n as a cubic polynomial in n.

40. $1/(1-x)$.

42. (a) $(x + x^3/3! + x^5/5! + \cdots)^k$; (b) $(e^x - 1 - x - x^2/2 - x^3/6)^k$; (d) $(1+x)(1 + x + x^2/2!) \cdots (1 + x + \cdots + x^k/k!)$.

43. $h_n = 4^{n-1}$ if $n \geq 1$ and $h_0 = 0$.

45. Hint: The exponential generating function is $(\frac{e^x + e^{-x}}{2} - 1)^2 e^{3x}$.

Chapter 8 Exercises

1. Let the number of ways for $2n$ points be a_n. Choose one of the points and call it P. Then P must be joined to a point Q such that there are an even number of points on either side of the line PQ. This leads to the recurrence $a_n = a_0 a_{n-1} + a_1 a_{n-2} + \cdots + a_{n-1} a_0, a_0 = 1$. Let $b_n = a_{n-1}, (n \geq 1)$. Then $b_n = b_1 b_{n-1} + b_2 b_{n-2} + \cdots + b_{n-1} b_1, (b_1 = 1)$. Hence, $b_n = \frac{1}{n}\binom{2n-2}{n-1}$ and $a_n = b_{n+1} = \frac{1}{n+1}\binom{2n}{n} = C_n$.

2. Hint: Consider the sequences $a_1, a_2, \ldots, a_{2n}$ of $+1$'s and -1's obtained by taking a_j to be $+1$ if j is in the first row of the array and -1 if j is in the second row.

5. Generalize the proof of Theorem 8.1.1.

6. $\sum_{k=0}^{n} h_k = 3\binom{n+1}{1} + \binom{n+1}{2} + 4\binom{n+1}{3}$.

9. Use induction on k.

10. Use the fact that $\binom{n}{k}$ is a polynomial of degree k in n. Thus, c_m has to be chosen so that $c_m/m!$ is the coefficent of n^m in h_n.

12. (b) $S(n, 2)$ is the number of partitions of an $n \geq 2$ element set into 2 indistinguishable boxes so that no box is empty. There are $2^n - 2$ partitions into nonempty *distinguishable* boxes.

13. Hint: The inverse images of an onto function give a partition into k nonempty *distinguishable* boxes.

15. Partition the partitions according to the number of boxes that are nonempty.

19. (a) $s(n, 1)$ is the same as the number of circular permutations of an n-element set.

26. (a) $12 = 4 + 3 + 2 + 2 + 1$.

Chapter 9 Exercises

4. Any bipartite graph in which there is at least one vertex that meets more than 4 edges.

5. Hint: Put the dominoes vertically column by column unless one is forced to put a horizontal domino.

7. Hint: The total number of edges is the number of edges that meet a vertex of X. It also equals the number of edges that meet a vertex of Y.

9. Apply Theorem 9.2.5.

10. For the bipartite graph on the right, the largest number of edges in a matching is 6. A matching of 6 edges is

$$\{\{x_1, y_7\}, \{x_2, y_5\}, \{x_3, y_3\}, \{x_5, y_2\}, \{x_6, y_6\}\{x_7, y_8\}\}.$$

A set of vertices which cover all edges of the bipartite graph is $\{x_1, x_2, x_5, x_6, x_7, y_3\}$.

12. Largest number is 5.

13. The number of different SDRs is 2 (for all n).

15. Delete x (if present) from each of $A_2, \ldots, A_n$ and show that the resulting $n - 1$ sets satisfy the Marriage Condition.

17. Hint: Suppose the number of black squares equals the number of white squares. Show that there are two consecutive squares, either in the same row or in the same column, such that removing those squares leaves a board of the type in the exercise. Now proceed by induction.

20. Hint: A woman's kth choice is a man whose $(n + 1 - k)$th choice is that woman. If $p < k$, then $n + 1 - p > n + 1 - k$.

22. In both cases, we get the stable complete marriage $A \leftrightarrow c, B \leftrightarrow d, C \leftrightarrow a, D \leftrightarrow b$.

23. Since $(n^2 - n)/n = n - 1$, it follows, that after $n^2 - n + 1$ proposals, some woman has been rejected $n - 1$ times and every man has received at least one offer.

24. Hint: Introduce *fictitious* women in order to have an equal number of men and women with each man putting the fictitious women on the bottom of his list.

27. Hint: Construct the bipartite graph $G = (X, \Delta, Y)$, where $x = \{x_1, x_2, \ldots, x_n\}$ and $Y = \{y_1, y_2, \ldots, y_n\}$, in which there is an edge between x_i and y_j if and only if $a_{ij} \neq 0$. Then show that its cover number is n.

Chapter 10 Exercises

6. Use Exercise 5 and the fact that $a - b = a + (-b)$.

9. $-3 = 17, -7 = 13, -8 = 12, -19 = 1$.

10. $1^{-1} = 1, 5^{-1} = 5, 7^{-1} = 7, 11^{-1} = 11$.

11. $4, 9$, and 15 do not have multiplicative inverses.
    $11^{-1} = 11, 17^{-1} = 17, 23^{-1} = 23$.

12. The GCD of $n - 1$ and n is 1.

14. (i) GCD$=1$.

15. The multiplicative inverse of 12 in Z_{31} is 13.

17. (i) i^2; (iii) $1 + i^2$; (v) i.

19. No: If there were such a design then $\lambda = r(k-1)/(k-1) = 80/17$.

21. Its parameters are $b' = v' = 7$, $k' = r' = 4$, and $\lambda' = 2$.

23. Each is obtained from the other by replacing 1's with 0's and 0's with 1's.

27. $\lambda = v$.

29. No.

33. There is a Steiner system of index 1 with 3 varieties. Now apply Theorem 10.3.2 $t - 1$ times.

37. Interchanging rows and columns does not change the fact that the rows and columns are permutations.

40. Take $n = 6$, $r = 1$, and $r' = 5$.

43. Use Theorem 10.4.3.

44. To construct 2 MOLS of order 9, one can use the construction in the proof of Theorem 10.4.6, or one can use the product construction, introduced to verify Theorem 10.4.7, starting with 2 MOLS of order 3. To construct 3 MOLS of order 9, one should first construct a field of order 9, starting with a polynomial with coefficients in Z_3, which has no root in Z_3 (e.g. $x^2 + x + 2$). Then apply the construction used to verify Theorem 10.4.4.

45. Take two MOLS A_1 and A_2 of order 3 and two MOLS B_1 and B_2 of order 5. Then $A_1 \otimes B_1$ and $A_2 \otimes B_2$ are two MOLS of order 15.

47. The n positions in A that are occupied in B by 1's give a set of n nonattacking rook positions.

55. One completion is

$$\begin{bmatrix} 3 & 2 & 0 & 4 & 5 & 1 \\ 2 & 0 & 3 & 5 & 1 & 4 \\ 0 & 3 & 2 & 1 & 4 & 5 \\ 4 & 5 & 1 & 2 & 3 & 0 \\ 5 & 1 & 4 & 3 & 0 & 2 \\ 1 & 4 & 5 & 0 & 2 & 3 \end{bmatrix}.$$

57. Take one completion. Another is obtained by interchanging the last two rows.

60. The 0's in the last $n - 1$ rows and columns pair up the integers $1, 2, \ldots, n - 1$. Hence, $n - 1$ is even.

Chapter 11 Exercises

1. 1, 2, and 4, respectively.

3. No.

4. No; Yes.

5. See Exercise 16 of Chapter 2. Not true for multigraphs.

6. Hint: Try loops.

7. Hint: Put in as many loops as you can.

8. Hint: For any set U of k vertices, how many edges can have at least one of their vertices in U?

11. Only the first and third graphs are isomorphic.

14. No.

15. No.

19. Neither connectedness nor planarity depend on loops or the existence of more than one edge joining a pair of vertices.

21. If G is connected, then surely G^* is. The two vertices x and y must be in the same connected component of G (Why?). Hence, if G^* is connected, then G must have been connected.

29. The second, but not the first, has an Eulerain trail.

32. 5.

39. Hint: First construct a graph of order 5, four of whose vertices have degree 3 and the other of which has degree 2. Now use three copies of this graph to construct the desired graph.

48. No, but yes if we delete the loops.

49. (a) In order for $\{a, b\}$ to be an edge, either a and b are both even, or else they are both odd. From this it follows that the answers are (i) No; (ii) No; (iii) No; (iv) No.

50. 4 (to get $K_{2,3}$, which has 6 edges).

54. Only the tree whose edges are arranged in a path.

55. Again, only the tree whose edges are arranged in a path.

56. There are 11.

57. Hint: Use induction on n. At least one of the d_i equals 1.

59. If there were more than two trees, then putting the edge back could not result in a connected graph.

64. Hint: Try a "broom."

66. Just one.

68. The graphs in Figure 11.49 give positive, neutral, and positive games, respectively.

71. Hint: Otherwise could the edge cut be minimal?

75. (c) A BFS-tree is a tree whose edges are arranged in a path with a root "in the middle."

76. (c) A DFS-tree is a $K_{1,n-1}$ with the vertex of degree 1 as root.

78. Hint: Consider a pendent vertex and use induction on n.

86. Hint: Consider two spanning trees of minimum weight and the smallest number p such that one of the trees has an edge of weight p and the other doesn't.

Chapter 12 Exercises

5. Hint: In a digraph without any directed cycles, there must be a vertex with no arc entering it.

7. Hint: There is a Hamilton path.

9. Hint: A strongly connected tournament has at least one directed cycle. Show that the length of the directed cycle can be increased until it contains all vertices.

11. Hint: Open trails.

16. If not, then t_1 would pull out of the allocation, and hence the allocation would not be a core allocation.

18. Just check the 6 possible allocations. The core allocation pro-
    duced by the algorithm is the one in which each trader gets the
    item he ranks first.

19. Otherwise he would pull out of the allocation.

Chapter 13 Exercises

4. C_n is not bipartite and it is easy to find a 3-coloring.

5. 2, 3, and 4, respectively.

8. (a) All of the null graphs obtained by applying the algorithm for
   computing the chromatic polynomial have at least one vertex;
   hence, their chromatic polynomials are of the form k^p for some
   $p \geq 1$. (b) G is connected if and only if one of the null graphs
   obtained has order 1. (c) To get a null graph of order $n - 1$, one
   edge has to be contracted and the other edges have to be deleted.

9. Use the results of Exercise 7.

10. $n - 1$.

12. $n - 1$.

13. $n - 2$.

15. Hint: Remove an edge and get a bipartite graph.

21. Hint: Put the lines in one at a time and use induction.

23. Hint: Examine the proof of the inequality (12.5).

26. Hint: Theorem 12.2.2.

27. Hint: Examine the proof of Theorem 12.2.2.

29. Hint: Choose a longest path $x_0, x_1, \ldots, x_k$. To which vertices can
    x_0 be adjacent?

33. Hint: A tree is bipartite.

37. 2.

38. $\lceil n/3 \rceil$.

42. Hint: If G is a graph of intervals, then any induced graph is the
    graph of some of the intervals.

44. Hint: A chordal bipartite graph cannot have a cycle.

51. $\min\{m, n\}$.

52. Hint: Assume that G is not connected. What does this imply about the degree sequence of G?

53. (a) $\lceil (n-1)/2 \rceil$.

Chapter 14 Exercises

1. $f \circ g = \begin{pmatrix} 1 & 2 & 3 & 4 & 5 & 6 \\ 2 & 5 & 3 & 4 & 1 & 6 \end{pmatrix}$; $f^{-1} = \begin{pmatrix} 1 & 2 & 3 & 4 & 5 & 6 \\ 4 & 3 & 6 & 2 & 5 & 1 \end{pmatrix}$.

5. The symmetry group contains only the identity motion. The corner symmetry group contains only the identity permutation of the three corners.

10. The symmetry group of a rectangle that is not a square contains four motions: the identity, a rotation by 180 degrees about the center of the rectangle, and the reflections about the two lines joining midpoints of opposite sides.

13. (a) (R, B, R, B, R, R); (b) (R, R, B, R, R, B).

14. 4 (10).

16. If $f(i) = j$, then $f(j) = i$. The cycle factorization of f contains only 1-cycles and 2-cycles.

18. $\frac{p^4 + 2p^3 + 3p^2 + 2p}{8}$.

22. $\frac{p^4 + 3p^2}{4}$.

23. (a) Label the two squares A and B. The number of marked-dominoes equals the number of nonequivalent colorings of $\{A, B\}$ with the colors $0, 1, 2, 3, 4, 5, 6$, under the action of the group G of the two possible permutations of $\{A, B\}$. Hence, by Theorem 13.2.3, the number of different marked dominoes equals $\frac{7^2 + 7}{2} = 28$.

24. (a) The group of permutations now consists of four permutations of the four squares to be marked. This gives $\frac{7^4 + 3 \times 7^2}{4} = 637$.

25. There are a total of 10 ways to color the corners of a regular 5-gon in which 3 corners are colored red and 2 are colored blue. Under the action of the dihedral group D_5 the number of nonequivalent colorings is $\frac{10 + 5 \times 2 + 4 \times 0}{10} = 2$.

26. $\frac{35+7\times3+6\times0}{14} = 4$.

27. $f = [1\ 6\ 3\ 2\ 4] \circ [5]$.

28. By reversing the order of the elements in each cycle of the cycle factorization of f.

31. $\frac{k^5+5\times k^3+4\times k}{10}$.

33. See Exercise 28.

36. $\frac{30+5\times2+4\times0}{10} = 4$.

41. If ρ_n^k, $(k = 1, 2, \ldots, n - 1)$ contains a t-cycle, then by symmetry the cycle factorization of ρ_n^k contains only t-cycles, implying that t is a factor of n. Since n is a prime, $t = 1$ or $t = n$. Since $t = 1$ implies that ρ_n^k is the identity permutation, we have $t = n$; that is, ρ_n^k is an n-cycle.

42. Using Exercise 41, we get $\frac{k^n+n\times k^{(n+1)/2}+(n-1)k}{2n}$.

43. The cycle index of the group of permutations is

$$P_G(z_1, z_2, \ldots, z_{10}) = \frac{z_1^{10} + 2z_1 z_4^2 + z_1 z_2^4}{4}.$$

Hence the number of nonequivalent colorings is

$$P_G(2, 2, \ldots, 2) = \frac{2^9 + 2^4 + 2^5}{4} = 2^7 + 2^2 + 2^3.$$

49. The cycle index for the group G of three rotations is

$$P_G(z_1, z_2, \ldots, z_9) = \frac{z_1^9 + 2z_1 z_3^3}{3}.$$

The generating function for nonequivalent colorings is

$$P_G(r + b, r^2 + b^2, \ldots, r^{10} + b^{10}) = \frac{(r + b)^{10} + 2(r + b)(r^3 + b^3)^3}{3}.$$

Bibliography

Many references have been cited in the footnotes in the text. Here we list some more books, primarily advanced, for further reading on many of the topics discussed in this book.

Ian Anderson: *Combinatorics of Finite Sets.* Oxford, England: Oxford, 1987.

Béla Bollobás: *Modern Graph Theory.* New York: Springer-Verlag, 1998.

Claude Berge: *Graphs and Hypergraphs.* New York: Elsevier, 1973.

Richard A. Brualdi and Herbert J. Ryser: *Combinatorial Matrix Theory.* New York: Cambridge, 1991.

Louis Comtet: *Advanced Combinatorics.* Boston: Reidel, 1974.

Shimon Even: *Graph Algorithms.* Potomac, Maryland: Computer Science Press, 1979.

L.R. Ford, Jr. and D.R. Fulkerson: *Flows in Networks.* Princeton, New Jersey: Princeton University Press, 1962.

Ronald L. Graham, Bruce L. Rothschild, and Joel L. Spencer: *Ramsey Theory*, second edition. New York: Wiley, 1990.

Frank Harary: *Graph Theory.* New York: Addison-Wesley, 1969.

Frank Harary and Edgar Palmer: *Graphical Enumeration.* New York: Academic Press, 1973.

D.R. Hughes and F.C. Piper: *Design Theory.* New York: Cambridge, 1985.

Tommy R. Jensen and Bjarne Toft: *Graph Coloring Problems.* New York: Wiley-Interscience, 1995.

C.L. Liu: *Topics in Combinatorial Mathematics.* Washington: Mathematical Association of America, 1972.

L. Lovász and M. D. Plummer: *Matching Theory.* New York: Elsevier, 1986.

L. Mirsky: *Transversal Theory.* New York: Academic, 1971.

C.A. Pickover: *The Zen of Magic Squares, Cicles, and Stars*, Princeton: Princeton University Press, 2002.

Herbert J. Ryser: *Combinatorial Mathematics.* Carus Mathematical Monograph No. 14. Washington: Mathematical Association of America, 1963.

Thomas L. Saaty and Paul C. Kainen: *The Four-Color Problem.* New York: Dover, 1986.

Richard P. Stanley: *Enumerative Combinatorics*, Volume I (1997) and Volume 2 (1999), Cambridge.

N. Vilenkin: *Combinatorics*. New York: Academic, 1971.

Douglas West: *Introduction to Graph Theory* (2nd edition). Upper Saddle River, NJ: Prentice Hall, 2001.

Index